# The New Larousse ENCYCLOPEDIA OF THE EARTH

LEON BERTIN

Foreword by Sir Vivian Fuchs

Introduction by Carroll Lane Fenton

**Editorial Consultants (English Edition)**
Norman Harris, B. Sc., Ph. D.;
Deputy Director of Geological Survey of Uganda, 1951-61
Carroll Lane Fenton, Ph. D., University of Chicago
Chalmer L. Cooper, Ph. D., University of Chicago;
United States Dept. of the Interior, Geological Survey
Henry Hill Collings Jnr, M.A., Harvard University
Olivia V. Haslau-Perry, M.A., Columbia University;
Instructor Hofstra College, New York

Hamlyn London
Crown Publishers, Inc., New York.

**Larousse Encyclopedia of the Earth**

Translated by R. Bradshaw M. Sc. and Mary M. Owen
from LA TERRE, NOTRE PLANÈTE
First published in France by
Librairie Larousse, Paris

Published by
THE HAMLYN PUBLISHING GROUP LIMITED
LONDON • NEW YORK • SYDNEY • TORONTO
Hamlyn House, Feltham, Middlesex, England
Distributed in the USA by Crown Publishers, Inc., New York

Copyright © The Hamlyn Publishing Group Limited 1961
Revised edition 1972
Casebound ISBN 0 600 35464 4
Paperback ISBN 0 600 35472 5
Printed in Singapore
by Times Printers Sdn Bhd.

# Contents

## The Present

## Earth in the Service of Man

# The Past

# Colour Plates

# Foreword

*by Sir Vivian Fuchs*

THE EARTH describes many wonders of the world in which we live — the home from which man may soon venture to other planets, but to which he will always owe the origin of ideas and, more important, his driving sense of curiosity. All knowledge is based on experience of what is, and what has been; how important therefore to know at least the facts about our own environment.

This book tells the story of the Earth itself, explaining the interplay of the oceans and atmosphere, the forces of mountain building and those which ultimately destroy them. All the natural forces at play, whether they are the violent explosions of volcanic eruption, or the slow deformation of the crust which has drowned inhabited regions and raised new surfaces for man to occupy, all control our existence and destiny.

At first man was forced to live where Nature allowed; with increasing knowledge we have learned to draw upon Nature herself to help us in our struggle to exist. At the beginning the natural resources of the animal and vegetable kingdom helped man to extend his occupation into areas of less favourable climate. Then came a gradual understanding of the materials available from the rocks themselves — this led to the use of many minerals and the harnessing of power from coal, petroleum and, more recently, radioactive minerals.

When the only material available to Stone Age man was a coarse unfavourable rock, the tools he made were crude and unrefined. Where he found materials like flint, or the natural glass obsidian, his manufacturing technique improved rapidly and the delicate stone knives, chisels and arrowheads that resulted set his foot another step forward in the evolutionary struggle.

Comparably, we are now but beginning to use the latest tools of the nuclear age — tools derived from the Earth, turned by man's mind to his purpose, and surely about to provide a great impetus to human advance in the battle with our present habitat — perhaps even extending it to other worlds.

Man in his present state is the end product of a long evolutionary line that extends hundreds of millions of years into the past. Yet he himself, as a genus, has existed for little more than a million years. Through the ages since life first appeared, living things have remained within the bounds of their own habitat, each limited by hydrological or meteorological factors. Salinity, temperature, altitude, light, humidity — all these have conditioned and controlled life forms, both in the sea and on the land. These physical factors created laws by which each community, and every individual, was compelled to live. To transgress such laws by moving into colder water, or into a region of greater aridity, meant death. Sometimes Nature itself changed the physical environment of an entire community, and this, too, for primitive life, sometimes meant death and extinction of the race. To continue successfully the thread of life and its slow development, it was necessary for living matter to establish a wide distribution and a considerable tolerance of inimical factors. Distribution ensured that local environmental changes destroyed only a proportion of the population, while the greater the tolerance, the less likely that seasonal or longer-term changes would affect the entire race. Certainly the less adaptable individuals died, but the rest survived to continue reproduction and to demonstrate that even when life began, physical laws permitted only the survival of the fittest.

As time progressed, more and more forms of life developed, each specialising to compete with its habitat and, in so doing, becoming more and more successful. But woe betide any living thing that specialised too greatly, for this brought inflexibility, and in evolution there is no sudden return to the past. Biological history reveals countless examples of races which became so well adapted that the slightest climatic change destroyed them.

As life spread and diversified upon the earth, different species had an increasing influence upon one another. A habitat could no longer be described in terms of physical factors alone for it now included many forms of life — one preying upon the other, whether as predator or parasite. This increasing complexity of the lifestream on earth can be traced in the past from the fossil remains. These tell a dramatic story of the struggle for existence by every individual and every race of plant and animal.

Of all the countless thousands of species living today, only man possesses the supreme advantage of developed reasoning power. He has found that matter cannot be created; but it can be transformed from one state into another. For this there must be a starting point, and for Earth-bound man that point must be within the limits of the hydrosphere, atmosphere and lithosphere.

The development of man's existence down the ages has depended upon his understanding of the planet on which he lives, and on his ability to turn to his own best uses the materials that are there for the taking. From it he has had to seek food, water, air light, heat, clothing, building materials, power to drive the machines of his inventive mind, and the substances necessary to his innumerable scientific and technical achievements.

Now that man is on the verge of visiting extra-terrestrial bodies, and perhaps establishing himself there, we must expect future generations to be influenced by new and previously unencountered factors. What these factors may be must be left for the future to show, but meanwhile there is still much to be learned about our own world. For the non-specialist a book such as this makes the origins of scientific work more readily understandable, and lends point to its aims.

V.E.F. 1961

# Introduction

by Carroll Lane Fenton

THROUGH 300,000 years, at the least, men of varied species have looked to their world for living space, food, shelter, and raw materials for industry. Age by age these demands have grown and multiplied. Today we ask and must get much more from our planet than did Stone-age men.

Do you doubt that? Then let us consider a few examples. First comes Pekin man, or *Sinanthropus*, who inhabited the hills and prairies of China during Mid-glacial times. This wide range suggests that Pekin man was successful, yet all he asked was a place to hunt, a damp cavern for shelter, and stones that could be shaped into simple weapons and tools. The best materials for this purpose were quartz and quartzite, but when these were scarce Pekin man got along with less durable rocks.

Our second example from the past is Egypt of 2600 B.C. — the time of great pyramid builders. Stone was still shaped into tools and weapons, but different rocks were chosen for different purposes. Flint was hard and took a sharp edge; it went into sickles, scrapers, knives, daggers, and battle axes. Coarse sandstone was best for files and grinders, and massive rocks such as limestone and granite were quarried for use in temples and pyramids. Dried mud made houses for common people, but clay baked into tiles was used by the nobility. The wealthy folk also had metals — copper for knives, daggers and fishhooks, and gold for ornaments. Some chisels and saws were made of bronze, for this alloy of copper and tin was harder than either metal alone.

William the Conqueror and his barons of A.D. 1066 were Normans who knew nothing of Pekin man or ancient Egypt, and would have scorned flint battle axes. These conquerors lived in stone castles and fought with steel weapons whose ultimate source was iron ore that had accumulated in an ancient sea. The Normans also had gold and silver coins, as well as jewels which were minerals extracted from rocks.

We moderns still smelt ore to get iron, but we change the raw metal with alloys and shape it into tools that range from pocket knives to automobiles. We turn limestone and shale into cement, mine vast amounts of sulphur and salt, and use a great number of lesser minerals, from actinolite to zircon. We get heat and energy from coal (a black rock) and petroleum (rock oil), as well as from gas, running water, and atoms secured from selected minerals. These things keep our complex, crowded civilization going — and when complexity becomes oppressive, the earth's surface offers the relief of seashores, lakes, deserts, and mountains.

We use the earth on an unparalleled scale, but we seldom pause to ask questions about what, how, when, and why. When such questions do arise we shrug them off or leave them to specialists.

The practical values of earth science become most evident when they are lost. Suppose you want to build your home on a hillside because it provides a beautiful view. The steep slope seems as firm as the rock of Gibraltar, yet plain geologic clues would warn you that it is creeping downward into a valley. The movement is much too slow to be felt, but it is fast enough to twist floors, distort doors and windows, and open cracks in walls.

This is one example; others are easily found. Bridges have been built where streams are sure to destroy them; fields waste away because their owners are ignorant of the action of rainwater as it runs over bare ground. Towns spread across level land built by rivers in flood, in surroundings which give certain warning that floods will come again. Other towns have grown up on eroding seashores, where waves and currents wash away sand and let buildings sag into the water. Mines are sunk that reach neither ore nor coal, and oil wells drilled that find no petroleum. Farmers have tried — and failed — to grow crops on deposits of brick clay or fuller's earth. In all these cases acquaintance with geology could have replaced failure with royalties or profitable sales.

Even city dwellers are more concerned with geological matters than they imagine. They walk through parks, drive into the country at week-ends, and take long trips during vacations. All provide contacts with the earth; contacts whose meaning can be enriched by some knowledge of geology.

What one sees depends upon where one goes. The New Yorker who strolls in Central Park winds among roots of ancient mountains, whose worn surface was scraped smooth by a glacier that melted barely 12,000 years ago. The San Franciscan sees layers of rock that formed in a sea, as well as cliffs that plainly have risen from the Pacific Ocean. Near Chicago are more ancient sea beds, while the week-end camper from Portland or Seattle is likely to visit a volcano that is six to ten million years old. If he visits Crater Lake, he sees the remains of a mountain that exploded and then collapsed some 7,500 years ago.

The more varied the scene, the more it offers to the person who knows how it has formed. Suppose that two people stand at Desert View lookout, on the rim of Arizona's Grand Canyon. The one who knows no geology sees a vast and mysterious gorge, with multicoloured walls and isolated cliffs that are larger than many mountains. He sees the gorge as a rift in the earth, torn open by some catastrophe of 'prehistoric' times. Steep walls abound in 'lovers' leap'; cliffs become fanciful castles and forts.

The second visitor too sees the canyon with its brightly coloured cliffs and walls. But to him this is no fanciful rift; it is a valley cut into an upraised plateau by rain, meltwater, and the silt-laden river that still flows through the Inner Gorge. The supposed castles and forts are parts of the plateau that remained when the high ground around them was worn away, without help from cataclysms or giants.

Thus far science has replaced fancy. That in itself is worth while, but it is no more than a prelude to a story that nothing except geology can tell. This story begins in the deep Inner Gorge, whose dark, twisted rocks reveal mountains that formed when the Southwest was squeezed into tremendous earth-folds some two billion years ago. In time these same mountains were worn down to a plain; it sank beneath a shallow sea which deposited layers of sand and fine mud. Then came uplift and more mountain building which broke the rocks into blocks that heaved upward and tipped into long, narrow ranges much like those of modern

Top
Capitol Reef, Utah, U.S.A.

Bottom
The Italian Dolomites

An island explodes. The Taal volcano off the coast of Batangas.

Utah. Then followed a long series of marine invasions, as well as an epoch of fern-covered lowlands and another in which persistent winds piled loose sand into dunes. Then one more sea, another lowland, and renewed uplift which turned both marine and terrestrial deposits into the Colorado Plateau. It must have approached its present height before cutting of the Grand Canyon began. There also was vulcanism nearby, producing lava flows, cinder cones, and the great volcano whose shattered and divided summit now forms the San Francisco Peaks. Mammoths and camels roamed their lower slopes as the Ice Age drew to a close.

This story is simplified and condensed, yet it shows how greatly earth science can enrich travel and scenery.

Since earth science enables us to use, know, and understand our planet, we might ask why the subject has all but vanished from twentieth-century education. Most people who want to understand it are forced to seek their information for themselves, from books.

Long shelves have been filled with volumes written to meet this demand. Some are old and quaintly sensational; they reflect a time when scientists were amazed and often embarrassed by their own discoveries. Next come works that state theories of the earth's origin and summarise its past, chiefly in terms of plants and animals that lived during successive periods. These books, too, were often sensational, though their purpose was to spread new concepts in place of the dogma that our planet was created less than 6,000 years ago yet passed through a sequence of cataclysms before reaching its present state.

By the 1920s geology had become an extensive and varied science, divided into highly specialised departments. This change was reflected in specialised books devoted to earthquakes, volcanoes, minerals, rocks, fossil plants, reptiles, or ancient members of the horse family. With this division came precision and detail. Thus a popular guide to minerals of the post-1950 period was a good deal more technical than a textbook widely used in 1921. A book on earthquakes said little about spectacular disasters and much about instruments, body waves, epicentres, and other elements of seismology. The once casual rockhound was growing into a serious amateur.

This brought the need for books covering the entire earth and summarizing our knowledge of it. Such works must be large, in keeping with their subject. They must also be modern and up-to-date, for geology is changing as rapidly as physics or astronomy. Finally, the volumes must be richly illustrated, for pictures will enable readers to translate statements into mountains, valleys, volcanoes, or rocks.

Of the books in this class, a substantial number were written in Europe, where they served so well that they then were exported to other continents. In theory this procedure was sound, for geologic principles are the same the whole world over. But details, emphases, and scenes do vary. Readers in Boston, Winnipeg or San Francisco should not be asked to interpret their world in terms of German shores, British lakes, Italian volcanoes, and earthquakes in Portugal. Even fossils found in the New World are not quite identical with those that have been named and described in the Old World.

THE EARTH began its career in Europe. But the publisher of this English-language edition has not been content merely to translate the original French work and reproduce its handsome pictures. The book has been revised from introduction to index; extensive additions have been made to deal with Britain, North America, Australia, and the rest of the world. New text is matched by new illustrations which cover rocks and fossils of the New World as well as the Old. Though THE EARTH is anything but a guidebook, it will tell Americans what to look for while they travel, and will often explain what they have seen after they return home.

Here we come back to the theme of interest and understanding. Geology does not make 'escape' reading; it is literally underfoot and around us, and frequently right above us. Yet the principles that explain a brook, a clay pit, or an eroded boulder can reach across great distances to account for a canyon, a coastal plain, or an ice-sheet in Antarctica. Thus geology meets one great need of our mobile age, with its long trips, its voyages in winter, and its luxury tours to regions once reached only by intrepid adventurers. Modern man can visit most parts of his world. If he uses THE EARTH to understand what he sees, the author, revisers, and illustrators of this volume will know that their work has been worth while.

C.L.F. 1961

# The Present

# Atmosphere and Weather

MAN has always been familiar with wind, storms, rain and snow, but where once he observed these phenomena only as independent forces whose origins could not be explained, today he knows how they are inter-related; their study is the science of meteorology.

Without going as far back as the discoveries of classical times, for the beginnings are to be found in the fifth century B.C., let us look at some of the important dates in the development of meteorology.

About 1500 Leonardo da Vinci constructed an improved wind vane and a mechanical moisture indicator.

In 1597 Galileo made a temperature indicator which, with the addition of an arbitrary scale, later became a measuring device. This marked the beginning of the transition from visual to instrumental observations, and was the first step toward precision. Progress toward the goal of complete precision, still continuing today, could not begin until the basic instruments (barometer, thermometer, anemometer and so on) developed.

In 1630 the Academy of Florence made the first thermometer.

In 1644, Torricelli invented the mercury barometer.

In 1675 the Royal Observatory was founded at Greenwich, England, for the purpose of improving standards of navigation. From it are derived the measurement of longitude and also the concept of Greenwich Mean Time.

In 1684 Louis XIV of France set up a rain-gauge to find out how much rainwater he could expect to work his fountains at Versailles. His measurements were as precise as those made today.

In 1686 Halley gave the first account of the trade winds and monsoons. He explained the significance of the Earth's different temperature zones in producing winds.

In 1710 the Fahrenheit scale of temperature was developed.

In 1736 the Celsius (Centigrade) temperature scale was introduced.

In 1749 Benjamin Franklin discovered a lightning conductor.

In 1775 Borda and Lavoisier made simultaneous observations in different parts of France and attempted the first weather forecast with barometer and hygrometer.

In 1805 the Beaufort scale of wind force was developed.

In 1855, during the Crimean War, severe storms caused considerable damage to the French fleet. At about the same time a French astronomer, Le Verrier, won world fame by suggesting the existence of a new planet. He had used Newton's laws of motion, observed the orbits of the planets and had concluded that another planet must exist in our solar system. He also calculated its approximate position for a certain time. Telescopes trained in the predicted position proved him correct. His 'new' planet was Neptune. Napoleon III seemingly thought that if Le Verrier could discover a new planet he would be able of forecast the weather, and accordingly set him this difficult task. Le Verrier established a meteorological service at the astronomical observatory in Paris, the first of a series of such institutions established in Europe and the United States during the nineteenth century.

With the advent of telegraphy and, later, of wireless, the results of observations could be sent daily to meteorological stations all over the world. During the 1914–18 war, the development of aviation brought with it the need for rapid forecasts of wind strength and direction, and produced the first real advances in meteorology since Le Verrier's time. Between the wars meteorology made new strides. Observations were made in mid-ocean by meteorological ships, and in mountain areas by high-altitude observatories; exploration of the upper regions of the atmosphere was carried out by means of radio-sonde ballons, aircraft and rockets.

Astronomical observations today often record data and develop instruments which aid meteorological studies. Among the better known observatories so engaged are those on Mont Blanc in France, on the Jungfrau in Switzerland, Mount Palomar in California (which has the world's largest reflector telescope), Mount Hamilton (Lick Observatory) in California, and William Bay (Yerkes Observatory) in Wisconsin. At Kitt Peak National Observatory, Arizona, the world's largest solar telescope examines the sun's surface, sunspots and solar flares. Similar observatories exist in countries all over the world.

A recent development is the setting up of the Jodrell Bank radio telescope in Cheshire, England. It is a bowl of steel 250 feet in diameter hung on lattice-work steel towers 180 feet above the ground; the total weight is 2,000 tons. It is fitted with electronic apparatus of various kinds which can be used to catch the faintest radio waves, to track earth satellites and space rockets, and to measure the size of the universe itself. Signals sent out from Jodrell Bank actuated the radio transmitters of the American sun satellite Pioneer IV at distances of up to five millions miles.

There is today a world meteorological organisation attached to UNO which co-ordinates the work of meteorologists of sixty-seven nations. This organisation is in touch with 10,000 ground posts and numerous radio-sonde stations. Readings are taken at exactly the same time of day all over the world.

In the present age exploration of the atmosphere is of prime importance. It is necessary for advance weather-casting, stratosphere flying, and also for space travel. The outer atmosphere is also a natural laboratory for the study of low pressures and cosmic radiations.

The earliest meteorological equipment to be used for studying the higher portions of the atmosphere was the balloon, carrying either automatic equipment or the meteorologist himself, who ascended to make his observations direct. In 1931 Auguste Piccard, the Swiss physicist, reached 53,000 feet in his stratosphere balloon, *Explorer* 11, and carried out investigations into cosmic rays. *Explorer* 11 was 180 feet in diameter and contained 4 million cubic feet of gas. This type of balloon, carrying the passengers in a spherical air-tight car, contains a relatively small amount of gas compared with the 10 million cubic feet capacity of some of the new plastic balloons used by the United States Navy. (Capacity figures are quoted for the reduced pressure conditions of high altitudes, not for ground level.) In August 1957, Major D.G. Simmons, U.S.A.F., reached 102,000 feet. At that altitude he saw the 'sun set as a white disc and rise as a green flare'.

Even in this era of rockets and man-made satellites, the balloon is far from obsolete. The radio-sonde, consisting of self-recording

*Left*
The tiros Operational Satellite undergoing tests at Cape Kennedy in preparation for launching. The satellite will transmit photographs of the weather conditions surrounding the Earth.

*Right*
Cumulus and cumulonimbus cloud formations.

thermometer, barograph, hygrometer and radio transmitter, is still used for exploring the atmosphere. Carried by a balloon filled with hydrogen or helium, it transmits the barometric pressure, temperature and humidity every thirty seconds (that is, about every 500 feet of the ascent). The details are picked up by recording apparatus on the ground where they can be put into immediate use. The speed at which this information can be obtained is of obvious value for relaying to aircraft in flight. Balloons are favoured too in many atmospheric and weather studies because they can be brought to a standstill in mid-air and will remain there for many hours.

But balloons and aircraft cannot reach the higher atmospheric layers, and here satellites take over. By 1967 more than 700 satellites had orbited the earth since the first *Sputnik* was launched from the U.S.S.R. in 1957.

EARTH SATELLITES. In 1959 America began her *Explorer* series of scientific satellites and *Explorer VI* sent back cloud-cover photographs. In 1962 the Russian series of earth satellites known as *Cosmos* began. By December 1963 twenty-four of these scientific satellites had been launched. The first international satellite was launched in 1962 with British instruments on board, and a Canadian satellite was launched in that year to study the ionosphere. America also sent up the *Anna IB* to measure more accurately the size and shape of the Earth and the tiny *TRS* to chart radiation belts in space. In July *Telstar I,* a communication satellite was launched, later followed by *Telstar II* which relayed television pictures across the Atlantic. The *Syncom II* was launched in 1963 and achieved synchronous earth orbit. In May 1963 American scientists released 400 million tiny copper dipoles in space for a controversial experiment in radio communications. At the end of the year *Explorer XIX* was launched, from California. It was designed to measure air density and temperatures in the thin atmosphere high above the North and South Poles. Early in 1964 Russia launched *Elektron I* and *II* to study the internal and external radiation belts of the Earth.

SPACE SATELLITES. The first attempts to explore outer space began with the Russian *Lunik I,* launched in January 1959, the first rocket to escape from the Earth's orbit. This was followed by *Lunik III* which took photographs of the far side of the Moon. America, too, began her Pioneer series of cosmic satellites in 1959. In February 1961 Russia launched a satellite towards Venus from a heavy satellite in a parking orbit around the earth, and in 1962 she launched *Mars I.* In August 1963 Russia announced that she might abandon the attempt to land a man on the Moon. America continued her space probes with *Mariner II,* launched in December 1962 towards Venus, which remained in contact up to 54 million miles, and in 1964 the *Ranger VII* spacecraft sent back thousands of close-up photographs of the Moon's surface.

MANNED SATELLITES. In 1962 Russian satellites containing dogs and other biological specimens were put into orbit and recovered safely, thus opening the way to manned space flights. The first of these took place on April 12, 1961 when *Vostok I* was launched. In May of that year came the first American sub-orbital flight, and in February 1962 *Mercury III* carried the first American into space. Russia put two men into space later that year and for the first time two spacecraft were able to achieve a rendezvous. America's *Mercury* series continued, and in 1963 *Mercury VI* made 22 orbits. In June 1963 Russia again put two spacecraft into orbit which were able to establish two-way communication with each other as well as with the Earth. In 1967 the American spacecraft Surveyor V produced evidence that the Moon's surface is chemically similar to that of the Earth.

So much information has been gained from the satellites that only computers can analyse it. In addition to the information obtained from the weather satellites, American satellites discovered the Van Allen radiation belts which encircle the Earth; found that the Sun exerts pressure on objects in space; determined that the Earth is slightly pear-shaped with the 'stem' at the North Pole; and collected new data on the make-up of magnetic fields in space. It also seems possible that hitherto unsuspected layers of helium, hydrogen and cosmic dust surround the Earth far beyond the oxygen-nitrogen atmosphere that men breathe. In 1967 the Russian spacecraft *Venus IV* discovered that Venus has a surface temperature of approximately 270°C and is therefore too hot to support life as we know it, and manned space flights have built up a small catalogue of photographs of the Earth's surface for use in geophysical studies. They are particularly valuable for Africa, where there has been relatively little geological exploration. The study of these photographs is also hoped to lead to the development of techniques for interpreting photographs taken by satellites flying close to the planets.

Electromagnetics offer another method of studying the upper atmosphere. We shall refer to this method later, when we discuss how Hertzian waves have revealed several layers of ionised gases in the upper atmosphere which make wireless transmission possible.

## THE LAYERS OF THE ATMOSPHERE

The atmosphere is the envelope of all gases enclosing the Earth. A specimen of dry, pure air contains by volume about 78 per cent nitrogen, 21 per cent oxygen and about 0.03 per cent carbon dioxide. The remainder includes traces of several other gases, such as neon, krypton, helium, ozone, xenon and hydrogen. This last group

*Top*
One of the satellites used to obtain global meteorological data on the entire sunlit portion of the Earth daily and to relay the information to the meteorological centre for analysis.

*Centre*
One of the satellites in the Tiros Operational System being prepared for launching.

*Bottom*
Mosaic view of North America received from Essa 1 satellite, which was launched into orbit on February 3 1966 from Cape Kennedy, Florida. Coastlines and other features have been outlined and storms and cloud cover are discernible.

**THE TIROS SYSTEM IN OPERATION**

2. TIROS TAKES CLOUD PICTURES OVER SPECIFIED AREAS.

3. PICTURES ARE 'READ OUT' TO GROUND STATION ON COMMAND.

1. GROUND STATION SENDS INSTRUCTIONS TO TIROS IN ORBIT.

4. DATA GO TO NASA, WEATHER BUREAU, AND OTHER AGENCIES FOR ANALYSIS.

5. UP - TO - DATE REPORTS GO TO SHIPS, PLANES, OTHER USERS.

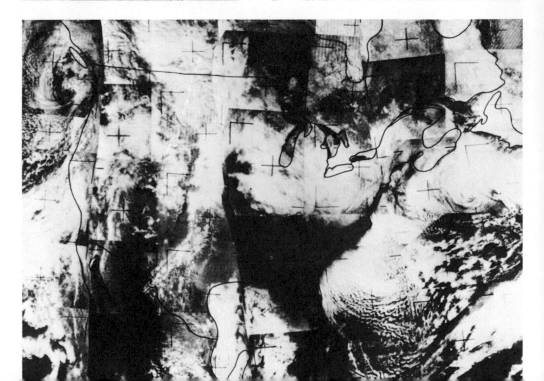

constitutes such a minor quantity that it is of no importance in meteorology. In addition to the regular constituents, impurities such as soot, dust and salt particles are present in variable amounts.

The carbon dioxide content of the air varies slightly, in general the animal world producing it and the vegetable world consuming it. While these processes are not always balanced, the oceans act as a regulator by absorbing the excess carbon dioxide, so that the balance is kept virtually constant.

The amount of ozone varies too. It occurs in minute amounts in the lower atmosphere but is present in larger and more variable quantities in the upper atmosphere.

Except for the carbon dioxide and ozone content and the varying amount of water vapour in the atmosphere, the composition of the air is remarkably stable throughout, even to great heights. The water vapour is in many ways its most important constituent. The maximum amount of water that the air can absorb depends on temperature: the higher the air's temperature the more water it can hold.

Although the chemical composition of air is fairly uniform, successive layers of the atmosphere display variations in temperature, pressure, mechanical and electrical properties.

The main layers of the Earth's atmosphere, working outward from the surface of the globe, are the troposphere, the stratosphere, the mesosphere and the ionosphere. Variations occur within these major 'spheres', so that each may be made up of zones of a different nature. The total thickness or ultimate limit of the atmosphere is possibly something like 20,000 miles, where perhaps centrifugal forces offset the pull of gravity and Earth's atmosphere is slowly escaping.

A description of the main atmospheric layers is slightly more detailed given below:

**Troposphere.** This zone extends from ground level to a height ranging from 18,000 feet over the Poles to 60,000 feet over the Equator. Pressure and temperature decrease upward regularly. Water vapour, winds and dust are present. Separating this layer from that above it is a level called the tropopause. A thin transitional layer, the tropopause varies in height with latitude, season and weather. It marks the level at which the steady temperature lapse rate ceases and beyond which temperature becomes constant even with increasing altitude. It is in this zone that meteorology conducts most of its studies, for it is here the region of weather lies.

**Stratosphere.** This extends upward from the tropopause to around 250,000 feet. A little above the tropopause it contains a thin ozone-rich layer where a special variety of clouds, called mother-of-pearl clouds, occasionally form in the normally cloudless stratosphere. The air between the ozone-rich layer and the tropopause is remarkably stable. Here, in the lower stratosphere, conditions are ideal for flying. Pressure decreases regularly upward through the stratosphere but temperature does not, remaining constant up to something like 120,000 feet. Between 120,000 feet and 180,000 feet it increases upward; beyond that it decreases again, up to a level of 250,000 feet. Horizontal temperature variation in the stratosphere is the reverse of that in the troposphere; it is warmer over the Poles and cooler over the Equator. Air currents are strongly marked. Dust and water vapour are absent.

**Ionosphere.** This is the layer above 250,000 feet and continues to the limit of the Earth's atmosphere. Pressure here decreases up to 300,000 feet and thereafter remains virtually constant. In the lowest part of the ionosphere noctilucent clouds occur. These very rare luminous night clouds whose essential character is still a mystery sometimes disclose the presence of strong winds at high altitudes. Temperature increases upward. Water vapour and dust are absent. Several electrically conducting layers characterise the ionosphere. It is in the lower portions of this layer that aurora borealis and related phenomena are seen most often.

## THE FUNCTIONS OF THE ATMOSPHERE

Beyond talking endlessly about weather, we take the atmosphere for granted, seldom pondering on the existence and nature of this

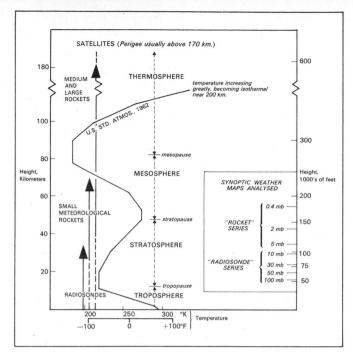

The layers of the atmosphere. Satellites are in orbit up to 292,000 miles above the surface of the Earth.

most necessary part of the Earth. Yet the type of atmosphere on which our lives are dependent is unique in the solar system.

AN INSULATOR. The Sun's radiation consists mainly of short wavelength rays. Because these short-wave rays pass easily through transparent substances, only about 15 per cent is absorbed directly by the air and stored as heat by the water vapour and carbon dioxide content; 40 per cent is reflected back into space by dust and clouds, though this amount varies with weather conditions. The solar energy to reach the ground is therefore about 45 per cent of the original total output. Of this, 10 per cent is reflected back into the air; only the remaining 35 per cent actually heats the upper surface of the land and sea. As the continents and oceans re-radiate this heat both day and night in the form of long-wave rays, which are readily absorbed by the atmosphere, the latter is heated mainly indirectly, by a process which allows the Sun's radiant energy to be absorbed in one form but not in another. The atmosphere, then, acts as an insulating blanket, keeping temperatures in middle and high latitudes from dropping to extremes during night and winter. This principle is also known as the 'greenhouse effect' because it finds application in greenhouses where the glass structure permits the entry of short heat waves but hinders the escape of the long re-radiation waves.

A FILTER. In addition to acting as an insulator, the atmosphere serves to protect us from those of the Sun's rays which are harmful. The ultra-violet end of the spectrum in particular, while important and necessary for life, is nevertheless fatal below certain wavelengths. It is fortunate for us that there is in the atmosphere a protective screen allowing only rays with a wavelength of more than 2,900 angstroms to penetrate it. (The angstrom is equal to one ten-millionth of a millimetre.)

The protective screen is a thin layer of ozone which encircles the globe at an altitude of about 90,000 feet to 100,000 feet. Its effectiveness is proved by comparison of the spectra obtained from rockets. For example, a V2 fired in New Mexico in 1946 showed a spectrum below the ozone layer hardly richer in ultra-violet rays than one near the ground. Above the ozone layer, however, it showed a spectrum with ultra-violet rays from 3,900 to 2,100 angstroms. As rays shorter than 2,900 A are harmful, it can be appreciated just how vital this ozone layer is. Life on earth depends on it. If it dispersed, life would cease; if it gained in thickness, it would absorb not only the harmful ultra-violet rays but also those which bring the vitally important vitamin D to our bodies.

HYDROLOGY. To this function the atmosphere adds that of a major and indispensable link in the hydrologic or water cycle of the Earth. The surface waters of the ocean undergo evaporation;

the water vapour produced rises into the atmosphere and is carried landward by the winds. The vapour condenses into minute droplets which form clouds and, later, fall as rain or snow. (The mechanism of precipitation is discussed on p. 30). Some of this precipitation evaporates as it falls and never reaches the ground at all. What does reach the ground may evaporate back into the air, be absorbed by the vegetation, seep into the ground to form underground water, or flow on the surface to reach streams and eventually return to the sea. The cycle is overlapping and endless.

## ATMOSPHERIC PRESSURE

The first thing to remember is that, despite all appearances, air has weight. A cubic foot of air at normal atmospheric pressure and at 32°F. weighs about 1¼ ounces. This may seem very little, but a bathtub that looks empty contains about 20 ounces of air, and a man at mean sea level carries a weight of 14·7lb. (the pressure of the atmosphere at sea level) on each square inch of his body. He escapes being crushed under the pressure only because his body fluids are maintained at a slightly higher pressure.

Atmosphere pressure decreases rapidly upward till, at about 3·4 miles, it is only about half as much as at sea level. In the lowest six miles of the atmosphere pressures are quite variable; above this they are more or less uniform at all times in a given place.

Most people notice abrupt changes in air pressure, whether produced in a high-speed elevator or in an air-plane's take off. At the approach of a low pressure storm centre sensitive or ailing people may experience actual physical discomfort, perhaps in the form of headaches. These are but rough indications of the vagaries of pressure and their effects. Accurate measurement of such changes are made by barometer. The barometer is a familiar instrument; its readings are quoted daily, even hourly, on the radio and in newspaper weathercasts. To most of us the numbers quoted have little significance, except that a rise in numerical value comes with fair weather, and a lowering of values foretells foul weather. These numerical changes are minute. In the malestrom of events and forces involved in, say, a hurricane, whose intensities are almost impossible to gauge, the barometer provides the best indicator of the storm's progress.

Barometric readings are most commonly quoted in inches of mercury. The newer unit, the millibar, is generally used in the laboratory and by meteorologist. (The millibar is a force equal to 1,000 dynes per square centimetre. A dyne is the force required to produce a velocity of 1 cm. per second to a mass of 1 gramme in 1 second). In its simplest form the mercury barometer consists of a glass tube about a yard long, sealed at one end. This is filled with mercury and carefully inverted into a small, open container partly filled with mercury. After inversion, the column of mercury in the tube will drop slightly, stopping at a level about 30 inches above the surface level of the mercury in the open container. Because atmospheric pressure acts on the surface of the mercury in the container, the weight of the column of mercury above this level must be equal to the weight of the air column above the same surface. Therefore the length of the mercury column at any moment indicates

*Left*
Radio technician tracking the balloon on his radar screen. The screen at the top records the elevation.

*Top right*
Stevenson screen showing thermometer (left) thermograph (centre) and hydrograph (right).

*Bottom right*
A barometric chart of the North

Atlantic showing depression (D) and anticyclones (A).

*Far right*
Weather ship. All information is sent within a few hours to the forecasting stations. Weather charts are then prepared first to provide a picture of recent meteorological events and then to permit forecasting conditions for the next 24- or 48-hour period.

the air pressure at that time. Measurements are read off against a calibrated scale on the glass tube.

The aneroid barometer—the type generally found in domestic use—is a cheaper, more robust and more convenient instrument than the mercury barometer. Air pressure in the aneroid barometer is balanced by elastic forces; by flexing a metallic surface the action of atmospheric pressure moves a scale indicator to read off the appropriate figure.

Variations in pressure are produced mainly by heating, cooling, and the amount of water vapour in the lower atmosphere. A meteorological service has constant need of the patterns of these variations for its weather forecasting. Charts or maps are drawn up each day showing only the air pressures — unlike the more familiar weather map which includes temperatures, wind direction and velocity, precipitation, and sky conditions. Some of the newer weather maps show analysis and distribution of air masses and fronts.

Pressures are normally shown on maps by lines drawn through points of equal pressure, called *isobars* (Greek *isos*–equal, *baros*–weight).

Let us take, for example, a barometric chart of the North Atlantic. Curved lines can be seen, some grouped in concentric circles; others continued to the edge of the map, lying between the circles. These are the isobars and form a pattern resembling that of contours on a topographical map showing the rise and fall of the ground. The comparison is more apt than one might suppose at first sight. Let us look at the amount of pressure, expressed in millibars. South of Iceland pressures decrease from the margin towards the centre. This is known to the meteorologist as a depression or low *(D)*, into which the air is moving from the surrounding regions. In contrast, off Spain, pressure increases from the margin inwards. This forms an anticyclone *(A)* out of which air flows.

Depressions and highs have lives of their own, and may be said to evolve. They appear, develop, travel, and disappear. These changes may be rapid, and are the cause of the more violent atmospheric disturbances such as storms and cyclones. Even when more gradual, they determine wind direction, the movement of clouds, and the development of rain from the clouds.

However varied, though, areas of high pressure and low pressure are not distributed haphazard over the Earth's surface. As a result of statistical studies, the Earth can be divided into climatic zones. These relate to the alternation of high and low pressure. For example, low pressure characterises equatorial and temperate regions, while high pressure prevails in tropical and polar regions, with certain irregularities caused by the contrast in temperature between continents and oceans. Climate classification is not based solely on varying pressure, of course; other weather factors, including temperature, winds, precipitation, and sometimes natural distribution of vegetation and solids, are taken into account when formulating climatic divisions. Because of the many variables, climatic zones are difficult to classify. Systems which include too many variables are confusing; those which are simple and clear-cut are apt to be misleading and inexact. There is no such thing as an ideal climatic zone system, and there probably never will be.

## AIR MASSES

Present-day techniques of weather analysis and forecasting are founded on the fact that the general air circulation of the world tends to produce enormous masses of air differing in physical properties but inherently of a uniform internal character.

Air that stays in contact with the Earth's surface will slowly develop properties typical of the underlying area. For example, if the air currents are favourable and the underlying region is uniform in character, the physical properties of the air mass (particularly temperature and humidity) will become uniform in a horizontal direction. The transition from one large air mass to another is relatively abrupt. The greatest differences in energy occur along the boundaries between neighbouring air masses, and it is in these areas that cyclones, or depressions, develop.

Air mass sources are the areas where air masses develop. When an air mass moves away from its source region, its physical properties begin to change, influencing local weather phenomena.

The important factors which bear upon the life history of an air mass are: the source, (whether marine or continental, dry or moist, warm or cold), the path it follows (toward moister or drier, colder or warmer areas), and age (the time it has spent en route).

The major air mass sources lie where the Earth's surface has fairly uniform conditions and where the wind systems are divergent. Convergent winds produce stronger temperature contrasts, while divergent winds reduce them and tend to make conditions more uniform. Air masses are grouped into six main categories on the basis of their source areas: polar (or arctic), continental, polar maritime, tropical maritime, tropical continental, equatorial and monsoon.

They are also classified according to the influences sustained by them as they travel out of their source areas. This thermo-dynamical classification consists of two main groups: cold air masses (those which are colder than the ground they travel over and absorb heat from below) and warm air masses (those warmer than the surface they travel over and give off heat to the underlying area).

Cold air masses normally originate in arctic or sub-polar regions but in the winter they may originate over continents down to a latitude of 30°N. In their source areas they are cooled from below and are characterised by stable layering, low specific humidity and low temperatures. The cold masses acquire different physical properties, determined by their movements over continental or maritime regions. Changes over these areas vary with season. A maritime cold mass invading a continent in the summer becomes increasingly unstable and showers are frequent and intense. The same type of air mass invading the continent in winter develops great stability and showers decrease.

From warm air masses the most important source regions are the subtropical high-pressure oceanic areas. They may also form in summer over the southern continents. In general, maritime subtropical air masses are fairly stable and have a high humidity. When they invade warm continents in the summer, instability is increased and heavy fogs form, often over large areas.

If an air mass travels slowly it changes its characteristics gradually, but if it travels rapidly enough it comes up against masses of quite different character. The plane along which the two meet is called the frontal surface. The line of intersection between the frontal surface and the Earth's surface is called a front. There are four types of front: cold front, along which colder air replaces warmer air; warm front, along which warmer air replaces colder air; stationary front, along which there is no replacement; occluded front, which results when a cold front overtakes a warm front.

It is along the moving fronts that clouds form, precipitation occurs. air turbulence develops. and cyclones arise. Cold fronts are usually accompanied by marked atmospheric disturbance, wind squalls and thunder showers; warm fronts have more stable air conditions, long cloud systems and steady rain. Thus the movements of these fronts bring about our changing weather conditions.

## AIR CURRENTS

Why does the wind blow? The answer is found in what we have already said about the distribution of pressure on the Earth's surface, and about the pattern of the isobars. The air is bound to move of its own accord from the high to low-pressure areas.

Theoretically, wind direction should be perpendicular to the isobars. Atmospheric pressure is not however the only factor influencing an air mass. Air tends to rise when it is warm and light, and to fall when it is cold and heavy. Where it is in contact with the ground, friction decreases its speed and affects its direction – though friction itself falls off with the height of the mass. Valleys and mountains, too, can direct the wind, and their effect is seen in certain winds with a constant direction, such as the mistral and the tramontane. Finally, there is another influencing factor, the Coriolis force, produced by the Earth's rotation on its axis. This force gives every wind in the northern hemisphere a deviation to the right, and in the southern hemisphere, to the left.

The combined effect of these various factors is that winds blow obliquely to the isobars, and under certain conditions even parallel to them.

Let us apply these results to the movements of depressions and highs, or anticyclones. The illustration shows a depression represented by three isobars. A marginal body of air at A would, ideally, move toward the centre C, following the line AC. The Coriolis force, however, diverts it to the right (assuming this depression to be in the northern hemisphere) so that it takes the direction AB. The same force operates throughout the depression with the result that the whole air mass moves anti-clockwise. It is evident that similar forces operate in an area of high pressure, though it will be seen that here air at the centre cannot move towards the circumference without being swung to the right, so that the currents of this air mass move clockwise. In the southern hemisphere, on the other hand, the Coriolis force causes deviation to the left, and the winds blow clockwise into a depression and anti-clockwise out of an anticyclone. This force will also be considered later as an influence on whirlwinds.

The position of the isobars can give information on the strength of the winds as well as pressures. Winds are strongest and fastest where the isobars are closest together. For example, on the map on page 20, the wind is strongest to the north and weakest to the south.

Wind speed is measured by means of an anemometer. A simple and widely used type is a sort of horizontal windmill fitted with little cups on its arms. Wind force is expressed in miles per hour or in knots, one knot being a speed of one nautical mile (6,080.20 feet) per hour.

The Beaufort scale, by which the strength of a wind at sea could be indicated, was introduced in 1805 by the British admiral Sir Francis Beaufort (1774–1857). It consists of a series of numbers linked to descriptions of the effects produced in the behaviour of certain objects (smoke, flags, sails) by the force of the wind. Thus by simple visual means a moderately skilled observer is able to estimate the force of any wind. The Beaufort force numbers range from force O (a calm) to force 12 (a hurricane 'which no canvas

The action of the Coriolis force.

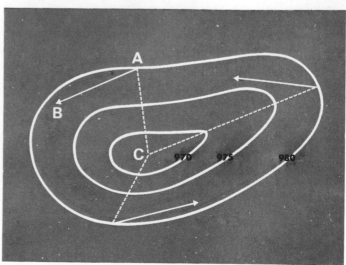

could withstand'). The Beaufort scale was adopted in 1874 by the International Meteorological Committee.

In June 1939 the same body set up a scale of wind velocity as an international standard of comparison, to be measured by an anemometer 20 feet above the ground. The velocity scale covers all forces from calms to hurricanes, ranging through light airs, breezes, wind squalls and tempests. The Beaufort scale is still in international use. Measured wind velocities are converted to Beaufort numbers to ensure a standard throughout the world.

## TRADE WINDS AND MONSOONS

We have seen that the equatorial zone is an area of low pressure, while the tropical areas bordering it to the north and south are high-pressure zones. In a room heated by an open fire, a draught of cold air can be felt entering the room by the cracks of the doors and windows and crossing the floor in the direction of the hearth This air current becomes lighter by expansion on heating, rises in the chimney, and subsequently is lost in the sky. By the same principle, the equatorial region is the largest chimney in the world. Here the ground heats up faster than anywhere else because the Sun's rays, striking the region perpendicularly, travel through the smallest amount of atmosphere. (In northern and southern areas, because of the Earth's curvature, the Sun's rays must travel further through the atmosphere, with the ultimate result that the ground does not receive as much heat.) As the humidity of the air in the equatorial belt is also extremely high, there is a continuous ascent of water-vapour and cloud growth until sunset. On the coast, a torrential downpour generally ends an equatorial day.

### BEAUFORT SCALE OF WIND FORCE

| Force Description | Equivalent speed at 10m. (33 ft) above ground | | | |
|---|---|---|---|---|
| | Knots | | Miles per hour | |
| | Mean | Limits | Mean | Limits |
| 0  Calm | 0 | 1 | 0 | 1 |
| 1  Light air | 2 | 1 – 3 | 2 | 1 – 3 |
| 2  Light breeze | 5 | 4 – 6 | 5 | 4 – 7 |
| 3  Gentle breeze | 9 | 7 – 10 | 10 | 8 – 12 |
| 4  Moderate breeze | 13 | 11 – 13 | 15 | 13 – 18 |
| 5  Fresh breeze | 18 | 16 – 21 | 21 | 19 – 24 |
| 6  Strong breeze | 24 | 22 – 27 | 28 | 25 – 31 |
| 7  Near gale | 30 | 28 – 33 | 35 | 32 – 38 |
| 8  Gale | 37 | 34 – 40 | 42 | 39 – 46 |
| 9  Strong gale | 44 | 41 – 47 | 50 | 47 – 54 |
| 10  Storm | 52 | 48 – 55 | 59 | 55 – 63 |
| 11  Violent storm | 60 | 56 – 63 | 68 | 64 – 72 |
| 12  Hurricane | 68 | 64 – 71 | 78 | 73 – 82 |
| 13  — | 76 | 72 – 80 | 88 | 83 – 92 |
| 14  — | 85 | 81 – 89 | 98 | 93 – 103 |
| 15  — | 94 | 90 – 99 | 109 | 104 – 114 |
| 16  — | 104 | 100 – 108 | 120 | 115 – 125 |
| 17  — | 114 | 109 – 118 | 131 | 126 – 136 |

The rise of heated air from the Equator causes a barometric depression at ground level, and an inward movement of air from the bordering tropical regions. The winds resulting from this are known as the trade winds. Deflected by the Coriolis force, they

blow from north-east in the northern hemisphere and toward north-west in the southern hemisphere. Between the north and south trade winds lies the equatorial belt of rising air known as the zone of equatorial calms, or doldrums, where the ships of Christopher Columbus were becalmed for so many weeks before reaching America. As the warm air rises and travels away from the Equator, it becomes progressively cooler and heavier until finally, heavier than the air beneath, it sinks downward. This takes place near the 30° latitude, forming another belt of calms, called the horse latitudes. The air then divides here, part of it flowing northward and part of it returning to the equatorial zone. The poleward-flowing portions (the zones of the westerlies) at about 60° latitude meet a front of colder, heavier air moving outward from the Poles (the zones of the polar easterlies) that inserts itself under the warmer air. The warmer air then rises to higher altitudes where part of it continues to flow poleward and part toward the south. A portion of the cold wedge is warmed and also rises to return to the Pole.

In contrast to these major wind belts, constant and reliable enough to earn the name of trade winds in the days of sailing ships, there are other winds which periodically change their direction, giving rise to violent storms during each change-over. These are the monsoons, a name derived from an Arabic word for 'season'. They occur in southern Asia where they produce a characteristic pattern of weather recognised even in the ancient world, though their cause was not understood until discovered by the astronomer Halley at the end of the seventeenth century.

In addition to seasonal temperature and pressure variations over land and sea which produce monsoons, there is a daily variant that develops a similar but local effect. It can, perhaps, best be explained by reference to the phenomenon of land and sea-breezes familiar to all of us. In summer the land heats up more quickly during the day than the sea. Warm air rises and causes a low-pressure area near the coastline, into which cooler air flows off the sea. This sea-breeze increases in strength until the middle of the day, then decreases. After sunset, when the land cools more rapidly than the sea, it gives way to a gentle land-breeze blowing in the opposite direction. This effect is particularly noticeable in a seaside town lying at the mouth of a valley opening on to the coast. A somewhat similar effect is found in mountain country, where on sunny days the mountain slopes are warmer and at night colder than the free atmosphere. Because cold air tends to sink and warm air to rise, a local system of winds develops, with an upslope trend in the daytime and a downslope motion at night. Except on cold winter nights in narrow valleys, these winds do not usually attain any great velocity.

Let us now enlarge this local explanation to encompass continents and oceans. In winter, continents are cooler than oceans. For example, because of the high plateaux and mountains of Yunnan, Tibet and the Pamirs to the north, India is much cooler than the Indian Ocean. The air over the continent becomes colder and denser. In other words, at this period India is an area of extreme high pressure, while the Indian Ocean is the centre of a barometric depression. Consequently the wind blows from north to south, or rather, influenced by the Coriolis force, from north-east to south-west. The direction recalls that of the trade winds of the northern hemisphere. This wind is the winter monsoon, and is is dry and cold.

After the month of March or April, the situation alters. The land surface in India heats up more rapidly than the sea surface, and into the resulting barometric depression blows a strong wind, this time from south-west to north-east. Though occuring where the northerly trades would normally blow, this takes the same direction as the southerly trade winds. That is perhaps why it is much stronger than the winter monsoon. It is characteristically warm and wet — warm because it comes from the Equator, and wet because the air has taken up water-vapour while passing over the sea. This is the summer monsoon which is replaced by the next winter monsoon, which occurs in September or October.

It is now clear that the winter monsoon corresponds to the nocturnal land-breeze, and the summer monsoon to the daytime sea-breeze. The explanation is the same for both; the difference lies in the size and importance of the monsoons, and their great influence on climate. For six months of the year the dry north-east wind parches the earth, bringing farming to a standstill. For the remaining six months, the south-west wind brings the nourishing rain to India. This is the season when the luxuriant tropical vegetation flourishes, the grain and fruit ripen and the cattle become fat again. Hardly any other country in the world offers such a clear-cut example of the influence of climatic forces on life.

In North America the monsoon effect is similar, but because of the continent's smaller size the winds are less extreme. In the summer the air tends to move generally from the Gulf of Mexico northward, and in the winter southward from sources in Canada.

## STORMS

Trade winds and monsoons follow well defined paths; there are other winds whose tracks are more complex and whose internal movements are spiralling. Revolving storms are variously called whirlwinds, tornadoes, willy-willies, waterspouts and hurricanes. On a larger scale are the cyclones and typhoons which from time to time cause devastation in the West Indies, the southern United States, India, Madagascar and China.

A storm is any disturbance of the atmosphere which moves its location and is attended by precipitation, clouds, and moderate or strong winds. Most storms are connected with areas of low atmospheric pressure toward which air is moving, generally with an in-spiralling motion. The meteorologist uses the word 'cyclone' for any low-pressure area into which there is a spiral motion of air, whether the wind is gentle or of storm velocity. To the layman a 'cyclone' signifies a terrific storm, and to many Americans the word is synonymous with tornado. In an attempt to avoid confusion we will use the meteorologists' terminology: 'cyclone' for low-pressure areas with gentle to moderate spiralling winds; 'cyclonic storm' (or a specific name, such as tornado) for low-pressure areas with storm velocity winds.

*Cyclones,* then, as classified by the meteorologist, are mild affairs that are experienced daily in the middle latitudes. Listed according to increasing severity but decreasing size they are: middle-latitude cyclone, tropical cyclone (called cyclone in India, hurricane in the West Indies and in parts of the United States, typhoon in east Asia), and tornado.

*Thunderstorms* represent another variety of storm, local in nature and associated with cumulonimbus clouds (see chart, p. 27). storms occur in large numbers in a single day in the middle latitudes, and occasionally tornadoes develop with them.

A spiralling wind originates in a barometric depression at the contact of two air masses of different temperature, for example, warm tropical air and cold polar air masses. A depression, or cyclonic area, is, as we have said, an area into which the surrounding air is moving. We have already seen that the Coriolis force turns the simple centripetal movement into a clockwise movement in the northern hemisphere. If the barometric depression is very deep (a condition which would be shown on the weather chart by the isobars lying very close together), this movement of air becomes more and more rapid, and the deep depression becomes a cyclonic storm.

*Cyclonic* storms develop when the mass of rotating air departs from its normal course. Without entering too deeply into this complex study of double movements, let us recall that a cyclone arises where a mass of warm air moving away from the Equator meets a cold air mass from the Pole in the middle of the tropical regions, or doldrums. This nucleus of low pressure has intruded into an air mass of high pressure, and makes its escape from it as quickly as possible. As if shot from a sling, it whirls along a curved course, turning north-eastward in the northern hemisphere, and south-eastward if it originates in the southern hemisphere.

The combination of the elements of movement in a tropical cyclone or hurricane — its rotation on its own axis plus its forward movement — causes a difference in wind speed at the margins. On the most dangerous side, the two movements are in the same direction, and the speed of the air may reach up to 200 miles per hour. On the other side the movements are in opposite directions and air speed is therefore less.

Near the centre of the tropical cyclone prevails an area of calm,

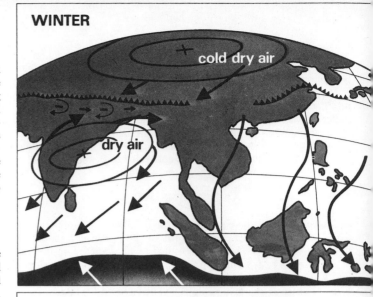

WINTER

cold dry air

dry air

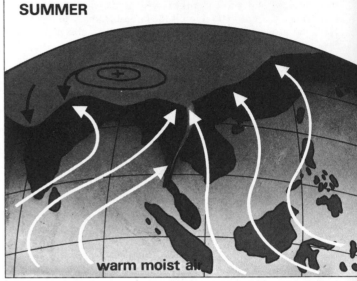

SUMMER

warm moist air

which may even be sunny, with light variable winds. This area is known as the 'eye' of the storm. A ship entering it soon finds her bridge and rigging covered with insects and birds exhausted by their struggle in the side of the hurricane they have just passed through, and seizing the chance to rest before being caught up in the fury of the other side. The eye is simply a hollow vortex formed by the spiralling of the air, and is comparable to the centre of a vortex of water going down a drain.

Progress of a tropical cyclone resembles that of a top spinning rapidly round its axis as it moves slowly over the ground. While the rotary speed may be as high as 250 miles per hour, lateral movement is generally between 10 to 20 miles an hour. Even so, cyclonic storms lasting for about ten days (an average duration in the Atlantic) may travel a total of about 3,000 miles.

There are wide differences in the size of tropical cyclones. They may be only a few miles wide, or by the time they are shallow and weakened just before dying out they may be as much as 500 miles across. The nearer the Equator, the smaller their diameter; as they move away from the Equator so diameter increases.

How does a cyclonic storm end? Simply by the progressive replacement of warm air by colder air coming in from its margins. Slowly the barometric depression fills up, and revolving motion and lateral movement grow slower. The storm finally dies of exhaustion, though not before it has left behind it a trail of demolished houses, uprooted trees and grounded ships.

The force of a cyclonic storm is almost impossible to describe. In 1935, during a historic Florida cyclone, a relief train was sent to the devasted area. The violence of the wind was such that it blew the train right off the rails, except for the locomotive and tender. In 1938, a violent hurricane devasted the east coast of the United States. In New York, the highest skyscrapers swayed, while in the surrounding countryside thousands of houses fell to the ground or were crushed by falling trees. Pylons and more than 20,000 miles of electric lines were torn down, and millions of

trees. In 1944, the U.S. Third Fleet was overtaken by a typhoon in the north-west Pacific. Within minutes, more than twenty ships were in trouble, battleships and aircraft-carriers among them. The heavy cruiser *Pittsburgh* had one hundred feet of its forward superstructure smashed and scarcely made the base at Guam only.

The following eye-witness account by de Chambertrand of the 1928 storm in Guadeloupe Island, West Indies, was published in Paris *(L'Illustration, 1928)*.

'Soon after noon on the 11th September, the wind began to blow from the north, while the barometer which stood at 1,013 millibars (29·9 inches), about normal, began to fall — unmistakable signs of an approaching storm. At 4 p.m. it had fallen to 1,010·6 millibars (29·84 inches), and as the sun went down we could see one of those ominous sunsets with copper-coloured clouds which are the forerunners of foul weather. At 9 p.m. pressure was 1,008 millibars (29·7 inches), and the wind was gradually increasing. At midnight, I decided to start barricading several of the doors in the house; the barometer was down to 1,005 millibars (29·6 inches). At 5 a.m. on the 12th it fell rapidly to 1,002 millibars (29·6 inches), and by 7 a.m. to 1,000 millibars (29·5 inches). After that the situation rapidly deteriorated. The wind was gaining strength every moment. A telegram came from Puerto Rico giving the centre of the depression as 300 miles from Guadeloupe, and moving towards it. My house, which lay in a high and isolated position seemed less secure, and at about 10·30 a.m. I left it to take refuge in a neighbouring house in a lower and more sheltered position, taking with me two children and my barometer. Towards midday the house collapsed; the barometer was then down to 960 millibars (28·3 inches). Even so, the wind was still increasing in violence and the cyclone growing more intense.

'Everywhere there was a fearful chaos of tiles and floors from ruined houses whose walls were crumbling. Waves poured over the quay, and wrecked the docks. The sea swept up to the town and the shops. Then the house where I had taken refuge

started to go. The roof was torn off piece by piece, and the ceiling of the upper storey collapsed on to the floor below. Everything was swimming in water and the wind had reached prodigious strength. At last, towards 2 p.m. the eye of the storm passed over us. The sky cleared for about ten minutes, which gave me time to note the lowest pressure I had yet observed, 942 millibars (27·8 inches). The wind, which had previously been blowing from the north, now swung to south and raged more furiously than ever. Several times we felt the house shake and shift. However, it remained standing, and by 4 p.m. to our joy, the barometer had risen to 970 millibars (28·6 inches), and by 10 p.m. it stood at 990 millibars (29·2 inches). By 4 a.m. next day, the 13th September, it rose to 1,005 millibars (29·6 inches) as the last gusts of the dying wind disturbed the ruined town. When we ventured out at 7 a.m. the barometer recorded 1,008 millibars (29·7 inches).

'What a spectacle awaited us! Houses wrecked and gutted! Streets blocked with every kind of debris, trees stripped to bare trunks — all, that is, that were not uprooted. The place was unrecognisable; it was a scene of devastation where rescue work was proving more than difficult. Appalling sights were on all sides. Corpses were being dug out of the ruins. And worst of all was the isolation. Communications were at a standstill; famine and epidemic preyed among the twisted girders, splintered beams, and flattened houses.'

The contemporary newspaper records of the Guadeloupe hurricane describe the successive phases of the storm in perfect detail. The stages are similar whether the cyclonic storm occurs over land or sea, and except for intensity they are the same in each and every storm. They may be roughly summarised as follows:

THE APPROACH OF THE STORM. The day before the storm breaks is usually calm with air pressures slightly above normal. The sky is streaked with cirrus clouds (*see* chart p. 27), sometimes veil-like or in streamers, sometimes giving a halo to the sun or moon.

A house blown down in the
Guadeloupe hurricane of 1928.

Map showing the path
taken by the 1928 hurricane.

and sometimes producing vividly red sunsets. Out at sea a long swell is felt; it is made up of dying storm waves that have out-distanced the slowly travelling storm centre.

THE ARRIVAL OF THE STORM. Atmospheric pressure drops — from 1,010 millibars (29·8 inches) to about 980 millibars (28·9 inches), and a dark cloud approaches, enveloping all in torrential rains, thunder, lightning, and tremendous wind gusts. Huge waves with high-blowing spray are generated at sea. This phase lasts several hours.

THE PASSAGE OF THE CENTRE. With the arrival of the centre, total calm comes abruptly; skies clear, the sun shines, and there may even be a steep rise in temperature. The barometer is at its lowest, around 965 millibars (28·5 inches). At sea, waves are still mountainous, rising in great peak-like masses. This stage may last a half hour or so.

ARRIVAL AND PASSAGE OF THE SECOND HALF OF THE STORM. This is the last stage. It is as severe as the first half of the storm — sometimes even more severe — but with reserved wind directions. The full fury of this portion lasts for several hours. Then winds gradually lessen and return to normal, clouds break, air pressure rises to normal, and fair weather returns.

*Tornadoes* are the smallest but most violent of all storms. They occur throughout the tropical and subtropical areas of the world but are most frequent and violent in North America and, to a lesser extent, in Australia. The number in the United States is about 670 yearly. The tornado appears as a dark funnel cloud suspended from a huge cumulonimbus cloud (*see* chart p. 27). At its narrow end the funnel may be a few feet to a mile in diameter. Its dark colour is caused by the density of condensing moisture, the debris, and the dust picked up by the powerful winds in the funnel.

This small, extremely violent cyclone has wind velocities higher than those known in any other storm. The super-hurricane velocities are estimated to run as high as 500 miles per hour, and in the central part of the funnel there is a strong rising air current. The tornado funnel, accompanied by heavy rains or hail, thunder and lightning, twists and turns as it moves across the country, travelling alternately in contact with the ground or rising above it. Where it touches the ground it leaves complete destruction in its wake; where it rises it leaves the area below unharmed. The region of destruction is narrow but complete. The combination of great wind velocities and the sudden reduction in air pressure is often sufficient to cause closed buildings to explode outwards. Storm cellars built entirely below the surface of the ground give effective protection if the tornado is noted in time to take advantage of them,

and if given due warning a car travelling at right angles to the storm's path can find escape.

Luckily tornadoes are fairly localised and short-lived, lasting only an hour or two, and travelling a maximum of 50 miles. In the United States they often form in series and follow almost parallel paths along the squall line of some types of strong cold fronts that cover much the same course each season. Then the annual course of the tornadoes follows a similar path.

*Waterspouts* are tornadoes that form over the ocean, usually in subtropical waters near continental areas. They are smaller and weaker than land tornadoes; although spray is carried high, the water is lifted only eight to ten feet into the air.

## ATMOSPHERIC HUMIDITY

Only passing reference has so far been made to the water that is always present in the air and produces such phenomena as dew, hoar-frost, clouds, fog, rain, sleet, hail and snow. The quantity of water required for this is hard to imagine. The total annual rainfall over the land areas of the world is about 26,000 cubic miles. To provide this vast amount great quantities of water must be lost initially by the oceans through evaporation. Evaporation rates range from very low in the cold polar regions to very high in the belts of the subtropical highs. The Mediterranean Sea, to take an example, loses a five-foot-thick layer of water annually. Of the total rainfall, one-fifth is returned directly to the oceans by streams. Most of the remainder sinks into the ground and is returned to the seas indirectly. A comparatively small amount is returned directly to the atmosphere by the evaporation of falling rain, run-off, and surface waters on the land.

The permanent presence of water-vapour in the air can be proved by simple experiment. A glass containing cold water mists over when brought into a warm room, demonstrating the presence of moisture which condenses on contact with the cold surface. Again, in the country on a clear night, ground and vegetation radiate the heat they have stored up in the day. Temperature drops during the night and by dawn there is a covering of dew (water-vapour which has condensed on the cold surface). In very cold weather, the water-vapour may pass directly from gaseous to solid state, and become frost, which covers branches and telegraph wires with a layer of ice crystals, which varies in thickness.

**Measuring humidity.** Atmospheric humidity is measured by the hygrometer and psychrometer. The hair hygrometer is based on a simple phenomenon: the variation in the length of human hair at different humidities. When hair is dry its cells pack closely together. When the air is humid, water absorbed by the spaces between the cells pushes the cells apart so that the hair is longer

The map legend reads:
- - - - - Hurricane of 12-23 August 1830
———— Track of hurricane of 12-20 September 1928

when moist. The hair hygrometer measures variations in the length of a bundle of hairs scaled to indicate not length but relative humidity.

The psychrometer consists of two thermometers. The bulb of one is wrapped in wet cloth. If the surrounding air is not saturated with moisture, evaporation of the water in the cloth takes place, cooling the wet-bulb thermometer which consequently registers a lower temperature than the normal, or dry-bulb, thermometer. The lower the atmospheric humidity, the more intense the evaporation, and the greater the temperature disparity between wet and dry bulbs.

There are standard humidity tables which give the relative humidity for any combination of wet and dry-bulb temperatures.

**Clouds.** Normally water-vapour rises in the atmosphere with the ascending currents of air. As the air rises and expands, it cools. This condenses the water-vapour into minute droplets which make up the clouds.

Clouds can be divided into layers: the lower clouds (below 10,000 feet), the intermediate clouds, (10,000 to 20,000 feet), the upper clouds (20,000 to 40,000 feet), and clouds of great vertical extent belonging to more than one of these zones simultaneously. The variety of cloud shape or form is endless. The number of basic types, however, is limited. The international classification includes ten main types, which for the sake of convenience are arranged according to height. It is given above with altitude zone and approximate elevations above ground.

In addition, to this classification by height, clouds are typed according to their form into two main categories: stratiform and cumuliform. The stratiform or layered types include cirrus, cirrostratus, cirrocumulus, altostratus, altocumulus, stratocumulus, nimbostratus and stratus. The cumuliform or massive, globular types include cumulus and cumulonimbus.

Clouds do not occur singly; they form long processions and masses which meteorologists call cloud systems. The systems are often over 100 miles wide and 1,000 miles long, and may extend upward from 6,000 feet to about 20,000 feet. Consisting mainly of stratiform clouds, they are generally associated with movements of large masses of warm air. Groups of cumulus clouds and showers are typical of cold air masses.

| Cloud type | Altitude |
|---|---|
| Cirrus<br>Cirrostratus<br>Cirrocumulus | High: approximately 20,000 to 40,000 feet |
| Altostratus<br>Altocumulus | Medium: approximately 8,000 to 20,000 feet |
| Stratocumulus<br>Nimbostratus<br>Stratus | Low: generally less than 8,000 feet |
| Cumulus<br>Cumulonimbus | Great vertical development<br>Base: average 3,000 feet or less<br>Top: up to and over 20,000 feet |

The forepart of the cloud system generally consists first of high clouds: delicate white feathers of cirrus, the highest of all clouds, veils of halo-forming cirrostratus, and grey blankets of altostratus. These clouds may be several hundred miles long and while they are passing overhead there is no danger of rain. Later they merge with the main body of the system that follows, and little by little the appearance changes. The mass of clouds grows thicker and the early light clouds give way to the dark layers of stratus and the shapeless, ragged layers of nimbostratus. The prefix nimbo-signifies that rain or snow is falling from the cloud. Precipitation lasts as long as the main mass of the formation, sometimes 200 miles in extent, is passing over. Beneath both the forepart of the

system and the main body, low clouds of the stratus or cumulus types may be present, partly obscuring the upper series.

Toward the end of the procession, the clouds begin to break up, and cumulus and altocumulus may form a sort of rearguard. Their thick, rounded shapes are a sign of the end of the low-pressure area and the passing of a weather front. Altocumulus clouds and cirrocumulus clouds are often arranged in groups or lines suggesting the markings on a mackerel's back and are called 'mackerel sky'. Cumulus clouds are normally associated with fair weather but can develop into cumulonimbus or thunderhead clouds, with great rising masses like mountains crowned by an anvil-shaped top and often ringed by false cirrus. Accompanied by heavy rain, thunder, lightning and gusty winds, cumulonimbus are often associated with the front of a moving mass of cold air.

**Fog.** Clouds that sweep down to ground are called fog if visibility is reduced to about five-eighths of a mile or less. If visibility is more than that, then in Europe the ground clouds are called mist. In North America the term 'mist' is often used as a synonym for drizzle or fine rain. Fog and mist form when the air at ground level is cooled below the dew point (the temperature at which moisture begins to condense on cooling). There are many types of fog, originating under different conditions and associated with a variety of weather phenomena and geographic conditions. Some of the more frequent types include: fogs associated mainly with cold oceanic currents (sea fogs), those produced by cold air passing over warmer ocean waters (steam fogs), those produced by cooling of the air while in contact with cold ground, those formed by radiative cooling of the ground, and those formed by air moving upslope to cooler altitudes.

Haze, which consists of dust or salt particles, should not be confused with fog or mist. The particles which form haze are microscopic in size, or nearly so, and even though they cannot be detected by the eye they affect visibility. Distant objects if dark in colour appear to be veiled in pale blue (the veil appears yellowish if the distant object is white). If distances are fairly long or haze is heavy, all details disappear and objects stand out like silhouettes.

**Cloud Formation.** It was once believed that clouds developed whenever air saturated with water vapour suffered a sudden drop in temperature. This alone, however, is not sufficient to explain the phenomenon. If the air is absolutely pure, condensation does not take place; the air simply becomes supersaturated with water vapour. The deciding factor in condensation is the presence of dust and microscopic salt particles carried by winds. Analyses of fog waters and cloud waters show that they contain sodium chloride (common salt) and various sulphates. It is thought that the salt particles are brought into the air through the evaporation of ocean spray, and that when the spray droplets evaporate, the salt particles, because of their minute size, remain airborne even in seemingly still air. Some salt particles may be introduced by the wind blowing over inland salt flats. The sulphur compounds are introduced by fuels burned for domestic and industrial purposes. Both substances, especially the sulphur compounds (mainly sulphur trioxide), have an affinity for water and absorb it readily, so that when the temperature of the atmosphere approaches the dew point their hygroscopic particles begin to absorb moisture.

Up to at least 35,000 feet the atmosphere always contains plenty of these nuclei of condensation. Naturally there is more dust over cities than in rural areas. In the latter there are several thousand nuclei to every cubic centimetre of air; above the cities there are about ten times as many. This explains why fogs are thicker and more persistent over urban areas (especially industrial areas) than in the country. The famous fogs of London are made up of droplets covered with a thin oily film which retards the process of evaporation.

Average diameter of a cloud droplet is 40 microns (1 micron equals 0·001 mm.). In one cubic centimetre of air there may be 500 to 1,500 droplets. To get some idea of what this means, let us take a large cumulus cloud of about 30 cubic miles. This would contain some 50 million million droplets, about 150,000 tons

Because they are so light and because they form mainly in rising air currents, cloud droplets remain suspended. They do not reach

*Left to right*
Cumulus.
Cumulonimbus.
Mammato cumulus.

*Far left*
1. Stratus.
2. Nimbostratus.
3. Altostratus.
4. Cirrostratus.
5. Cirrus.
6. Cumulus.
7. Cumulonimbus.
A. Head of cloud system.
B. Body.
C. Tail of system.

*Left*
Upward deflection of strong air
currents by distant mountain
ranges caused this banner
cloud formation.

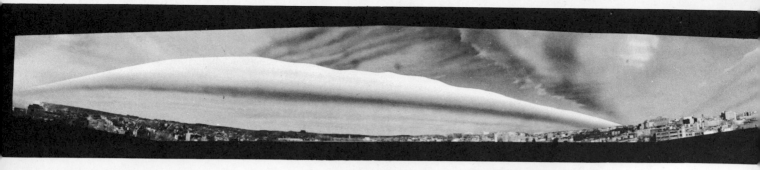

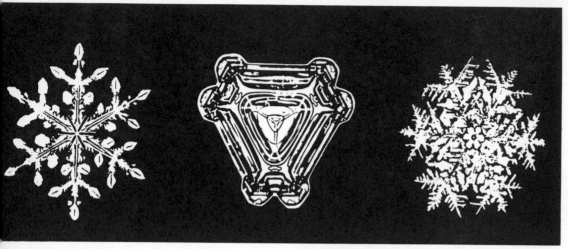

large proportions because as they grow the hygroscopic content becomes diluted and consequently less active. The average composition of a cloud droplet is 1 part hygroscopic material to 10,000 parts water, by weight. Average diameter of a droplet is, as we have already learned, about 40 microns. An average condensation nuclei is about 1 to 2 microns in diameter. The smaller cloud droplets (less than 40 microns) grow rapidly till all are relatively large (around 40 microns in diameter). The growth of condensation nuclei to average cloud-size particles takes about 100 seconds. It would take a cloud droplet 24 hours to grow to raindrop size (500 to 4,000 microns in diameter). Because raindrops, as such, cannot be kept in the air for such a long formation period, it follows that they must form through coalescence of several cloud particles and not through continued condensation.

Raindrops measuring more than 5 mm. in diameter break up during the course of their descent. The speed of their vertical fall is limited to 25 feet per second, without taking into account any extra speed imparted by a driving wind.

**Rain.** Exactly how cloud droplets coalesce into large raindrops is not known with certainty. Many mechanisms which might produce coalescence have been studied; most of them have been discarded because they are not of sufficient significance. The factors studied have included electrical charges (these are too minute to affect the issue, and the distance between the droplets too great), varying size of droplets, variations in temperature, variations of particle size plus air turbulence, turbulence of air alone, and the freezing of ice. Only the last is significant.

The Norwegian meteorologist Bergeron states that droplets increase in size only when the upper part of the cloud is high enough to be made of grains of ice, not water droplets. The droplets of water immediately below this ice-layer are thus in a supercooled condition, that is, they are still liquid although at a temperature below freezing-point. Under these conditions, the falling ice particles gather up the supercooled droplets, just as a cold object gathers frost. As they become heavier so they fall to the ground as hail (that is, still in the solid form) or they melt on the way down and are finally precipitated as rain.

It has been suggested that rain is always produced in this way. This is not so. Showers can be observed to form at high temperatures in cloud that have no portion in sub-freezing areas. These,

however, are generally very light showers and can be attributed to turbulent air mixing warmer drops with colder drops in a cloud.

Confirmation of Bergeron's theory is offered by artificial rain which can be produced by 'seeding' clouds. The first experiments in 1946 consisted in sprinkling dry ice (frozen carbon dioxide) into the clouds. The crystals cool the water droplets which turn into ice and start the rain-producing reaction. Rain falls within a few minutes. In later experiments, ice was replaced by fine crystals of silver iodide, the properties of which resemble those of ice and induce precipitation in the same way. Currently the experiment is usually carried out from the ground by burning a substance that releases silver iodide. The day may not be far off when, provided clouds are available for 'seeding', we shall be able to produce rain artificially whenever the farmers want it, or thick snow whenever the winter resorts want to ensure good skiing conditions.

**Drizzle.** This is a fairly uniform precipitation of very tiny drops (less than 0·5 mm.) almost floating in the air and following every slight air movement. It falls from low stratus clouds or fog. Sometimes when rain falls from high clouds and partly evaporates en route, raindrops are as minute as drizzle. The difference is that the tiny raindrops are more widely spaced than the drizzle drops.

**Hail.** Hail is an impressive precipitation of irregular lumps or balls of ice from 5 to 50 mm. or more. Real hail generally falls only during violent thunderstorms and is rare at ground temperatures below freezing-point. An ice crystal falling through a cloud gathers any droplet it touches in the form of frost, as has already been explained. An ascending air current may carry it back into the upper part of the cloud, from which it falls yet again and acquires more frost. The hailstone thus goes up and down, yo-yo fashion, growing bigger and bigger with each successive layer of ice deposited on it. Often alternating concentric layers of clear ice and snow reveal repeated movement from the liquid to the snow area of the cloud. The largest hailstone ever verified fell in Nebraska, U.S.A., in 1928. Its circumference was nearly 17 inches and it weighed 1½ pounds. Even larger hailstones are not beyond possibility.

**Snow.** Frozen water usually falls not as hail but as snow. Crystallisation must occur slowly, in calm air and at a temperature near freezing-point. These conditions produce the beautiful hexagonal

crystals revealed in detail in photomicrographs. Their delicate shapes are the skeletons of incomplete crystals. Tiny needles of ice form along the crystal axes, but in most cases they never get as far as joining up to form the complete crystal, a hexagonal plate. A scientist can study the form of a snow crystal and determine what air currents and what variations in temperature it has gone through to reach its existing shape. Indeed, no two snow crystals are exactly alike. As for snowflakes, they are aggregates of snow crystals. Some have been known to reach eight inches in diameter.

Snow clouds are not special clouds; they can build up at any altitude. A calm atmosphere and low temperature are the only essential conditions for the formation of snow.

Finally, it should be mentioned that the highest clouds, cirrus and cirrostratus (both recognisable by their wispy shape) are made up of minute ice crystals (5–20 microns in diameter) that will never reach ground level. Reflection and refraction of light by these ice crystals cause the halo which sometimes appears around the Moon.

## CLIMATE AND WEATHER

The climate of a region is the aggregate of the atmospheric conditions which characterise it in the course of an average year: temperature, pressure, wind, humidity, and precipitation. The word refers not simply to the average conditions but to the overall factors, the extremes as well as the means. Weather, the target of daily comment and humour, is the condition of the atmosphere at a given moment or for a short term. Meteorology is the science of the atmosphere and weather; climatology the study of climates.

Climate is one of the most powerful controls imposed on man by nature. It governs to a great extent the soils and vegetation native to a region; it is reflected in the form of the topography, man's clothing, foods and buildings, and in countless other things.

The classification of climates is a complex problem, and no single existing system is perfectly adequate for the purpose. Any single major factor (temperature, pressure, vegetation, soil, etc.) or a combination of them may be used as the basis of a classification system. For example, a classification based on rainfall can be made: arid (0 to 10 inches), semi-arid (10 to 20 inches), sub-humid (20 to 40 inches), humid (40 to 80 inches) and very wet (over 80 inches).

Another classification divides climates roughly into belts matching up with the pressure zones. An area of low pressure runs from 5°N to 5°S, and is about 600 miles long. It is the hottest zone of the Earth and has the most constant temperatures. Within this tropical climatic zone, temperature averages close on 80°F and the seasons are determined by the variations in humidity and not, as in temperate regions, by temperature differences. Rainfall is heavy throughout the year, but somewhat heavier during the period December to March. The average amount of rainfall is 80 inches or more a year. Situated between the trade winds of the two hemispheres, the equatorial zone is, by comparison, a zone of calms with light variable winds, and very little pressure variation, the pressure being only slightly less than at sea level.

The zones of high pressure occur between 5° and 40° of latitude. There is a variety of climatic types in these zones which are crossed by the trade winds carrying few clouds and having little tendency to form them. The customary clearness of the air contrasts with the disturbed skies of the equatorial zone. The result of the absence of clouds is that many of the land masses of the tropical high-pressure areas are deserts, and here we find the driest regions of the Earth. The average annual rainfall is nearly always less than five inches, though some areas in this belt, particularly between 5° and 15° have tropical climates with alternating wet and dry

seasons. Coupled with the monsoons the rainfall is tremendous.

In a zone roughly between 15° and 35° lie the areas with remarkably uniform dry climates. The vast deserts of the world are found here: north Africa, Arabia, Iran, north-west India, south-west United States and south central Australia. In these areas embracing the Sahara, the Sonora and the Kalahari, annual rainfall is always less than five inches, and number of years may go by without any measurable rainfall at all. Occasional brief, heavy rains may occur when tropical air masses penetrate the zone and form single storms restricted to small areas covering only a few square miles. Such rains may be brief but they may form the major part of a year's rainfall and may result in huge and damaging mudflows. Daily and seasonal temperature variations are often great in this zone. The Sonora Desert of the United States has perhaps greater temperature variations than any other.

Low-pressure areas are bounded approximately by the parallels 40° and 60°. Depressions are frequent here but never turn into truly devastating typhoons. West-coast Europe owes its rain-bearing westerly winds to these depressions. Markedly different seasons characterise this belt: warm summers and cold winters, and highly changeable weather with moderately violent storms. Although variation is great, extremes of temperature and weather are rare.

High-pressure areas occur in latitudes above 60°, the areas of prolonged days and nights, long and very cold winters, short and very hot summers, brief spring and autumn seasons, constant winds, and a relatively small annual precipitation concentrated in the warm season.

To this rough enumeration of the successive climatic zones must be added local variations due to altitude, seasonal winds, influence of the land versus the sea, and so on. The influence of altitude is obvious: the climate of a mountainous region, in whatever part of the Earth it may occur, is always allied to the polar climate and supports the same sort of vegetation. In regions with seasonal variation in wind, we have already seen how the monsoon regime completely changes the climate twice a year in southern Asia.

In addition, each of the climatic zones can be divided into a continental and a marine variant. The water of the sea is an immense heat storage unit which mitigates extremes of temperature in neighbouring land areas. This explains why localities at the same latitude may have such different temperature variations; for example, a range of 44°F. in the Faeroes against 143°F. at Irkutsk; 46°F. in the Azores against 78°F. at Saint Louis, U.S.A.; 48°F. in Brittany and 60°F. in Paris.

Another action of the sea upon the climate which has also been mentioned, is its creation of warm and cold currents, of which the best examples are the Gulf Stream and the Labrador Current respectively.

## ATMOSPHERIC ELECTRICITY

Pressure, temperature and humidity are the only variable factors in the atmosphere. The presence of electricity must also be taken into account.

**Lightning.** It is common knowledge that the air can be momentarily intensely electrified. Proof is found in storms which are frequently accompanied by thunder and lightning.

The exact mechanics of the phenomenon of lightning are not known. Most theories put forward concerning thunderstorms claim that falling particles (rain, snow or hail) become electrically charged while the air and smaller droplets, carried upward in it, become oppositely charged. This segregation of electrical charges continues until the difference of potential is large enough to produce a lightning stroke.

The most popular theory (G. C. Simpson's), substantiated partly by experimental data, suggests that when raindrops fall at a rate greater than eight metres per second they break up and the resulting smaller droplets fall more slowly. In a cumulonimbus cloud where the ascending currents exceed this velocity, the large drops will be spilt and the smaller droplets will be carried upward in the cloud. Because the ascending current is gusty rather than constant the drops may form, split and reform, rising and falling repeatedly. As each drop splits, the positive and negative electrons are separated, the air taking up a negative charge and the drops a positive one. The air moves rapidly upward and imparts its negative charge to the main body of the cloud. Where air currents exceed the 8-metre-per-second velocity, generally in the forward portion of the cloud, a positive charge accumulates. This fact has been checked by balloon, which has shown that the bulk of the cloud is negatively charged, and that the upper zone (temperature of 10° to 20°F.), where ice crystals occur, is positively charged. As the cloud charges build up, the ground beneath becomes positively charged by induction until a discharge takes place. This produces a gigantic spark (lightning) accompanied by thunder. It is believed that the thunder is produced partly by the pressure increases resulting from the heat generated along the path of the lightning stroke.

As sound travels much more slowly than light (1,120 feet per second against 186,000 miles per second) the thunder clap reaches us later than the flash of the lightning. The time lag between perception of the two phenomena allows calculation of the distance separating the observer from the storm. If the difference in time is less than 10 seconds, meteorologists report the storm as occurring 'at the station'. If the difference is more than 10 seconds then it is in 'the neighbourhood of the station'. When lightning is seen but the thunder is inaudible, it is called 'distant lightning'.

Lightning discharges may be from cloud to Earth, from Earth to cloud (about 80 per cent of the lightning strokes on New York City's 1,275 feet high Empire State Building are intiated by the building), from cloud to cloud, and even from one section of a cloud to another. There are many forms: streak lightning, the kind most commonly seen; sheet lightning — not much is known about this type where discharge is hidden in the cloud and only reflections are seen; forked lightning, which is streak lightning splitting into two or more paths; heat lightning, which is simply lightning too far away (15 – 18 miles) for thunder to be heard; beaded, chain, zig-zag and other uncommon minor types or varieties of streak lightning. Lightning balls, or ball lightning, are an unsolved problem; even their existence is not yet fully authenticated.

The damage done by lightning is well known. When it strikes it can burn and kill. It can destroy houses and damage electrical installations. It is capable of shattering rocks or making holes in them, and has a similar effect on the slates or tiles of roofs, and bricks or stones of walls. It has even been known to make holes in church bells. On the hard rocks of high peaks, lightning scars occur in star-shaped bursts well known to climbers. Another effect is to fuse the surface of siliceous rocks, leaving them with a vitreous coating. Striking the Earth and burying itself below the surface, lightning may sometimes excavate a branching passage like a mole-run. When it penetrates sand, its path is marked by lines of fused silica, given the name of fulgurites, structures up to 2 inches in diameter and 40 feet in length. The temperature necessary for their formation must be in the region of 3,000°F.

**Polar auroras.** Ordinary storms take place in the lower layers

*Left*
Icicle formations.

*Right*
Zigzag lightning. The electricity seeks its way along conductive particles in the atmosphere.

*Far right*
Lightning striking a tree.

of the atmosphere. The electric phenomena called polar auroras are produced much higher, above the 40-mile level. To understand how they are produced, let us first examine conditions at this height.

Atmospheric pressure is determined by the weight of the surrounding air, and by the pull exercised by the Earth's gravity. It therefore decreases regularly with a rise in altitude. At about 3·4 miles above the Earth's surface air weighs about half as much as at sea level. As pressure decreases at the same rate, at 300,000 feet altitude it would be 1/750 millibar. The molecules of nitrogen, oxygen, helium, hydrogen, etc. occuring at this height are spaced further and further from one another, and the forces of gravity applying at lower levels are no longer effective in the upper atmosphere. Above 300,000 feet pressure decreases extremely slowly, and not only the molecules of gas in the atmosphere are widely separated, but even the atoms which compose them. For example, a molecule of oxygen, which is normally formed of two atoms, splits in two; nitrogen and other gases behave in a similar way. In other words, the air becomes more and more rarefied.

It is in areas of rarefied gases between 40 miles and 700 miles, and occasionally between 950 miles and 1,250 miles, that polar auroras — the aurora borealis or 'northern lights' of the northern hemisphere and the aurora australis of the southern hemisphere — appear. Their splendid arcs, rays, curtains, bands and patches are seldom seen in the middle latitudes.

Coloured auroras are most commonly yellow-green, but reds, blues, greys or violets are not infrequent, while the sky beneath is contrastingly black. The complete absence of sound during an auroral display is a startling feature to most observers.

Scientific study of auroras began in 1716 with a display that was visible over the whole of Europe. The English physicist Halley found no difficulty in demonstrating their connection with Earth's magnetism. Auroras occur most frequently in the areas around the Earth's magnetic poles, and the highest part of an auroral arc is usually in this direction. The bands extend generally from east to west and are nearly perpendicular to the east-west compass line. Their rays and flutings are inclined from the vertical parallel to the line a freely balanced magnetic needle would assume in the same place. Auroras are most pronounced during magnetic storms, that is, during the time that the Earth's magnetic field is most disturbed. They tend to coincide with the presence of sunspot eruptions, seeming to wax and wane with the eleven-year sunspot cycle. The displays also have a tendency to repeat with a period equal to the lunar rotation relative to Earth's (27·6 days).

Zollner, in 1870, compared auroras to the light produced by luminous tubes, such as neon lighting, and offered as evidence the similarity of their spectra. Neon tubes produce light by means of electrical discharges through the rarefied gas with which they are filled. This would certainly give a possible explanation of auroras when charged particles emitted by the Sun enter the highest layers of the rarefied atmosphere and illuminate it.

## THE IONOSPHERE AND RADIO

The subject of atmospheric electricity leads us on to other phenomena of great economic importance. The gases of the upper atmosphere are subject to ionisation as well as to molecular

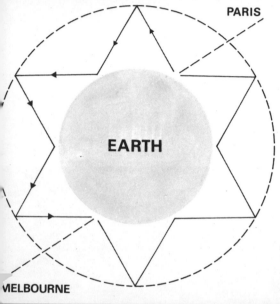

PARIS

EARTH

MELBOURNE

*Top left*
The drapery of a polar aurora showing the path of the stars.

*Centre left*
The passage of radio waves round the Earth.

*Centre right*
The new G.P.O. station in London is designed as a 619-foot tower, because its microwave transmitters need clearance over any buildings likely to be built in the foreseeable future.

*Bottom*
A radar screen showing the weather situation in and around London. The main rain areas and showers—the white patches —are clearly discernible.

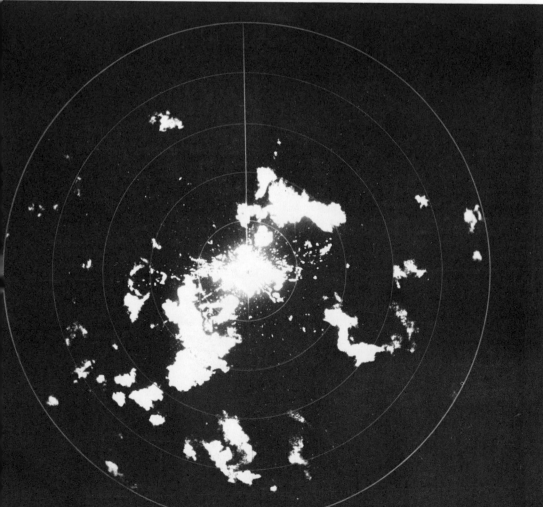

dissociation. In this region the molecules and the atoms of the atmospheric gases are electrically excited by radiations from the Sun, and gain or lose electrons. Those which gain an electron acquire a negative electric charge and become negative ions, and those losing an electron acquire a positive charge and become positive ions. Atoms may be singly ionised, doubly ionised (that is, lose two electrons) or multiply ionised; they may even lose all their electrons. The resultant ionised gases have the property of reflecting the electro-magnetic or Hertzian waves utilised in radio transmission and reception. If there were no ionised gases in the upper atmosphere, the radio waves would simply proceed in a straight line to lose themselves in space. However, alternately rising to the upper atmosphere, and being deflected again to the Earth, the waves can travel round the world and even reach the side of the world directly opposite the transmitter. Indeed, the pioneers of radio never imagined that from their experiments would come the discovery of these ionised gases.

Ionisation of the upper atmosphere, or ionosphere, is not, however, uniform, as has been said before. There are ionised layers separating layers which consist simply of rarefied gases. The most important is the *D* layer, which occurs approximately between altitudes of 180,000 and 240,000 feet in the daytime; then follows layer *E*, known as the Kennelly-Heaviside layer, occurring between 300,000 and 450,000 feet and becoming thicker and denser at night. Finally there is layer *F*, or the Appleton layer, occurring between 690,000 and 1,500,000 feet and doubled during the day into two layers known as *F*1 (day) and *F*2 (day-night).

Electromagnetic waves are reflected by different layers. The long waves are reflected from layer *D*, which explains why they are more audible during the day. The medium waves are chiefly reflected by layer *E* and are therefore received more clearly at night than in the daytime. The short waves are reflected by layer *F* and are scarely affected by the time of day.

Another layer, *G*, is thought to exist around 250 — 450 miles. It is possible however that the long period reflections on which its existence is based may be caused by multiple reflection between the *F*1 and *F*2 layers.

The total height of the ionised layer has not yet been determined for it merges outward with the exosphere, that still unexplored zone beyond. Soon, because of the ever-widening exploration the character of even these zones will be known.

This brief outline gives, of necessity, an over-simplified account of the reflection of the Hertzian waves. Here we have paid no attention to the fact that the atmospheric layers vary in degree with the time of day, the seasons, the latitudes, and even with sunspots. The cause or causes of these variations is not known. It may be because ionisation is a product of solar radiation, and sun–produced magnetic storms may arise suddenly to upset the ionosphere and indirectly influence radio transmission. These difficulties, known technically as fading effects, are causing a return to submarine cables for transatlantic telegraph and telephone communication. Satellites can now relay several telephone calls at a time and this is capable of further development.

Up to now we have mentioned only radio waves. We must not, however, forget to examine the shorter waves, those with a wavelength of several feet used for television, and those of only a few inches used for radar. These very short waves travel freely through the atmosphere without being reflected back to the Earth. From the U.S.A. and Australia, for instance, the Moon has been successfully reached by powerful radar transmitters. It took only two seconds for the reflected beam to reach the Earth again. Reflection by any solid substance in its path is, of course, the principle of radar detection of aerial objects. Of advantage in radar techniques non-reflection of the shorter waves by the ionised layers hampers television broadcasting. Television wavelengths of only a few feet are not strong enough to use the Moon as their reflecting surface, and for effective results transmitters must be closely spaced (every 200 miles or so) on high ground.

## ANOMALIES OF THE UPPER ATMOSPHERE

We have already mentioned that today rockets, satellites and electromagnetic waves are making possible the study of the upper

layers of the atmosphere. Thanks to these methods of investigation, information previously undreamed of is furnishing scientists with explanations of phenomena that were once a mystery.

For example, it was once imagined that, by analogy with the lower layers, pressure and temperature decreased steadily up to the farthest limits of the Earth's atmosphere. We now know that these hypotheses have been completely disproved. It has already been explained how Earth's pressure, instead of continuing to decrease regularly above 300,000 feet, steadies or, as a result of the Sun's radiation, even increases during the day.

Anomalies of temperature are even more marked than those of pressure. Let us take an average temperature of 60°F. at ground level. If we make an ascent in a balloon or an aircraft, or climb a mountain, the temperature decreases steadily. At 30,000 feet it is −65°F. Although we are getting nearer the Sun, the layers of the atmosphere which envelop the Earth and protect it from the cold of interplanetary space are thinning out. We might therefore expect this decrease of temperature to continue above 30,000 feet. In fact, it does not. To begin with, temperature remains constant (−65°F.) between 30,000 and 120,000 feet, then increases up to 180,000 feet (140°F.), and decreases again up to an altitude of 240,000 feet (−95°F.). Above this height, there is a continual increase as far as the boundaries of the atmosphere, starting with temperatures of 60°F. at 300,000 feet (this is the same as the average ground temperature) and continuing upward to 300°F. at 450,000 feet, and 1,800°F. at 600,000 feet and above.

The outer layers of the atmosphere have a very extreme 'climate', and display differences of several hundred degrees between day and night and between summer and winter temperatures. We must not, however, be too impressed by these figures. To state that the upper atmosphere is at 1,800°F. or 3,600°F. does not mean that any object traversing it must be burnt up or melted. Temperature acquires a calorific value only in relation to the amount of matter raised to this temperature. To take an example nearer the ground, a radiator of 120°F. will heat a room–its calorific value is considerable; a burning cigarette with a heat of 1,500°F. raises the temperature only fractionally–its calorific value is negligible. Similarly, the extremely rarefied gases of the upper atmosphere are calorifically weak.

The change of conditions affecting atmospheric pressure above 300,000 feet has important consequences on matter. Let us imagine an object coming from outer space and, passing first through the upper atmosphere with its constant pressure before reaching the lower layers of the atmosphere where pressure rises rapidly. On passing from one layer to another this object would come into contact with what may be called 'the ether wall'. This resistance, would produce a sudden frictional check which would heat the object to incandescence. This, incidentally, is what happens to shooting stars.

There exist at altitudes of 50 to 190 miles violent winds which may be produced by the Sun's radiation and tidal effects. These powerful winds may have some influence on the winds we feel at the Earth's surface. Of course, winds are a good deal more complex, and endless horizontal and vertical movements combine to keep the atmosphere in perpetual motion. All that has so far been discovered about the atmosphere is but a shadow of future knowledge.

An important problem at present occupies the attention of the physicists: are we losing our atmosphere? Hitherto there was no doubt in the scientists' denial of this. For a body to detach itself from the Earth's orbit, it would have to travel at a speed higher than 7 miles per second. The atoms of gas in the upper atmosphere have speeds in the order of half a mile per second. Thus the Earth must maintain, practically intact, the quantity of air around it. In the light of recent investigation and experiments, however, we may have to reconsider this problem. It has now been suggested that the Earth may have a 'tail' composed of atmospheric gases which is being blown away by the Sun's radiation. If this is true, some of the atmosphere must be steadily escaping into space. The planet Mercury and the Moon have already lost any atmosphere that they may have had because they were too small to retain it, or too near the Sun. It is probable that this condition will one day overtake the Earth and after some few million years life as we know it at the present time will cease to exist.

# Atmosphere and Soil

ATMOSPHERIC conditions, as we have seen, vary over the face of the Earth, producing different climatic zones. To the geologist the primary importance of the zones is their effect on rock decay, soil formation and erosion, for every natural or man-made object exposed to the atmosphere is affected in some way. Unless continually repaired and restored, all man's structures become weather beaten, and in time weaken and fall in ruins. Exposed rocks suffer in the same way, and eventually are reduced to different physical and chemical forms. Weathering is the term given to all rock changes, physical and chemical, caused by exposure and its attendant variations in humidity and temperature.

Weathering is of two kinds: disintegration (mechanical weathering), and decomposition (chemical weathering). Often the two are difficult or impossible to separate in nature because they are overlapping and integrating. In many areas, however, one or the other predominates, and for the purposes of preliminary study it is convenient to review them separately.

**Disintegration.** This is the mechanical weathering process by which rocks are gradually broken down into smaller and smaller fragments. The main factors involved are temperature and humidity. Changes of temperature alone, if they are great enough, may produce breakdown of rocks. For example, uncovered rock without a protective covering of soil may be fractured by the heat generated by a forest fire or by direct lightning stroke, when rapid heating of the outer surface of the rock causes it to expand and break off in flakes, chips or large pieces.

Whether the daily or seasonal temperature changes causing expansion and contraction in rocks are pronounced enough to produce mechanical breakdown is questionable. Until quite recently many weathering results were attributed to them. For instance, it was thought that in an area free of vegetation and subject to considerable daily variations in temperature (in the Sahara, for example, it is common for the thermometer to rise to 150°F. during the day, and fall to within several degrees of freezing-point at night) the rocks alternately expanded and contracted until they could no longer stand up to the mechanical stresses. The theory was that disintegration would be particularly marked in granite, because of the different colours of the minerals composing it: the black mica, an amphibole, would absorb heat, while the quartz and feldspar would reflect it. The resulting differential expansion would cause the mineral grains to lose their cohesion, and the rock to crumble and fall apart.

This seems a reasonable theory, and theoretically it should cause breakdown. But does it? Granite and similar rock subjected to alternate heating and cooling such as would be experienced naturally over a period of about 250 years showed no such effects, not even microscopic ones. Possibly if continued for much longer periods, the variations would produce changes, but at present the

effects of normal changes of temperature are not fully known.

When moisture or water is present, temperature changes do, however, produce breakdown, especially when freezing temperatures are involved. When water that seeps down into the pores, joints and other fractures of a rock freezes, it increases almost 9 per cent in volume. The expansion produces pressures which are directed outward from the central part of the rock body and which may be as high as hundreds or even thousands of pounds per square inch. Such pressures loosen mineral particles and fragments of rock, splitting them away from the main body. Frost action is most prevalent in mountain regions and moist areas where temperatures vary daily or seasonally across the freezing-point.

Plants and animals facilitate this and other types of weathering. Plants growing in cracks and crevices in rocks widen them as they develop, sometimes splitting sections of rock. Burrowing animals, rodents, ants and worms, also loosen material and expose more rock surface to the action of the atmosphere.

Exfoliation, spheroidal weathering and granular disintegration are all classified as types of mechanical weathering, although in many instances the mechanical pressures which cause the breakdown develop as a result of chemical activity.

Exfoliation produces large dome-like hills by spalling off concentric layers or plates from a large rock mass. Stone Mountain in Georgia, Half Dome in Yosemite Park, California, and Sugar Loaf Mountain in the harbour of Rio de Janeiro are outstanding examples of exfoliation domes. Exactly how their forms are developed we do not know. Roughly paralleling the outer surface of these rock bodies there exists a series of concentric joints or fractures whose origin is unknown. The slabs of rock along these cracks become loosened and slough off, eventually producing a broad, dome-shaped hill. The curved fractures are thought to have developed through expansion of the rock under decreased pressures following erosion of overlying rocks. In New England quarries measurements are being made to ascertain the amount of expansion and the kind of fracturing that develops as the confining pressures of overlying rocks are released.

Exfoliation of a somewhat different sort produces smaller, spheroidal forms. Rounded rocks or boulders develop through the sloughing off of concentric layers of rocks, but here the pressures causing exfoliation stem from chemical activity. The outermost surface of a rock is normally dry, or dries rapidly after wetting. Beneath the surface and extending for some distance into a rock moisture is held longer and sets up a chemical action. Certain common rock-forming minerals are more active, chemically, than others and combine readily to produce new substances which occupy a greater volume than the original minerals. This expansion creates the pressures which produce spheroidal weathering and rounds the boulders.

Sometimes the individual mineral grains cohere badly and the rock crumbles grain instead of peeling in layers. This is known as granular disintegration or granular exfoliation. In this type of disintegration, where breakdown is caused by chemical activity, there is a considerable decomposition of the grains.

Examples of spheroidally weathered boulders can be found in

*Top left*
Marram grass helping to resist the movement of wind-blown sand in the Sahara.

*Bottom left*
Shattered rock formed by frost action in the Aletschwald, in Switzerland.

*Top*
The spectacular amphitheatre of Malham Cove in Yorkshire, rises 300 feet from the Aire Valley. In the foreground the process of chemical decomposition has produced irregular blocks in the limestone pavement.

*Bottom*
Concentrically weathered granite near Virginia Vale, Colorado. The process is known as exfoliation.

many areas of the Rocky Mountains. In the Pikes Peak region huge, round boulders piled in great stacks are found. Near Baltimore, Maryland, the oval or round masses of black gabbo called 'niggerheads', are produced by deep weathering along rock-joints.

Extensive fields of weathered boulders, generally above the timber-line in mountain regions, are called *Felsenmeer* (German, 'boulder-sea').

**Decomposition.** Rock decay or chemical weathering is the result of reactions between the chemical elements of the atmosphere and the rock constituents. The commonest chemical reactions include oxidation, hydration, carbonation and solution. In oxidation, the oxygen content of the air, generally in the presence of moisture, combines with some or all of the minerals of a rock. The feldspars usually decompose rapidly, increasing in volume as they do so and causing granular disintegration and exfoliation. Some rocks (limestone and gypsum are examples) are soluble and yield relatively easily to moisture and rain. Growing vegetation, too, takes chemicals from the rock and when it decays, produces acids which increase the solvent action of water.

Mechanical weathering simply breaks rock down into smaller fragments; chemical weathering produces new substances. Decomposition is therefore a much more complex process, influenced by many factors.

As substances react chemically only on contact, the size of the individual rock grains is significant in decomposition. The smaller the particles in a rock the greater the surface areas open to chemical attack. The different minerals, too, which form a rock determine in good part the rate of weathering. Feldspars, olivine and calcite decompose much more readily than, say, quartz or muscovite.

Weathering, both chemical and mechanical, affects the rock face and a considerable depth below surface, penetrating fractures and pore spaces. Granites and gneisses in Uganda, for example, are weathered to depths of 100 feet or more; limestones in Georgia, U.S.A., are penetrated to about 200 feet, and some shales in Brazil are affected to 400 feet.

Weathering is a slow process on the whole, though there are exceptions. A one-hundred-foot layer of volcanic ash deposited during the eruption of Krakatoa in 1883 had forty-five years later developed a fourteen-inch top layer of soil. The indecipherable inscriptions on tombstones indicate clearly the sometimes speedy results of weathering.

The type and extent of weathering is controlled by climate. Moist and hot climates speed chemical weathering; in dry climates decomposition is very slow and mechanical effects predominate, though special forms of chemical weathering do take place. The comparative weathering effects of climate can be illustrated by the action on an obelisk which had stood in Egypt for centuries almost without change. Soon after its erection in Central Park, New York City, it began to disintegrate rapidly and had to be given special protection against its new climatic conditions.

Common soluble rocks, such as limestone and marble, which weather rapidly in humid areas, do so very gradually under arid conditions to form prominent ridges. In general, arid zone topography is angular with abrupt and steep bare slopes, and flat valley floors. In contrast, topography developing under humid conditions tends to be rolling, and most of its slopes are covered with a layer of soil held in place by the vegetation; there are few bare cliffs.

WEATHERING PRODUCTS. Covering bedrock over a great portion of the Earth's surface is a layer of unconsolidated rock debris and soil which grades downward into solid rock. This material is produced mainly by weathering. It is called mantle: residual mantle if it lies where it formed, and transported mantle if it has moved from its place of origin.

Soil, which is simply rock debris sufficiently altered to support vegetation, is the most important product of weathering, from man's point of view.

DIFFERENTIAL WEATHERING. This is the process by which portions of rock bodies decompose or disintegrate at different rates. Areas of greater resistance project as ridges above the faster weathering zones. Often, rock that is homogeneous in its resistance

will, in fact, weather differentially because of its position; for example, shaded zones which are slower to dry out will break down more rapidly than the remainder of the rock mass.

There are a number of characteristic features developed by weathering, especially by differential weathering. Though not, perhaps, useful, these features are sometimes beautiful and fantastic in form. Some of the finest examples of differential weathering are to be found in the west and south-west United States: they can also be found abundantly everywhere in the world. Some demoiselles, hoodoos, earth pillars and perched or balanced boulders are the work of differential weathering. Most of them, however, are sculpted by differential erosion and will therefore be described in the chapter on running water (p. 77).

MASS WASTING. Wherever rocks are exposed to attack by weathering process, loose material forms, sometimes in large quantities. In time, under the influence of gravity, the material starts to move downslope. Slow or fast, gravitational movements of surface materials are called mass movements. The complete process, known as mass wasting, although often overlooked because the moving agent is invisible and most of the movements are slow, should not be underestimated. It is an operative and important factor in erosion wherever there is a slope, however gentle.

Materials are moved by gravity only when they can overcome their inertia, and any factor which helps to overcome this internal resistance promotes mass movements. Mass wasting is almost inseparable from weathering and the many other agents of gradation. Water, for example, aids its work considerably. Water-soaked materials move more easily than dry because the water adds weight to the mass, as well as disrupting its cohesion and lubricating it. For example, landslides often develop in water-soaked rock debris. Rivers and glaciers may start mass movements by steepening a slope until the mantle moves under the force of gravity. Earthquakes, blasting, quarrying and mining activities may also trigger off the process.

Gravity movements are frequently divided into two categories: slow and fast. The slow ones include creep, solifluction and rock glaciers; the fast ones include rock slides, debris slides, mudflows and earthflows.

Any steep, exposed slope of rock is subject to weathering. In mountain areas daily freeze-and-thaw action, or frost wedging, plays its part. Fissures in the rocks fill with water which freezes and expands at night. Under the pressure of these innumerable wedges of ice, the rock cracks. Next morning the ice melts in the sun and no longer supports the rock fragments, many of which roll down the slope to join other rocks and debris at the foot. The accumulation of fragments which pile up at the foot of hillslopes and cliffs is known as talus.

The screes of the English mountain areas are talus slopes. Much of the District is formed of slate. This has weathered and split into flakes to form unstable sliding slopes on the sides of such mountains as Skiddaw. The mountains of Wales and Scotland are also largely composed of fissile slates and schists and have extensive developments of scree.

The effects of freeze and thaw are important also in the polar and sub-polar regions, where the ground is permanently frozen to a depth of several hundred feet, and for this reason is called permafrost. The frozen soil is, of course, impermeable. The summer's warmth succeeds in thawing only the top few inches. The water from the melted snow and ice saturates and softens the soil until it is sodden and semi-liquid, and unable to support even the weight of a man. Huge sheets of soil and rock debris are then transported, flowing slowly down the gentlest of slopes under the influence of gravity. This movement is known to geologists as solifluction. The stability of buildings, roads and railway lines is threatened in areas where solifluction occurs. Communications and transport are made difficult by the pressure of innumerable lakes and temporary ponds. At this season drinkable water is completely lacking or very scarce, and that scourge of polar regions, the mosquito, swarms at large for several months. With the returning winter, the mud becomes solid as ice forms and the rocky material remains fast in muddy, contorted beds–a clear indication of the method of their formation.

*Far left*
The obelisk of Thothmes III shortly after erection in Central Park, New York City, in 1881.

*Left*
Weathering in New York's climate quickly brought deterioration to the centuries-old 'needle'.

*Right*
Curious weathered rocks in Goblin Valley State Park, U.S.A.

*Far right*
The Needle's Eye in the Black Hills, South Dakota. The pegmatitic granite has weathered along fractures.

Closely related to solifluction are soil structures or patterns such as stone polygons, stone nets, polygonal ground or stone rings. These are extremely striking in appearance, making it seem as if a gardener has separated the stones from the finer earth and carefully laid them out in circles or polygons, up to 30 feet in diameter. If the ground is steeply sloping, the polygons elongate until the stones are arranged parallel to one another to form stone stripes reaching downward into the ground for as much as two feet. These interesting features, whose exact origin is unknown, seem to be the end product of a complex series of actions controlled by freezing and thawing in heterogeneous material, which includes everything from clay to boulders. Fine examples occur in the English Lake District.

A slow down-slope migration of soil and rock fragments under the influence of gravity is called creep. It is operative in temperate and tropical climates even on gentle, vegetation-covered slopes. Its effects may be seen in tilted poles and fence posts, warped and cracked buildings, broken or displaced walls, road and railroad tracks, and in curved tree trunks which have been inclined down-slope by creep and returned to vertical by growth.

A multitude of factors aid creep: the presence of water, frost heaving, burrowing animals, the prising action of tree roots. A network of roots itself tends to slow creep, but beneath this protective layer the process continues. Rate of movement is slow (with annual speeds of 5 to 10 feet), yet rapid enough to make creep a factor to be considered when permanent structures are being planned.

More spectacular than creep and solifluction are landslides and mudflows. Although it seems a contradiction in terms, mudflows occur mainly in semi-arid to arid climates. Here there is scant vegetation, and loose rock and sand lie exposed. In such climates what little rain that does fall, is often concentrated in short, heavy, local cloudbursts. Within minutes of a storm starting, water flows down the bare rock slopes, gathering up all loose material until it becomes a viscous mass. The consistency of the moving mud and rock rubble is that of newly mixed concrete. The forward margin of the mudflow becomes more heavily burdened, slows its progress, and actually forms a slow-moving dam against the more rapidly moving, less heavily loaded mud that follows. On reaching the mouth of the valley down which it has been travelling, the dam bursts. Its flow, often carrying huge boulders weighing up to 85 tons or more, spreads rapidly in all directions across the plain or larger valley, often blocking it. The Slumgullion mudflow of Colorado, for example, dammed the Lake Fork branch of the Gunnison River to form Lake San Cristobal.

Mudflows are not confined solely to arid regions; in alpine areas they may follow the rapid melting of winter snows. One mudflow of this kind in Southern California was estimated to have moved about one-and-a-quarter million cubic yards of debris over a distance of 15 miles. Slopes of active volcanoes, with their large accumulations of volcanic ash and dust, are also vulnerable to mudflows. Herculaneum, at the base of Vesuvius, was buried by a large flow following the heavy rains that accompanied the eruption in 79 A.D.

While mudflows are rare in humid areas where soil and rock debris are held in place by vegetation, debris slides occur on deep slopes in mountainous areas, such as north-western United States and in the Appalachians.

A landslide, or landslip, is the rapid sliding of large masses of bedrock, soil and rock debris wherever there are steep slopes. They therefore occur mainly in rugged mountain country. Switzerland, Norway, and the Rocky Mountains in United States and Canada are areas of frequent landslides. Here they are a

constant menace to the inhabitants of steeply sloped and mountain areas, sometimes bringing only the loss of small patches of forest or pasturage, at others carrying away houses and crops. Landslides generally result from unstable conditions built up slowly over periods of tens or hundreds of years and upset in a moment, perhaps, by a minor earth tremor, an abnormally heavy rain or local mining activity.

A famous landslip occurred at Rossberg in Switzerland, on September 2nd, 1806. A mass of rock, estimated at 400 million cubic feet, took only several minutes to move down the plain, burying three villages under a hill of debris and causing many deaths.

Another well-remembered landslip was that at Elm, also in Switerland, which occurred on September 11th, 1881. A large slate quarry had been cut obliquely into a steep mountain slope above the town. The rock above the quarry was progressively undercut until one day it sheared off, and a block about a quarter of a mile wide slid to the quarry floor, shattered, and flew across the valley in a more or less level path to strike the opposite side. The impact deflected the flying mass down valley, where it moved with increasing speed as a sheet of impressive dimensions: 4,200 feet long, 1,500 feet wide and 60 feet thick.

In Wyoming, south of Yellowstone in the Gros Ventre Valley, a huge slide took place in 1925. The slide mass, consisting mainly of sandstone, slipped along a layer of water-soaked clay, descended 2,000 feet, plunged across the valley and travelled 350 feet up the opposite side. It then settled back to form a dam 250 feet high and half a mile long, behind which a lake formed. A rancher in the path of the slide managed to escape on horseback, but the cattle in the valley were engulfed.

In mid-August, 1959, an earthquake in the Yellowstone National Park area set off landslides which dammed the Madison River.

Probably the best known of all North America's landslides occurred in 1903 in Alberta, Canada. Seventy people in the coal-mining town of Frank were killed when 40 million cubic yards of rock broke loose from the Turtle Mountain. This mountain rises steeply to 3,000 feet above the valley floor and is formed of strongly jointed limestone, with the joint fractures inclined down into the valley. Many factors added slowly to the effects of the jointing and inclination: the steepness of the mountain slope, the coal-mining activity at its base, frost wedging, heavy rains, melting snow, a minor earthquake two years before, and plastic deformation and settling of shaley rock beneath the mountains. Finally the accumulating stresses reached the critical point. The rock cracked and a mass of it slid down to the base of the mountain, shattered, shot across the valley and 400 feet up the opposite wall. A total distance of two and a half miles was covered in less than two minutes.

Examples of landslides involving large and small quantities of rock, and often causing great damage occur on every continent and even below sea level. Submarine landslides or flows, such as that which fractured the Atlantic cables off the Grand Banks in 1929, do take place and, apparently, attain high velocities, with effects comparable with those of their land-born relatives. It would seem that warning signs of imminent landslides are perceived mainly by animals, whose behaviour reflects nervousness. The few survivors of the 1806 slide at Goldau, Switerland, recalled that although they themselves had no reason to suspect that a slide was threatening, cattle and horses were very nervous for several hours beforehand, and even bees left their hives. Perhaps because of their greater sensitivity the animals noted some slight preliminary disturbance on the slopes.

Though sand lacks cohesion, when dry it can be fairly firm as a foundation. When thoroughly wet, however, its character changes and it becomes almost fluid, moving down gentle slopes and even

A huge rockslide formed when a series of earthquakes shook the

Yellowstone National Park in 1959, transforming the landscape.

Land reclaimed from swamps and marginal coastal areas is also susceptible to subsidence. Again pilings are used as a foundation.

Sliding, slumping and flowage were a major problem in the Culebra Cut during construction of the Panama Canal, while a slow-flowing mass of saturated silt and sand hampered work on the Grand Coulee dam on the Columbia River until engineers circulated a freezing mixture through pipes and froze it until the necessary work was done.

Preventive measures to slow down mass movements of all types include reforestation of denuded areas, reinforcement of stream banks and steep slopes, and the development of proper construction techniques. The most important measure is, of course, to ensure future selection of the safest possible sites for permanent structures by the study of mass wasting phenomena.

Much of the cost involved in repairs of tunnels, railroad lines, highways and aqueducts which are damaged by mass movements could be avoided by educated site selection and proper planning and construction. The remainder of the cost is the fee we pay for dwelling on an unsteady planet.

## SOIL

Soil is the uppermost layer of the rock mantle on which vegetation lives. It is composed of organic and inorganic matter in three states: solid, liquid, gaseous. Weathering, which produces the inorganic particles that make up most of a soil's volume and weight, plays a crucial part in its development. The organic portion of the soil includes both living and decayed plant and animal substances. The water content, called soil water, is actually a dilute but complex solution containing a wide range of chemicals including bicarbonates, chlorides, nitrates, sulphates and phosphates.

Although colour is a minor factor in soil composition it is an excellent characteristic by which to distinguish different soil layers. Colours range from white through brown to black, depending on the amount of humus (decomposed plant material) present. Reds and yellows, which are common, generally indicate the presence of small amounts of iron oxides and hydroxides. Texture, varying from fine, impervious clay to coarse permeable sand and gravel, is important because it regulates the water content of soil.

Composition varies with depth. A soil grows thicker as it develops and falls into layer-like horizons which may differ in colour, texture, structure and consistency. A complete succession of these horizons from the ground surface downward to the parent bedrock is known as a soil profile. The major parts of a soil profile consist of: Horizon A, the topmost zone, often dark in colour and rich in organic matter; Horizon B, lying directly beneath A, rich in clay and poor in organic material. Together, Horizons A and B make up the true soil or solum. Beneath lies Horizon C, or subsoil, consisting of partially decomposed rock material; Horizon D is the bedrock on which Horizon C rests. In most areas soil horizons may be further subdivided.

Many factors are involved in soil development: climate, parent rock, transportation, topography, biologic activity, and time. The composition of fully mature soils (those which have a full or complete profile) reflects mainly the climate or environment in which they formed; only immature soils reflect strongly the parent rock. Classification is into about twelve major soil groups which, under much the same types of climate and topography, have almost world-wide distribution. Some are listed below:

**Podzols.** These are the most widely distributed of the humid climate soil groups. The podzols need cold winters and an adequate precipitation spread throughout the year. They are therefore generally closely linked with cool to cold climates. Horizons A, B and C are well developed. The surface material consists of a layer of partly decayed vegetation, while the upper portion of Horizon A is rich in humus. The lower portion of the A horizon is light in colour — whitish or greyish — because the iron oxide colouring substances have been leached out and carried down to B. The B horizon is brownish and clayey. Both A and B horizons are strongly acid.

Conifers and mixed hardwood forests grow on these soils.

on the flat if it is under pressure. Quicksands are often formed where the sand holds a large amount of water. Serious catastrophes may result if sand breaks into underground workings. One of the many difficulties in the exploitation of the great coal-basin of North France and Belgium is set up by the flowing sands through which the workings must pass before reaching coal-beds which lie below.

Water can cause the movement of both clays and sands; both, therefore, are bad subsoils for building operations, necessitating the construction of special foundations. The subsoils of Paris, for example, caused difficulties in the building of the basilica of the Sacre-Coeur on the heights of Montmartre. This steep hill is composed entirely of Tertiary marls, sands and gypsum beds (used in the making of plaster-of-Paris). The foundations of the church, which rest upon 83 pilings, each 130 feet long, had to be driven right through the sands and marls to reach the compact gypsum.

The Sacre-Cœur was completed in 1919 and still stands firm. There are many ancient monuments which have not been so fortunate, and have leaned over or collapsed with the movement in soft subsoils: the Campanile of San Marco in Venice, which collapsed in July 1902, and the leaning towers of Pisa and Bologna are famous examples.

In New Orleans, Louisiana and other areas in the United States where subsurface materials are unstable pilings are driven into the ground to support large buildings and other structures. In New York City La Guardia Airport is built on such foundations.

Waterspout observed at San Feliu de Guixols, Spain in 1965.

occur further inland. Both soils are cultivated to produce cotton, groundnuts and tobacco. The natural vegetation of the yellow soils consists of pines; that of the red soils, oak and pine. Soils of this type are not developed in Britain.

**Latosols or Lateritic Soils.** Latosols form under warm, humid climates and are closely associated with equatorial rain forests and tropical savannas. They have a thin organic layer above a strongly leached reddish layer, which is followed by a deeper red layer. Except that leaching is much less, the latosols are similar to the red-yellow soils of the southern United States. A good growth of hardwoods or grasses is supported under natural conditions but under cultivation the latosols quickly lose their fertility because, except in a thin surface layer, the plant nutrients are removed by leaching.

Some types of valuable mineral deposits (aluminium, iron manganese) occur in lateritic soils. The ores, which are not soluble in soil water, accumulate in the latosols as the parent rock is eroded away. Soils of this type have a wide distribution in East Africa, India and Australia.

**Chernozems or Black Earth.** The chernozem is a distinctive and widely distributed soil type associated with a semi-arid climate. Horizon A is black, shows no leaching and is high in humus. It grades into Horizon B, which is brown or yellow-brown and rich in calcium carbonate. Aridity–hot summers, cold winters and pronounced dry periods–is an important factor in chernozem development. The black earths occur in a north-south belt in the central Great Plains region of the United States. Steppe and prairie grasses are the natural vegetation of the chernozems. Grains such as wheat, oats, barley and rye are grown in the black earth zones. Black cotton soils of India are of similar origin; so are the Canadian black soils and the black earths of south-eastern Australia.

**Prairie Soils.** In the United States, the prairie soils lie between the belt of the chernozems and the grey-brown podzolic soils and are transitional in type between the two. They have the general chernozem profile and appearance but lack the large amount of calcium carbonate. The prairie soil region in the United States has a rainfall of 25 to 40 inches, diminishing westward toward the chernozem belt. The natural vegetation is the tall grasses which are able to survive the dry, hot summers. Prairie soils are very productive if moisture is present; in the United States, the corn belt and the prairie soil belt coincide. The soil of the Wimmera of Victoria, Australia, a very fertile wheat area, is said to belong to this group.

**Chestnut and Brown Soils.** The chestnut soils and the brown soils are distributed in the semi-arid middle latitude steppe regions of North America and Asia. In the United States they lie to the west, on the arid side, of the chernozem belt. They have a profile similar to the chernozem but with less humus and therefore are a lighter colour. They border the grain belts, and in times of rainfall and under irrigation are fertile. Similar soils are found on the South African veldt.

Chestnut soils in the United States grade westward under increasing dryness into the brown soils which are still lighter and have even less humus. They occur in the Rocky Mountain basins and in parts of the Colorado Plateaux, Utah, Arizona and New Mexico. The natural light growth of grasses on the brown soils is suitable for grazing, while under irrigation the soils can be farmed.

**Grey Desert Soils (Sierozem) and Red Desert Soils.** The soils of the middle latitudes and tropical deserts usually fall into the red or grey desert soil categories. The grey soils are well developed in the Great Basin area of Nevada, western Utah, southern Oregon and southern Idaho. A soil profile does exist but is difficult to see because the horizons differ only slightly. With very little vegetation — only sagebrush and bunch grass — there is very little humus. A lime crust, or caliche (calcium carbonate or sulphate), is often present. This develops in the long dry periods, during which subsurface water moves upward under capillary attraction, evaporates near the ground surface and deposits salts.

Examples are to be found in parts of the Cheshire Plain and under heaths in Surrey and Kent. Similar soils occur in the coastal regions of Australia, normally under sclerophyll eucalyptus forests, and in the Pine Barrens area of New Jersey. Addition of fertilizers and lime to counteract acidity will make a podzol relatively productive.

**Grey-Brown Podzolic Soils.** The grey-brown podzolic soils are richer and less acid than the podzols and have been less intensely leached. Deciduous forests (maple, beech, oak) grow on these podzolic soils. When cleared they make excellent diversified agricultural and dairy land. In Britain, such soils are characteristic of most of the eastern and southern countries and of the better drained lowland areas of Wales. In the United States, grey-brown podzolic soils occur in the cooler, 35-inch to 40-inch rainfall area, and underlie parts of southern New England, New York, Pennsylvania, Maryland, north Virginia, West Virginia, Ohio, Kentucky.

**Red-Yellow Podzolic Soils.** To the south of the grey-brown podzolic soil area, in a warmer area with a similar rainfall, the red-yellow podzolic soils are found. These humid, subtropical climate soils extend in the United States from the Atlantic coast through east Texas. The A horizon of the red-yellow soils is leached as in the podzols but because the warm moist climate speeds organic decay, the humus content is low. Iron hydroxides produce the red and yellow colours. The yellow soils are more strongly leached and occur along coastal plain belts; the red soils

45

Some authorities place the Wimmera of Victoria in this group.

Red desert soils form when desert conditions are hotter and drier. Their horizons are poorly developed, and fragments of parent rock are often present throughout. Colour ranges from pale reddish grey to deep red. The limey deposits common in the grey desert soils are present in the red too, and again sparse vegetation of cacti and desert shrubs means minimal humus. Both soil types are cultivable under irrigation, if they are of fine texture.

The soils just described are all of relatively wide-spread occurrence. There are many other types which are more local in distribution. Probably the most common of these are the tundra soils of poorly drained, arctic regions. The intensely cold winters of the tundra keep the water frozen for much of the year. Decomposition is slow because of the low temperature, and many fragments of parent rock are present in the soil. The ground surface is covered with a two- or three-inch layer of lichens and mosses. The soil consists of thin layers of clay and humus; there is no distinct profile.

In desert basin areas, where playas (temporary lakes) occur, saline soils are sometimes found. Light-coloured and of poorly developed profile, these soils consist of a mixture of mud, silt, sand and salts (soda, borax, calcium carbonate, sulphates, etc.) Saline soils are local in occurrence, and where the salts are very thick it is impossible for a true soil to be formed.

Bog soils form on marshy bottomland in humid regions. Because the ground is water saturated most of the time and the water circulation is poor, plant decay is retarded, and partly decomposed material accumulates into a layer of peat three or four feet thick. Beneath this is the gley, a sticky, grey-blue impervious clay horizon.

The gley also occurs beneath the meadow soils which develop in humid mid-latitude regions along flood plains where drainage, though better than in swamps, is still poor. Above the gley of meadow soils lies a thick humus layer. Grass grows rapidly and densely, giving useful pasture land.

Calcareous soils are found in the downland regions of south and east England, where the bedrock is chalk.

## TRANSPORT AND EROSION BY THE WIND

The wind's carrying capacity is considerable. Those who have experienced a desert sand-storm can testify to this, or those who have been in the sort of dust-storms which can occur in the United States when clouds of dust are thick enough to stop traffic and bring darkness in the middle of the day. In the 'Dust Bowl' during the 1930's unusually heavy quantities of dust were transported. Some of the material from one of the storms was carried

*Left*
The Great Sphinx of Egypt
eroded by the desert winds.

*Top right*
The approach of a tornado.

*Bottom right*
Crescent dunes in the
Mauritianian desert.

as far as New York City, 2,000 miles away, in amounts large enough to produce twilight conditions. Fine material may travel considerable distances; 'sand' from the Sahara has been carried as far as Spain, France and Italy. The same 'sand', blown into the valley of the Nile, mixes with the river alluvium and increases its quantity and its fertility. The material which produces the 'black blizzards' of United States and the Nile valley is dust not sand. It has been estimated that 4 million cubic yards, or enough to fill forty daily trainloads of thirty cars each, fall annually on the Canary Islands. The trade winds similarly shower sand upon ships passing the coasts of Mauritania and Senegal.

Volcanic ash is carried much greater distances. Its extremely fine texture enables it to remain airborne for months and even years. After the 1906 eruption of Vesuvius, ash was recorded in Istanbul, Paris and Brussels. In 1883, ash from Krakatoa, situated between Java and Sumatra in the Sunda Strait, was spread throughout the entire Earth's atmosphere, and caused glorious sunsets for several years afterwards — 'Chelsea sunsets' as they were called in England. In the same way, radio-active clouds produced after the explosion of atomic bombs are carried round the Earth's atmosphere, and the time may not be far distant when they will exert an influence on human life as well as on climatic conditions and seasons.

Salt is another substance transported by the wind. A deposit

estimated at 55 million tons was brought by the wind from the Rann of Kutch in India to the arid plateau of Rajputang, 375 miles to the south.

The wind's transporting power is, of course, increased during a cyclone or tornado and in waterspouts at sea. Apart from the dust they carry during these disturbances, they bring most unusual occurrences. Showers of small fish, frogs or other aquatic animals were once almost legendary. It has now been proved that these creatures are drawn up from their ponds and marshes by whirlwinds and deposited elsewhere. Waterspouts have been known to gather up crabs and sea-shells from the shore and drop them several miles inland.

In recent studies it has become apparent that, although the wind is an important transporting and depositing agent, very few major topographic forms are produced by its activity. Even in desert areas few of the characteristic features are wind-created. The wind moulds and shapes details of larger features but is itself incapable of producing through abrasion an entirely new feature of large extent.

Yardangs are perhaps the only topographic form that can be attributed to wind abrasion. These are large knife-like ridges separated by furrows or troughs following the direction of the prevailing wind. Originally described in Chinese Turkestan by

Hedin, yardangs are almost invariably excavated in weak materials, such as silts.

The main effects of wind abrasion are polishing, pitting, faceting and shaping. Many surfaces and features, such as pedestal rocks, caves, arches and alcoves, once thought to be wind-produced, are now known to be caused by other agents (rainwash, differential weathering and the like) and are only slightly modified by the wind. Ventifacts, highly polished, faceted pebbles with flat or curved surfaces, fluted or pitted and sharp-edged, are wind-produced. They are named according to the number of their facets — einkanter, zweikanter, dreikanter, etc., ranging from one to twenty.

The bedrock of desert areas, too, may become polished to a high degree. That of the Sahara is an example.

Abrasion by the wind is similar to abrasion by the artificial sand-blast used for cleaning buildings. In some areas of the United States glass windows have been frosted to opacity in a single storm, and where the wind carries particles of a resistant substance, such as quartz, damaged telegraph and fence poles often have to be replaced. Within about three feet of ground level, the wind is a powerful force. Even foundations may be deeply eaten away.

It goes without saying that man-made monuments do not escape the wind's attention. In Egypt, the softer layers of the stones of the Sphinx and the Pyramids have been deeply etched. The damage that sand can do is apparent from the industrial use of a sand-blast to engrave glass. Window-panes of houses by the sea are often scratched by the action of sand in the local wind. There is a sheet of glass in the United States National Museum, Washington, which was removed from a light-house at Cape Cod after a storm lasting forty-eight hours. Sand had so scratched the glass in the first few hours of the storm that it was opaque.

In areas of loose or weak materials wind erodes depressions called blowouts, or deflation basins. Most are small in size. These, too, are products of the wind's transporting capacity rather than its abrasive powers. The process of deflation involves rolling or dragging particles along the ground as well as carrying material in suspension. Except during severe storms the material so transported is generally confined close to ground level. Fine dust may rise higher, but 93 per cent of the coarser material is kept within a few inches of the ground. In most true sand-storms, the head and shoulders of a man standing upright will be above the level of the blowing sand. Sand particles are carried short distances and then dropped where they may dislodge other particles, throwing them into the air on impact. Except under unusual conditions the specific gravity of air (about one eight-hundredth that of water) is not great enough to produce sufficient energy to support coarse particles.

The wind therefore carries mainly the finer sediments, often sweeping desert areas clean of fine substances and leaving behind only coarse materials, or lag deposits. Such expanses of bare rock rubble and bedrock are called desert pavement, or desert armour. They are typical of the arid regions of the world, such as the hammadas and regs of the Sahara, and contrast sharply with the smaller sand-covered areas, the ergs.

Contrary to popular idea, most desert regions are areas of barren rock and rock rubble. Of the Sahara's vast area of 3,500,000 square miles, less than 500,000 square miles is dunesand; most of it is

*Top left*
Crescent dunes creeping forward with the wind, Mauritania.

*Bottom left*
This wind-carved window known as Delicate Arch is one of the features of the Arches National Monument, in Utah, U.S.A.

*Right*
The original springs near Alice Springs Northern Territory, Australia. This dependable natural occurrence of surface spring water is caused by the level of the land being below that of the water table a striking feature in this tropical region.

bare rock. In the United States less than 10 per cent of the arid region is sand-covered. The pebbles, cobbles and bedrock of desert pavements may be very shiny and smooth, partly the effect of polishing by the wind but probably mainly due to a black or brown enamel-like coating, called desert varnish. This glossy film, which often covers very extensive rock surfaces in dry regions, consists largely of iron and manganese oxides, though the exact chemical composition is not unanimously agreed upon. Desert varnish appears to result from deposition by evaporating moisture which has risen to the surface of the rock by capillary action. Lichen growth, too, may be an important contributory factor in the development of the coatings in some areas.

## EOLIAN DEPOSITS

In dry regions the wind may be a carrying agent of considerable importance, shifting large quantities of very fine material and, on mountain tops where velocities may rise to 200 miles an hour, even large rocks. Aridity coupled with lack of vegetation, abundant sand and moderate to strong winds produces an ideal situation for eolian activity — the process of wind transportation and deposition.

Sooner or later sand particles carried by the wind are dropped. There are two major categories of such deposits: sand and loess, fine silty material. Sometimes the two are found interbedded, but generally the finer loess deposits are carried farther and lie beyond or to the leeward of sand areas. In Nebraska the sandhills region lies to the west and the loess to the leeward east.

Eolian deposits do not necessarily originate in deserts nor are their deposits restricted to these areas. In fact, they occur most often just outside desert regions, along shorelines, along stream courses (especially in semi-arid regions), in areas of glacial debris or any region where poorly indurated sandstone is disintegrating.

One of the characteristic features of wind deposition is the development of sand-dunes. In its simplest form a dune is a hill or ridge of sand with unequal slopes. The gentle slope is convex, upward in profile and facing the wind; the steep and concave side is the sheltered side. The shape is determined by the wind, which impinges on the slope nearest to it, sweeps the sand up and moves it forward. This side of the dune is being continually removed and can never acquire a slope of more than 10°. When the wind reaches the summit, it blows straight on, and the sand is left to fall and roll down the leeward side and assume its normal angle of repose, about 30°–40°. Thus the internal structure of a dune consists of a series of inclined layers or beds of sand. This type of lamination is known as eolian cross-bedding. At the edges of the dune, where the wind encounters less resistance, it is able to blow the sand away completely, shallowing the ends. In this way the crescent shape of the barchan dune develops.

Dunes are seldom as simple as this; they generally occur in roughly parallel chains, colonies or complexes. Most are ripple-marked and are made up largely of particles of quartz, the commonest material found in sand-size (0·06 mm. to 2 mm. diameter) sediments. Other substances may form dunes too. The dunes of Bermuda are built up of sand-size particles of calcite, formed by weathering and erosion action on limestones, coral and shell fragments. The White Sands region of New Mexico is an area of

*Left*
The rocky desert scenery of Ras en-naqb (Naqueb) in Northern Sinai, situated west of Eilat on the Gulf of Aquaba.

*Right*
The sand dunes near Stovepipe Wells Death Valley National Monument, in the California-Nevada desert.

500 square miles covered with snow-white dunes of gypsum grains. The largest surface deposit of its kind, this region has become a rocket proving ground and a tourist recreation centre. Its sand is softer, whiter and therefore cooler than quartz sand. Being so soft, it rubs to powder between the fingers.

Local dune formation may consist of clay or silt, generally in aggregates, each about the size of a sand grain and composed of several particles. Yellow-green olivine, pink coral fragments, black magnetite, and even ice crystals or snow can form dunes.

**Coastal Dunes.** Three things are necessary for the formation of coastal dunes: a low coast, a wide area of sand uncovered at low tide, and a fairly constant prevailing wind. Under these conditions the sand moved by the wind accumulates at the foot of the smallest natural or artificial obstacle and forms the first dune which grows, layer by layer, as fresh sand is added.

The action of the wind on a dune is, as we have seen, continual transport of sand grains from the windward to the leeward slope. Thus the dune shifts along in the direction of the wind.

Suppose a stake is driven into the ground on the lee side of a dune 30 feet wide, and that the dune passes it in a year and a half; simple calculation will show that the dune is moving at a rate of 20 feet in a year. Movement as fast as this along a coast can become a danger. Dune migration rates vary with local conditions, usually averaging only a few feet annually but reaching or exceeding 100 feet in some places. Many settlements have, in fact, been rendered uninhabitable by sand, and many old villages around Europe's coasts have been swallowed up by advancing dunes. In Cornwall, early Christian churches have been found half buried in the sand, and in a number of eastern and western coastal areas in the United States migrating dunes present a difficult problem. Culbin sandhills, near the mouth of the Findhorn on the southern shore of the Moray Firth, Scotland, are a classic example of the destruction of cultivated lands and habitations by advancing sand. Prior to 1694 the sandhills had reached the fringe of the Culbin estate. In that year a great storm started a phase of accelerated encroachment which led to the complete obliteration of houses, farms and orchards, and even destroyed fir plantations.

Many attempts are made to control sand-dunes. First of all fences are erected as wind-breaks, and then an attempt is made to establish vegetation in the sand to give it more stability. For this operation sand-loving species which put out dense sand-holding root systems must be used. By far the best plant for the purpose is marram-grass (*Ammophila arenaria,* derived from words meaning 'sand-loving' and 'sand-living').

**Desert Dunes.** The three conditions for the formation of dunes (absence of vegetation, abundance of loose material, a moderately constant wind) are at their most effective in the desert regions of the world. Here the dunes are much larger than coastal dunes, and may be as much as 600 feet high. They may develop parallel or, alternatively, transversely to the direction of the prevailing wind. Areas where there are two or three fairly constant wind directions may also be marked with longitudinal dunes. The great seif dunes of North Africa and the Middle East, too, belong to the longitudinal class. With jagged, knife-like crests they reach a height of 200 to 600 feet; width is about six times the height. Chains of seifs may stretch as far as 150 to 200 miles. They seem to grow in length while the prevailing wind blows parallel to them, and in height and width when there are cross-winds; they may merge with each other or be completely separated by expanses of bare rock.

Even larger than seif dunes are the whaleback or sandlevee dunes. These, too, form parallel to prevailing winds but are more rounded in form than the seifs. A single whaleback may be 100 miles long, 2 miles wide and 150 feet high, and may carry seifs upon its back. These giants seem restricted mainly to the Egyptian Sand Sea. They appear to keep fixed positions, unlike the mobile seif or barchan, and are named and mapped to serve as landmarks for desert travellers along the permanent caravan routes which cross them.

The dunes of the Sahara have been thus described:

'The traveller crossing these dunes imagines that he is in the middle of an inextricable maze; but if he climbs to the summit of one of them, his labour will be rewarded by the impressive view of the landscape. Dunes surround him on every side, like the waves of the sea, stretching away to the horizon. It is as if a stormy sea had been solidified.' *(G. Rolland)*

Although occurring most often in barren desolate regions, dunes form some of the most spectacular topography in the world. Along much of the east coast of the United States and the eastern shores of Lake Michigan, they are highly developed. Unusually large dunes are found in the Great Sand Dunes National Monument, near Colorado. The sands of this area west of the Sangre de Cristo Range are coarser than average.

Dunes are a characteristic feature of many of the arid basins, such as Death Valley, in Nevada, and California. Some of the best-known in the United States are found in the Imperial Valley, not far from Yuma, Arizona, and in the Colorado and Mohave Deserts.

**Ancient Dunes.** In many past geologic periods arid conditions prevailed even as they do today; dunes developed. These ancient dunes with their typical cross-bedding are preserved still in the rocks of many regions. A good example is the Penrith Sandstone, an eolian deposit of Triassic times which is exposed in the River Eden valley, in the north of England. Outside Zion National Park, in southern Utah, and in many sections of the Colorado plateau the Jurassic Navajo sandstone with its 'fossil' dunes is exposed. In Arizona's spectacular Canyon de Chelly, too, there are exposures, 800 to 1,000 feet thick, of the beautifully cross-bedded Coconino dune sandstones laid down in the Permian age.

**Loess.** Loess is yet another wind-borne deposit, composed of very fine silty, unconsolidated material, generally grey to buff in colour. It is permeable and made up of particles that are mainly angular to sub-angular grains of quartz, feldspar, calcite and dolomite, bonded together by a clay — generally montmorillonite. Most loess deposits seem to be derived from glacial deposits. Some, however, — desert loess, for example — have no connection with glacial sediments and are simply accumulations of fine material picked up by the wind in arid regions.

Loess deposits lack the layering that is so characteristic of most sediments. Their particles, though uncemented, show considerable cohesion and the mass tends to break in a vertical direction. Vertical cliffs are therefore a natural development in loess areas under erosion, and unpaved roads crossing them often become bounded by these cliffs.

In China, central Europe, and a few other areas dwellings have been carved out of loess deposits. When the loess becomes moist, the water evaporating on the walls leaves a film of lime to form a 'case-hardened' surface.

In the presence of water loess makes an unusually fertile soil. Many of the world's rich agricultural regions lie in loess zones. These districts include parts of the plains of south Russia, the Argentine pampas, the grain region of China, and parts of Iowa and Illinois in the United States. Many more of these areas are being reclaimed by irrigation.

*Top*
The rocky desert, part of the
Algerian Sahara.

*Bottom*
Sand-dunes and tidal sand-flats
planted with marram grass.

# Subsurface or Ground Water

THE terms subterranean water, ground water, underground water, and subsurface water are all used to refer to the water present in the pore spaces, cracks, crevices and channels of the solid or un-consolidated rocks below the surface of the lithosphere, or solid crust, of the Earth.

The major source of ground water is precipitation in the form of rain, snow, hail, and so on. The proportion of the total precipitation which finally becomes ground water depends on many factors. Generally, the more rain that falls, the greater the volume of ground water. The more meagre the rainfall — in arid regions. for example — the deeper the surface water lies. The rate of rainfall is important too. The more rapid the fall, the more quickly the surface of the ground is saturated and the greater the amount of water free to run down the slopes to form streams and surface run-off. This is true of thawing snow too; the more rapid the thaw, the more water runs over the surface of the ground and proportionately less sinks into the subsurface, especially into frozen ground.

The nature of the terrain is another factor governing the fate of precipitation. Steep slopes are conducive to surface run-off; level or gentle slopes retard flow, giving the water longer to sink into the ground. The presence or absence of vegetation cover also helps to determine the amount of precipitation turned to ground water. Barren slopes hasten run-off; vegetation hinders it by holding the water and thus giving the ground time to absorb larger quantities. Other factors include the characteristics of the rocks (permeability and porosity, and their structural position relative to the slope of the ground), the amount and kind of vegetation, and the humidity.

## SPRINGS

Few things are more refreshing than sparkling spring water, cool and untouched by the summer's burning heat. Throughout the centuries man has turned to springs for refreshment, and in earlier ages the origin of their seemingly inexhaustible bounty caused a great deal of speculative thought. In their attempts to shed light on the mystery, the more keen-witted of the early philosophers soon perceived that there was a connection between springs leading into streams which in turn lead into rivers and, finally, to the sea. One fact, however, brought them to an abrupt stop. The sea, although constantly receiving water from the rivers of the world, never rose in height. Was it, like the Danaids' barrel, bottomless? The explanation seemed unlikely, and it was suggested instead that the sea water returned to the springs through deep subterranean cavities — a theory not far removed from that of the Tartarus into which Plato had earlier imagined all the water of the Earth was precipitated. The ecclesiastic Kircher went even further, and in his extraordinary work, *Mundus subterraneus* (1678), described water being driven underground from the sea by whirlpools and returning to the sea by rivers.

According to Van Helmont, in his *Initia physicae inaudita* (1655), there was in the centre of the Earth a mine of water in the form of inexhaustible 'living sands'. A similar idea is supported in the book of Genesis: 'The Lord God had not caused it to rain upon the earth, and there was not a man to till the ground. But there went up a mist from the earth, and watered the whole face of the ground.'

We may ask why none of these early men of learning hit on the idea of rain being the medium by which water from the seas was returned to the springs. Two facts seem to have stood between them and the truth: the sea is salt, yet springs are fresh; rain falls from the sky, yet springs rise from the ground. The first barrier opposed the theory which supposed sea water to reach springs by underground channels. The French philosopher Descartes (1596 — 1650) imagined great underground alembics distilling the sea water to make fresh water. But then, it was objected, the Earth would become nothing but a mountain of salt. We can see the quandary in which the early scientists and philosophers found themselves. Today their problems seem not a little naive, for it has been incontestably proved that rain is evaporated sea water, and therefore distilled and free of salt; that, after falling on the ground, it sinks below surface to reappear in the form of springs.

It is curious that the first man to realise what actually happened was not a scientist but an artist. He was Bernard Palissy, the potter perhaps more famed for throwing his furniture into the furnace to maintain the temperature of his kiln as he tried to discover the secret of producing enamel; he was notwithstanding a discerning observer of natural phenomena. His *Discours Admirables de la Nature et Fontaines* (1580) contains the following extract, 'When I had well considered the sources of natural fountains, and the places in which they arise, I at last understood that they proceeded from, and were engendered by, none other than the rainwaters.' This conclusion he patiently and carefully demonstrated by means of his own observations.

Another two centuries passed before Palissy's opinion was widely held and gained authority. One or two of the early encyclopedias elaborated the theory and to support it produced further evidence, that of the pluviometer. When, finally, it had been proved beyond doubt that the amount of rain falling in an area was amply sufficient to supply the springs in it, a great advance had been made.

Today we know that springs are the outflow of ground water; that aquifers (Latin *aqua* – water. *ferre* – to carry) are rocks which are saturated with water and form underground reservoirs; that springs appear wherever subsurface water flows out at the surface of the ground; that the outflow often occurs along rock fractures or at a point where an aquifer intersects the slope of a hillside.

In order to make this clear, see top the diagram opposite showing a fairly common arrangement of strata. The hill consists of a series of inclined rock layers, permeable and impermeable beds alternating. The beds outcrop on the surface of the hill, where falling rain sinks into the permeable layers to form a series of underground reservoirs in which the water flows slowly down the slope of the strata, later emerging as lines of springs lower down the hillside.

Water from aquifers should, theoretically, emerge as continuous seeps along the line of contact of the permeable strata. More commonly, it flows out at a number of separate points along this

**Top**
A section showing water
supply collection.
1. Rain; 2. Bead of sand and
loose stones; 3. Bed of clay;
4. Bed of limestone;
5. Second bed of sand;
6. Bed of gypsum; 7. Well;
8. Reservoir; 9. Conduit;
10. Water-tower; 11. Rising
column of water.

**Centre**
Chamber of the Grotte de
Choranche Isère.

**Bottom**
The water cycle, illustrating the
process of evaporation,
condensation and precipitation
and the eventual return of
groundwater to the sea.

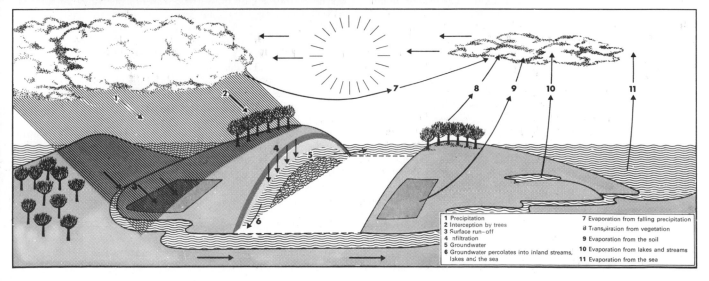

| 1 Precipitation | 7 Evaporation from falling precipitation |
| 2 Interception by trees | 8 Transpiration from vegetation |
| 3 Surface run-off | 9 Evaporation from the soil |
| 4 nfiltration | 10 Evaporation from lakes and streams |
| 5 Groundwater | 11 Evaporation from the sea |
| 6 Groundwater percolates into inland streams, lakes and the sea | |

line: these points are known as springs and the lines spring lines.

The appearance and origin of the points of emergence are very varied. Their exact location may depend upon the existence of more porous zones, slight variation in the slope gradient, the presence of cracks or joints in the aquifer, or on some other factor. Sometimes the seepage is scarcely noticeable; sometimes it is clearly marked by a flowing spring; sometimes there is a running stream, flanked by poplars and willows that can be seen from miles around, while water-loving plants—reeds, rushes, iris, forget-me-not, water-cress and mosses—spring up round about. Some springs are large enough to supply whole villages with water, and settlements have grown up around them, especially in arid areas.

The quality of a spring depends on the thickness and chemical composition of the soil and rocks through which its waters pass. Let us turn again to the diagram on p. 55.

The first permeable bed of rock is near the surface and can be polluted easily by animals (farmyards, dungheaps) or by human agencies (cesspools, drains and cemeteries). The water of the upper layer is therefore suspect, for it may carry Salmonella typhosi (typhoid fever) and other water-borne bacteria.

The second permeable layer is formed of a porous limestone, and the water with which it has become impregnated has dissolved some of the calcium carbonate. The additional calcium carbonate produces a type of hard water. The taste of heavily mineralised water (hard water) is generally unattractive, and soap lathers in it only with difficulty. Nevertheless, the water supply of many areas of the world comes from limey rocks and consists of lime-bearing waters, perfectly drinkable but not ideally palatable.

The third permeable bed is sandstone, an excellent aquifer from all points of view. Sand is generally capable of holding greater quantities of water than most other porous rocks and allows it to flow freely over long distances. Water that has been filtered through it is entirely free from all traces of organic matter.

The fourth permeable layer is composed of gypsum. The water contained in this rock dissolves some of the calcium sulphate and is therefore unfit to drink.

It is clear, then, that far and away the best spring is one whose water has been filtered slowly through a thick bed of sand. To preserve its purity such a spring must be enclosed within a protected area where it is illegal to deposit sewage or manure, or to establish slaughter-houses or chemical works. A watertight tank should be built round it, roofed over to keep the spring free from air-borne dust. A pipeline supply will then be fed to the water-tower or reservoir from which the water will finally be distributed.

Springs were harnessed by many ancient peoples to provide their towns with drinking water. The Roman aqueducts, including France's famous Pont du Gard, built to supply Nimes with water, give ample evidence of this. Such was the veneration of earlier days for the springs' supplies that each was thought to have its own gentle water-nymph or group of nymphs and naiads. Sacrifices

were made nearby and in the caverns from which the springs issued.

Many an old town was named after the god who was patron of its springs. And in countries speaking a language derived from Latin, there is an enormous number of place-names based on the Latin word for spring, showing how this type of supply determined areas of settlement. In France, such places as Fontainebleau, Fontenay, Fontvielle come to mind; Fontanarejo and Fontanosas in Spain; Fontane and Fontanelluto in Italy; and Fontana in Switzerland.

## WATER TABLES

Rain that has soaked into the ground travels slowly and almost vertically downward until it reaches a level below which all the pores and crevices are already filled with water. The upper surface of this zone of saturation is called the water table. It is not a plane but an undulating surface, reflecting very roughly the topographic profile of the area. Therefore, although under valley bottoms it is usually at a lower elevation, it lies nearer the ground surface than it does beneath the higher hilltops.

The depth to which the zone of saturation, or phreatic zone, extends varies with the nature of the rocks. In some areas water exists several thousand feet down; at the same depths in others there is only dry rock. Water in the saturation zone moves slowly because of the extremely tiny spaces open to it. Its passage can be measured in feet per day or even feet per year. General direction of movement is from higher to lower levels of saturation, paralleling the topographic slopes of the water table.

The zone above the water table, which is wet or dry, depending on the rainfall and on the vertical travel of the water, is called the vadose zone, or the zone of aeration. Here, except after rain, only small amounts of water are found, held by capillary attraction.

Actually there is wide variation in subsurface conditions. Impervious layers or lenses of tight rocks can stop or deflect flow. When such layers occur in the vadose zone, a temporary pool of water, a perched water table, forms above them. The water collects above the impervious layer and then migrates laterally, eventually spreading out beyond its edges to resume its downward travel.

Generally sloping toward the rivers, the water table controls or replenishes most streams. Permanent rivers are actually surface streams receiving ground water from seepages and springs, as well as from surface run-off. In some areas, mainly arid regions, the reverse may be true.

## WELLS

We have just seen that springs are the natural outflow of ground water, occurring at a point where the aquifer is intersected by a slope. This reservoir of water can also be obtained by

digging a hole deep enough to reach the water table, thus creating an artificial spring, or well.

The water table level fluctuates with seasonal precipitation but always lags somewhat behind. A fundamental principle in drilling thus evolves. When possible wells should be dug in the dry season to ensure that they reach beyond the lowest level of the water table. Of course, the well-digger must also make certain that he has reached drinkable water, and then he must protect it from infiltration of polluted water. The well facing must be made completely watertight. Only the lower part is constructed of stones laid together without mortar. This is to allow the pure water to seep in.

Nowadays it is more usual to sink a tube-well, a narrow steel tube driven down to the underground water. The end of the tube is cone-shaped, pierced with holes; above ground a pump is fitted.

Well waters, as all ground waters, have temperatures and qualities dependent on the rocks through which they have passed and on the aquifer in which they have been stored. Differences in quality are, in fact, the same as those found in spring waters.

A frequently debated puzzle is the occurrence of fresh water near the sea coast. How can a well dug in such a reservoir below the level of the sea be free from salt water? For infiltration as the sea water passes the permeable rocks will not remove salt from it in the way that it removes bacteria, and when sea water is put through such a filter, it still remains very definitely salt. If, therefore, a coastal underground reservoir gives fresh water, the pressure behind the fresh water must be greater than the pressure of sea water, and the source of the fresh water is not the sea but fresh water aquifers.

In many areas along the Atlantic coast of the United States

fresh water is obtained from aquifers actually extending beneath the ocean and generally separated from the sea water by impervious rock layers. In many Pacific coast sections, however, heavy withdrawals of ground water have produced an influx of sea water in the subsurface for a considerable distance inland. The wells in these areas now yield salt water.

## ARTESIAN WELLS

If an aquifer is enclosed above and below the impermeable beds of rock, and if the strata are downfolded into a basin, or syncline, so that the level of the water table in the rocks round the edge of the basin is high enough to give a good head of water, the water will be under enough hydraulic pressure to flow up a well-shaft drilled into the lower portion of basin. The water enters the bottom of the well and flows upward until it equalises the pressure. If pressure is great enough, the water will overflow at the top. All flowing wells are artesian wells, though many deep wells which do not flow are incorrectly called artesian.

The first step in the construction of an artesian well is to drill the ground and line the hole with sectional steel tubes. The water from these wells is abundant and, as it has been filtered for so long while passing through the rocks, it is pure. Underground temperatures are higher than surface temperatures, so artesian water is also especially well suited for industrial purposes.

Artesian wells owe their name to the Artois region of France, where they were first noted by Carthusian monks in the twelfth century. Since that date many important artesian basins have been tapped for water all over the world.

The London basin contains a layer of porous chalk which forms the main aquifer, the rainwater entering the chalk along its northerly outcrop of the Chilterns, and the North Downs along the southern rim of the basin. This water is sealed in by the Gault Clay below the chalk, and by the London Clay above it. One of the earliest records of this deep water being tapped is found in an account of two wells dug through the London Clay at Paddington in 1725. Each was 300 feet deep; water rose in them as far as the surface, indicating that the borings must have reached the chalk.

In London the water does not gush out of the wells, for the catchment areas of the chalk downs of Surrey are less than 600 feet high. On the western side of the London area, artesian water exists also beneath the chalk, in the Lower Greensand. Its outcrop in the wooded hills of Surrey forms a catchment area; where the greensand passes underground into the basin, it is covered over by the Gault Clay directly underlying the chalk.

The abundant supply of water in the chalk under London was so widely exploited about a century ago by breweries, laundries and factories looking for cheap water that the amount drawn up by the numerous wells began to exceed the amount of recharge slowly seeping down into the centre of the basin from the catchment area; hydrostatic pressure fell steadily. In 1844 one of the wells near the National Gallery supplied consumers. The water at this period rose to about 68 feet below sea level. By 1911, the water level had sunk to 195 feet, and soon afterwards the wells were abandoned. The fountains once supplied by these wells are now operated on the continuous circuit principle, with the same water being used over and over again. In some of the London parks, ornamental waters such as the Serpentine and the lake in St. James's Park, which were started by the damming of streams and later supplied by deep wells, are now fed from shallow wells in the underlying gravels. Depletion of water reserves is so serious that the Government has forbidden the construction of new wells.

Where the estuary of the River Thames cuts into the chalk on the eastern side, the pressure of water in the aquifer has been so much reduced that brackish water is now appearing in some inland wells sunk below the level of the estuary.

The structure of the Paris basin is similar to that of London. The most important aquifer here is not the chalk but the greensand which underlies it. Above the greensand is a thin layer of Gault Clay; below it lies another layer of clay which seals in the water. These formations outcrop some distance from Paris. The edge of the chalk basin, for example, forms the dry plains of Champagne where, overlooking the low Gault Clay vale, we find the famous

vineyards. The Gault Clay gives a damp, low-lying stretch of land with abundant pools and marshes that is succeeded to the east by the thick forests of the Argonne. These forests occupy infertile sandy soils with alternating dry hilltops and waterlogged valleys, marking the outcrop of the greensand.

The greensand encountered at 1,800 feet below Paris outcrops 125 miles from the city, at a height of 390 feet. The rain falling in this region soaks down into the sands, and at the end of several months reaches Paris. We have already seen that the aquifer is folded into a basin and roofed in by an impermeable layer of clay. Water entering the sand at 390 feet is now 1,800 feet below the surface, and cannot regain its former level. If a well is drilled to a depth of 1,800 feet, the artesian water tries to find its original level, and gushes into the air. It is exactly as if the water of a reservoir 390 feet high was being piped to Paris.

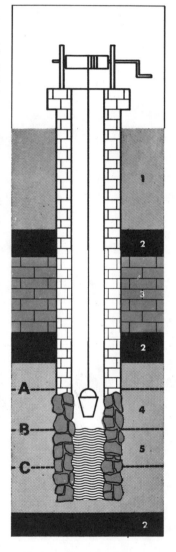

*Left*
A well sunk through a succession of rock.
1. Superficial sand; 2. Clay;
3. Limestone; 4. A deeper sand;
5. Aquifer formed by these layers.
A, B and C are the levels of the water table.

*Right*
The laying of a 72-inch diameter steel water main on the Newton-Prescot Aqueduct.

## OCCURRENCE OF GROUND WATER

The London and Paris basins occur in areas of good rainfall and large rivers, so subsurface water is simply a useful supplement to the main supply; neither city is by any means dependent on it. In some arid regions of the world, however, ground water is the only source of supply.

Some of the oases of the Sahara and other deserts occur where artesian water reaches the surface naturally. Thick sandstone underlies much of the Sahara, and where it outcrops on the high ground the very rare torrential rains of the region have formed a considerable reservoir of water under the desert. The natural oases occur where the ground-surface is low enough to intersect the aquifer. Hollows eroded by the wind may be deep enough to produce this result; the water-bearing beds may be brought to the surface by folding or fracturing. (Many springs, especially thermal springs, are associated with fault fractures.) The artesian water may, alternatively, escape through fissures in the rock. Where no natural springs exist, wells have been dug for countless centuries. Traces of them still exist in Africa and the East, pointing to man's utilisation of artesian water long before the 'discovery' of such properties at Artois. With great difficulty the primitive Arab inhabitants of the Sahara managed to dig wells 200 feet deep; modern methods of drilling have greatly changed the land by making the water much more readily obtainable. The flow of underground water is of prime importance here, where except for areas round the oases the land must be irrigated by man-made wells. The most typical of these underground watercourses is the Saura, which decends the Sahara Atlas mountains and flows southward until it is finally lost in the Tanezrouft. Its subterranean

course is marked by a string of oases and date-palms from Beni-Abbès to Zaouiat Reggane, the southern limit of cultivation.

The largest artesian basin in the world is the Great Artesian Basin of eastern Australia, 600,000 square miles in extent. It measures 1,270 miles from the Gulf of Capentaria in the north to Darling River in the south, and 900 miles from Lake Eyre in the west to the Great Dividing Range in the east. The water-bearing strata are Jurassic sandstones, up to 6,000 feet thick in the centre of the basin, and capped by impervious shale. The sandstones curve up steeply on the eastern edge of the basin to form the flat-topped plateaux of the Great Dividing Range which absorb the adequate rainfall of the coastal area. The water flows westward into the basin, and on the eastern edge, beyond Lake Eyre, mound springs are its natural outlet. These springs are highly saline, rimmed with rings of salt left by evaporation. Blown sand adheres

*Top left*
Llanforda Reservoir and the slow
sand filters at Oswestry,
in Shropshire.

*Bottom left*
The Elan Dam, part of Birmingham
Waterworks in the Cambrian
Mountains, mid-Wales.

*Right*
A desalination plant at Weir
Westgarth, Shuaiba, Kuwait. This
is one of the three flash distillation
units built in the area to obtain a
constant supply of fresh drinking
water from the sea.

to the salt rings until the spring, no longer able to overflow the rim, becomes a pool inside the mound. Some mounds reach 100 feet high and 300 yards wide.

Although artesian water is a boon to the eastern area of Australia, where yearly rainfall is seldom above 25 inches and often as low as 10 inches, its high salt content makes it useless for irrigation or agriculture. Cattle, however, need salt, and in some areas the stock-holding capacity of the land has been doubled by the use of artesian water, sometimes led for miles over pastureland.

In the early days, most people in the United States, except those residing in the arid sections, were relatively unworried about water supplies, for the rainfall recharge matched or exceeded withdrawal. Today the position is changing slightly. Wells and springs currently supply about 32 billion gallons a day. About 63 per cent of the ground water used goes into irrigation, 17 per cent is tapped by industry, 11 per cent supplies the needs of municipalities, and the remaining 9 per cent is used in rural areas for stock and in the home. The rate of consumption continues to grow rapidly. Ground water supplies in many areas are running low or are exhausted, and large lakes and rivers are being tapped. With proper conservation measures, however, the present water supply can be made to suffice.

Subsurface water is obtained more readily from some materials than others. Sediments with a high yield include alluvial sands and gravels, glacial outwash, and sandy coastal plain deposits. Rocks giving a good yield include permeable sandstones, cavernous limestones, vesicular lavas, and joined and fractured rocks of all types, including artificially fractured rocks.

In Britain, the most prolific water-producing rock is the Chalk. Its contribution to the water supply of the London area has been already described. In other parts of southern England, conditions suitable for an artesian or a sub-artesian supply do not exist and the copious amounts of water available in the Chalk are tapped by wells. Frequently 'headings', that is, large-diameter horizontal tunnels, are driven outwards from the bottom of the well so as to intersect water-bearing fissures in the rock. At Eastbourne, for example, one large well at Friston, only 110 feet deep, produces 3 million gallons of water a day through a system of

headings two miles long. The water requirements of Brighton are met from five wells having five miles of headings into the Chalk.

In Cambridge, the aquifer is the Lower Greensand which yields water to wells averaging 150 feet in depth. Farther north, good supplies are obtained from Triassic sandstones and marls, for example, at Ormskirk, where boreholes reach a depth of 1,000 feet.

Variations in climate, geography, and geology across the United States mean that sources of ground water differ considerably from one region to the next. The country can be divided into about twenty-four main ground water provinces. Subsurface water conditions within these areas vary slightly, but on the whole are similar. The twenty-four areas can in turn be grouped together into four major regions: the Coastal Plain, the East Central Old Rock Region, the Great Plain Region and the Western Mountains.

COASTAL PLAIN REGION. This region covers a belt of varying width running southward from Long Island, New York, to the Texas-Mexico border. Here, sedimentary rocks laid down in Cretaceous, Tertiary and Quaternary seas dip slightly seaward.

Water is found in the porous sandstones and limestones lying between impervious shaley beds. In some areas the inclination of the beds from the intake areas inland, where these beds outcrop, is great enough to produce artesian water conditions. In this belt, especially in the Mississippi Valley embayment, ground water is also obtained from stream fill and terrace deposits.

EAST CENTRAL OLD ROCK REGION. In this region the surface rocks range in age from Pre-Cambrian to Recent, and ground water supplies come from a variety of sources. In parts that have been glaciated, good supplies come from shallow wells in glacial gravels and sands; in sections underlain by very old igneous and metamorphic rocks, small supplies of excellent water are drawn where fracture zones lie near the surface. Over a large portion of the area underlain by Palaeozoic rocks, water—sometimes artesian—is obtained from almost level-lying sandstones and limestones and from fracture zones in all types of folded rock. Deep wells in this region, and many shallower wells in the northern and western parts, tend to be highly mineralised and unfit for use.

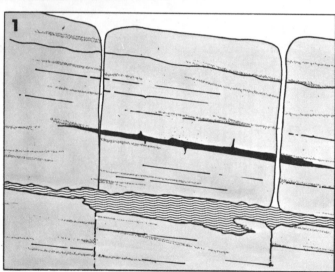

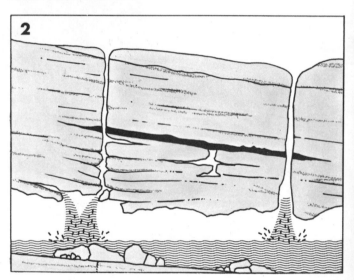

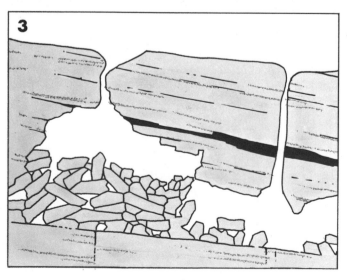

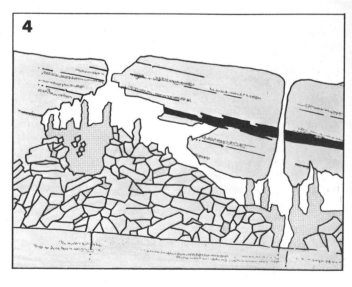

GREAT PLAINS REGION. Sedimentary rocks of the Mesozoic and Tertiary underlie much of this area. The bulk of the water comes from wells intersecting Cretaceous sandstones, especially the Dakota sandstone. A great deal of the water is artesian; most of it is highly mineralised but usable. Outside this source there is much variation in supply from place to place. In the northern part of the area there is abundant shallow-well yield in some of the glacial deposits. In the central part and toward the south, Tertiary river gravels produce good local shallow yields. In the south-western section, shallow production is highly saline and unusable, but deep wells penetrating Palaeozoic limestone rocks give excellent supplies of artesian water.

WESTERN MOUNTAIN REGION. This region embraces one-third of the different ground water areas. Here, rocks of all ages and types produce water under a great variety of conditions. Two great plateaux lie in this section. In the Columbia Plateau and some other parts of the north-west, porous Tertiary lavas as well as some of the interbedded sediments produce water. Over much of the Colorado Plateau water is scarce. Some is found in shallow wells close to streams, and a little artesian water comes from deep wells penetrating sandstones. In the mountainous sections a variety of springs occurs and provides a great part of the ground water supply. Good surface waters, too, are available in many places, and in these areas the ground water remains untapped. In the south-western area, near the coast, the need for water has led to the exploitation of the alluvial cones, fans, and fill of large valleys, and in some coastal regions water is piped over long distances from inland mountain sources.

## WATER SUPPLY

What qualities do we look for in drinking water? Obviously it must be well oxygenated, pure, and as odourless and tasteless as possible—or at any rate without any unpleasant taste or smell. Ideally, the quantity of calcium and magnesium contained in solution (mainly as bicarbonates, chlorides and sulphates) must not be high enough to make the water unpalatable or to make it difficult for soap to lather. Any chemical reaction must be as near neutral as possible. Finally, and most important of all, it is vital that it should be entirely free from pathogenic bacteria. These specifications apply to ground water and surface water.

Some of the defects in domestic water can, of course, be remedied. Deodorisation, for instance, can be effected by ozone produced electrically from the oxygen in the air. The amount of alkaline salts in solution can be controlled too. Hardness of water is measured by mixing it with pure castile soap in an alcoholic solution, with degree of hardness based on the amount of the standard soap solution needed to produce a five-minute lather in a fixed volume of water. Calcium bicarbonate hardness is a temporary hardness; on boiling, carbon dioxide is discharged, and the bicarbonate becomes insoluble carbonate. All other types of hard water are considered permanent, and must be treated chemically with softening sodium salts which bring about a base-exchange with the carbonates and precipitate them.

The acidity or alkalinity of water is expressed by its $pH$ value, the proportion of acid ions $(H+)$ to basic ions $(OH-)$. Water is chemically neutral when its $pH$ value is 7, acid when the value is lower and alkaline when it is higher. Water that is too acid can be treated with lime; if too alkaline, acid can be added.

The final point is of the most fundamental importance, and concerns the presence of pathogenic bacteria which may be revealed by bacteriological analysis. Any micro-organisms which might endanger the health of man or his domestic animals must be scrupulously removed by one of many different methods. If the purity of the water is doubted, it must first of all be filtered through a layer of sand resting on a bed of gravel or on slabs of porous concrete. Coagulation of undesirable materials is achieved by the addition of chemicals that produce sediments or settling. The filtered water is then collected into covered basins to await pumping to the necessary elevated reservoirs. While awaiting distribution it must be subjected to final purification by ozonisation or chlorination before it is fit for delivery to the consumer.

The most common sources of contamination are sewage, broken pipes, and drainage from privies, barnyards and cesspools. Strangely enough, ground waters in heavily populated areas are frequently of excellent quality. For example, on Long Island, New York, one community of 70,000 in an area underlain by sand and gravel disperses sewage at depths of 15–30 feet from septic tanks (averaging one per house) through the zone of aeration to the water table. The entire water supply for this community comes from eight ordinary wells, none less than 100 feet from a septic tank. Still the water supply exceeds legal purity requirements without chemical treatment. This remarkable situation is the result of the passage of the water through sediments with a high aggregate internal surface area and, therefore, a large molecular force that attracts and holds water while purification takes place.

Purification is accomplished by mechanical filtering out of bacteria, followed by their destruction by other organisms and, finally, by chemical oxidation. In some areas sewage-polluted river water is simply pumped into the ground and becomes part of the local subsurface water supply. In one United States city, lateral percolation through less than 500 feet removes all impurities and produces drinkable water.

Supplying a city with water sets many problems, and it might be of interest to consider the development of the supplies to London and Paris. The earliest inhabitants in their simple settlements on the banks of the Thames and the Seine were presumably content draw the water they needed from the rivers. In those far off times the waters were still pure and fresh. When the Roman cities were founded on hills overlooking the rivers, a closer source of water was needed. In London the hill forming the city is of gravel, and the Romans had no difficulty in digging wells that reached the abundant supply of water in the subsoil which was to provide Londoners with their water for many centuries to come.

In Paris, the Roman town of Lutetia was situated on a limestone bluff on the left bank of the Seine, and wells would have had to be very deep indeed to reach the water. Towards the end of the third century, then, the construction of an aqueduct was put in hand to bring spring water from Arcueil for the baths and the Emperor's palace. Still later, in the twelfth century, the religious houses of Saint-Lazare and Saint-Martin-des-Champs took water from springs flowing from the hillsides of Belleville and Romanville, and brought it to Paris by conduit. The course of their supply can still be traced in various parts of Paris, where some of the inspection chambers still exist.

London was not so fortunately placed for supplies of spring water, though every use was made of any issuing from gravel cappings on neighbouring hills, the water being led in by conduits or in oaken pipes. In 1613 the Lord Mayor of London, Sir Hugh Myddleton, constructed the New River, an open trench bringing pure water from the chalk springs at Amwell and Chadwell, in Hertfordshire, down the Lea valley to London.

As technological knowledge advanced, improvements were made. At first very large pumps were built to raise river water for the cities. In London the rapid flow of the Thames through the narrow arches of old London Bridge (later destroyed in the Great Fire) was used to turn water-wheels that worked a pump for forcing the Thames water through wooden conduits to the City. This pump was constructed in 1582 by Peter Morrys. A similar pump was established in Paris in 1715 on the Pont Neuf to supply water to the Louvre and the Tuileries.

At the end of the eighteenth century cast iron came into use, and the widespread distribution of piped water became possible. Water could be sent for longer distances and at greater pressures than in the leaking old wooden conduits. In London several water companies were founded which drew their supplies from the river and stored them in reservoirs. The city's great expansion over the areas of clay subsoil in the London Basin now became possible, for the population was no longer confined to the areas with a gravel subsoil for wells.

Unfortunately, as water thus became more abundant for the citizens of London and Paris, it also became more and more foul as the river suffered pollution from the outpourings of factories and sewers. Epidemics of water-borne diseases were widespread in the nineteenth century. Typhoid and, particularly, the dreaded

Asiatic cholera raged. Though the actual vibrio, or bacterium, of cholera was not to be discovered until 1883 by the German scientist Koch, it was gradually realised that the disease was spread by water, and purer supplies were sought.

After the Paris cholera epidemic of 1832, new sources of water were urgently needed. Work on the deep well at Grenelle was started in 1834. Later, in 1854, an extensive programme of aqueduct construction was begun and is still being extended.

London's problems were less easily solved; aqueducts could not supply the purer water needed. An effort was made to draw water from higher up the Thames, and filter beds were laid out on flat land alongside the river. In spite of this and other developments, cholera epidemics still lingered on into the 1860s, with an outbreak as late as 1871. It was only after this date that drastic improvements were at last made, and London's water was no longer regarded as 'poisonous'. Reservoirs were constructed along the broad, trench-like valley of the River Lea which flows into London from the north, bringing abundant water from the chalk. The bulk of the water supply of the eight million inhabitants of Greater London still comes from these two rivers, only 16 per cent coming from other sources.

The present supplies of two cities are today by no means adequate for their rapid rate of expansion. Both the London and Paris Basins continue to increase in population, and maximum water withdrawal from both Thames and Seine has been reached. Paris is already looking as far afield as the valley of the Loire; London's future sources remain a problem.

The position is not so acute in the industrial north of England. The city of Manchester obtains its water supply from two lakes in the English Lake District, Thirlmere and Haweswater, through a pipeline some 95 miles long. Similarly, Liverpool's water supply is largely obtained from Lake Vrnywy in North Wales.

Most towns and cities come to feel the pinch of an inadequate water supply as they grow in size. Except in cities as large as New York, careful planning and sound technical advice can generally produce an adequate supply without too great an effort. But many pitfalls exist. In Bloomington, Indiana, for example, two dams were constructed on Karst topography against the advice of geologists. Leakage made the first useless, and the second only partly useful. Finally a third dam was built, as advised, on impervious rock. This produced a reservoir with an adequate supply.

Louisville, Kentucky, taps the Ohio river for its main municipal supply, augmenting this with wells in glacial deposits buried in preglacial valleys. In summer the ground water supply is preferable, particularly for air conditioning purposes, because its temperature of 57°F. (14°C.) remains constant while that of the river water rises to 80°—85°F. (26°—28°C.) When increasing consumption made a shortage of ground water supplies imminent, it was decided not to pump the wells in winter but to use the river water exclusively, and to recharge the glacial gravels from the municipal river supply. In this way, a plentiful supply of cool ground water is ensured in the summer.

New York City is looking forward to tapping the Hudson River to augment its present supply, which is brought in by pipeline from north of the city. The city's ever-increasing demand has made it necessary to consider the possibility of having to use ocean water in the distant future, in spite of the expense involved in purification.

## UNDERGROUND CAVITIES

So far we have considered almost exclusively the water which runs off the surface of impermeable soils or sinks through porous soils to form underground reservoirs. Certain regions of the world present us with another phenomenon: that of water penetrating the rock fissures and gradually excavating them into a network of subterranean cavities. We are entering the territory of fissures, pot-holes, caves and caverns, which the exploits of speleologists make more familiar every year.

If water is able to open up these countless channels and galleries, the rock must at once be impermeable, fissured and, of course, soluble. To explain this: porosity is the amount of void space

*Top*
The collapsed chasm of Padirac, 96 feet wide at the top, and more than 225 feet deep.

*Bottom*
Stalagmites piled up like plates in the Aven d'Orgnac, in Ardeche.

present in a rock; permeability is the ability to transmit a liquid. A porous rock is not necessarily permeable. Pore spaces must be interconnected if a rock is to be permeable, and they must be large if capillarity is not to prevent flow.

While many limestones are very compact and dense, and therefore have a very low primary porosity and are impermeable, most are jointed or cracked. The joints and cracks create openings in the rock, and the fissured limestones have an excellent acquired, or secondary, porosity lending them permeability. An unfissured, impermeable rock merely supports a superficial run-off of water, in the manner of clay soils. Soft or incoherent deposits would wash away, just as sand and clay does, and would not be strong enough to form the solid walls of pot-holes, galleries, and underground chambers. Finally, in an insoluble rock, fissures are not enlarged by water running through them.

For cave formation the solubility of the rock must not be too great. It is obvious that a mass of rock salt, for example, would disappear rapidly and entirely under the dissolving action of the water, instead of being hollowed out into underground cavities. On the other hand, a granite massif would not be sufficiently soluble to allow the formation of caves to proceed at all. In fact, caves in granite generally form only on the coast, where the dual action of mechanical and chemical erosion by the waves succeeds in hollowing them out.

The most suitable rock for the formation of subterranean cavities is a compact limestone, preferably with a somewhat varying composition, ranging from pure limestone to dolomite. Composition of soluble and relatively less soluble substances results in cavernous development strong enough to stand without extensive collapse. Besides alternating beds of limestone and dolomite, which are ideal, limestones grading into or alternating with shale can also result in cavern development.

Limestone, as we have said, may be an impermeable rock but generally it is cut by countless joints and fissures, some of which developed from rock shrinkage caused during drying-out after the initial deposition period, and others from later deformations of the rock. The fissures range from simple cracks passing through several layers to faults which may mark the movement of a large section of rock. In most cases the clefts are narrow, and the water which flows slowly down them has ample time to carry out its work of dissolving the limestone. Little by little, the fissures become enlarged into corridors and galleries; at the points where they intersect, larger chambers and caverns appear. The various networks of passages join up as erosion proceeds, forming enormous subterranean cavities, or pot-holes, hundreds of feet deep and usually leading into miles of galleries.

We see from this that it is useless to search for cave-systems in areas of hard, igneous rock or in lowland areas of soft clays and sands. It is to the limestone areas of the world that we must turn. The age of the rock is of little importance; a limestone very young by geological standards may be as full of caves as an older one.

Before dealing with the principles underlying cave formation, let us consider the caves of a famous tourist area, say, the Causses limestone area of south central France. Here in the departments of Dordogne, Ardèche and Lozère, thick Jurassic limestones give rise to a Karst topography pierced by abundant caves, some of the best and earliest explored in the world.

It matters little which we examine first, but if we choose the cave of Padirac, we shall find a chasm of medium size, but large enough to impress us with the dimensions of the subterranean world. The cave was one of the first in the neighbourhood to be discovered and opened up for tourists. The first serious exploration of it was made by E.A. Martel (1859—1938), the French pioneer.

From the town of Saint-Denis-près-Martel, the road and railway climb up to the Causse de Gramat, formed by the arid grey limestone which stretches away to the horizon among sparse clumps of oak and ailanthus. The soil is covered with stones, with short coarse grass striving to grow between them. Poor habitations are scattered here and there on this upland desert. This is the true Karst landscape in all its sombre grandeur. The surface of the barren soil is deeply furrowed, pierced by clefts and holes, and rent here and there by gaping hollows revealing beds of clay left

after the solution of the limestone. The clay supports a more abundant vegetation. At odd points large crevices, sink-holes, and chasms break the landscape. Aridity reigns everywhere, not because the rainfall is low but because it vanishes instantly down the crevices in the bedrock.

Several miles away, there is the curious spectacle of the town of Rocamadour perched dangerously on the side of a steep, riverless gorge. Below the dry river bed, the Alzou follows an underground course to join its waters with those of the River Ouysse. Only in winter does it contain enough water to rise into view, and after very heavy rains it forms a sizeable torrent.

The Gouffre de Padirac, with the tops of trees emerging from it, is an oasis in the barren land around it. Here there is a museum of objects found in the cave by explorers. Here the visitor may rest and take refreshment in the restaurant before descending into the abyss by elevator.

From the bottom it is easy to appreciate the immense depth of the abyss, 96 feet in diameter at the entrance, 165 feet in average depth, and 225 feet at the deepest point. The elevator does not descend to the deepest point, but only as far as a slope of debris left after the collapse of the old roof of the cave. Beyond, there is a further 100 feet to the deepest point, and then an entry to a lateral corridor for another 100 feet. Finally, at 300 feet underground, a stream runs along a sombre gallery that is 15 to 30 feet wide and topped with a dark roof hanging 30 to 150 feet above.

Constant seepage all round swells the stream into a river, filling the gallery from side to side. At the water's edge, small flat-bottomed boats wait, ready to float the visitor down the river between steep walls merging into the darkness above. A quarter-of-a-mile downstream the water widens out into the Lake of the Raindrops, so called because the walls sparkle with countless tiny crystals of calcite. Huge stalactites hang down from the roof. One, the Great Pendant, is several yards long, 10 feet thick, and narrows to a point 6 feet from the ground. Centuries of time and innumerable drops of water must have gone into the formation of this colossal needle of rock.

Apart from the Great Pendant, the ornamentation of the caverns of Padirac, like that of Kentucky's Mammoth Cave, is spectacular rather than beautiful. For stone lacework and pretty masses of stalagmites the caves of Aven Armand and Aven d'Orgnac at Lacave, or the Carlsbad Cavern of New Mexico are better examples. At Padirac everything is on a much grander scale. No inspection is complete without a visit to the Great Dome, 180 feet long, 75 feet wide and 280 feet high. Beyond this, the underground river runs on for another three miles, passing through several lakes and over several dams of stalagmites.

The chasm of Padirac, as the heap of debris shows, was formed by collapse. The chasm was formerly an underground chamber whose roof finally became too thin to support its own weight. This is a common factor in many cave histories. Other chasms and pot-holes have been completely excavated by water; these usually show deep solution-channels and are narrow in relation to height.

Some galleries are wide and low-ceilinged; others are extremely narrow and so high that the roof vanishes into the darkness. The physical differences are explained by the type of fissure in which the gallery has developed; the wide, low-ceilinged type develops usually along the horizontal bedding-planes which separate the layers of rock; the high, narrow type along the vertical joints cutting through the layers. When the water has few cracks in the rocks to guide it, it may erode narrow winding galleries, called meanders because of their resemblance to those in surface streams.

It is worth noting a few details of nomenclature at this point. Although some of the terms employed by speleologists and pot-holers may not be very precisely defined and may vary from one cave to another, the more common ones are fairly widely used. A bare limestone surface eroded by surface run-off, carved into holes and pits and grooved and fluted is called a limestone pavement (known also as a clint in northern England, a *lapiaz* in France, and a lapies in United States). The large, shallow, funnel-shaped depressions are called *entonnoirs* (funnels) in France, and are also known by many local dialect names (sink-holes in the United

States and shake-holes in Yorkshire). In most areas the larger depressions are called dolinas, after those in the Karst of Yugoslavia. A vertical hole or succession of straight-sided holes occurring in steps is a pot-hole or a ghyll in England, and a *gouffre, abîme, aven* or *chouroum* in the dialects of southern France. Horizontal galleries lead into chambers, rooms or halls. A chamber with an outside opening is a cave or cavern, a *grotte* or *baume* in France, and a *Höhle* in Germany. The point at which a river vanishes underground is called a swallow-hole or swallet in England and United States, a *perte, goule* or *embout* in France, a *Saugloch* in Germany, and a *ponor* in Yugoslavia. The point where the river reappears from the ground is a resurgence or rise-pit, or a *guier* or *gillard* in dialect French.

Cave-systems is the expression most generally used for an underground network of galleries and vertical gulfs. The French terms *gouffre* and *grotte* are also widely applied, the first to systems where vertical drops predominate over galleries, and the second where the galleries are more important. This usage is not recommended: *gouffre* should imply merely a chasm, and *grotte* an open cave-chamber.

## STALACITITES AND STALAGMITES

Apart from roof-collapse, it is certain that almost the whole of any underground cavern has been hollowed out by water, which introduces mechanical erosion as well as chemical attack or corrosion. In cave formation, chemical attack is the more important of the two, for the limestone rocks, being formed of calcium carbonate, are only slightly soluble in pure water. When the water contains carbon dioxide, they break down readily under the action of carbonic acid. The resultant calcium bicarbonate, which is soluble, can be readily removed in the dissolved state.

The capacity of water to dissolve limestone is considerable. The River Lesse, for example, in its passage through the caves of Han, in Belgium, has been estimated to dissolve some five tons of limestone a day.

Just as the calcium carbonate can be taken into solution as bicarbonate, so the reaction can be reversed and insoluble calcium redeposited by driving off the carbon dioxide, as in evaporation. It is redeposited carbonate that builds up the stalactites and stalagmites on the roofs and floors of underground cavities.

The continual seepage of water down the cave walls and through the roof produces constant dripping and evaporation at certain points. Stone icicles form on the cave roof, slowly growing with the addition of successive layers of calcium carbonate. Each drop of water that evaporates leaves its contribution of calcium salt. The word stalactite has a Greek derivation meaning 'drop by drop'. There is, too, a general term 'dripstone' which is used to cover all formations. In German a similar term, *Tropfstein,* is used.

The deposition of calcium carbonate generally proceeds fastest at the circumference of the drop, where a hard ring develops and gradually elongates to form a tube. A hollow stalactite of this kind grows from inside and outside as the evaporating water deposits its mineral matter. Little by little, the cavity fills up, and the stalactite becomes solid.

Where water trickles out of a narrow fissure in the cave ceiling and not just from one small hole, a hanging curtain of stone will form instead of a conical stalactite.

If the water flows too quickly to evaporate entirely on the ceiling of the cave, it falls to the floor, evaporates there, and deposits its calcium carbonate. The small domes of stone that begin to rise are called stalagmites, and may grow up to join the stalactites above as single columns. Some cave floors are covered with stalagmites; in others, the walls are lost behind a dripstone covering. Some cavities are so blocked with calcareous formations of this kind that access to them is completely barred.

Made up of innumerable tiny calcite crystals, stalactites and stalagmites, columns and stone draperies sparkle as if lit with millions of tiny lamps. The odd shapes they take on suggest comparisons with trees, flowers, candelabras, chandeliers, organs, statues, obelisks, mosques, minarets, cathedral spires, domes and carvings. The cave of Dargilan in France has a 'belfry' 60 feet

*Top*
Stalagmite joins stalactite to form unbroken columns in the cave of Betharram.

*Bottom*
The magnificent drapery of stalagmite and stalactite in the Abbé-Glory chamber in the Aven d'Orgnac.

66

high, formed of numerous small columns resembling the spire of a cathedral. In the Aven Armand there is a spectacular array of dozens of columns, some more than 50 feet in height, called 'The Virgin Forest'.

It is not always possible to find a ready explanation for the shape of these dripstone formations, as there may have been various interruptions in their development. Currents of air can cause stalactites to form out of the vertical. Curtains and gargoyles of calcite can change their shape with a slight change in the source of the dripping water. A cave whose moist air is saturated with water vapour will not provide suitable conditions for crystallisation by evaporation, and in this case the cave may become covered with a soft paste of calcite, sometimes fancifully called 'mountain milk' or 'moon milk'.

Calcite particles may be seen floating on the waters of some underground lakes. The celebrated 'cave pearls' are spherical calcite particles. At the centre of each pearl is the nucleus around which it has formed in a process very similar to that of pearl formation in the oyster.

Some stalactites form filaments as fine as spiders' webs; others, 'eccentric' stalactites, have strange, twisted shapes. Some young stalagmites resemble flowers, while larger ones may look very much like piles of plates. Each 'plate' is formed by the deposition of extra calcite over irregularities on the outside of the formation.

Finally, calcium carbonate in another mineral form, aragonite, may appear side by side with calcite, and be even more attractive in appearance, forming as elaborate groups of crystals, brightly coloured by traces of metallic salts.

## UNDERGROUND RIVERS

The cave of Padirac, as we have seen, contains an underground river which is navigable for a certain distance before it disappears into the depths. In other caverns, such as Saint-Marcel d'Ardèche, there is no river to be seen. It should not be assumed, however, that a river had no part in the formation of such a cavern: its river may be flowing at a lower level, and the galleries may be the old stream bed, now dry and abandoned. At Dinant, in Belgium, guides conduct visitors down to successive river beds more and more recently occupied, until the rushing of the underground water can be heard far below on a level with the Meuse.

Numerous underground streams exist in the limestone cave and sink-hole region of the Pennines in northern England and of Kentucky and Tennessee in America.

The water of these rivers comes principally from the swallow-holes which carry the surface water below. In addition, there is the ground water which seeps through cracks in the roofs of the caverns and trickles down the walls, and there is condensation from the water vapour of the warm air which penetrates the caverns and there becomes chilled. Finally, in mountain areas there is the meltwater from snow accumulated in the hollows of the limestone pavements.

Water behaves underground in much the same way as it does on the surface, except that it may abandon one course for a successively lower one, taking advantage of the fissures in the rock. The change to a lower level may be marked by underground waterfalls, or by apertures so narrow that speleologists call them chimneys, squeezes, or letter-boxes. Sometimes these apertures are siphons through which the roof dips down below the surface of the water, requiring considerable courage of the cave explorer if he is to dive 'into the dark water and into the unknown', as the famous speleologist Casteret puts it.

Like other streams, an underground river may carve pot-holes, this time in the form of rock-mills, not caves. The cave of Cuves de Sassenage, Isère, contains a striking example.

The course of shallow underground rivers may be ponded by little dams of stalagmite, called rimstone pools or gours.

Again like rivers above ground, underground courses are subject to floods and periods of low water. For cavers their sudden floods all too often spell disaster.

When a river flows through a limestone area, especially when the limestone is fissured, it sometimes disappears partly or completely underground. Here we have a swallow-hole whose nature

*Top*
Gours in the Grotte de Coufin, Isère.

*Bottom*
The resurgence of the River Loue (Doubs).

changes with the seasons and with the amount of water coming down the river. The water which disappears into the swallow-hole may reappear further down the same valley, or it may feed a tributary which rejoins the main stream. It may even rise in the valley of another river. The Danube, for example, after flowing above ground for some miles, sinks partially or totally underground above Immendingen in Baden, and flows into the river Aach, seven miles to the south. The Aach in its turn flows into Lake Constance and thus helps to feed the Rhine. It is curious to think that the waters of one river can terminate so far apart—in the case of the Danube in the Black Sea and the North Sea.

The terms resurgence, rise, or rise-pit, are given to the place where an underground river emerges from the earth. Underground rivers vanish into sink-holes, exsurgences, or swallow-holes. These should not be confused with a true spring, and it must be borne in mind that the river has merely flowed along an underground bed; its waters have not been purified in any way by gradual filtration through porous beds of rock. Numerous 'springs' in Missouri, Indiana and Kentucky, as well as the famous Silver Springs of Florida, are resurgences of this sort. A well-known British example is that of the River Aire which emerges at the foot of a spectacular limestone amphitheatre, with cliffs nearly 300 feet high, at Malham Cove in north-west Yorkshire. Generally, very large springs indicated the rise of underground streams; most smaller ones are simple gravity springs. The waters of rise-pits are usually clear, except during the rainy season.

One of the most famous resurgences in the world is the Spring of Vaucluse, mentioned in the verses of Petrarch. A rocky cliff enclosing the river valley gives Vaucluse its name (Latin *vallis clausa*—closed valley). The river rises from a siphon in a cave at the base of the cliff. The flow is extremely variable, and can double in an hour or two. An underwater research team from Toulon tried to plumb the depths of the siphon's outside arm, but proved little except that at a depth of 192 feet they still had not touched bottom. Beyond this everything is unknown; we have no idea where the water of this large and important underground river comes from.

## MAJOR CAVES OF THE WORLD

The longest cavern in the world is probably Mammoth Cave, near Bowling Green, Kentucky. It owes its name simply to its vast size; the series of connected galleries at various levels cover several hundred miles in length. It does not, however, contain many crystallisations. The French speleologist Martel stayed down it for three days and described it as 'a monstrous and interminable immensity'.

The cave is located in one of the most spectacular cave and Karst areas of the world. A region covering over 8,000 square miles in the states of Kentucky, Indiana, and Tennessee is underlain by more or less horizontal beds of limestone, several hundred feet thick. Mammoth Cave covers an area of five square miles, and although parts have been mapped in great detail, it has never been fully explored. The galleries vary in height from two feet to over one hundred feet, and contain streams, lakes and waterfalls.

The Carlsbad Cavern of New Mexico, besides being 30 miles

long, boasts some enormous chambers with magnificent decorations. Its original discovery and early exploration, in 1901, are attributed to a cowboy. James Larkin, like many other less curious ranchers in the neighbourhood, had been puzzled by sky-darkening flights of bats. Close investigation brought him the discovery that they spiralled from a cave mouth. Today an elevator takes visitors 600 feet down to view his discovery.

Carlsbad Cavern lies in a region of limestone and gypsum in the Guadaloupe Mountains. Spacious chambers of great variety and beauty characterise the system. The largest lies about 1,300 feet below ground. The Big Room is 4,000 feet long, with a width of 625 feet and a height of 350 feet. Stalactites, stalagmites, pillars, curtains and frescoes of onyx, at once fantastic and magnificent, lend an atmosphere of splendour. Bat Cave, the source of Jim Larkin's curiosity, is seldom shown to the half million people who visit the system. Here, as many as eight million bats sleep by day, and at night leave to feed outside at a rate of three hundred per second. Their exodus, filling the air with a sweet musky odour, may take all of four hours.

In Britain, good examples of caves occur in the Carboniferous limestone rocks of the Pennines in the north, and of the Cheddar Gorge area of the south-west. The Pennine range of hills, often called the 'backbone of England', extends in a north-south line from the Scottish border to the southern boundary of Derbyshire and separates the coalfields of Cumberland and Lancashire from those of Northumberland, Durham and Yorkshire. The caves at Clapham and Ingleton are particularly noteworthy. In the Clapham area is the almost perfectly cylindrical swallow-hole

Gaping Ghyll, 365 feet deep; at the foot of the main shaft is the Main Chamber, 500 feet long, 110 feet wide and 90 feet high, from which 3½ miles of surveyed passages and caverns radiate. Water entering the top of Gaping Ghyll reappears at Beck Head in Clapdale, a mile away. Also in this area is Malham Cove, a good example of Karst topography. In the Cheddar area, Wookey Hole is outstanding.

In Europe, the longest cavern is the Holloch, near Schwyz, in Switzerland, where more than 33 miles have been explored. In the Austrian Alps, not far from Bischofshofen, the Eisriesenwelt offers a labyrinth more than 25 miles long, with two galleries occupied by interesting ice formations.

The cave of Postojna (formerly known as Adelsberg, then as Postumia), in Yugoslavia, has a total length of 14 miles. Parts of it have been known since the Middle Ages, and today it is one of the best organised caves for tourists, offering a light railway and a concert-hall. There is also a valuable underground laboratory for the scientist.

In France, the underground network of the Dent de Crolles in the Grande-Chartreuse area is more than 11 miles long. The mountain 6,186 feet high, is riddled with a labyrinth of pot-holes and galleries to which there are five entries: an abyss at 5,808 feet, the Chevalier and Annette caves, the Trou du Glaz at about 5,100 feet, and, finally, the exit of the underground river at 4,000 feet which forms the 'spring' of Guiers-Mort. It took a team of explorers twelve successive years to complete investigations from the upper limestone platforms to Guiers-Mort, 1,809 feet below.

The world's deepest pot-hole is the Gouffre Berger, in the

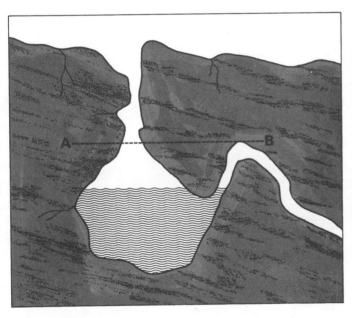

*Left*
The main sink hole of Gaping Ghyll, Yorkshire. This depression has been caused by the solvent action of rain water containing carbon dioxide from the atmosphere and has been enlarged by the sinking of the ground.

*Above*
Some underground siphons flow at regular intervals. As seen here the water empties as soon as it reaches the level AB.

*Right*
The underground lake in the Grotte de Gournier.

*Top left*
'Straw' stalactites in the Grotte de Coufin at Choranche.

*Top right*
Eccentric stalactites in the Aven d'Orgnac.

*Bottom left*
'Fir cone' Stalagmite in the Aven d'Orgnac.

*Bottom right*
'Balcony' formations in the Grotte de Gournier at Choranche.

A new cave of stalactites
discovered in 1956 branching off
the main system of Stump Cross
Cavern, Yorkshire.

Isère province of France. Speleologists have already reached 2,955 feet, and hope ultimately to penetrate to the full depth of 3,300 feet. The stream flowing through this cave reappears in the Cuves de Sassenage.

We must not forget the size and beauty of individual chambers within these famous cave-systems. The largest chamber is that of the Giant Cave near Trieste. It measures 720 by 390 feet and is 420 feet high.

Decorations of stalactite and stalagmite are at their most beautiful in the Aven d'Orgnac. Joly, who first explored it, was able to trace the stages of its formation which was twice interrupted during late Tertiary times by earthquakes. The early stalagmites, broken and thrown down, served as a foundation for the growth of new concretions which have remained intact, and date from the calm period of the Quaternary. Even though the earthquakes lasted only a short time, the hollowing-out stages and the subsequent accumulation of the concretions have taken not hundreds of thousands of years but millions.

The Eisriesenwelt of Austria contains a mile of the chambers and galleries occupied by glacial formations. Ice-caves are also found in other parts of the world—the Alps, Jura and Pyrenees. Those discovered by Casteret in the neighbourhood of the Cirque de Gavarnie, in the Pyrenees, are among the most remarkable.

In the Dèvolouy area of the Alps, the cave called the Chouroum de Parza also conceals an underground glacier. Its neighbour, the Grotte de la Tunnette, possesses a marvellous drapery of ice descending from a narrow fissure in the roof above.

## CORROSION LANDSCAPES

Though limestones are particularly soluble, they are seldom homogeneous and offer unequal resistance to the corrosive action of rainfall and run-off. This is especially true of dolomitic limestone, which is partly calcium carbonate and partly the double carbonate of calcium and magnesium. This last substance is less easily converted into the soluble bicarbonate, and can resist erosion until after the disappearance of the calcium carbonate. Shaley limestones, too, offer resistant layers that are less soluble than the pure calcium carbonate. In every case, the result of differential erosion is an extraordinary carving of the ground into deep twisting furrows, narrow and separated by sharp walls. If the bedrock is suitable, water may disappear and flow underground here.

Corrosion of the surface forms limestone platforms and sink-hole or Karst plains. Perhaps the best known of these areas in the world occurs in Europe, the Karst (or Carso) lying north-east of the Adriatic coast, between Trieste and Cattaro. The Karst landscape has given its name to all such deeply eroded areas of limestone. Pre-requisites for the development of Karst topography are almost the same as for cave formation, except that for Karst topography the soluble layers should be at or near the surface, while for caves they should be at depth. If the surface limestone is thick, as in Kentucky, then both will develop.

This point can be illustrated with an example from the Pyrenees whose deeply eroded limestone caves are world-famous. The first photograph on page 72 shows a limestone pavement with its characteristic aridity, its upstanding blocks fretted and channelled, and its loosened blocks clogging the deep fissures.

The next photograph (top right) shows a closer view of the vertical channelling of the limestone. This is caused by the solution of calcium carbonate by the rainwater, which is slightly acid after absorbing some of the carbon dioxide from the air.

The third photograph (centre left) shows an area only twelve feet long, and yet it is traversed by a miniature valley-system. Adjacent valleys are separated by crests, and two valleys converge at the spot marked by a cross. In front of this may be seen the vertical channelling along the wall of a longitudinal fissure.

The last photograph of this series (centre right) shows ridges of limestone completely separated by fissures. The tops of the ridges are worn into a series of cup-shaped hollows separated by sharp edges; the sides are, as usual, vertically channelled; their crests are jagged. It must have taken many centuries of erosion to incise and sculpture the rock so deeply by the chemical action of the rainwater.

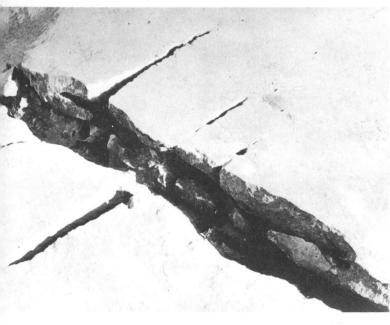

The photograph on the bottom left shows how fissures begin in a calcareous surface crossed by veins of softer rocks. A deep fissure only a few inches across has been excavated along one of the lines of least resistance. Smaller fissures, at right angles to the first, are opening out to the right and left along other lines of weakness in the rock. This process is indicative of corrosion.

Let us finish with an aspect of erosion in limestone pavements. The last photograph looks rather like an aerial view of the valley of a meandering river, not unlike that of the Meuse or the Wye. It is, however, simply the channel carved by a trickle of rainwater on a gently sloping surface of limestone. The 'valley' is only a finger's length in width, and a few inches deep.

Even so, all the features that we shall consider later as characteristics of meandering rivers are there: steep concave banks, convex slopes more gently curved, and so on. There is even what looks like an abandoned meander, now forming an ox-bow.

Famous Karst areas, besides those already mentioned, occur in Spanish Andalusia, Greece, north Yucatan, north Puerto Rico, northern England and central Florida. Northern Yucatan possesses some unique Karst features. If a limstone region consisting

of a gentle, northward sloping plain and a higher, hilly southern portion. There are no permanent surface streams. Sink-holes, galleries or caverns are the major source of water, most of which is obtained from cenotes opening downward into the caverns. (Cenotes are chimney-like sink-holes or caverns in which the roof has collapsed.) Near the coast the water level in the cavernous limestone is very near the ground surface. Because of this, in coastal lagoons and sometimes even in the Gulf waters just beyond the coastal reef, springs of fresh water occur.

Most of the topographic features characteristic of a Karst region have already been described in the preceding section about caves. They include: disappearing streams or sinking creeks; dolinas; solution pans; numerous sink-holes and sink-hole ponds, perhaps numbering hundreds to the square mile; swallow-holes; compound sinks, and blind or closed valleys.

Eventually, as most of the limestone is removed from the area, only isolated hills or group of hills will remain. These residual hills are analogous to the monadnocks produced in the last stages of a fluvial erosion cycle. In the Yugoslavian Karst they are called *hums;* in Puerto Rico, where they are very common, they are known as *pepinos* or haystack hills; and in the Causse area of France they

The Cirque de Gavarnie, France.

are *buttes témoines*. The *pepinos* of Puerto Rico vary in height from 200 to 300 feet; a few in Cuba stand over 1,000 feet and are honeycombed with caves.

## SPELEOLOGY: SPORT AND SCIENCE

From 1888 onwards Martel and, later, Casteret, made sensational discoveries underground with the most primitive equipment. By the light of candles or carbide lamps, sliding down ropes or heaving themselves up rope-ladders, swimming across lakes and underground rivers, edging along galleries and squeezing through fissures, they travelled, as Casteret writes, 'ad augusta per angusta: towards splendid things through narrow places'. Courage to meet every hazard, backed by athletic ability and sustained by ardent enthusiasm, made them the pioneers of a sport and a science: speleology, 'the study of caves'.

To get a more exact idea of the conditions under which the earliest speleologists worked, and of the progress made over the last twenty years in equipment and the results obtained, we cannot do better than to read their own books. Some of the most interesting are *My Caves* (1951), *Ten Years Under The Earth* (1952) *More Years Under the Earth* (1962) by Casteret, and *Underground Adventure* (1952) by Arthur Gemmell. Excellent books and articles about caves and their origins include *Origin of Limestone Caverns* by Davis (1930), *Caverns, ice-caves, sinkholes and natural bridges* by Henderson (1932—33), *Landscape* by Cotton, and *Geomorphic History of the Carlsbad Caverns Area* by Horberg.

Casteret's second book contains a chapter specially devoted to and in praise of those who began exploring caves before 1928, without the help of the modern tackle and equipment. Hemp ropes have now been replaced by rot-proof nylon; rope and wooden-ladders replaced by light, flexible metal ladders; windlasses facilitate the ascent and descent of men and equipment; collapsible metal poles hoist ladders up to the roofs of caves; pneumatic canoes are used to navigate underground streams; frogmen's suits make diving through siphons possible; telephones are used for communication. With these aids today's speleologist, or 'spelunker', can study the effects of living underground continuously for three months or more.

'Spelunking' is at once a sport and a science. Perhaps it would be more accurate to call it a sport that is of service to science. Like underwater sea exploration, speleology has many scientific aspects. It is naturally hydrogeology that first attracts the attention of the speleologist, for this explains the varied shapes of the subterranean cavities, and the link between the cavities and the rocks and water which hollowed them out. It explains, too, the curious formations of stalagmites and stalactites.

Water coming from underground is often used for purposes which make it essential to know its exact provenance and the amount of outflow. Here the speleologist plays a useful part in determining the water's character. In trying to trace the origin of water in a resurgence, however, it is not always possible to follow the course of a river underground as far back as the swallow-hole. Sometimes the only method is to put fluorescein in all the swallow-

*Far left*
Ice formation in the Chouroum
de la Parza.

*Left*
An ice curtain in the Grotte
de la Tunnette.

*Right*
The British expedition to the
Gouffre Berger, 1967.

holes that could possibly give rise to the flow in question. Then the explorers must wait for the tell-tale colour to re-appear above the surface.

An amusing story of involuntary water colouration is told of one region in France. In 1901 fire broke out at the Pernod factory at Pontarlier. The contents of the stills and more than a million bottles of absinthe were lost down the drains leading into the River Doubs. The waters were turned green with absinthe for several days before the colour dispersed. It was then noticed that the waters of the River Loue, which rises some eleven miles from Pontarlier, were turning green—a strange demonstration of the underground connections between these two rivers.

Cave exploration can also benefit industry. In the Pyrenees, the development of hydro-electric schemes has led to new speleological exploits. In 1948, a stream flowing into the Grave d'Ossau was dammed underground, 480 feet above its resurgence in the Grotte des Eaux-Chaudes, and its waters led by a tunnel to a hydro-electric power-station. This was the first river to have its waters diverted while still underground.

Other branches of science find useful data underground. The waters and floors of caverns are often inhabited by animals of great interest to the zoologist. Their characteristics have been described by Blatchley in *Indiana Caves and their Fauna* (1897), and by Eigenmann in *The Homes of the Blind Fishes* (1917).

The ancestry of many cave-dwelling animals is a long one, for they have been breeding in the darkness for millions of years. The surroundings, cold, dark and poor in food, have caused certain changes in their physical structure and mode of life. Crustaceans, insects, millipedes, spiders, worms, molluscs, fish and bataracians (such as the *Proteus* of Yugoslavian caves)—all display similar characteristics. For example, decolouration of the skin and blindness are common. To compensate for the lack of sight, the legs and antennae are elongated into tactile organs, giving a certain resemblance between cave-dwelling and deep-sea creatures.

The interest in cave fauna, and to a lesser degree their flora, has led to the construction of underground laboratories. About 1920 the first was set up in the cave of Postojna, Yugoslavia, and the most recent in 1950, in the Grotte de Moulis near Saint-Girons, in the Pyrenees.

During glacial periods caves not infrequently provided prehistoric man with a ready-made house, for not all caves are damp, and their remarkably uniform temperature, often varying only a degree or two from summer to winter, made them excellent natural dwellings. In fact, most of our information about early man comes from the caves he used for hundreds of generations. In them he left behind artifacts, tools, kitchen implements and refuse, and works of art. As civilisation and technology developed, man abandoned his cave dwellings for buildings of his own construction. Some caves, such as the Grotte de Lascaux in the Dordogne, are positive prehistoric picture galleries; others boast carved and sculptured walls, as in the Grotte de Montespan in the Pyrenees. The Stone-Age art of Aurignacian and Magdalenian times will be discussed later, in the chapter on the Quaternary Period.

# Running Water

On a steeply sloping and impermeable land surface, the greater part of the rainwater runs down the slope at random in countless little threads. This flow is known as the run-off. On smooth slopes the water appears as a thin film moving more or less evenly downhill, and is called sheet flow. In many regions run-off removes a tremendous tonnage of soil and is highly damaging. Each raindrop striking the soil surface on uneven slopes make a little cup-shaped depression. Several depressions merge to become tiny streamlets, which in turn make little channels for themselves by loosening and carrying away particles of earth, until the run-off is concentrated first into irregularities, then rills, then gullies, and eventually streams. In sheet flow on smooth slopes differential erosion removes weaker materials more quickly, and in time this too produces rills, gullies and streams.

A familiar effect of mechanical erosion, particularly in soft, unconsolidated materials such as clay and sand is the furrowing of unpaved roads and paths after a thunderstorm. The amount of the erosion depends on the gradient of the slope, the velocity and volume of water flowing, and the nature and quantity of material. Then, to the mechanical action must also be added the much slower chemical erosion, or corrosion, of the soluble rocks.

It is easy to see that a bare soil offers little resistance to erosion. If steady rains are disastrous, alternating intensely wet and dry seasons are even worse. The work of erosion begun by rainwash is continued and modified by the wind, weathering and mass wasting. The rock waste deposited by these agents is carried in suspension, in solution, or by tracing along the channel bottom until it reaches still water.

In many arid or semi-arid regions rainwash and differential weathering accompanied by wind action produce what is called badland topography. The badlands of South Dakota are a classic example. Throughout the western and south-western areas of the United States and in other arid regions of the world there are numerous similar badlands. Examples of differential erosion on a small scale are to be found everywhere, sometimes both picturesque and fantastic.

Glacial deposits known as boulder clay sometimes occur in mountainous areas, composed of soft clay plus stones and boulders of various sizes. Under the action of weathering and rainwash, the clay surrounding the surface stones is carried away. That beneath them, however, is sheltered and continues to support the boulders at their original height while the surrounding ground is etched deeper and deeper. Finally, the stones are left perched at the top of columns of unaffected clay. Usually, smaller stones jut out of the sides of the column, acting as further protection against rain. Many such pillars are pointed at the top, and have never themselves been capped by an accumulation of stone or boulder.

Various fanciful names are given to these features; the French call them *demoiselles;* in America they are commonly known as hoodoos.

Earth pillars have been carved out of the glacial deposits of many areas. The most perfect examples in the Alps are found in the Tyrol. In France, the highest *demoiselle* of the Savoy Alps touches 120 feet, while in the Vallauria valley there is a remarkable group called 'The Ballroom' whose pillars form long lines following the valley sides. Some have already lost their capping stones and after every rain become a little shorter, but as erosion of the valley sides is going on all the time, so new *demoiselles* are constantly being formed.

Earth pillars form in other soils too. An impressive example is found in the volcanic tuffs of Cappadocia in Asia Minor. Half-way between the coasts of the Black Sea and the Mediterranean there is a plateau 6,000 square miles in extent. The base is composed of volcanic ash consolidated into tuff, and the summit is capped by weathered and fissured lava. The tuff is 1,000 feet thick in places, and the bed of lava about 150 feet. From December to April rain is abundant in Cappadocia, and surface run-off carves its way into the purple lava and down into the white tuff, carrying away the loose, weathered material to produce a fantastic landscape of scattered rocks and pillars. Despite the apparent inhospitality of the area, many towns grew up from the tenth century onward. Religious communities settled there, and the rocks are honey-combed with homes, churches and cathedrals, often ornamented with frescoes and sculptures. Even the earth pillars have been inhabited. Some still are — proving how very slowly weathering and erosion of the soft material takes place as long as it is sheltered by a protective cap of lava.

In Colorado, colossal earth pillars have been eroded out of conglomerates of 'pudding-stones', and striking examples can also be observed on the Zuni Plateau of New Mexico, where pillars of soft calcareous sandstone are protected by blocks of harder Dakota sandstone fallen from the edge of the escarpment 1,000 feet above.

Spectacular examples of differential erosion on a much larger scale are to be found in many arid and near-arid regions. Isolated blocks of resistant rocks form the unusual 'Cathedral Spires' in the Garden of the Gods, Colorado, while near Douglas, Arizona, lava has been eroded into colourful pillars. Probably the most famous area in the world exhibiting these boldly sculptured forms is Bryce Canyon National Park, Utah. Besides variations in the resistance of the beds, strong vertical jointing and good stratification are characteristic of the rocks in this region, and a gloriously coloured maze of fantastic towers and pinnacles has been produced by many centuries of differential weathering, rainwash and mass wasting.

Sandstone beds, often well jointed and friable, are sometimes more readily eroded than limestone. A notable example occurs to the south-east of Dresden. Here the sandstone lies in well-marked horizontal beds, cut by vertical jointing into numerous rougher rectangular blocks called Quadersandstein (German *Quader*—freestone). Escarpments of this sandstone overlook the Elbe at their highest point, the 600 feet high 'Bastion'.

*Top*
Laucaut Loop, a meander of the River Wye.

*Bottom*
Erosion has removed the topsoil in Ceara, Brazil.

Not every example of differential erosion is attractive. And not all are natural phenomena; many are the result, at least in part, of man's own actions. During the 1930s the inhabitants of towns in the mid-west United States became accustomed to yellowish-brown clouds of dust blowing past in the dry season, sometimes thick enough to bring darkness at midday. The dust was the soil of the Great Plains being blown away and lost for ever. It was calculated that a single dust-storm in 1934 carried away from the area 500 million tons of earth. The result is that fertile soil has been removed from many hundreds of square miles, and hundreds of thousands of settlers have been virtually driven from their homes by the rapid transformation of once-fertile farmland into a desolate dust-bowl. The wind's carrying capacity has already been mentioned in this connection. The major factor, however, was improper cultivation methods and deforestation which stripped the ground of its protective vegetation cover, enabling running water, wind and other agents of erosion to take hold.

In Tennessee, toward the middle of the last century, trees were used to fuel the smelting plant of the copper ores worked there. The sulphurous vapours given off in the process killed all vegetation, even grass, and erosion quickly took hold on the bare ground. Within a short time, over a radius of more than 20 miles round Ducktown, the copper city, the ravaged and sterile soil lay bare and red. Despite considerable reforestation, only one-third of the area has been restored to fertility after more than twenty years' work and considerable capital outlay.

In Brazil, the whole southern part of Minas Gerais, an area of over 1,000 square miles, has been devastated since the original forest was replaced by coffee plantations, which offer less resistance to the development of sheet-flooding after heavy tropical rainfalls.

Similar results of soil erosion can be quoted from Africa, Asia and Australia. Where settlers have cleared the land for cultivation or where bush fires have destroyed the forest-cover, the soil is soon carried away by wind and rain. Experiments conducted in the United States have shown that soil decreases in volume by 1 per cent in five years if it is wooded, by 10 per cent if it is planted with cotton, and by 20 per cent under cultivated vines. To look at this problem from another angle, a cloudburst can remove from bare ground in a few hours one inch of soil which has taken roughly one thousand years to form. The situation is a serious one. Man's numbers are increasing while the amount of tillable land decreases. Every possible means is being used to arrest this decline. Vast reforestation schemes are being put into operation, areas are being contour-ploughed, crops are being planted in terraces. The improvement of the Russian steppes is a model in this field and an encouraging example of what can be done, but we have yet to see who will win the race between erosion and man, between survival and extinction.

## STREAM CHARACTERISTICS

Rainwater flowing off the surface of the ground, as we have seen, forms rainwash. This type of run-off simply seeks out the easiest and, generally, the most direct and steepest slope, flowing from high to low ground. In streams, on the other hand, the water is concentrated in more or less well-defined channels which lead eventually to the oceans. These characteristics are accentuated in young streams, most of which are alpine streams with a steep bed and irregular flow, heaviest when the spring snows melt or after a summer tempest.

A young stream is a powerful agent of erosion, as can clearly be seen if its course is explored from end to end, from the mountain summit where it rises to the level plain below, where its velocity is checked.

The head of a stream valley is a funnel-shaped hollow down which the run-off flows. In winter the snow piles up in the hollow; in summer the rain pours into it as if into the funnel of an enormous rain-gauge. Thus the head of the valley is continually deepening at the expense of the surrounding mountain-side, and each year another stretch of pasture or forest is lost.

All the water running into the funnel converges upon a deep, narrow gorge where, with the help of the rocks strewn here and there, the stream enlarges its channel still further, for the erosion

*Top*
Rainwash, differential weathering and wind action have combined to produce the Badlands of Dakota. The different strata indicate the marked stages in the process of erosion. Some of the gorges are so deep that one can hardly see the bottom.

*Bottom*
Close-up of "Cap Rock", the Badlands, showing the effects of erosion on alternating bands of hard and soft rock.

*Right*
A demoiselle in the Vallauria of the French Alps.

force of a river depends as much on the weight and resistance of the materials it carries as on the weight of water.

The channel is widened by the waters steadily undermining the banks until they collapse. Violent and sudden floods may speed the work, sometimes detaching large chunks of forest and pasture from the river banks and carrying them downstream like floating islands, often complete with their population of animal and insect life.

Deepening by abrasion occurs as the bed-load of rock debris is dragged, pushed and rolled downstream, a process called traction. Further erosion, though on a smaller scale, is produced by solution along the channel and by saltation (the bouncing of smaller rock particles along the bed of the stream).

Sometimes depressions in the channel create whirlpools which suck in sand and pebbles, swirling them round to hollow out pot-holes. In the early stages of formation centrifugal force is strongest, and corrosion of the rocks by the swirling rock fragments is concentrated at the margin of the pot-hole. The hole therefore deepens at the edge, leaving a conical hump in the middle, and horizontal grooves round the sides. As deepening progresses the hole is occupied by a greater volume of water, and the centrifugal force decreases until the central section begins to wear away too. Where several pot-holes form close to one another, the rock walls separating them are finally worn away, and the bed of the stream is correspondingly deepened.

As boulders and pebbles are being carried down to the mountain-foot, wearing away the stream bed as they go, they are at the same time broken up by striking and abrading each other. Experiments have been made with a rotating cylinder in which blocks of sandstone as large as a man's head have been ground down into sand in the time the rock would take to move one mile along the stream-bed. Under the same conditions a boulder of granite would cover the equivalent of 25 miles before being completely reduced.

The materials removed and carried along by the stream are deposited lower down its course. Sometimes extensive deposits are made along the whole length. Velocity and volume exert a strong control here: reduction of volume, such as follows the flood stage, results in deposition; so too, does reduction of velocity. Young streams are generally of small volume and clear, carrying little sediment except during heavy rains and high water stages, when the waters are turbid and huge blocks of rock may be moved. Mature rivers are proportionately larger and slower flowing. They carry a good deal of fine material throughout the year, and coarse material mainly during the flood period. As the flood waters recede, the coarse sediments are dropped along with much of the fine material, but in a large stream there is always enough of the latter in suspension to cloud the waters all through the year.

**Sources of streams and stream gradients.** While the law of gravitation demands that a river should always flow from higher to lower levels, there are big variations in the rate and character of its motion caused by differences in the gradient of its channel. A vertical or steeply inclined face of rock will produce a waterfall, and a rocky declivity will produce rapids, while a flat plain will allow the stream to flow slowly and smoothly on its way.

Young streams or the head portion of large rivers in mountains may have high gradients. The Yuba River in California, for example, drops 225 feet per mile, and the Uncompahgre River in Colorado about 350 feet per mile. In large rivers, on the other hand, gradient may average only 2 feet per mile, while in many navigable rivers the slope may be 1 foot per mile, or less.

Unlike young streams (almost always mountain watercourses), mature streams or large rivers flow, at least over some of their length, on a plain. They therefore have a reduced gradient and rate of flow. In general, they never run dry, although they show alternation of high and low water according to the seasons. The wadis of desert regions with their intermittent flow may be considered the exception to this rule.

A river and its tributaries can be compared to a branching tree, each tiny tributary emptying its contents into a larger one until the main trunk of the river is reached. This is known as a river system or drainage system, a gigantic natural network which

removes surface water and a vast amount of mineral material
from the land and carries them out to sea. The area drained by
a stream system is called the watershed, or drainage basin, and is
separated from adjacent systems by a series of ridges called divides.
The Continental Divide is a series of high ridges and mountains
separating two major drainage networks: that of the Mississippi
Basin and that of the Pacific Region. Another major divide
occurs in western Uganda between rivers draining to the Atlantic
and the Mediterranean (the Nile).

The sources of rivers are numerous: springs, rills, brooks,
creeks, resurgences of underground water, glacier meltwater, or
a combination of any or all of them.

The development of valleys and the erosion of land by rivers–
the fluvial cycle–is one of the most important aspect of geo-
morphology, the science which deals with the study of land forms.
The depth of the valleys, downcut over centuries, determines the
relief of an area, while their shape is the most evident manifestation
of the type of erosion and the stage of the erosional cycle prevailing.
The amount and distribution of rainfall is the controlling element,
and erosive work begins when the first raindrop falls. Annual
rainfall varies from less than 10 inches in the Sahara to more than
500 inches in parts of India. In some regions it is concentrated
into one short season; elsewhere it is distributed relatively evenly
throughout the year. The pattern of its fall controls to a great
extent a stream's regimen, its normal fluctuations, variations and
adjustments during the year.

**Velocity.** The velocity of a river is the distance its water travels
in a given unit of time, and is generally quoted in feet per second.
A velocity of half a foot per second (0·3 miles per hour) is a low
velocity; 25–30 feet per second (17–20 miles per hour) is high.
As we have already seen, the speed is very variable; it depends on
gradient, the amount of water, and the form taken by the stream-
bed at the point of measurement. It is easy to observe that speed
increases where the channel is narrow and decreases where it
widens. Some erratic rivers pass through an alternating series
of rapids and still waters.

If several people set at intervals along a bridge over a river
were to throw sticks into the water at the same time, the sticks
would be swept into a curve, shaped with the river's concavity
upstream and demonstrating that speed is greater in midstream
than near the edges, where the banks act as a sort of friction brake.

Speed varies also with distance from the bottom of the river-
bed. Variations can be measured with apparatus consisting of a
propeller rotated by the flow of the river, which produces a sound
or light signal for each rotation. Relative speeds can then be
worked out.

Let us take as an example a river that is 18 yards wide. If a
cable is stretched across it, and the speed measured yard by yard
from one bank to the other, at the levels $A$, $B$, $C$, and so on, these
points can then be plotted on a scale and curves drawn, such as
those given here for points, 8, 12 and 15. Curve 8 shows that the
speed increases from the surface as far as level $B$, then decreases,
increases again to level $E$, and finally decreases toward the bottom.
Curve 12 is quite different; speed decreases immediately below
the surface, increases between levels $E$ and $G$, drops to nothing
between $H$ and $I$, and then rises again at $J$. Lastly, in curve 15,
the direction of flow is completely reversed below point $F$, doubtless
due to an obstacle in the stream-bed checking the flow and producing an eddy. In general, the water flows fastest in the central
and upper areas where friction is lowest.

This example serves to show that the speed of a river is not
uniform throughout its depth. It must also be remembered that
the speed at any given point is not always constant, but alters
with the amount of water flowing down the river. Thus the velocity
is dependent primarily on its gradient, then on its volume, and
to a lesser extent on the shape or configuration of its channel
and the amount of material or load it carries along its course.

**Profile of equilibrium.** Let us take a mountain slope which has
not yet suffered the action of stream erosion and, for simplicity,
let us suppose that it is composed entirely of a soft, impermeable
and homogeneous rock. Water will run off the sides of such a
mountain, following the steepest gradient until it reaches the foot

of the slope along the edge of the plain which is its local base level. At the upper level the stream's gradient is steepest and produces its maximum speed. It is here, that the stream starts to cut into the mountain side and form gorges which work headward to the mountain crests. Stream erosion, therefore, proceeds from the mouth of the stream or from base level up to the summit of the mountain, and is called retrogressive or headward erosion. On the other hand, sedimentation progresses on the plain by the constant addition of new layers to the alluvial cone, alluvial fan or alluvial plain, or to the river's flood-plain, delta and levees. This is progressive deposition.

Erosion in the upper part and deposition in the middle and lower parts of its course give a stream a graded slope, or profile of equilibrium, a smooth curve that is tangential to the vertical at the summit, tangential to the horizontal at the base, and concave toward the sky. As long ago as the seventeenth century, Italian investigators studying the Po and its tributaries had noted this phenomenon. Galileo believed it to describe an arc of a circle, but more recent work has shown the curve to be more complex in shape, steeper at the summit than at the base. As soon as the river has eroded its bed to this profile of equilibrium, further active erosion or deposition ceases; the load of material in suspension is almost exactly adjusted to the carrying power of the river.

To find out how far a river has advanced toward the state of equilibrium, its longitudinal profile must be ascertained. This is a lengthwise diagram of the valley obtained by plotting on paper heights measured along the valley. The nearer this profile comes to a profile of equilibrium, the closer to full maturity the river is. Any stream with an irregular longitudinal profile, flowing in a deep narrow valley, is considered to be in an early stage of development. Once it has attained equilibrium it is described as mature. This method of assessing a river's age refers to the stage of erosive work it has reached, not its absolute age in years. A river which has been in existence for only a short time may appear older than a longer established river if, rising from a lower source, it finishes its work of erosion sooner.

It will be understood that an equilibrium profile curve represents a theoretical limit rather than the form actually attained by a stream-bed. Unequal resistance of rocks and irregularities of flow interrupt the development of a smooth or graded profile. Therefore most streams are eroding slightly or depositing at any given place along their channels. A stream may even be rejuvenated if it regains its erosive power by a lowering of its base level or a raising of its upper course. Work is then resumed until a new profile of equilibrium is attained. This is a fairly frequent phenomenon and takes place when, in relation to sea level, the region through which a river flows is raised by earth movements. It accounts for the development of entrenched meanders like those of the Meuse and the Wye. Meanders on a flood-plain are a sign of maturity or old age in a river; in deep valleys they are a sign of rejuvenation. River terraces may also be a further proof of successive rejuvenations, as we shall see later.

**Transverse profile.** The study of a river valley is incomplete without consideration of its transverse profile, the outline of the cross-section. This is determined partly by the nature of the rocks. In an area of soft and homogeneous rocks, the transverse profile is a V, and erosion is proportional to the speed of the water, diminishing from the middle to the banks. In hard, fissured rock there is a continual removal of base blocks and caving in from the top. The banks therefore remain vertical, and give the transverse profile more of a U-shape. This is especially noticeable in valleys formed by a young river where vertical erosion of the stream-bed is more evident than lateral plantation, or horizontal erosion, which widens the valley floor. This type of valley develops as a narrow corridor between two cliffs, and is called a gorge or canyon.

At the other extreme are the shallow valleys of the rolling chalklands of England and France, such as Salisbury Plain and the plains of Champagne. Here the ground is bare of trees, short grass or poor cultivation prevails, and the white chalk often shows through the thin soil. These areas are sparsely populated and are often used for encampments and military manoeuvres. The

*Top*
The stream above Cauterets in the Pyrenees today follows a corrected course. This diagram shows the three parts of the stream: 1. Funnel-shaped hollow; 2. Ravine, and 3. Alluvial fan.

*Bottom*
The dried up summer bed of the River Coe, Scotland, showing the surfaces eroded by the grinding action of pebbles.

wide and shallow valleys are marked by the appearance of marsh, peat and clumps of trees. Slow streams with numerous lateral channels wind through the marshy floor of the valley.

Lastly, let us have a look at the form of stream valleys in an area of both hard and soft rocks, say, where limestone and clay alternate. In such an area erosion affects each rock differently. The beds of clay give rise to slopes, and the layers of limestone to vertical cliffs. A series of tiers results—numerous if the beds of limestone are thin. Thicker beds of clay and limestone may produce a V-shaped valley carved entirely in one thick stratum of clay or marl, with a cornice of limestone jutting out at the top of the valley wall. Much of the drama and beauty of a landscape depends on differential erosion of this kind. The Arizona Grand Canyon is an excellent example of such an area.

**Discharge of rivers.** The discharge of a river is the quantity of water which passes through a given cross-section of its channel in a given unit of time, usually quoted in cubic feet per second. It is calculated from the speed curves already dealt with in this chapter.

Volume is essentially variable, for a river goes successively through stages from high water, or flood, to low water; it may even cease to flow altogether for weeks or months on end. Between these extremes is the river's average height, known as mean water. The fluctuations are caused by a multiplicity of factors: relief and nature of the ground, the local pattern of rainfall, and variations in temperature. These last, if high, may produce rapid evaporation of surface water or, if low, may freeze the water so that it is held in reserve.

Rivers and river systems can be classified in many ways: according to their erosional age, their drainage pattern, their mode of origin or history. One classification is based on the nature of seasonal discharge. Its system is outlined here:

WINTER MAXIMUM. This type is characterised by the predominance of rainwater over the other sources. High water therefore occurs in winter. Because the lower temperatures cut

*Top*
Wadi patterns in the Sahra el-'Arabiya, Egypt, during the dry season.

*Bottom*
Gorge below the Godafoss on the Skjalfand River, Northern Iceland.

*Right*
A water-wheel in the Sudan. This primitive method utilises the waters of the Nile to irrigate the parched semi-desert land of the Sudan, thus facilitating the cultivation of crops.

down evaporation this holds good even if the rainfall during this period is less than at other seasons.

SPRING MAXIMUM. These rivers come from upland regions where the melting of the snow in spring, often augmented by spring rains, causes maximum flow.

SUMMER MAXIMUM. This type can be observed in rivers rising in the glaciers of mountain regions, and which have most water in summer when the ice melts rapidly.

SPRING AND AUTUMN MAXIMA. These are characteristic of any area of the world with a 'Mediterranean' climate, that is, with a hot dry summer, cool dry winter, and rain in spring and autumn. The rain is often sudden and heavy, running quickly off the dry ground so that the typical rivers of these areas are torrential.

Among the rivers of the winter maximum type, a further distinction can be made between those that receive most of their water directly from run-off and those that are fed by ground water and springs. The flow in the second is obviously more dependable than in those relying on run-off. Similarly, lakes regulate the flow of the rivers emerging from them. The most important to be governed in this way is the St. Lawrence, Canada, whose discharge is absolutely uniform. This constancy it owes to the Great Lakes.

The figures of the mean discharge of several of the world's rivers make interesting and useful comparison. The values are expressed in cubic feet per second: Seine, 1,200; Nile, 7,000; Mekong, 15,000; St Lawrence, 30,000; Yangtze, 45,000; Mississippi, 50,000; Rio de la Plata, 70,000; Congo, 110,000; and Amazon, 250,000.

In smaller rivers, such as those of Britain, whose basins lie in a region enjoying uniform general features of climate, the discharge is regulated by local rainfall. A wet season swells the streams, a dry one diminishes them. So, in estimating and comparing the geological work of rivers, we must take into account whether or not the sources of supply are liable to occasional heavy augmentation or diminution.

In some rivers, there is a more or less regularly recurring season of flood, followed by one of drought. The Nile, fed by the spring rains of Abyssinia, floods the plains of Egypt every summer, rising 30 to 35 feet in Upper Egypt, 23 to 24 feet at Cairo, and in the seaward part of the delta about 4 feet. The Ganges and its adjuncts begin to rise every April, and continue to do so until their plains are one vast lake as much as 32 feet deep.

In other rivers, sudden and heavy rains occurring at intervals swell the usual volume of water and cause flash floods, freshes or spates. This is markedly the position of the rivers of Western Europe. Thus the Rhône sometimes rises $11\frac{1}{2}$ feet at Lyons and 23 feet at Avignon; the Saône from 20 to $24\frac{1}{2}$ feet. In the middle of March, 1876, the Seine rose 20 feet at Paris, the Oise 17 feet near Compiègne, and the Marne 14 feet at Daméry. The Ardèche at Gournier rose more than 69 feet during the floods of 1927.

**River deposition or sedimentation.** This is the process by which a stream builds up its channel and flood-plain by depositing some of its load. It is, in fact, the opposite of erosion. The deposits

*Top left*
The Waitaki River, South Island New Zealand. The river is 95 miles long and is seen here 15 miles from its mouth. The extremely braided course with ox-bow formations is a result of the deposition of gravel and sediment, causing the original stream to divide and form new channels.

*Bottom left*
The Kaieteur Falls on the Potaro River, Central Guyana, fall 741 feet from a sandstone tableland into a wide basin. From here they fall a further 71 feet. The river at this stage is 350 feet wide.

*Right*
The source of a mountain stream. Here it is dry, but as it begins to flow it will eat away more and more of the thin layers of peat, working its way farther back.

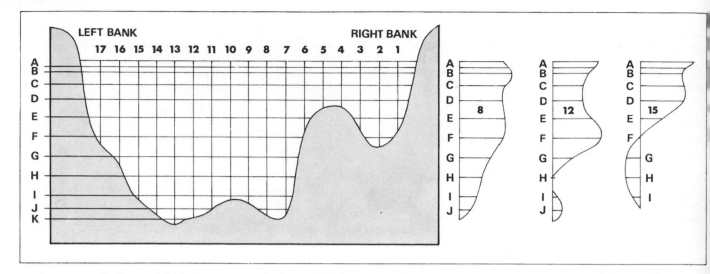

are known as alluvium and include stones, mud, silt and debris.

The load's amount and kind is controlled by the stream's velocity. its volume. and the nature of the rocks over which it flows. Sedimentation will occur whenever and wherever volume or velocity are reduced – most commonly after flooding, at the foot of mountain slopes, and where a stream enters a lake or a body of standing water. The size of a particular load usually decreases from the headwaters to the mouth, because although the volume of water is greater at the mouth, velocity is less. During normal or low water times only sand and silt manage to reach the sea. Silt is a clayey or limey mud that is fairly rich in fine sand. Its yellowish or reddish colour is given by the presence of oxides of iron.

River deposits build up to form alluvial fans, alluvial cones, alluvial plains, flood-plains, levees, deltas, bars and channel-fill. Bars of sand or gravel and channel-fill usually develop where there are irregularities or obstructions in a stream's channel. Sometimes heavily loaded streams with a marked low water season will deposit these features so extensively that the river is split into numerous branching and reuniting channels. Such a river is called a braided stream. The South Platte River in Nebraska (now nearly dry as a result of extensive tapping for irrigation) and the Matanuska River in Alaska are examples.

A flood-plain is constructed at times of receding flood-waters and natural levees are built up when the river's velocity is checked as it overflows its banks and deposits the coarsest part of its load. Natural levees of a large river may build up so high that tributary streams are prevented from joining it directly and are forced to flow parallel for long distances over the flood-plains until they find a lower point at which to join the main stream. The classic example of this kind of stream is the Yazoo River, a tributary of the Mississippi. Other deferred juncture streams, such as the Big Sioux, are called yazoo rivers after this notable example.

At the break of slope where the stream emerges on to the plain, the sudden check in the speed of the water diminishes carrying power rapidly, and the load is deposited in order of relative weight. First the boulders are dropped, then pebbles, gravel and sand, and finally mud. The rock debris is built out on to the plain as an alluvial cone or fan, predominantly coarse at the head and mid-portions, and very fine in the outer margins. When the change in velocity or volume is abrupt, sorting of materials is poor; when conditions change gradually, sorting is good.

Streams often run so close to each other their fans may coalesce at the foot of a mountain to form a piedmont alluvial plain. The region of Italy called Piemonte was once a gulf of the Adriatic; today it is a plain formed by stream and glacier deposits from the Alps, and has given its name to all similar features. An example occurs in the United States at the western foot of the Sierra Nevada Mountains where, at the Great Valley, huge merging fans are built out to form a continuous alluvial plain many miles wide and hundreds of feet thick. Similar fans are found in the Basin and Range of Nevada, and along the Front Range of the Rocky Mountains.

Vast quantities of alluvium are transported annually by the rivers of the world. Experts have worked out that the Mississippi river system carries over 500 million tons of material every year, building its enormous delta out into the Gulf of Mexico. A more detailed breakdown of the Mississippi's annual discharge and load shows that this great river carries 22 million million cubic feet of water, 340,500,000 tons of material in suspension, 40,000,000 tons rolling on the bottom, and 136,400,000 tons in solution–a total of 516,900,000 tons of rock waste. Yet vast as these quantities are, the Mississippi and its tributaries, which drain a watershed of 1,265,000 square miles, lower the drainage basin at the very low average rate of about one inch in 9,000 years.

The Colorado River ranks second in the United States in depositional powers. In China, the Yangtze pours 500 million tons of mud into the sea yearly. It is nevertheless the Hwang-ho that carries the heaviest load of any river as it flows across huge areas of loess in northern China. This material, which gives the waters their yellow colour, is easily eroded and carried along to produce a total of 2,000 million tons of delta alluvium annually.

**Estuaries and deltas.** Estuaries are narrow, elongated bays, deeply hollowed out at points where the sea penetrates at high tide and reverses the direction of a river normally flowing into the bay.

The mouth of a river is subject to the heaviest deposition because it is here that the bed's gradient is reduced, here that the water suffers a check in speed on encountering the resistance of tide-water, here that the salt content of the latter exercises a precipitating action on the clay. A simple experiment will demonstrate this last point. If two glasses, each containing the same amount of mud, are filled with water, and then a little common salt is added to one before the contents of both glasses are stirred and left to stand, it will be seen that the salt water clears in a few minutes, leaving a deposit of clay on the bottom. In the glass containing no salt, the clay remains in suspension for several hours.

The mouth of the river, then, is the scene of active deposition of alluvium. Sand and silts form temporary bars and sandbanks, deposits which may rise above wave-level or stop short a few feet below the surface of the water, causing considerable inconvenience to shipping, and demanding constant dredging.

Where a river carries large quantities of alluvium into a sea area, conditions are suitable for the growth of a delta. Its form is controlled by the shape of the water body entered, its currents and its waves, and by the quantity and quality of the load carried. In quiet seas, almost every river has a delta. The Ebro, the Rhône, the Tiber, the Po, and the Nile have deltas in the Mediterranean; the Danube in the Black Sea; the Volga in the Caspian; the Mississippi in the Gulf of Mexico; and the Colorado in the Gulf of California. In seas with strong waves and current action, and in the great oceans, only the large rivers have deltas — the Niger, the Ganges, the Mekong and the Yellow River, for example. Some rivers, of course, have no deltas because they carry little or no sediment; the Niagara is one outstanding example of this.

When a stream reaches a body of standing water, it loses its

*Left*
River profile speed of flow at different depths can be plotted.

*Right*
The Lower Falls of the Yellowstone River, Yellowstone National Park, Wyoming.

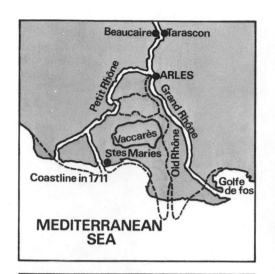

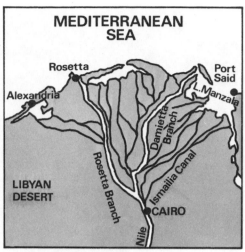

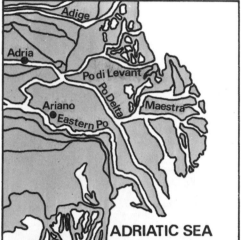

*Top*
The deltas of the Rhône, Nile, Po and Mississippi.

*Bottom*
Braided course of the Eyja Fjordur River, Iceland.

*Right*
The All-American Canal carrying the Colorado River water to several valleys through-out California.

carrying power and drops the bulk of its load, retaining only the finer material for some distance off shore. The deposits form three groups. The first is a series of nearly level, topset beds at the mouth of the river. Next comes the thickest and coarsest group, the foreset beds, extending at a seaward-sloping angle from the topsets to the floor of the sea. These form the bulk of the delta. Finally comes the third group, the bottomset beds, which are again nearly level and grade outward from the foreset beds along the sea floor.

The sediments settle and reduce the slope of the river, causing it to deposit more rapidly. Soon the channel becomes choked and breaks up into a number of smaller channels or distributaries. These in turn become choked, and new outlets form. As the stream travels over its own delta deposits, it drops a cover of topset beds. In flood, it may sweep away the topset beds partially or even completely, eroding into the foreset beds. The rate of delta growth depends on the amount of the stream's load and on the capacity of the ocean waves and currents to carry away deposited material. The Mississippi delta grows at the rate of about one mile in sixteen years along its chief distributary.

There are three main types of delta: arcuate, estuarine and bird's-foot.

Arcuate deltas are built where the stream's load varies from fine to coarse (mainly sands and gravels), and where the currents are relatively strong. A delta constructed of such deposits is porous, and streams flowing over them develop many small, communicating distributaries. The outline is triangular, with the apex pointing upstream. Many of the world's rivers (the Nile Rhine, Hwang Ho, Irrawaddy, Po, Ebro, Volga) build deltas of this type. The Rhine, the Meuse, the Scheldt and the Ems form a combined arcuate delta covering an area somewhat larger than that of the Nile. Much of the polderland of Holland which has been reclaimed by the building of dykes lies within this delta region.

An estuarine delta occurs where a stream's mouth has been submerged to form an elongated bay, and the delta deposits take the form of a long, narrow estuarine filling, some of which is flood-plain and bar deposit. Rivers with such deltas include the Mackenzie, Elbe, Vistula, Susquehanna, Seine and Hudson.

A bird's-foot delta is built by large rivers carrying heavy loads of fine material, with a good deal of it in solution. These deposits are fine and relatively impermeable, and the river's water is concentrated mainly in a few large distributaries. This combination is very rare. The bird's-foot delta of the Mississippi is unique. It grows by addition of material along the four main distributaries, or passes, which extend outward from the same place to form the bird's-foot shape.

Let us examine several well-known deltas more closely.

**Rhône delta.** This river carries 550 million cubic feet of sediment into the Mediterranean every year, enough to fill 150 trains of 40 cars every day. Lack of vigorous currents favoured the formation of its arcuate delta. The present distributaries are the Grand Rhône, or eastern branch, which is the only one navigable, and the Petit Rhône, or western branch, which is gradually disappearing. A third branch, leading off the Petit Rhône, has almost silted up. Others disappeared earlier, notably the Vieux Rhône which flowed from Arles and entered the sea to the west of the Grand Rhône.

These variations in the position of the Rhône's mouth have naturally caused variations in the form of the coastline. It is obvious that the coastline near an active distributary must advance outward into the sea, while that on each side of a dead channel will be subject to wave-erosion and retreat. This pattern of development can be clearly seen when the present-day outline of the Rhône delta is compared with that shown on a map made in 1711.

The distributaries of the Rhône divide the delta into two triangular islands, the Grande and Petite Camargue. Largely areas of marsh, their grassland stretches are devoted to stock-raising. Today, that part of the Rhône delta where the Grand Rhône opens into the sea is extending by 150 feet a year. The town of Arles, situated where the river forks into the Grand and Petit Rhône, was six miles from the sea in Roman times but today it is twelve.

**Nile delta.** It was to this most regular and typical example that the name delta was first given (delta being the Greek letter D, triangular in shape). The Nile divides into two main branches and numerous smaller branches extending from Cairo to Alexandra and Port Said, with a gradient of less than one foot per mile–about half that of the Hwang Ho and three times that of the Mississippi. The delta is slowly advancing into the sea at the rate of about twelve feet per year. The cause is now clear; the Nile has abandoned a great part of its alluvium on its flood-plain in Egypt and carries only a reduced load to its mouth. Its flood-plain is an area of extensive cultivation where dams built across the river to collect water for irrigation purposes prevent all but a few disastrous floods.

**Po delta.** This river, unlike the Nile, is now embanked over the greater part of its course and cannot deposit much alluvium in the plains of Lombardy. Its load is carried as far as the Adriatic, where it forms a delta advancing 200 feet a year. The town of Adria, once a seaport, today lies 14 miles inland, showing an annual advance varying between 80 and 200 feet. Looking further afield, the whole basin of the Po, comprising the provinces of Piemonte, Lombardy, Emilia and Venezia, is one enormous series of deltas which has filled a depression between the High Alps and the Apennines. Borings in the Lombardy Plains have reached a depth of 550 feet without penetrating the river alluvium. It has been calculated, therefore, that the delta of the Po began to fill up the depression as soon as the Alpine chains were upraised.

**Mississippi delta.** The Mississippi carries such an enormous amount of alluvium down to the sea (over 500 million tons per year) that despite the tides of the Gulf of Mexico it has built an enormous delta which advances at more than 200 feet a year. The sediment carried is very fine, the bulk consisting of a mud deposited as impervious clay along the banks of the channels. Thus the distributaries of the river are confined within levees or

embankments, similar to those further upstream on the flood-plain and extending in a seaward direction raised well above water level.

Besides extending its delta seaward, the Mississippi also keeps pace with a steady subsidence of the Earth's crust at this point. Borings have revealed 2,000 feet of alluvial deposits, so that the actual volume of sediment in the Mississippi delta is far greater than might be imagined from the surface.

**Colorado delta.** The headwaters of this great river lie in snow-covered mountains, while the greater portion of its course is through arid country. Across the Gulf of California it has built a delta separating the head of the gulf from the main body, producing a basin in which evaporation has caused the water level to drop. This depression, nearly 275 feet below sea level, is occupied by a relatively large, extremely saline lake, the Salton Sea. Occasionally the Colorado River forces its way into this basin along irrigation canals, with disastrous results.

**Lake deltas.** In addition to its delta in the sea, the Rhône has formed a delta where it enters Lake Geneva, laden with alluvium from the melting glaciers. The eastern end of the lake has been filled in for more than twelve miles, one mile of which has been formed since Roman times. The River Durance, which enters the lake on the south between Thonon and Evian, has also formed a small delta.

The Lake of Thun and Lake Brienz in Switzerland once formed a single sheet of water; this has been cut in two by the growth of a delta. The town of Interlaken (meaning 'between the lakes') has been built on this isthmus of alluvium.

In the Lake District of England the lakes of Buttermere and Crummock Water follow each other along the same deep valley blocked by glacial deposits. It is obvious that they were once a single lake, now split in two by a delta deposited by a stream flowing down the eastern side of the valley. The alluvial flat at the head of Lake Buttermere is also a result of delta formation by the river feeding its load into the lake.

As mentioned earlier, rivers also deposit in artificial lakes. The Colorado, for example, is building a large delta where it enters Lake Mead.

**Freezing of rivers.** In the cooler countries of the world, the rivers freeze in winter and thaw in spring. The rhythm of life in these regions is regulated by the two events. During the short summer, the rivers are an important means of communication, and even when frozen solid in winter they sometimes serve as paths for sledges with their loads of furs.

Freezing of surface water in winter follows different patterns. Lakes and ponds, where the water is still, freeze over uniformly; in rivers, the current keeps the middle of the course clear while ice forms along the banks and in odd patches in areas of quieter water. The patches, sheets or slabs are carried downstream, increasing in size as they go until the river is full of ice-floes. If freezing continues, the floes may be joined together by still more ice until one continuous sheet forms across the river, or they may pack the river and override one another until they freeze together in a single block. At points where ice gathers against natural or artificial obstacles in the river's course, there the blocks of ice sometimes pile up in dramatic masses. On the great rivers of Siberia and Canada, the spectacle is magnificent.

With the spring comes the break-up. The ice cracks and melts, adding to the waters of the melting snows and heavy rains which often prevail at this season. This is, therefore, the time of high water in these rivers, often accompanied by heavy and disastrous flooding.

Ice plays an important part in river development by acting as an agent of transport. Large and heavy rocks may be carried along, frozen into the ice. They are often so large that river forces alone would be powerless to move them. Mud, gravel and boulders are frozen into the 'anchor-ice' which forms on the bottom of the river-bed and then rises to the surface. When the ice breaks up, these deposits are dropped in deeper water downstream or stranded on some other part of the shore. The rocky islands in the St

Lawrence are fringed with blocks carried downstream in this way. The carrying capacity of ice is displayed on a much greater scale by glaciers.

Ice action also causes great destruction along river-banks – especially where the bank is composed of soft rocks.

**Stream classification and drainage patterns.** Although ice has sometimes played a part in remoulding them, valleys are formed by the action of the streams flowing in them. Observation verifies this. Valleys vary in size according to the size of the streams they contain; the larger the valley, the larger the river in it. The valleys of tributary streams are always smaller than those of their main rivers.

Let us consider an ideal example of valley development on a completely fresh land surface, that is, an area which has newly emerged from the sea, for every part of every continent has been submerged by the sea at some stage of geological history.

On being warped up into dry land again, parts of the shallow sea-bottom receive their first rainfall, and the usual pattern of subaerial drainage becomes established. The raindrops loosen the soil particles and are carried down in rivulets. Where the rivulets join together, the increased volume of water creates a gully. Several gullies converge to form a ravine. Finally a valley develops.

The slope, newly arisen from the sea, on which the valleys are forming is called an initial slope, and the streams following it are called consequents. If the rocks in the area are homogeneous, the network of streams resembles the venation in a leaf. This arrangement is called a dendritic pattern. In the beginning it is made up of consequent streams; as erosion progresses and the drainage pattern grows, insequent streams develop—streams whose course is governed by the slope of the secondary or erosional ground surfaces.

The land surface on which streams form is not, of course, always as simple as this. It may be a previously formed valley or a structural valley formed by movements of the Earth's crust. The Jordan Valley, the Rhine Valley, the Great Rift Valley of East

Africa, and the Owens Valley of California were all formed by the down-faulting and subsidence of long and narrow blocks of crust. The Great Valley of California was formed by the uplifting of land on either side of it.

Other rivers flowing in structural valleys are those of the Jura region in Europe and the Appalachian Mountain area in the United States, where the mountains are formed of smooth, rippling folds of rock, and the rivers flow along bands of weak rock exposed by the folding. The bending of the upfolds, or anticlines, stretches and fractures the rock at the crests of the folds, and lowers its resistance to erosion. In this way new valleys often form in the summits of the anticlines.

Streams develop more quickly along geological faults in areas of weak rock, or along belts of softer or weaker rocks. Such courses are called subsequent streams. The drainage pattern which normally evolves in areas of folded rock resembles a lattice, and is called a trellis pattern.

An interesting situation occurs in North Germany, where the directional trend of the valleys is determined by the front fashioned during the Glacial Period of the continental ice-sheet which extended across Hanover, Brandenburg, Pomerania and Prussia for about 50,000 years. The rivers are for the most part parallel to the margin of this ancient ice and, consequently, their flow is directed from east to west. For example, the course of the Elbe can be followed from Hamburg to its meeting with the Havel, which flows on as far as Berlin. From this city, the course of the Spree can be traced upstream toward Frankfurt. Next, the Oder can be traced to the point where the Warta and Netze lead toward the River Vistula. Similarly, the Vistula and its extension, the Bug, continue toward the east, again influenced in their course by the early ice-sheet. The lower parts of the Oder and the Vistula, so clearly perpendicular to the general direction just outlined, are but a later addition to the rest of the network, formed after the melting of the ice.

There are other drainage patterns: there is the radical pattern which develops on the slopes of a volcanic cone, where streams

stream flowing through the wide valley it had carved for itself earlier. A former river-valley through a ridge may be left with no river running through it. This is called a wind-gap.

River capture, or stream piracy, is of frequent occurrence in the development of a river system in an area where alternate belts of hard and soft rock outcrop at right angles to the main rivers, as in the trellis stream pattern. Here the tributary streams develop quickly along the softer beds. The tributary of a strong river may gain upon one belonging to a smaller river, and may even succeed in reversing its flow and diverting the water of the weaker into the larger.

An excellent example of this can be seen, on a small scale, at Farnham in Surrey. Here, the River Wey flows north-east toward a gap in the chalk ridge of the North Downs known as the Hog's Back, but before it enters the gap it makes a sharp turn south-east, and flows away in an easterly direction. It is not until nine miles further east along the hog-back that the Wey crosses the North Downs through a deep valley at Guildford. The sharp bend in the river at Farnham is a suspicious sign. Careful observation through the low wind-gap shows a wide valley running past Aldershot northward to the Thames. This valley is occupied by the small River Blackwater. It is obvious that the bend in the river is an elbow of capture and marks the spot where the Wey successfully appropriated the waters of the upper part of the Blackwater.

There is a further example in northern England where, from the eastern side of the Yorkshire Pennines, several rivers flow east toward the North Sea. The Swale, Ure, Nidd and Wharfe, however, all join the Ouse and flow southward to enter the North Sea with the Humber at Hull. It is thought that the Ouse has worked its way northward by retrogressive erosion along the north-south outcrop of the soft Triassic rocks, and captured all these rivers.

A notable example in the United States is the capture of the headwaters of the Chattahoochee by the Savannah. The Chattahoochee drainage flows into the Gulf of Mexico, and the Savannah drainage into the Atlantic. The Savannah, with a shorter distance to cover and therefore greater erosive power, has cut back into the Chattahoochee valley, and an elbow of capture at Tallulah Falls, Georgia, shows where the waters have been diverted. Stream capture has occurred also in many places in the folded Appalachian Mountains of eastern United States.

## YOUNG RIVERS

Although young rivers are not navigable, they have been made to serve man in other ways. In Canada, Sweden and Austria, rapidly flowing torrential rivers are used to transport huge rafts of timber from the clearing camps to the sawmills. It is here and in other parts of the world that mountain rivers have been 'captured' and dammed, and their potential energy harnessed in the form of hydro-electric power.

Characterised by steep and irregular gradients and marked by numerous rapids and waterfalls, young rivers have high velocities, downcut actively, and deposit very little. Their valleys are narrow, V-shaped and steep-walled. Their courses are relatively straight and poorly adjusted to the rock structure. The gorge of the Niagara, the valley of the Rhine and the canyon of the Colorado are all typical examples of young river valleys.

**Gorges and canyons.** Gorges, or canyons, which present some of the world's most spectacular scenic wonders, are deep, narrow, young river valleys with precipitous sides, generally eroded in hard rocks. They form in the upper parts of the valley of any river that is young and active and flowing from very high ground, when the valley floor is being cut down so rapidly that the sides have no time to widen. The most extensive development of this feature occurs in the upland areas of resistant rocks that have been recently and slowly uplifted across the path of a large mature river, which is thereupon rejuvenated and starts to cut down actively again. The rejuvenated, or antecedent, river is able to carry a large load of rock fragments and to wear down its bed quickly, in this way carving a gorge across the upland area.

The most spectacular gorges are produced where some factor

*Above*
Via Mala, Switzerland.

*Right*
Yosemite Falls in Yosemite National Park, California. The waters plunge 1,430 feet over a hanging tributary valley. The Yosemite Valley is a V-shaped canyon formed by glaciation and erosion by the River Merced. Viewed from Glacier Point, 3,254 feet above the valley floor.

radiate outward from the central high point; there is the rectangular pattern which develops in areas strongly fractured by faulting or jointing, the streams following the zones of weakness produced by the fractures; and there is the annular pattern, a modified form of trellis, which develops in areas where the rocks have been domed.

The drainage pattern of an area is governed by the type of rock structure found in it. Thus geologists are able to determine the major as well as many of the minor geologic structures from studying air photographs and topographic maps of a region.

**Stream Piracy.** It is not unusual to come across a valley disproportionately large compared to the river running through it. Obviously a miserable trickle of a stream could not form a large valley. In many cases, the miserable trickle is the weakened remnant of a once larger river, or of a river that has been entirely diverted from its original course by a neighbouring one. Basically, all that is necessary to produce diversion is two rivers eroding opposite flanks of the divide between them. The river descending the steeper slope will carry on its work of erosion faster than the other, and may finish by cutting through the divide into the course of the second, thus diverting its waters. The lower course of the unsuccessful river, deprived of its headwaters, becomes a misfit

retards lateral erosion of the valley sides. For example, a strong and swift river, actively downcutting and abundantly fed from the snows of the mountain region in which it rises, may flow through an arid region lower down its course. Here there will be little rainfall to erode the valley walls, which will be affected only by an occasional thunderstorm and the action of the atmosphere and the wind; a canyon will be formed.

Other rivers flow through highly jointed or fractured rock such as limestone. Here the rainwater sinks into the crevices so that surface erosion is accomplished slowly.

The largest canyon in the world is the Grand Canyon of the Colorado River in North America. It is more than 200 miles long, from 7 to 15 miles wide at the top, and reaches a depth of 6,000 feet. This tremendous gash in the Earth's crust has been carved out by the waters of the Colorado River and its tributaries. It is a most impressive example of the enormous power of river-erosion.

The enormous quantity of rock material removed and carried away by the Colorado has been deposited in the great delta at its mouth, while the finer sediments have been carried out into the Gulf of California. The Canyon has been formed during the present period of geological time and is still being steadily deepened.

The Colorado is far from being graded; it carries a heavy load of material and lies 2,500 feet above sea level at one point. During the Tertiary period, it was a large river flowing on a low land-surface not far above sea level. The surface was subsequently and slowly uplifted 7,000 to 8,000 feet; the river was rejuvenated, gaining a steepened gradient and power to form the canyon.

It is not mere size alone but the form and colour of its rocks that make the Grand Canyon one of the wonders of the world. The alternation of horizontally bedded hard and soft rocks out-cropping on the sides of the valley presents a series of red, grey and greenish terraces. The sides of the canyon are carved into fantastic shapes by ravines and deep tributary valleys, with buttresses of rock jutting out between them. At the foot of the lowest plat-form, the river can be seen flowing far below, at the bottom of an inner gorge. This final trench alone is 2,000 feet deep and is cut into the crystalline Pre-Cambrian rocks, which are much older and harder than the stratified rocks above and are probably some of the oldest rocks on the Earth's surface.

Gorges are found in most limestone areas throughout the world. The largest in Europe is the gorge of the Verdon, in the southern part of the French Alps. Twelve miles long and 200 feet deep, it begins at Rougon and forms a giant cleft through great thicknesses

95

of white Jurassic limestone. Its sides are so precipitous that it was not explored until 1905, by the French speleologist Martel.

The gorges of the Fier, in Savoy, are spectacular though small. They consist of a narrow passage, hardly more than an enlarged fissure, where it has just been possible to insinuate an overhanging path with a hand rail for the use of tourists. At certain points, the opposite wall can be touched with the hand. The roaring torrent at the bottom of the gorge has been known to rise up to 100 feet in a few hours. Deep furrows, gouged out of the walls by the running water, and overhanging ledges are evidence of erosion. So, too, are the round hollows, vestiges of the countless pot-holes which played their part in the early excavation of the gorge. Here the work of a torrential river can be observed at first hand.

The narrowness of the Fier gorges is rivalled by those of the Vernaison, in a neighbouring limestone area. Similar narrow, steep gorges with numerous right-angle bends, stream-cut along joint fractures in limestone, can be found in Watkins Glenn, in New York.

More widely known are the gorges of the Tarn. This river of central France leaves the uplands of impervious rock on which it rises to plunge into deep gorges through the bare limestone country of the Causses (causse being a local word for a limestone plateau). The Tarn gorges can be explored in a flat-bottomed boat. The limestone walls reach an average height of 1,500 feet; at one point they approach each other very closely, and then widen out into a large amphitheatre several miles in diameter. The high, unusual-coloured cliffs are fissured, sculptured by the water, and riddled with caves. The Tarn leaves this part of its course by a narrow defile, passing through a chaotic mass of blocks fallen from the overhanging sides of the valley. These produce rapids in the river, and boats have to be abandoned from this point on; any further exploration must be on foot.

A special type of gorge is the clue or cluse of the mountain areas marginal to the European Alps. These are gorges cut into folds of rock lying perpendicular to the course of the river. The large,

wide cluses of the Jura are classic: less well-known, but more extraordinary, are those in the upper region of Provence. For example, the Clue d'Aiglun cut by the River Esteron is 1,200 feet deep, only a few feet wide, and over a mile long. Gorges or water-gaps of this type are common in the folded section of the Appalachian Mountains in the United States.

The rivers of the limestone areas of Great Britain, too, have developed gorges in some parts of their courses. Those of the Pennines offer many examples, especially around Buxton and Matlock, in Derbyshire. The Avon Gorge near Bristol has been cut into limestone of the same age. The Cheddar Gorge, unlike the example just given, has no river in its upper part, and is really a ravine carved by the underground stream which emerges near Gough's Cave, at the exit from the gorge. The bottom of this narrow valley winds steeply upward for four miles between buttresses of white limestone and below towering crags 500 feet high, until it emerges on the level surface of the Mendip Hills. The Cheddar River flows through a series of caves below the present floor of the valley. It is to the collapse of old cave-systems that the gorge owes its origin.

Although young river gorges are not normally highways for travel or commerce, occasionally they offer the only possible passage through a mountain area, and must be utilised despite construction and maintenance expense. Arkansas River's Royal Gorge (1,000 to 1,500 feet deep) cutting through the Front Range of the Rocky Mountains in Colorado is an example. For through-way rights in this gorge, a remarkably bitter struggle was waged by rival railroads between 1870 and 1880, until finally a compromise was reached. Today the gorge is spanned by the world's highest suspension bridge, slung 1,052 feet above the river.

As might be expected, some of the deepest gorges in the world are to be found among the highest mountains, in the Himalayas. The River Arun, which rises in Tibet, passes between Everest (29,140 feet) and Kanchenjunga (28,146 feet) in a tremendous gorge. Many tributaries in the headwaters of the Indus and the Ganges also run through gorges in the Himalayas. The Indus,

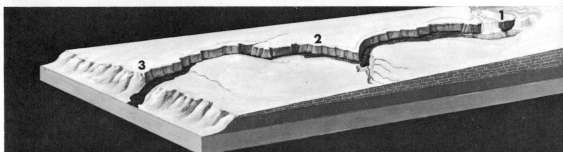

*Far left*
The largest gorge in Europe, the Gorge du Verdon, cleft in the Jurassic limestone of the southern French Alps.

*Left*
Cheddar Gorge is really a ravine formed when the original cave systems carved by the Cheddar River collapsed.

*Top right*
The Victoria Falls in Southern Rhodesia lie in the middle of the River Zambesi, hurling their waters into an almost vertical chasm. The Falls divide into four sections: Eastern Cataract, Rainbow Falls, Main Falls and the Boiling Pot. The view here is of Main Falls.

*Bottom right*
Diagram showing:
1. The present Niagara Falls;
2. The gorge cut by retrogressive erosion from the original site;
3. The original site.

in Kashmir, is only 3,000 feet above sea level, but its precipitous valley sides rise to nearly 20,000 feet. This river has managed to cut through 17,000 feet of rock while geological forces of uplift were raising the Himalayas to their present height. Rivers have also cut vast gorges in another of the world's great mountain chains, the Andes.

**Rapids and waterfalls.** Young and torrential rivers often develop rapids, that is, stretches of river-bed with a steep and rough gradient that present a serious obstacle to navigation. The water is shallow at such points and cascades over the rocks. Many of the great navigable rivers of the world, among them the Congo, the Niger and the Mekong, have stretches of rapids along their courses. The so-called cataracts of the Nile are really rapids. The lowest of them is the site of the important Aswan Dam: five others lie upstream from it.

When a river passes over a vertical rock face it is known as a cascade, cataract or waterfall. The feature is usually caused by bands of hard rock which resist erosion. Like rapids, they mark an irregularity in the smooth longitudinal profile of a river. The irregularities, however, are condemned to disappear when the river finally becomes graded. Waterfalls are continually cutting back upstream as erosion proceeds.

Some of the highest waterfalls are found in North and South America, and in Africa. In the last, the high plateaux of the interior form an escarpment some 12,000 feet high in Natal, overlooking the coastal plain. The rivers rising on the plateaux plunge over high waterfalls to reach the sea. One of these, the Tugela River, though small, has a fall 1,800 feet high, composed of five minor falls with cascades occurring between them; it is one of the highest in the world.

The Niagara Falls on the Niagara are undoubtedly the most famous. The amount of water flowing down the river from Lake Erie to Lake Ontario has been estimated at 212,000 cubic feet per second. It is, therefore, a river of some size that plunges over the vertical fall, 160 feet high and 3,500 feet wide.

Once over the falls, the river enters a seven-mile gorge which opens on to the shores of Lake Ontario. Niagara is an Indian word meaning 'thunder of the waters', and the name is well deserved. Today, large hotels stand close by this natural marvel, boats carrying tourists ply to and fro near the foot of the enormous cataracts, and during a certain period the entire scene is illuminated at night by coloured floodlights. It is by moonlight, in the calm of the night, however, that the grandeur of nature makes its greatest impression.

Let us look for a moment at the origin of Niagara Falls. The rocks of the region are inclined gently downward to the south, and consist of a series of sediments including sandstone, shales and dolomite. The most resistant layer is the hundred-foot-thick dolomite (calcium and magnesium carbonates). The shales and a little sandstone sandwich it below and above. Carving away at these rocks, the river wears down the soft shale more easily than the hard dolomite. The falling water leaps up again as it strikes the river-bed below the fall, and ceaselessly attacks the shale below and behind the lower part of the falls. Finally the dolomite overhang that is left collapses into the river-bed below, leaving the wall of the fall vertical. The erosion of the shale continues, and the whole process is endlessly repeated.

It is by this repetitive cycle that the Niagara Falls have steadily worked back to their present position, seven miles away from their initial emplacement—a particular case of retrogressive erosion producing headward migration. Measurements show that the falls are currently retreating at an average rate of 3·8 feet per year. Had this rate been constant during the formation of the entire gorge, its formation would have taken 10,000 years. As it is, the rate of erosion is probably higher now than in the past because the lakes above Ontario, which once had different outlets, all now drain through the Niagara. Taking this and other factors into account, it has been estimated that the falls developed over a period of 20,000 to 39,000 years. Recent drilling operations invalidate all previous estimates, however, for they have revealed that a middle portion of the Niagara gorge is filled with glacial debris, and

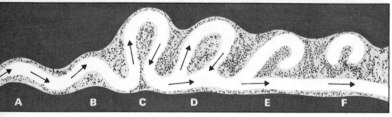

Top
Depleted course of the Zambesi River seen from the 420 feet high wall surrounding the Kariba Dam.

Centre
Formation of an oxbow lake.

Bottom
The Welsh Afon Llugw flows for many miles along an open valley before plunging abruptly into a steep, tree-covered gorge in which it becomes the Swallow Falls.

therefore pre-dates the last glacial epoch in western New York. Thus it is impossible to estimate accurately the time involved.

If Niagara Falls are the best-known waterfalls and the best utilised for the generation of hydro-electric power, it must not be forgotten that in size they are surpassed by the Victoria Falls. Situated near Livingstone, in Southern Rhodesia, where the Zambesi flows in a wide shallow valley through an arid basalt plateau, these falls were not discovered until 1885. From a distance, nothing betrays their existence except 'the smoke that thunders', as the natives describe the spray. The river above the falls is so gentle that elephant and hippopotamus bathe in it. Suddenly it plunges 300 feet over a fall 4,500 feet wide, into a deep and narrow gorge hollowed out of the basalt.

There are many falls greater in volume and in height than either Niagara or Victoria. That with the largest mean annual flow is Guaira (Sete Quedas) on the Alto Paraná River, between Brazil and Paraguay. It discharges 470,000 cubic feet a second, and is almost as high as Niagara. One of the most variable of the high waterfalls is Cauvery, which is transformed during the monsoons from a mere trickle to a flood of 667,470 cubic feet per second. In Yosemite Valley, California, there are many narrow but lovely, veil-like falls of unusual height: Upper Yosemite Fall is 1,430 feet high, and Ribbon Fall is over 1,600 feet.

The highest perennial fall on record is Angel Falls, in Venezuela. With a total height of 3,212 feet, it consists of several leaps, the highest dropping 2,648 feet down a sheer cliff. The only other falls to approach Angel Falls in height are Kukenaan (2,000 feet), also in Venezuela, and Tugela Falls (1,800 feet) in Natal. This last is also a multiple-leap, but its highest fall is only 1,350 feet, some 1,300 feet less than the highest drop of Angel Falls.

Though they cannot be compared with the world's great falls, the waterfalls of North Wales (Conway Falls, Swallow Falls and Aber Falls) attract many visitors from all parts of Britain and have, in addition, earned a measure of universal reputation, as have those of the Columbia River in the United States.

**High water damage of young streams.** Man does not hesitate to get the most he can from the mountains, destroying forest and replacing it with pastures for his sheep and goats. Plants hold the soil firm with the binding network of their roots and impede the devastating action of erosion. When grasslands disappear and the slopes are denuded, the topsoil is left without protection.

In mountain regions, rainwash and streams constitute a major threat to cultivable land. Each year the stream head gains some ground as the soil slips down into the funnel-shaped hollow until, finally, whole villages may disappear.

In the plains at the foot of the mountains, the damage is no less. At the melting of the snow or after violent storms, the mass of water coming down from the headwaters of the river is sometimes so great that it is hurled along the gorge with tremendous speed and strength. Vast blocks of rock are carried down in the roaring waters. Temporary dams are sometimes built up, behind which the waters pile up until they amass strength enough to sweep away the barrier and send a liquid avalanche to bury crops and villages on the piedmont plain below.

Mountain streams have been described by Surell as 'a fatal scourge, like a cancer on the side of the mountain; they eat their way into the soil and eject it as debris on to the plain beneath. After long-continued accumulation these enormous cones continue to grow, and threaten the invasion of the plain, condemning to sterility the soil buried under their coarse deposits. Every year they swallow up another farm, interrupt communications, or prevent a good road system from being established.

'The ravages are all the more deplorable in that they fall on poor areas without industries, where cultivable land is rare and the only resource of the inhabitants. Fields which have taken years of fatigue to cultivate may be destroyed within an hour by one torrential flood.

In the three centuries between 1471 and 1776, the western Alps of Europe lost three-quarters of its cultivated land, and its inhabitants were forced by starvation to take refuge in more hospitable regions. In other parts of the world, however, where the mountainous regions are less populated than in Europe, flood damage may not be so apparent to man, as he is less often affected by it.

The amount of destruction brought by young streams may seem excessive when the size of these small watercourses is compared with the volume of water in a wide, mature river, but with a moderate increase in speed the transporting capacity of water increases out of all proportion. Civil engineers have determined that the transporting power of running water varies as the sixth power of the velocity. That is, a river flowing fast enough to move blocks of rock weighing one pound can, if its speed is doubled, transport rocks not of two pounds, as one might expect, but of sixty-four pounds. Another factor to be considered in a study of a stream's carrying capacity is, of course, the buoyant action of the water. A rock of average specific gravity loses about two-fifths of its weight when it is immersed in water. Heavy material is transported by traction (rolling and dragging along the stream's bed), not in suspension.

It has been found that water with a speed of one foot per second can transport fine gravel, while currents of four feet per second can move pieces of rock weighing two pounds. When the speed rises to thirty feet per second, or more, the watercourse is able to transport huge boulders. As we shall see later, the ravages of floods in lowland areas are severest where the water is moving rapidly. In floods caused by the breaching of dams and similar sudden surges of water, massive bridges may be swept away, and even vehicles as heavy as locomotives have been known to be carried considerable distances. When the St Francis Dam in California broke in March 1928, the water rushing down the valley moved blocks of concrete weighing 10,000 tons through distances of 2,500 feet or more.

The first stage of any counter-measures against flooding, in young as well as mature rivers, consists in building a series of dams across the course of the stream. These halt transport of the load and provide quiet, ponded waters in which it can accumulate. The water pours in a cascade from dam to dam, each representing a step in a staircase which breaks the force of the torrent by reducing the original steep slope to a series of horizontal flats.

Replacement of forest and turf already eroded must then follow in order to repair the damage done, possibly over centuries. This long and costly operation consists in sowing grass, then positioning saplings and other plants on the edge of the watercourse, on the slopes, and at the head of the stream. The re-planted areas need protecting with fencing until they are well enough established to resist the slope on their own. When vegetation is firmly re-established, the forest area will resist erosion for several seasons. The branching roots distributed throughout the soil hold it together and resist downhill washing away, while the surface mat of living and decomposing vegetation acts as a sponge when rain falls, and allows it to seep off slowly instead of washing directly on to bare earth. Lastly, the slight shade of the trees helps to prevent over-rapid melting of snow in spring.

The problem of high water is only part of the general struggle against soil erosion. Once an area is bared of its forest cover, erosion of all kinds proceeds more rapidly. This is especially true in mountain areas.

## MATURE RIVERS

When lateral planation becomes dominant over downcutting then a stream enters the mature stage of its history, characterised by meanders, moderate gradient and velocity, and broad, flat-bottomed, U-shaped valleys with flood-plains and levees; tributaries are numerous. Except during flood time, a mature river's load consists mainly of silts and sands, the longitudinal profile is smooth, and the stream is well adjusted to the rock structure over which it flows. Deposition is the principal activity of a mature river as it crosses its flood-plain. Regular meanders (seen very clearly in the Mississippi, the Seine, and the Severn) are further traits of maturity.

It is interesting that meanders form, move and multiply according to the dictates of the river. A meander at its beginning is a simple detour taken to avoid an obstacle. In such circumstances the water is deflected in the direction of the arrow (A) on the diagram,

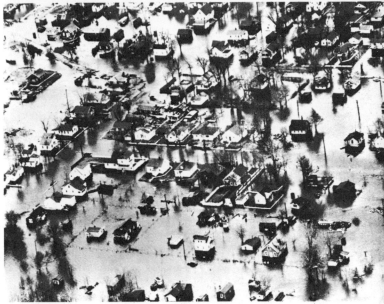

Top
Rainbow Bridge, Utah.

Centre
Floods at La Crosse, Wisconsin when the Mississippi overflowed in 1954.

Bottom
The overflowing of the banks of the River Po, Italy. Large teams of workers are engaged in defence work, erecting dykes along the sides in order to halt the advance of the floodwaters.

The U-shaped valley and
waterfall at Yosemite,
in California.

in the centre of page 98. This means that the flow of the entire
river swings against the concave bank. This bank is hollowed away,
and its concavity increases *(B)*. Conversely, the main current
avoids the convex bank and leaves an area of slack water where
alluvium is deposited. Little by little the meander becomes a loop
enclosing a peninsula of land. The neck of the peninsula becomes
narrower and narrower until the river breaks through it and flows in
a straight line again. The loop becomes by-passed by the main current, silted up, and overgrown with vegetation *(D)*. Favourable

atmospheric conditions produce a peaty marsh. Consequently the downstream end becomes dry land *(E)*, and finally the loop becomes completely separated from the river, and is known as an abandoned meander, a meander scar, or ox-bow lake. In southern England the term mortlake is also used. The former meander, sometimes unseen from ground level, is often distinguishable on air photographs. This is because soil and drainage along its erstwhile course remain different from the surrounding alluvium of the flood-plain and produce differences in colour and type of vegetation.

Meanders are very clearly developed in the lower Mississippi, where the river winds on a flood-plain 20 to 75 miles wide. In this region, long sections of abandoned meanders, or meander scrolls, are called bayous while they still contain water.

For purposes of navigation, the course of a river can be shortened by cutting through the neck of the meander if the severance has not occurred naturally. Such measures were taken in 1841 in the case of the Connecticut River near Northampton, Massachusetts. During the decade 1927–37, the Mississippi River Commission was able to shorten a 331-mile stretch of the river by 116 miles, facilitating navigation and affording better flood control.

The importance of meanders is considerable. By means of these successive deplacements a winding river manages first to erode and then to cover with alluvium a wide area quite out of proportion to its own size.

### Incised meanders.

Gorges are a sign of youth in rivers, meanders a sign of maturity and old age. But how can we explain that many gorges, such as the Wye in England, the Meuse and the Semois in France, the Neckar and the Moselle in Western Germany, and the San Juan and North Fork of the Shenandoah River in the United States, also reveal meanders? It is not really a contradiction for both features to be displayed, for a senile river can be rejuvenated. As we have seen, the region through which the river flows has only to undergo slow uplift for the meanders, at first formed on the surface, to bite deeper, little by little, and become incised or entrenched meanders. The Wye area and the Ardennes, for instance, were upraised during the earth movements that produced the Alps during the middle of the Tertiary period. The rivers kept pace with this by cutting downwards all the while.

Throughout many of the most picturesque areas in Europe, the phenomenon of incised meanders will be frequently observed, and also the work of the rivers in eroding the concave slopes while depositing alluvium on the convex. The concave slope always appears as a steep wall plunging vertically into the water. The convex slope, on the other hand, is a gentle slope, generally wooded or meadow-land.

The tributaries of the Rhine, such as the Main, the Neckar, the Lahn and the Moselle, are all incised in the Rhine highlands, with meanders deeply entrenched below the surfaces of the upraised plateaux. The Moselle, flowing between the Eifel and the Hunsrück, is a famous wine-producing area where vines are grown on the terraced south-facing slopes in the shelter of the deep valleys. The meander-cores almost surrounded by water formed easily defended sites for settlement in early times, and many a castle rises from their conical hills.

Meanders, whether incised or not, have always been important in the siting of towns. The city of Durham springs to mind, for here the cathedral was built inside a loop of the River Wear, on a meander-core with steep cliffs falling away to the river. The landward side was protected by a castle. New Orleans, Louisiana, called the Crescent City, is similarly located on a meander of the Mississippi.

The River Wye flows in entrenched meanders across the hills of the Welsh borderland, between Ross-on-Wye and Chepstow. The curves are up to three miles wide; several, such as Symond's Yat and the Wyndcliff, describe almost complete circles *(see p. 76)* overlooked by cliffs. The gorge from Monmouth to Tintern is not so winding, but the sides of the valley show that there have been other meanders, now abandoned. At St Briavel's Castle the river flows well to the west of a wide semi-circle of cliffs which must have been carved out by a meander swinging against it. Further upstream there is an even more interesting feature of the landscape, between Redbrook and Newland. A loop-shaped channel, now occupied only by two small tributaries, lies 400 feet above the level of the Wye, proving that the meander which occupied it must have been abandoned long before the river had succeeded in cutting down to its present level.

A very curious phenomenon occurs at a point of the Ardèche river, France, where the isthmus of the peninsula formed by the incised meander was so narrow that the river eventually bored its way through and established a new course through the archway. Numerous natural bridges or arches have been formed in this manner throughout the world. A superb example, unfortunately in relatively inaccessible terrain, is Rainbow Natural Bridge in Utah.

### Flooding of mature streams.

From time to time, a river is subject to exceptionally high water. Abandoning its normal channel, it spreads out over the level valley floor and deposits a flood-plain as it recedes.

The flood-plain may be extensive. In the valley of the Mississippi, for example, it covers about 30,000 square miles, extending 600 miles from the mouth of the Ohio River to the Gulf of Mexico, although the numerous curves and meanders of the river itself cover twice that distance. The alluvial soil formed by the silts brought down by the river is extremely rich, and the area is well populated despite the recurrent danger of disastrous flooding. Floods at the end of the last century and at the beginning of this were especially severe. In 1903, 65,000 people were driven from their homes when great damage was done to property and traffic suspended for three weeks. Still worse was the great flood of 1912, in the lower Mississippi Valley, when the cities of Memphis, Vicksburg and New Orleans all suffered severely. North of Louisiana, 30,000 people lost their homes. In 1951, floods again spread over a large portion of the same area, resulting in severe damage to livestock and property, and in some loss of life.

Perhaps the most disastrous recurring floods in the world are those of the Hwang-Ho, in China. The fertile soil of its vast alluvial plain (400 miles long and 100–300 miles wide) has been densely populated for several thousand years. When the floods spread out over the plain, untold damage is done to property and crops. Those of 1887 destroyed hundreds of villages, and more than a million people are estimated to have died by drowning or starvation. In 1892 the floods covered such an enormous area that when the waters finally subsided the Hwang-Ho had changed its lower course completely, and had eroded a new entrance to the sea, 300 miles north of its old mouth. It now empties into the Gulf of Chihli; before 1892 it flowed into the Yellow Sea. Much change of mouth has occurred in the last few thousand years.

Side by side with 'tidal' waves, cyclonic storms, earthquakes, and volcanic eruptions, floods bring home to man the realisation of his weakness in face of natural forces. In most instances, the utmost he can do to control them amounts only to minimising their effects.

The first step to be considered in face of flood danger is the embanking of rivers, practised since the time of the Pharaohs of Egypt. Dykes and levees, however, are a remedy often worse than the evil they seek to cure. Built for immediate protection, with no consideration for neighbours upstream and with no thought for future results, the artificial levees allow the swollen river to build up to a greater height than usual, and if they are overtopped or burst, the resulting flood is even more devastating than it would have been without the dykes. Moreover, they raise the crest level upstream and cause flooding there.

In Italy, in 1952, the Po was flowing 18 or 20 feet above the surrounding countryside when the dykes broke. Crops were inundated over an area of 400 square miles, 4,000 villages were destroyed, and several hundred people were drowned.

In 1735 the French settlers on the Mississippi constructed three-foot-high levees from about 30 miles above New Orleans to some 12 miles below it. The people upstream built levees, too, so those round New Orleans had to be raised still higher. This necessitated raising the upstream levees still more. The result is a levee system which, in the mid-1950s, had an average height of 24 feet and a total length of 2,846 miles, containing about 1,392 million cubic yards of earth.

Another current method of flood control is by the construction of dams. However, expense here would be great, as normally these dams would be empty and of no use for irrigation or hydro-electric schemes. Moreover, the capacity of such reservoirs would have to be enormous to deal with the largest floods that could be expected. What use would a normal reservoir's capacity of 500 or 700 million cubic feet be in the face of the discharge of several thousand million cubic feet of river in flood?

Several small dams constructed on tributaries seem more useful than one large dam, especially if the excess water can be diverted into overflow channels leading into a marsh or grassland of low fertility alongside the river. Flooding of chosen areas to protect others appears to be one of the best ways out of the problems raised by river overflow. Along the Mississippi and other United States rivers permanent overflow channels or spillways of this kind have been constructed. The Bonnet Carre Spillway, 23 miles above New Orleans, was put into operation in January 1937, when the greatest flood waters on record developed in Ohio. It was calculated that the crest when it reached New Orleans would exceed 20 feet. The spillway diverted at maximum some 200,000 second/feet of water into Lake Pontchartrain and held the crest at New Orleans at 19·3 feet.

Artificial cut-offs to lop off the meanders of a river and thus steepen its gradient and deepen its channel through increased erosive power are used in the United States as another flood-control measure. In some areas such cut-offs have lowered flood waters as much as twelve feet or more.

The best method of dealing with floods is, of course, to prevent their occurrence by reforesting the region of the river's headwaters. It must again be stressed that forest cover is the best regulator of run-off. In practice, however, the extent of pasture or cultivation, and varied ownership in the uplands, may make the operation impossible. Reforestation may have to be confined to wasteland and the steeper valley slopes, but even this will produce some amelioration of the situation in the lower regions of the river's course.

## OLD AGE

If the Earth's crust remains still for a long period of time, rivers will eventually reach that stage of development called old age. Its characteristics are very similar to those of maturity, but meanders are very numerous, gradient is very low, and the water flows sluggishly. There are no waterfalls or rapids. The valley is extremely broad and shallow, with low divides. Tributaries are few and large. Deposits are thick, and the stream's load is fine in texture.

A large river is never in one stage throughout its length; most, like the Mississippi, are young in their headwater region, mature in the middle and old near the mouth, where they enter the sea.

*Below*
Levees built along the
Atchafalaya, a tributary of
the Mississippi.

Moensklint, a limestone cliff
469 feet high on the east coast
of Moen Island, one of the
smaller Danish islands situated
in the Baltic.

A meander on the River Lot
between Calais and St Circ-
Lapopie (Lot) France. On the
inner sides of the bend the river
has carved steep cliffs leaving a
small flood plain opposite.

# Oceans and Lakes

It has been observed that Earth is a singularly inapt name for a planet covered over 71 per cent of its surface area by oceans. Add to this the fact that the oceans are, on average, 12,450 feet deep, while the average height of the continents is only 2,600 feet, and the disproportion between land and sea mass can be seen even more clearly.

The area of sea water covering the Earth's surface has been estimated at 143 million square miles. This can be roughly divided into five major areas: the Pacific, Atlantic, Indian, Arctic and Antarctic Oceans. Various other seas, including the Mediterranean and the Black Sea, can be looked on as annexes to the oceans. Others, including the North Sea, the English Channel, the Baltic, the Caspian, the Gulf of Mexico, and Hudson's Bay, are shallow coastal waters which temporarily cover the margins of the continents.

The important function of the hydrosphere — the oceans and seas — is its regulation of temperatures, winds and water on the surface of the globe. Rachel Carson, in *The Sea Around Us,* has described it as 'the earth's thermostat'. It is the source of all the Earth's water, for without its immense surface open to evaporation the atmosphere could not take up the moisture needed to form rain clouds. Without marine currents and the movement of warm oceanic water, the climates of the world would not be what they are.

The observer on sea-shore or ship's deck sees little difference between one ocean and another. The Atlantic looks much the same as, say, the Pacific. There is a difference of size certainly; the swells, the waves, and the storms differ in strength. But what are the structural differences? How do the hollows, the abysses, the submarine volcanoes differ from one sea-floor to another? These are questions that cannot be answered from the shore or the deck, and it is only recently that, by means of photography, dredging, sounding, seismic refraction, gravity measurements, sedimentary sampling and coring, comparison between oceans has been possible.

Detailed mapping of the ocean floor presents two major problems: the determination of exact geographic location at which any observation is made, and measurement of depth. Out of sight of any land to give a bearing, the first problem is solved by methods ranging from dead reckoning and astronomical fixes to various radar systems. Depth is ascertained by echo-sounding with a device that measures the time required for sound to travel to the bottom and its echo to return. Velocity of sound in sea water is about 4,800 feet per second. This figure is subject to correction for variations produced by differences in temperature, salinity and certain underwater topographic features. Thus a small margin of error is inevitably introduced into undersea maps.

**Pacific Ocean.** The Pacific is the largest and the deepest of the oceans, covering almost one-half of the globe, with an average depth of 14,000 feet. Its broad, deep basin, remarkably smooth for great distances, contains many irregularities as well. High and abrupt ridges, deep elongated trenches volcanic mountains, and numerous other features characterise its broad expanse. Some of these topographic forms project above sea level, but most do not. Notable among typical Pacific features are the volcanic

islands and a variety of coral reefs. Recent sonic sounding in the Pacific and elsewhere has revealed great numbers of submarine 'islands', or hills, on the sea-floor. Many such sea mounts have been mapped in the Gulf of Alaska, ranging in height from 3,000 to 12,400 feet above the surrounding sea-floor. Most of them seem to be volcanic in origin. Those with truncated flat tops are called guyots. Some are 60 miles wide at the base, with flat tops as much as 35 miles in diameter. Over seventy-seven have been noted off the California coast. The cause of their truncation is still a puzzle. Perhaps the coastal erosion millions of years ago is the answer.

The Pacific Ocean is almost completely encircled by mountains which form near-continuous chains in the Americas, just as they do in Asia and Oceania. On the American coast these are the Pacific Coast Ranges and, a little farther inland, the Sierra Nevada, the Cascades, and the Rockies, while to the south there is the Andean Cordillera. The last is preceded throughout its length by submarine ridges and furrows, which appear to be a sort of Pre-Andean chain in course of formation. On the Asiatic side of the Pacific, arising from a similar cause, is a somewhat different arrangement of mountains. 'Island-arcs' succeed one another from north to south: the Aleutian Arc, the Kuril Islands, the Japanese islands, the Marianas Islands, the Philippines, New Guinea, New Caledonia and New Zealand. The Indo-Malayan arc, containing Java and Sumatra, separates the Pacific from the Indian Ocean.

Island arcs are arcuate submarine ridges extending above sea level. They are best developed in the western Pacific and West Indies areas. On their convex side there are deep trenches, troughs, or fore-deeps, which constitute the deepest portions of the oceans. All the island arcs mentioned are separated from Asia and Australia by shallow seas. They are, in effect, chains of mountains in the course of emergence. The Pacific is also ringed with volcanoes, along its shores and even within its boundaries. Earthquakes are often recorded from various parts of it.

Should it be concluded, then, that the Pacific is a recently-formed oceanic area? Geologists believe not. Filling a sea-basin of such magnitude would require enormous quantities of water, and there is no evidence of great and continual lowering of sea water on other coasts of the world. The Pacific is therefore probably a very ancient feature of Earth's relief, and scientists have even suggested that it marks the scar left by the Moon when it became detached from the Earth. Others have compared it to the 'mares' of the Moon, and think that it was formed at a time when Earth's crust was sufficiently thin and unconsolidated to be affected by colossal earth movements followed by a subsidence with uplift of the surrounding areas.

**Atlantic Ocean.** The smaller, shallower Atlantic presents a different picture. This ocean forms a winding corridor between the Americas and the continents of Europe and Africa. Following its long axis, and running from Iceland to Antarctica, there is a submarine ridge, the Mid-Atlantic Ridge, whose S-shaped curve is similar to that of the coastlines bordering it.

Recent studies made by members of the La Mont Geological

Observatory and others show that the ridge is a very complex area, with at least three distinct types of topography. The Main Ridge is a high central belt of several sub-parallel, north-east to south-west ridges, closely spaced, separated by deep, narrow valleys, and covered with relatively shallow water (about 800 fathoms). The second zone, the Terraced Zone, 200 to 300 miles wide, is found on both sides of the Main Range and consists of a series of flat terraces at a depth of 1,600 to 2,500 fathoms. The third and deepest of the three zones lies between the terraced areas and the actual sea-floor, in waters more than 2,900 fathoms deep. This region, called the Foothills of the Mid-Atlantic Ridge, is mountainous belt with many peaks.

The true geologic nature of the Mid-Atlantic Ridge complex is still unknown, but theories are many: that it is a great fold in the rocks, or an elongated block pushed up between major fractures, or a zone of volcanic extrusions, or a complex mountain chain similar to the Andes.

Plateaus, extensive areas with flat tops, are found in the Atlantic as well as in other oceans. The Azores plateau with its ridges and broad, shallow valleys extending at right-angles to the Mid-Atlantic Ridge is an example. So is the Albatross Plateau off Central and South America in the Pacific, and Seychelles Plateau in the Indian Ocean. Scattered groups of sea mountains occur in the Atlantic also. South-east of Cape Cod one of them towers almost 11,000 feet above the sea-floor.

In front of the shores of the Atlantic a continental shelf extends, a gently curving submarine surface beyond which the sea-floor drops to the abysses. Judging by the way in which submarine channels such as that of the Hudson River are carved into it, the shelf has not been long under the sea.

The great submarine canyons which exist on the continental slope and even extend into the shelf and into the ocean-floor at depths of more than 12,000 feet should not be confused with submerged stream channels. The canyons are found off all continents and in all oceans. Many of them extend across the shelf opposite the mouth of a large river; others are not coincidental with stream channels. Their origin is still a mystery. Most of the canyons are cut deep in sedimentary rocks of fairly recent geologic age, though a few off the California coast are cut in granite. Where continental slopes are gentlest, the canyons are rare or completely lacking. The slope of their floors is extremely steep, being about ten times steeper than the average continental canyon floor.

Many and complex theories have been evolved to account for these remarkable gorges. Earthquakes, landsliding, slumping, mudflows, entrenching by great rivers, and turbidity currents are some of the suggested causes. The last, turbidity currents, is one of the more popular theories for the origin of the stream channels.

Turbidity currents caused by variations in the density of water are induced by the differences in temperature, salinity, and the amount of sediment (heavier or denser waters sink through the ocean or lake and move along the bottom). The presence of such distinct turbidity currents are recorded in Lake Mead, Lake Geneva and Lake Constance. The strange submarine canyons deeply incised into the continental slopes appear to be the result of erosion by streams, most probably flows of muddy water produced when earthquakes, storms or landslides disrupt the unconsolidated material at the heads of the gorges. The turbidity currents flow rapidly down the continental slope because their specific gravity is higher than that of the ocean water, scouring the valleys deeply into the slope. As they move into the floor of the ocean basin they spread out and slowly abate, depositing extensive layers of sediment.

The complex structure of the Atlantic Ocean provides plenty of scope for different schools of thought. One claims that the Atlantic is the result of relatively recent subsidence. According to this theory there was, originally, a North Atlantic continent embracing North America and Europe, and another continent uniting South America and Africa; and for a long period a sea called the Tethys, of which the Mediterranean is the last vestige, stretched between these continents, parallel to the Equator. Then came subsidence of the middle portions of the continents, which left the Atlantic Ocean.

Upholders of the theory of continental drift maintain that there was no subsidence, but drifting apart of the land masses. The American and Eurafrican blocks, they claim, were one until median fracturing freed them and allowed them to separate. This conception will be referred to again in the chapter on earth movements. The principal argument here rests, of course, on the remarkable coincidence of the western and eastern shores of the Atlantic. Europe could be fitted into the coastline of North America and, more accurately, Brazil into the Gulf of Guinea, as one might fit together the pieces of a jig-saw puzzle.

Other schools of thought have it that the Atlantic basin has existed as such throughout the ages. Some believe, however, that it has contracted (to produce the line of volcanoes along the axial ridge) and that contraction has been accompanied by an enlargement of the Pacific. Still others believe that, far from contraction, expansion or enlargement of the Earth's crust has occurred along the Mid-Atlantic Ridge, material being added along fractures from below the actual surface of the Earth.

*Left*
Rocks subject to constant
erosion on the edge of the sea.

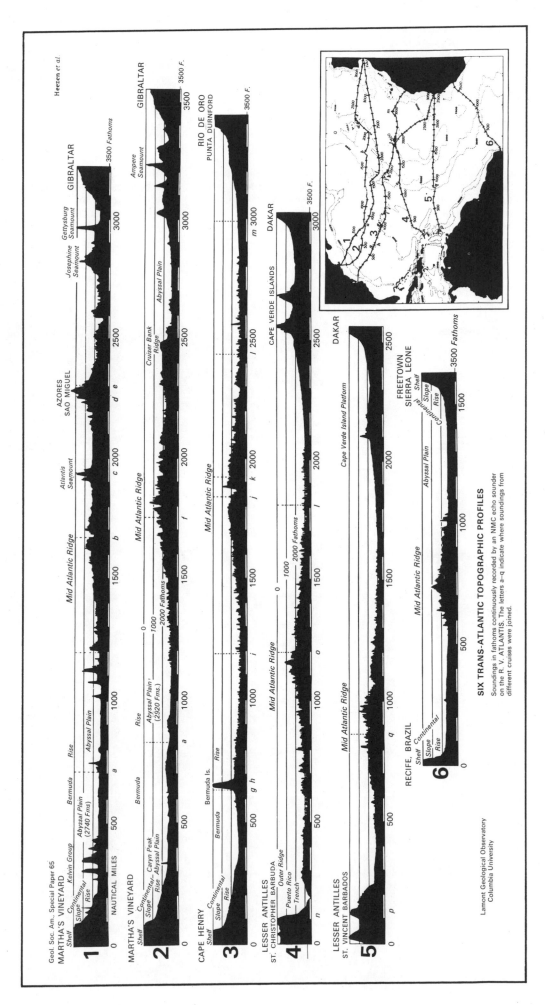

SIX TRANS-ATLANTIC TOPOGRAPHIC PROFILES

Soundings in fathoms continuously recorded by an NMC echo sounder
on the R. V. ATLANTIS. The letters a–q indicate where soundings from
different cruises were joined.

Lamont Geological Observatory
Columbia University

Research by echo-sounding has produced evidence in favour of the Atlantic's antiquity. The thickness of the floor sediments seems to be in the order of 2 to 2½ miles. With deposition of ocean sediments estimated to form at one-fifth to two-fifths of an inch in a thousand years, the existing thickness represents the accumulation of some 400 million years.

**Indian Ocean.** This ocean shares the characteristics of both the Pacific and the Atlantic. It is small in size but has an average depth of 13,000 feet. Its eastern part is an immense basin of abundant volcanic activity. There is a Mid-Indian ridge which runs for several hundred miles in a general southward direction from the coast of India. The ridge is similar to the Mid-Atlantic Ridge, but broader and flatter. Throughout one-half of its length it rises to within about 9,000 feet of the ocean's surface. On the western side, the ocean bottom has been compressed into parallel ridges. Part of one ridge has emerged above the surface as the island of Madagascar.

Nothing will be said here about the Arctic and Antarctic Oceans. The first is still insufficiently known; the second is part of the other three great oceans. While the Arctic Ocean is a true ocean-basin, albeit with an average depth of only 7,000 feet, the bottom of the Antarctic Ocean is simply the shelf around the vast Antarctica.

## SEA WATER

Except for a small number of salt lakes, seas contain the whole of the world's supply of salt water. The amount of salt in solution has been estimated at 56 thousand million million tons. Spread over the surface of the continents, this would form a layer more than 400 feet thick.

Although the seas and oceans are salty, the degree of salinity

varies from region to region and under different conditions, depending on fresh water being added by rivers and glaciers.

In the world's great oceans, maximum salinity (3·6 to 3·7 per cent) coincides with the tropic areas, excluding those of the equatorial belt. Here insolation is greatest and rainfall lowest. On the Equator, despite the rise in temperature, salinity decreases to 3·5 per cent as a result of the frequency of rain. In the temperate regions, too, the average is about 3·5 per cent. In the polar regions the effects of low temperatures and glacial meltwaters reduce salinity to its minimum of 3·2 per cent.

The seas of the great oceans follow different patterns, the temperate ones (the Channel, the North Sea and the Baltic) differing from the subtropical (Mediterranean, Red Sea, Persian Gulf). In the first group the inflow of fresh water is dominant; in the second, evaporation is the prevailing factor. Therefore the northern seas are reduced in salinity, while subtropical seas show an increase.

An extreme case is the Baltic, where salinity falls to 0·1 per cent in certain places (Gulf of Bothnia is one). The Red Sea is another; here, on the contrary, the swing is in the other direction, and salinity is never less than 3·7 per cent sometimes reaches 4·1 per cent.

The Black Sea, although in the same latitude as the Mediterranean, has low salinity (1 per cent to 2·5 per cent) because of the enormous amount of fresh water carried into it by the Danube, the Dniester, the Dnieper, and the Don. As for the landlocked seas (actually large lakes subject to high insolation), here salinity ranges from that of the Caspian at the lower end of the scale, to that of the Great Salt Lake and the Dead Sea at the other. In the last two, neither heavy rains nor large rivers modify the effects of the intense evaporation taking place in their climatic zones. The Dead Sea, in fact, is so salt (27·5 per cent) that it is impossible for a man to dive into its waters, and difficult to swim in them.

Salt and fresh water meet along coasts at points where streams enter the sea. In a bay, fresh water will flow seaward above the salt water because it contains less mineral matter and is less dense than sea water. A double current is thus formed, in which sea water predominates as the tide rises, and fresh water as the tide goes out or in times of river flood. Eventually mixing takes place, to produce the brackish water typical of estuaries and of coastal lakes and lagoons near the mouths of rivers. The current of fresh water can, if it is large enough (as in the Congo and Amazon), extend several hundred miles out to sea. Low tides, too, permit rivers to carry their brown muddy waters far out into the sea; the Mississippi is one to act in this way.

Over one-third of all the known elements are to be found in sea water, mainly in the form of chemical compounds. Common salt, or sodium chloride, obviously takes first place, forming four-fifths of the total. The next salt, which forms only one-eighth in quantity, is a magnesium chloride, and after this come magnesium sulphate and the sulphates of calcium and potassium. Such a list could be almost without end, since so many salts of so many different metals could be added. It is no exaggeration to say that every metal can be found in the sea, even silver, gold, platinum and radium. Indeed, if all the gold could be extracted from the oceans, such an enormous quantity would be available that it would become worthless.

Today, the practical and economic utilisation of seawater salts is limited mainly to sodium chloride, magnesium, potassium and bromium.

The mineral matter of the oceans is being constantly augmented by the rivers, glaciers, rainwash, ground water, and volcanic action. At the same time, organisms are extracting material (mainly calcium carbonate) from it. More mineral material is lost in deposition of limestone and salts in the shallow lagoons or salt pans along shorelines. It is not yet certain if withdrawal equals intake, or if a balance exists in this great chemical system.

## MOVEMENTS OF SEA WATER

The main movements of the sea waters are the tides, marine currents and waves.

**Tides.** The tides are caused by the attraction of ocean water to the Sun and the Moon, and so they have periodic movements of rise and descent, or ebb and flow. The highest are the spring tides, coinciding with the full and new moon, when the Sun's and the Moon's tide-producing forces complement each other; the lowest are the neap tides, which correspond with the first and last quarters, when the Moon's and the Sun's forces tend to counter-balance each other. Neap tides are about 20 per cent lower than average tides, and spring tides about 20 per cent higher. Because the Moon's orbit is an ellipse, it is at times closer to the Earth than at others. When it is closest, in perigee, tides are highest and about 15 to 20 per cent above average; when the Moon is farthest away, or in apogee, they are about 20 per cent lower. Apogean or perigean tidal effects may complement or detract from spring and neap tides; when they coincide, the tidal range is abnormally great.

Another factor to be taken into account is that each oceanic basin, each gulf, each bay, and each section of the coast modifies in some way the tides that occur along it. In the English Channel,

*Left*
Norway's Lofoten Islands.
Svolvaer, viewed from
Goat Mountain.

*Above*
Dyke-building operations
to protect the site of
Flevostad, Holland.

for example, when the tide is rising 20 feet at Cherbourg, it rises 30 feet at Calais, and 40 feet at Saint-Malo and the Mont-Saint-Michel. In the Mediterranean, the height of the tides is no more than a few inches; at the head of the narrow Adriatic a rise of 3 feet is produced, while in the Bay of Fundy an increase of 50 feet is recorded. Where such heights are reached, velocities soar too, with recordings as high as 14 miles per hour in narrow estuaries.

In spite of their normal regularity, tides can produce some unexpected and unfortunate surprises. This was proved on the night of 31 January–1 February, 1953, when enormous waves suddenly invaded Holland and parts of the North Sea coasts of England and Belgium. In Holland alone, 6,250 square miles were submerged, 1,800 people died, 50,000 more were homeless, and 50,000 head of cattle were drowned. Dykes were broken, and Dordrecht and Rotterdam menaced.

The cause of the tragedy? The tides in the North Sea travel southward and then, after reflexion from the coast of the Pas de Calais, move northward again. In certain areas direct and reflected waves cancel each other out, in others their effects are combined. On this particular night the tides worked together at the latitude of England and Holland, and the rise was higher than usual. The already high tides were reinforced by a wind which blew from the north at speeds up to 100 miles an hour. The water piled up in the southern part of the North Sea, resulting in what meteorologists call a surge. This combination of circumstances is fairly frequent but does not normally produce major catastrophes. Disaster comes when both winds and tides are exceptionally high.

There are many other dangerous currents. Their direction and force depend on the particular conditions of the coast. In the Lofoten Islands of Norway, dangerous whirlpools appear twice a day during the rising of the tide.

The 'bore' is a phenomenon of the contact of the river current with that of the rising tide. As the tide rises in the estuary, the converging shores intensify its height and speed. On encounter with the river water, speed is checked, the tidal wave breaks, and a near-vertical wall of tumbling and foaming water roars up the estuary. The narrow estuary of the River Severn at the east end of the Bristol Channel produces a bore which may be 3 or 4 feet high at spring tides. In some areas the bore may be a startling and destructive force. For example, tidal range is between 30

and 50 feet in the Bay of Fundy, between Nova Scotia and New Brunswick, and in this area a tidal bore about 6 feet high marks the front of the advancing water. In the larger rivers of the world, such as the Amazon and Tsientang, the phenomenon appears as a wall of water, 15 to 20 feet high, travelling rapidly upstream.

**Ocean currents.** One of the most interesting examples of a marine current is the Gulf Stream of the Atlantic, so important in giving western Europe and parts of eastern North America a temperate climate. It originates from westward-drifting equatorial waters travelling through the Gulf of Mexico. Here it is augmented by waters from rivers and rainwash, to emerge through the straits of Florida as the Gulf Stream. Its characteristic colour is, in part, the result of this excess of fresh water. Sweeping along the coast of Florida and northward along the continental shelf, it is deflected to the east by the Coriolis force, moves away from the shelf, and travels north-east across the Atlantic. As it travels it spreads out, gives off eddies, becomes less definite, and gradually disappears.

The warming effect of the Gulf Stream and the North Atlantic Current is well known on the coasts of western Europe, and explains why the coasts of Scandinavia, Spitzbergen and Novaya-Zemlya are so much warmer than, say, Siberia or Canada. Spitzbergen is especially spectacular. Situated between latitudes 74°N and 81°N, it is one of the most northerly countries of the world, and yet it enjoys summer temperatures of 55°F. The Gulf Stream is always warm, although its temperature varies a few degrees.

Ocean drifts or currents such as the Gulf Stream and Labrador Current are the product of many factors. Primarily they are caused by drag, or friction, exerted on the waves by prevailing winds. This is particularly noticeable in the belts of the westerlies and the trade winds. In addition, unequal heating of the water and differences in its density play their part; the colder, denser water of the Poles migrates toward the Equator, while from the equatorial region where the water is warm and where heavy rainfall raises the ocean level slightly, the drift tends to be poleward. Coupled with the wind direction, this produces a more or less circular movement in the surface waters of the oceans, modified in its turn by the configuration of the coasts and ocean basins, the topography of the sea-floor, the rotation of the Earth, the deeper water currents of the ocean, and other variables. Cold water currents travel

Sluices on the Afluitsdyk which, in 1932, turned the Zuider Zee into the Lake of Ijsselmeer.

A pilot scheme to reclaim the marshy flats along the Kafue River in Zambia following the Dutch example.

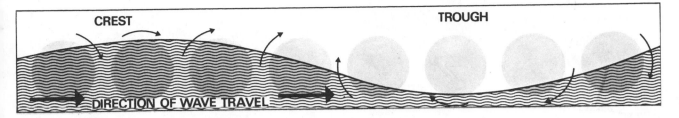

Orbital motions in waves.

along the ocean floor. Above these, but still beneath the surface currents, are 'intermediate' or 'depth' currents. The whole circulation system of the oceans is highly complex.

## IRREGULAR MOVEMENTS OF SEA WATER

In contrast to the periodic and constant tides and marine currents, which are always regular, the sea is also influenced by swells, waves, and local currents which are irregular and often unpredictable.

**Swell.** Swell, or free waves, is an undulatory movement of the water similar to that caused by a stone falling into a pond. It exists only in locally calm weather and is the manifestation of the dying waves of a storm raging several hundred miles away. Swell causes the two most unpleasant motions of a ship–rolling and pitching–and produces strong wave action on calm days.

**Waves.** Unlike swell, waves are produced by the direct influence of the wind. For example, in the morning the sea may be calm. Later, a breeze springs up, and miniature waves ruffle the surface. These small waves move slowly. As the day wears on, the wind gains strength. The ripples of the water deepen and become billows (wind waves, sea waves or forced waves) pushed along by the wind. As this action continues, it gives an asymmetrical shape to the wave. The crest is narrower than the base, and less retarded by friction with the underlying water, so that it becomes inclined in the direction of travel and, finally, bends over and breaks, to produce foam and white horses on the surface.

When the direction of the wind is variable the waves cross one another and interlace to form the familiar 'choppy' sea, cut with furrows and chequered with lines of foam. Waves which develop freely in open ocean waters are waves of oscillation. The main direction of movement of the particles in such a wave is orbital, moving forward on the crest, up at the front, backward in the trough, and down at the back. They move forward because velocity is great in the crest, where there is less friction, and the particles in motion return to a point slightly ahead of starting place.

As they break, waves of oscillation become waves of translation, flowing on to the beach as wash, and running back down the beach slope as backwash. Oscillation waves develop in groups; waves of translation are single, independent units which move forward without compensating backward movement. Most waves of translation are formed through the offshore breaking of oscillation waves, although some are generated at sea. They do not have pronounced crests or troughs, and the movement of their particles is in a parabolic path at the surface, flattening downward until it is almost in a straight line at the bottom. Waves of translation transport a great deal of fine material near the shore.

It is difficult to appreciate the height and speed of waves at sea. Sailors' descriptions of waves 200 feet high must be regarded as grossly exaggerated. Maximum height is about 15 feet in the English Channel, 25 feet in the Mediterranean, 65 feet in the Atlantic and Pacific, and 100 feet in the South Seas. This last is roughly the height of a seven-storey building, a terrifying size to tower above the mariner caught in a typhoon. Wave speed depends on wind speed; it can reach 100 miles an hour.

The factors producing and controlling wave development are: the velocity of the wind; the duration of the wind; the fetch of the wind, that is, the distance of water over which it is blowing; and, lastly, the depth of the water. The greater the wind's velocity and the longer its duration and fetch, the larger the waves.

When waves overturn on reaching the shallower water or shore-line they are said to break. Surf is the rough water formed on the surface by the breaking waves.

On a beach, the waves travel slowly inshore, the base slowed down more and more by friction while the crest rolls over into a graceful curve. The wave then breaks into a patch of swirling and boiling foam. It has been suggested that once the water has been carried up the beach by the breaking wave, it runs out to sea again, forming underneath the waves a current flowing in the opposite direction. This is known as the undertow, and is generally believed to be a danger to the strongest swimmer. Measurements made by current meters, however, fail to prove the existence of undertow. It would seem that mixing takes place inside the line of breakers. That the water returns seaward is obvious, but the exact manner in which it returns is difficult to determine. It runs back to the sea in complex and erratic fashion, not as a simple sheet along the bottom but in localised, narrow channels, some at depth and others at or near the surface.

Rip-currents are strong currents of water which flow seaward and usually occur in areas of breakers. They may attain speeds of two miles per hour, excavating channels in sand and moving much sediment. Because of their turbid nature they are clearly visible from a low-flying aircraft. It is probable that where they occur, the water has been concentrated into channels by the topography, and the currents have become confused with 'undertow'.

Waves breaking along a shallow, sandy coastline can be a great hindrance to coastal shipping, forcing vessels to unload into small boats which can pass through the surf, or at jetties built out into calm water.

Against cliffs where offshore water is deep, breaking is more violent than on flat beaches. When the waves are suddenly brought to a stop, all the force they have gathered in their coastward movement is directed like a battering-ram against the cliff face. In addition, the air in the rock joints and fissures aid erosion. Abruptly compressed under the impact of the waves, it acts as a wedge, forcing the rocks apart, and is then released explosively as the wave recedes. This produces very efficient mechanical erosion, and in soluble rocks the pressures speed up the solution activity.

Water dashed high up the cliffs carries spray into the wind. Such is the wave force at times that in storms lighthouse doors have been smashed in and windows broken, sometimes to the height of the lantern. It has been found that during winter storms the wave force can reach 6,000 pounds per square foot. A rough sea can move rocks and concrete blocks weighing several thousand tons.

**Longshore currents.** When waves strike the shore obliquely (the normal angle), they produce longshore or littoral currents, also called shore drift, which move great quantities of sediment. These currents will be discussed later in connection with the erosion they accomplish. Turbidity currents, mentioned earlier, are yet another type of irregular water movement with considerable geologic importance.

The spectacular but misnamed 'tidal wave', too, is an irregular movement. These waves, whose correct name is tsunamis (from the Japanese), are caused not by tidal action but by submarine or shore zone earthquakes and volcanic eruptions. The exact nature of their origin is not known because although they are always associated with submarine 'quakes and volcanic eruptions, they are not common to all earthquakes. Many a strong submarine shock has produced no tsunamis. In the Pacific speeds of 450 miles per hour have been recorded for such waves, and they are capable of travelling eight miles inland. Their high velocity causes a tremendous amount of damage along coastal areas.

# WAVE EROSION

Swells, waves tides and longshore currents are the great destroyers of shore-lines. The action varies in intensity over the whole length of the littoral. A steep, deep-water shore-line open to the ocean will suffer greater erosion than an embayed and shelving one, because in the first the force of the waves will be suddenly checked against the land and all their power exerted against the cliffs. The nature of the rock forming the cliffs — its composition, its structural position, and the amount of fracturing present — will favour or retard the sea's action. The amount and kind of material carried by the water also helps to determine the rate of wave erosion.

**Retreat of cliffs.** The coastline on either side of the English Channel presents a line of abrupt chalk cliffs, between St Margaret's Bay, Dover and Folkestone on the English coast, and from Le Havre northward through Fécamp, Dieppe and Le Tréport on the French coast. This is an area with optimum conditions for marine erosion. The waves and tides of the channel are average in height.

Let us analyse what happens on a stormy day. From the top of the cliffs waves roll in to attack the rocks. Driven on by the wind off the sea and by the rising tide, they overtop the shingle banks of the beach and begin to attack the foot of the cliff with earth-shaking blows. The noise of the attack reverberates throughout the hollows and clefts of the cliffs. Under the incessant bombardment of wave-carried rocks and pebbles, the cliffs retreat. The sea water is thrust into the clefts with terrific violence, forcing rocks apart and washing out clay and softer calcareous layers. The more solid rock is loosened and removed in blocks which are hurled along the beach and smashed into fragments. Little can be seen of this demolition work through the swirling spray and dashing waves, but the sea is so full of debris that it is dark and discoloured as far as the horizon.

Once the storm has abated, the work of the sea and of countless earlier storms can be measured. The base of the cliff is pitted with excavations, while higher up fissured walls appear insecurely balanced. Sections of the cliff have collapsed, and blocks of limestone or chalk, some weighing up to 1,000 tons, may litter the beach. Along the whole length of the coast the cliffs are retreating through undercutting, for the erosive action of sea water is similar to that of a river.

By comparing old and new maps of coastal areas it is possible to calculate that the chalk cliffs of the English Channel are retreating at a rate which varies from ten inches to one yard in a year. The speed of retreat is considerable, and yet it is exceeded by the speed of destruction of clay cliffs and sandy deposits in the Channel and on the East Coast of England, even though these softer cliffs are not vertical but sloping and deeply channelled by run-off rainwater. In the United States, the cliffs of Nantucket Island retreat as much as six feet in a year on average. At the other end of the scale are the cliffs of hard rocks, such as the granites, slates and quartzites of Cornwall and Brittany. Over centuries these show little appreciable regression.

Apart from the chemical nature of the rocks, other characteristics, such as stratification and fissuring, affect the rate of erosion. Fissures are the planes of least resistance, weak points enlarged little by little by infiltrating water, which predestine the rock to break down.

Heligoland, an island in the middle of the North Sea, is composed of beds of sandstone which dip from south-west to north-east. On the south-west coast, the waves attack the ends and the undersides of the beds, and are therefore able to demolish them more quickly than on the north-east coast, where the strata offer smooth upper surfaces to the waves.

The chalk cliffs of the Channel offer another aspect of erosion. The numerous layers of flint nodules, lying horizontal, are much harder than chalk. Uncovered by erosion, they fall to the foot of the cliffs and themselves become the weapons of the sea's attack.

The examples quoted help to show the many peculiarities of a coastline and its rocks that must be taken into consideration if the method and rate of its destruction is to be understood.

*Top*
Deeply eroded coast caves in Jersey, Channel Islands.

*Bottom*
A curious pillar on the Kent coast withstands the sea's constant attack.

*Right*
The 1,250 mile long Great Barrier Reef, Queensland, the longest in the world. The close-up photograph is of fish and coral of the Great Barrier Reef. Calcium from the skeleton of the coral polyp forms reefs and islands.

Stacks and natural arches eroded
from the soft sandstone cliff
by the sea at Loch Ard Gorge,
Pt Cambell, Victoria.

*Left*
Crater Lake, National Park, Oregon, was formed by the destruction of the summit of the Mt Mazama volcano. More than

21 square miles in extent and 1,932 feet deep, the lake is fed entirely by the snows from the high encircling cliffs. Wizard Island is part of the original cone.

*Above*
The Seven Sisters, Sussex. Soft chalk cliffs and forceful erosion of the Channel tides produced this coastal scenery

of truncated spurs and hanging valleys. The chalk cliffs of the English Channel are retreating at a rate of ten inches to one yard in a year.

**Shore-Platforms.** However strong waves are, their action is limited to shore regions. It is only between the levels of high and low water that cliffs are cut back. In this way an abrasion platform, bench, wave-cut terrace, or shore-platform, forms over the area at the bottom of the cliffs which was once the base of the destroyed part of the cliffs. Bare at low tides, the shore-platform is the temporary resting place of all the fallen blocks from the cliffs. It is covered with shingle and pebbles and overgrown with seaweed. The pools left by the retreating tide shelter a variety of creatures: fish, crabs, mussels, starfish, and sea-urchins. In places there are deposits of sand or mud in which worms and molluscs have their burrows. Nothing is more fascinating or instructive than a careful study of the marine life in these beach pools.

The abrasion platform forms a gently inclined plane which the waves must cross before reaching the foot of the cliffs. If the platform is extensive, all their strength is used up in friction as they cross it. At this point the cliffs almost cease to retreat, and platform becomes no wider; marine erosion is at a standstill. This state of maturity is comparable to that of a river which has attained its profile of equilibrium.

As with rivers, however, agents of rejuvenation exist which restore erosive power. A rise in the level of the land can lift the shore-platform sufficiently above the sea for renewed attack by the waves. A second platform is then formed at the foot of the first, which becomes a marine terrace or raised beach. On the other hand, a sinking of the coast can bring the shore-platform far enough below the sea for it to hinder wave-action no longer. Erosion and retreat of the cliff then continue with vigour, and the shore-platform grows. Possibly this is the origin of the continental platform fringing the shores of the Atlantic. Even when the land remains stable, rejuvenation along a coast can be accomplished by a pronounced change in sea level, produced for instance when large-scale melting of continental glaciers adds great amounts of water to the ocean.

**The Beach.** In contrast to cliffs, beaches form the level parts of a coastline, whether they occur at the foot of the cliffs and form

part of the shore-platform, or whether they border a low coast entirely without relief. They are built up of cobbles, pebbles, shingles, sand and mud, in decreasing order of size. This material comes mainly from the land, brought down by streams, slope wash, and mass wasting. Lesser amounts are produced by wave erosion, and a little is brought inshore by waves from the sea. The general name given to the material is littoral deposit.

When a cliff is composed of weak rock, this breaks down to form a calcareous or argillaceous mud which is carried out to sea on the tide. The harder materials, on the other hand, form an accumulation of blocks and pebbles which are rolled about by the waves on the shore-platform, and at times are thrown up over one another. This action is repeated time after time at each tide. Little by little this are abraded into shingle (flat pebbles and cobbles) and sand, and are spread out on the lower part of the wave-cut bench.

Compared with sandy beaches, shingle beaches seem more stable. The pebbles are rolled up to the highest part of the beach, again forming a storm-beach reached only during exceptionally high water. The shape of the pebbles is interesting: from their shape can be deduced their origin. The rounded pebbles come from hard rocks — flint, granite or quartzite; the flattened ones are usually formed from broken down laminated rocks such as shales or schists. Round or flat, they are well smoothed off, a characteristic distinguishing them from the angular pebbles produced by stream erosion.

A shore or beach is made up of two parts. Nearest the land is the back-shore area with its storm-beaches, or berms, the highest ridges thrown up by the storms and reached by the sea only when the water is highest. If the back-shore is formed of sand, it is generally dry and subject to being blown by the wind into dunes. Lower down is the foreshore, which is reached regularly by the tide. The sand here is moist and firm, and many minor features can be observed on its surface. At low tide, the sand may be covered with fine parallel ribbing or ripple-marks caused by the motion of the water during the retreat of the tide. Sometimes these ridges are arranged in a network, and their sand may be a

117

*Left*
The 'King' and 'Queen' rocks at Flamborough on the Yorkshire coast. Only three feet of the King's original fifty are left; the rest have succumbed to North Sea breakers

*Right*
A gently inclined abrasion Platform at Broad Haven, Pembrokeshire.

different colour from the rest of the beach, indicating that grains of different composition have been sorted out according to their densities.

On the upper part of a beach, or back-shore, there is often a different arrangement of sediments. Large, very shallow, crescent-shaped hollows appear, open to the sea and separated by ridges running perpendicular to the coast. These beach-cusps are formed by the retreating waves which carry some of the sand out to sea again. In a heavy storm a beach may be swept clean of sand in this way. Sometimes its height is reduced by several feet, and the underlying rock, mud or peat on which the sand formed only a temporary covering is laid bare. This phenomenon is common in winter, but in compensation, as soon as the fine weather returns, the sand reappears from the extension of the beach under the sea.

To minimise the effects of excessive aspects of erosion and sedimentation, groins or jetties are constructed. These range from simple lines of stakes driven into the beach to walls of thick concrete or rock, but all run more or less perpendicular to the shore-line to break the force of the long-shore currents and to offer maximum hindrance to their sand-shifting action. The groins, or breakwaters, are effective until their foundations are weakened by the removal of sand from beneath their base, or until they are overtopped by sand.

## SHORE OR MARINE DEPOSITION

Muds and silts are the scourge of seaside resorts and harbours. In many countries once flourishing seaports are now shallowed

by silt deposits. Many harbours are kept in use only by constant dredging to maintain the required depth.

Mud—a black, sticky, malodorous mud into which boats may sink at low tide—is prevalent in estuaries. A rarer deposit is shell-sand, a fine, iridescent sandy mud composed of innumerable fragments of pulverised shell. It occurs abundantly in the bay of Mont-Saint-Michel and in the Camel estuary in Cornwall. Before the days of artificial fertilizers, shell-sands were dug out and sent inland for the marling of acid granite soils. Beach sands are composed mainly of resistant quartz sand. Occasionally other types occur. The pink beaches of Bermuda are formed of calcite, those of the Bay of Naples of olivine. Sometimes valuable minerals such as native gold, zircon, magnetite (black sand), rutile and cassiterite occur in beach sand in mineable amounts.

The protection of ports against silting-up presents an extremely difficult problem. A port is usually protected by one or more jetties whose lay-out has been very carefully studied, taking into account all the available information on waves and tides, prevailing wind, offshore currents and so on. While these are adequate protections in theory, it may prove that in practice they are insufficient, for no port is ideally situated. Recourse, then, is to some remedy such as dredging, either by suction in the case of mud, or with a bucket-dredger for sand. The material so removed is dumped at sea. In the great estuarine ports of the world the operation is a more or less unending one, and adds greatly to their operational costs.

Setting aside deltas, which were described in an earlier chapter, many other parts of the shore show that if the sea is gaining on

119

the land in certain areas. Sometimes, however, the reverse is true.

We have seen that shingle or sand accumulates on the upper part of a beach. These materials can also be moved along the shore by the currents and redeposited to form spits, which often become recurved shoreward and are then called hooks. A fine series of these is displayed along the North Sea coast of England, particularly in East Anglia, and along the Atlantic coast of the United States, from Sandy Hook, New Jersey, in the north to the Gulf Coast in the south. Some of the beaches of Long Island (the Fire Island beaches are an example) are developed along spits.

Spits or sand bars which grow until they link a neighbouring island with the coast are called tombolos. The Isle of Portland on the Dorset coast of England is an example of this feature. Here, the west-to-east shore current in the English Channel has brought about the formation of the well-known Chesil Bank. For six miles south-east of Bridport the fine-shingled beach runs along the shore-platform like a normal beach, but for the next ten miles it is 20 feet high and separated from the land by a lagoon, called The Fleet. By the time Chesil Bank joins the Isle of Portland it

they extend with minor breaks (tidal inlets) for hundreds of miles.

A river arriving in a lagoon silts it up, and the amount of alluvium accumulating may bring about its total disappearance. The process is often hastened by the intervention of man, and the resulting marshland is dyked and drained to become marsh pasture (still flood-prone) or polderland, protected from the salt water.

The classic area of polderland is Holland, where land lying adjacent to the lower courses of the Rhine, the Meuse and the Scheldt is several yards lower than the level of the sea.

## THE CONTINENTAL SHELF

So far we have confined our attention to the tidal or littoral zone, the marine zone lying between the high and low water marks in which shore-platforms, beaches and spits are found. This is the zone continuously under the influence of the tides and waves. Conditions are rigorous, and if they are to survive, organisms must be well protected and firmly attached to their surroundings, or burrow into the sand. Many find safety in tidal pools, while others

*Top left*
Breakers hurling themselves on to the coarser pebbles of Chesil Bank, Dorset.

*Bottom left*
Jetties, breakwaters and concrete reinforcements help to minimise the effects of coastal erosion and sedimentation.

*Right*
Lagoons formed the mouth of the River Var, France.

is made up of material the size of large cobbles. At this point the height of the beach is 40 feet. Near Toulon, on the Mediterranean coast of France, the peninsula of Giens clearly appears to have been one of the Hyeres islands which has been attached to the coast by two tombolos. The western one is composed of shingle, and the eastern of sand. The lagoon which formerly existed between them has been drained and converted into marsh pasture. The peninsula of Quiberon on the south coast of Brittany has been linked to the mainland in the same way.

Sometimes a spit of shingle or sand may grow across an estuary or bay and cut it off from the sea, forming a baymouth bar. Silt and mud may then be deposited behind the bar, and an enclosed lagoon formed. Numerous lagoons formed in this way occur along the coast of California near Big Sur and Point Reyes.

Offshore and unattached to the shore in shallow coastal waters are barrier bars, offshore bars, or barrier beaches. These long, discontinuous ridges of sandy material are probably built through deposition by breaking waves and are modified by longshore currents. Offshore bars characterise the shallow water areas of the southern Atlantic and Gulf Coasts of the United States where

possess protective structures which permit them to survive exposure to the air at low tide.

Further out to sea, the next marine environment is the neritic zone. This coincides with the continental shelf, the submerged margin of the continent lying between low tide level and depths of about 600 feet. The width of this margin varies. In some areas it may be as much as 750 miles, in others it is very narrow or even non-existent. On the Atlantic coast and parts of the Pacific coast (the China, Java and Arafura Seas) the neritic zone deepens gradually in a gentle slope for scores of miles. The shallower seas appear as part of it; the North Sea, the Channel, the Irish Sea and the Baltic on the European side of the Atlantic, and the Newfoundland Banks on the American side fall within it.

Life abounds in the neritic zone, especially in depths of less than 400 feet where the sun's light penetrates and food is plentiful. This is probably the most heavily populated life zone on Earth, and vast numbers of plant and animal organisms live on substances found in solution in the seawater and also on each other.

Many hypotheses have been put forward to explain the origin of the continental shelf or platform. One theory is that it has

*Above*
The promenade at Rottingdean,
Sussex, acts as a barrier to
coastal erosion.

*Right*
Natural arches hollowed out
by the current along the
North Irish Coast.

been formed by the coalescence of deltas, and is gradually advancing towards the ocean deeps. Another is that it represents the enormous distance the shore-platforms have been cut back. More credibly, it can be visualised as the result of the progressive sinking of the continents, with marine erosion and sedimentation playing only a complementary role.

Originally the continental shelf was pictured as a smooth surface, gently inclined seaward. Actually, it has its hills and valleys, especially in sections near glaciated or mountainous coasts. Of the many theories concerning its origin none takes absolute precedence, and as shelf areas differ from region to region it may well be that they stem from different sources: the broad gentle slope off the eastern coast of United States, for example, through marine erosion and gentle downwarping, and the narrow, deep shelf zone off California through crustal movements along numerous fractures.

The continental shelf covers about 10 million square miles (about 8 per cent of the oceanic area). It is therefore an area of some importance, and is being more explored and more exploited by man. Fishing was the first and oldest industry to turn its attentions in this direction. The petroleum industry is currently interested in following coastal oilfields where they extend beyond the land. Even it the main reason for exploration is economic, the topography and constitution of the continental platform are in themselves important features. Several methods of investigation are used: simple physical sounding, ultra-sonic sounding, seismic exploration, and finally drilling and coring.

The last method, coring, is an old one which has the big advantage of presenting the geologist with an actual sample of the sea-floor. Coring drills are capable of raising cylinders, or cores,

of rock up to 70 feet in length. The value of a core cannot be under-estimated when it is realised that it reveals sediments built up over the last million years or more. These are real records of the ocean brought ashore to be studied by the geologist. From them he has already discovered that in the last few thousand years, for example, beds of volcanic ash have alternated with glacial deposits, clays, sands, and so on. These can be dated fairly precisely and form a key to geological events.

Generally speaking, the farther from the coast the principal sediments of the continental shelf lie, the finer they become, ranging from gravel, coarse sand and fine sand to sandy or calcareous mud, in that order. However, there are many exceptions to this general picture. Where currents are strong, numerous areas of the continental shelf have very thin deposits or none at all. Other areas reveal no tendency to grade sediments from fine to coarse; here sand, gravel and cobbles are found in patches and layers even on the outermost edge of the slope. On continental shelves of glaciated areas coarse material often makes up the bulk of the sediments. Muds (fine materials which, in theory, should lie far offshore) can be found in all regions, especially near bays and stream mouths on the shelf. Where currents are vigorous, limestone deposits are rare, occurring more often in quieter and warmer waters. This last is curious in that the geologic record indicates that most extensive limestone deposits were laid down in ancient seas.

Most of the shelf deposits are derived from the continent, carried to the coast by rivers, and then by ocean currents. Some sediments form from the remains of animals. Coquina, fossiliferous limestone, and coral reef sediments are examples. The presence of great shellbanks in the shelf area gives its waters the name of neritic zone derived from the Greek *neritos*, meaning a mussel.

## THE CONTINENTAL SLOPE

The continental slope forms the juncture between the continental shelf and the floor of the ocean basin. It is a narrow strip about ten to twenty miles wide. Average height is 12,000 feet, but in places it rises to 30,000 feet. Gradient of the continental slope is anything from 1° to 27°, or more. The slope shows more marked cutting by numerous canyons and furrows, and is ribbed with more hills and ridges than the continental shelf. The origin of the slope, too, is uncertain. Probably it is linked with that of the adjacent shelf, and the same theories are put forward to explain its development.

The bathyal zone, the marine water zone in this belt, lies between depths of about 600 and 6,000 feet. The floor supports a prolific animal population but because of the lack of light, plant life is very scarce. Sand, gravel and mud are deposited along with organic debris; sedimentation rates may be faster than on the shelves.

## THE OCEAN DEEPS

Exploration of the bathyal zone and the deeper portions of the ocean is carried out by sounding and sampling, and by use of the bathyscaphe. In 1960 the U.S. Navy bathyscaphe, *Trieste*, descended approximately seven miles to the floor of the Pacific and recorded a depth of 37,800 feet in the Marianas Trench. This was about 4,200 feet deeper than soundings had previously recorded for this point, and the presence of life (fish) and water currents was noted. This is the deepest point recorded for the ocean floor. The next deepest, 36,198 feet, was recorded in 1959, also in the Pacific.

The deepest parts of the oceans known today lie in the narrow trenches, or deeps, bordering the island arcs of the Pacific: Swire Deep (35,433 feet) lying north-east of Mindanao, one of the Philippine Islands; Emden Deep (34,125 feet); Kermadec Deep (35,445 feet) north-east of New Zealand. Next in order are Milwaukee Deep, 30,246 feet, north of Puerto Rico; and south of Cuba, Bartlett Trough, 1,000 miles long and almost 22,000 feet deep.

The oceanic water zone lying below 6,000 feet is the abyssal zone. If we recall that the oceans average 12,000 feet deep, we see at once that their greater part is abyssal: deep, cold and inhabited by a strange specialised fauna. At these depths there is no sunlight, temperature is permanently near freezing-point, and pressures exceed a ton to the square inch. The animals living here are carnivores or scavengers, depending on the animal and plant material that sinks down from sunny waters above.

The pelagic zone is that which includes all the surface waters outward from the littoral zone. Life in this environment includes floating planktonic organisms and many types of free-swimming animals, as well as plants. Algae and diatoms are the commonest of the sea plants. Their hard parts and those of the animals settle down through the waters to the ocean floor, contributing abundantly to the sediments of the continental shelf, slope and ocean basin.

## SEDIMENTS

Sediments deposited in the abyssal zone consist largely of fine materials which have sunk slowly through the deep waters from above. Blue mud is widespread on continental slopes. On one side it is found with the mud of the continental shelf, and on the other grading into abyssal oozes and red clays. It is a clayey mud which owes its colour to iron sulphide, and when stirred gives off a strong smell of hydrogen sulphide — the product of bacterial decomposition. Seaward of the great rivers in the tropics, red mud is deposited following erosion of laterite soils which are rich in iron oxides. Elsewhere, as off the steep coasts of Peru and Chile, the blue mud may be replaced by mud coloured green with the silicates of iron. In some places a calcareous white mud is found, the product of coral reef breakdown.

The dominant sediments of the abyssal zone are organic muds or the accumulation of carapaces, shells and skeletons which once belonged to the creatures of the pelagic zone. Two calcareous oozes are common: globigerina ooze in which limey shells of the single-celled *Globigerina* are abundant, and pteropod ooze in which shells of minute molluscs form the major part. These two cover large areas of the Atlantic, Pacific and Indian ocean floors.

Two other sediments are found in the abyssal zone: radiolarian ooze and diatomaceous ooze, both siliceous. The former is made up of the lovely, fragile, complex shells of minute marine protozoans (Order Radiolaria), while the latter consists of the siliceous shells of minute one-celled marine algae, called diatoms. Radiolarian oozes occur in the Pacific in an east-west belt just north of the Equator, and diatomaceous oozes in the north Pacific and Antarctic oceans.

Also in the abyssal zone are terrigenous (land-derived) sediments: wind-borne volcanic ash, silts and sands rafted by icebergs, and muds wasted down from the continents. Much of the ocean bottom, especially in the Pacific, is covered with a thin layer of very fine red clay thought to be produced by chemical changes in the muds, particularly by dissolution of their lime content. This red clay of the great deeps also contains volcanic ash, insoluble parts of plankton shells, and even the dust of meteorites and other bodies which have come from space, disintegrated into dust in the atmosphere, and settled into the ocean. The only large fossils found are sharks' teeth and the ear-bones of whales. The quantity of red clay in the Pacific is without doubt due to this ocean's greater depth and its greater power to dissolve completely any organic material normally present.

Sedimentation rates are very slow in the abyssal zone, except perhaps in deep basins close to the continents. Estimates of the deposition rate of red clays is about one centimetre every 1,200 years, and for some of the organic oozes one centimetre in 265 to 1,700 years.

We cannot leave the oceans without explaining why we have dealt only with erosion of coastline, and why it is stated that marine sediments are derived solely from the continents and from living organisms in the sea. The reason is simple. Submarine surfaces are free from erosion. Marine erosion takes place only in the zone of wave action, whose influence is strong only to a depth of a few yards. Maximum range for the highest storm waves is 600 feet, and even then motion at these depths is very slight. While a subaerial mountain is steadily destroyed by erosion, a submarine mountain remains almost untouched for indefinite periods of time. Weathering and mass wasting, which rapidly subdue continental features, are ineffectual here, or at any rate work very slowly. The layers of sand and bedding graded from fine to coarse are thought to be products of local turbidity currents set in motion by submarine landslides or volcanic eruptions.

## CORAL REEFS

In tropical seas, we find the well-known coral reefs. The term is a misnomer for a feature that is not produced solely by the reef-building activity of corals. Many other organisms contribute to its construction and commonly outnumber the corals. Algae are the major reef-builders, while snails, echinoderms, worms, foraminifers, molluscs and others may play a part. Bioherm, meaning 'life reef', is a better name for these structures.

Coral reefs are classified according to their relationship to the mainland: fringing reefs, which border the shore; barrier reefs, which lie some distance offshore, parallel to the coast; atolls, which are circular reefs enclosing a lagoon of water; and table reefs, which are roughly circular in form and lie just below sea level, without a depressed area in the centre.

Reefs normally form initially as fringing reefs. The organisms which construct them grow most vigorously in clearer waters where there is little or no sedimentation, and where agitation ensures a supply of food and oxygen. These conditions prevail on the outer or seaward side, and so the reef grows outward and away from the shore. Deposition from the shore reduces the organic population on the inner side of the reef. Gradually, the fringing reef becomes a barrier reef.

Reefs form only in shallow water, principally because the corals live symbiotically with various types of calcareous algae which need light for growth. Because light is absent from deeper waters, the algae and corals are rarely found in depths greater than 150 to 200 feet. Corals thrive best in warmer waters, in temperature ranges of 77°F. to 86°F. (25°C. to 30°C.), although some kinds can tolerate temperatures as low as 67°F. (19·4°C.). A normal salinity of 2·7 per cent is best suited to their existence, with the result that they do not develop near large river mouths where salinity is lower and sedimentation higher.

The reef dwellers are mainly sedentary organisms or directly attached to the ocean floor or to each other. They are therefore largely dependent on the waves for their food supply and for part of their oxygen supply. They prefer a firm foundation and do not thrive on unstable or muddy bottoms.

Because of these restrictions which control their environmental zone, it has been difficult to explain the origin of the many circular coral reefs whose bases extend into very deep waters. It was once thought that they might be formed round the rims of submarine volcanoes. This is true of some atolls, but not of all. Charles Darwin, author of *The Origin of Species,* conceived the idea of an island being surrounded by a fringing reef or barrier reef, and then subsiding very slowly, leaving the coral reef to form an atoll above the sea. This conception is illustrated by the series of diagrams on the right. Here, the island (black) disappears progressively under the waves while the reef (white) extends its area and finally covers the island completely. The first stage of the island with its fringing girdle of coral has its parallels in Tahiti, Wallis and Kusaie; semi-atolls can be seen at Truk and Gambier; the perfect atoll of the third drawing appears at Kwajalein, Eniwetok and Ulithi. Midway Island is an example of the fourth type, and the last is represented by islands composed entirely of coral, such as Nauru and Marcus.

There are many other theories of barrier reef and atoll formation, yet Darwin's is still the most favoured theory today.

**Geological information from atolls.** Darwin's theory implies progressive subsidence of the sea-floor, yet in many areas there has been stability and sometimes considerable uplift. The probability is that coral reefs have originated in a different way in different parts of the Pacific Ocean. Some unquestionably originate around submarine volcanoes and volcanic islands, others through normal reef growth on stable platforms, but most of them seem to have formed in subsiding areas, and new geophysical tests and borings appear to yield more and more data in favour of Darwin's subsidence theory. The best proof of subsidence is furnished by borings made in various atolls. At Bikini, for example, a boring which reached about 2,750 feet passed through nothing but reef coral. At Eniwetok, borings reached nearly 5,000 feet without encountering anything but the same coral limestone. Because coral reefs are built so slowly, such a thickness of dead coral indicates that the reef is very old. In fact, the base of the Eniwetok reef began to form about 100 million years ago, and the process has continued since. Many other examples of sinking reefs exist, even outside the Pacific. The coral island of Abulat in the Red Sea is one.

In contrast to Eniwetok, certain of the mountainous islands of Indonesia bear reef limestone at heights up to several hundred feet: 1,200 feet on the isle of Rotti, 4,000 feet on Timor, and so on. These raised reefs date only from the Quaternary and are not more than one million years old. Recent mountain-building movements in that part of the Pacific have brought about this comparatively rapid elevation.

**Destruction of reefs.** Many of the plants and animals inhabiting reefs contribute to their destruction. Bacteria, fungi, seaweeds, sponges, sea-urchins, molluscs, and annelid worms eat their way into the reef from all directions and weaken its resistance to the attack of waves which cut into the outside edge of the reef to make a shore-platform, uncovered at low tide and backed by a cliff. This part of the reef is dead and covered in broken fragments of coral. Percolating water cements the fragments together to form coral brecchia.

Reef approaches are marked by stretches of white sea water – 'coral milk'–produced by the suspension of fine particles of calcium carbonate. Deposited in calm water, the particles form

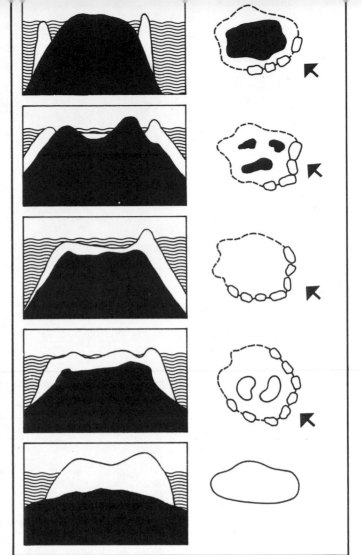

Formation of atolls, after
Charles Darwin.

coral muds. In more disturbed waters they may accumulate successive concentric layers of lime as they are rolled about, to become oolitic limestones.

Observations made in the Red Sea during the expedition of the *Calypso* (1951–52) revealed a curious method of coral formation. Parrot-fishes, one of the brightly coloured groups of fish inhabiting coral seas, have projecting jaws. They bite off parts of the coral, digest the organic parts, and excrete the remainder as fine calcareous powder, which scatters and eventually settles on the sea-floor.

The complex of deposits found in association with a reef zone is divided into three intergrading and interfingering groups: the back reef, the reef itself, and the forereef. The landward, or lagoonal, side is the back reef zone and includes interbedded limestone, dolomite, sandstone, red shale, and evaporites, generally anhydrite. These interfinger with the massive, porous dolomites and limestones of the reef itself. The reef rocks in turn grade seaward into the forereef rocks consisting of sand and reef debris, often very coarse in texture. Still further oceanward these slowly give way to the shales and limestones normal to ocean deposition.

**Ancient reefs.** Reefs not unlike those of our own day have been found in many parts of the world and in all geologic periods, from the Pre-Cambrian to the Recent. These ancient or fossil reefs were greatly developed in some periods (the Silurian, Devonian, Permian and Triassic) and poorly developed in others. In places they are uncovered in rocks, just as they grew originally; elsewhere they appear as mixtures of interbedded original and detrital organic debris and sediments. Fossil reefs vary in size from formations only a few feet thick and a few feet square to considerable structures hundreds of feet thick and several hundred miles long.

A reef very like the modern barrier reefs is Capitan Reef in the Permian sediments of west Texas and New Mexico. This

# THE PRESENT

*Top*
Great Barrier Reef, Australia.
Pool with flattened sheets of
living coral on the summit
of the reef.

*Centre*
Stagshorn coral exposed on
the surface of Pixie Reef,
Australia.

*Bottom*
Coral *(Madrepora abrotamoides)*
from Singapore.

*Right*
The Capitol Reef, Utah.

giant of all ancient reefs is over 1,200 feet thick and more than 400 miles long. Other extensive reef deposits occur in Silurian rocks in the Great Lakes Region in United States, in England as 'ballstones', in Gotland, and in Estonia. Large flat-bottomed, knob-like reefs occur in Carboniferous rocks in the Craven area of northern England. Fossil reefs such as the Capitan have to date yielded a substantial portion of the world's petroleum supply, their high porosity offering ideal reservoir conditions for oil accumulation.

## LAKES AND SWAMPS

Lakes and inland seas are expanses of water without access to the sea, or whose only access is by means of rivers. (The Rhône serves Lake Geneva in this way.) Lakes are agents or erosion, very similar in activity to oceans but on a much smaller scale. Waves develop in them, controlled by the factors that govern ocean waves. Longshore currents form, erode and deposit, just as they do in the oceans. Tides, too, form in lakes but are too slight to accomplish any significant erosion except gradational work. Beaches, bars, spits, wave-cut cliffs, stacks, and so on are all to be found along lake shores.

Ice ramparts come closest to being an exclusively lacustrine feature although these, too, appear occasionally along rivers and partially land-locked bays. The 'ramparts' are low ridges of rock debris that occur around lakes frozen in winter. As the water freezes, it expands and pushes debris along its margins on to the shore. Acting in the way of other solids, ice expands and contracts with variations in temperature. When a thick sheet of ice covering a lake expands the overall effect is to exert a very strong horizontal push in all directions. In this way the ice margin moves further up the shore, pushing debris still higher. Repeated movements of this kind during a winter may build up high ridges of loose rock material encircling the lake.

Among the world's largest lakes are the Caspian Sea (169,000 square miles), Lake Victoria (26,826 square miles), Lake Superior (31,820 square miles), the Sea of Aral (27,700 square miles), Lake Huron (22,322 square miles), Lake Michigan (21,729 square miles), Lake Tanganyika (12,586 square miles), and Lake Baikal (12,740 square miles). The largest lake in western Europe is Lake Geneva (223 square miles).

There are many types of lake, classified according to the origin of the depression in which they occur. The principal types are tectonic, glacial, volcanic, and sink-hole.

**Tectonic lakes.** These lakes have become established in natural catchment basins produced by differential movements of the Earth's crust. Lakes Albert, Malawi and Tanganyika in East Africa, and Lake Baikal in Siberia are good examples. Their elongated form, steep sides and great depth reveal at once that they are

located in valleys produced by the downward movement of a segment of the Earth's crust between fractures. Lake Baikal, which is situated at an altitude of 1,500 feet, has a depth of 4,500 feet. The bottom of the lake is thus 3,000 feet below sea level. The Dead Sea and Lake Tiberias in Israel lie in similar areas of structural depression. These and others, including Lake Leopold and the Sea of Galilee, form a long chain of lakes whose location is governed by the presence of a major rift, a series of fractures which extends from Zambia to the Holy Land.

**Glacial lakes.** Some lakes are entirely glacial in origin. In Russia, Finland, and Canada innumerable moraines have produced whole networks of winding lakes. In Russia and North America, morainic lakes, such as Lake Onega and Lakes Superior, Huron, Erie and Ontario, have enormous dimensions. The Great Lakes of North America owe their origin to the depositional activity of the glaciers, and to glacial erosion and modifications of the region's earlier drainage system by the great continental ice sheet. Jenny Lake at the foot of the Teton Mountains in Wyoming is the product of moraine deposits; the Finger Lakes of New York were produced by glacier erosion and deposition. At the opposite end of the scale, there are many moraine-dammed lakes no larger than ponds, many of the 'tarns' of the English Lake, including District.

An interesting feature results when a valley-glacier melts in stages and deposits several morainic dams at progressively higher altitudes. Not one, but a series of lakes forms. An example can be seen in the ancient glacier valley leading from the Col de la Schucht in the Vosges. Three lakes occur at intervals: the Gerardmer at 1,980 feet, the Longemer at 2,200 feet, and the Retournemer at 2,340 feet.

In the deep semi-circular heads of glaciated valleys tarn lakes frequently occur. These small lakes are common in the Sierra Nevada and Rocky Mountains.

**Volcanic lakes.** A lava flow can dam a valley and form a lake which will have the same shape as the valley itself. The lakes of Chambon, Aydat and Guery in France, and Lake Chelan in the United States have been formed by valley dams of Tertiary lavas. Lakes Bunyonyi and Mutanda in Uganda, and Lake Kivu in East Africa are of similar origin.

Also of volcanic origin are crater lakes, sited in the craters of extinct volcanoes or in cavities caused by subsidence after underground volcanic activity. They are circular and steep-sided. A well-known example is Crater Lake, Oregon, which fills a crater about 6 miles wide and 2,000 feet deep, and is surrounded by cliffs 500 to 2,000 feet high. Wizard Island, a small round island rising out of the lake, is an extinct nested or inner volcanic cone.

The famous 'maare' of the Eifel region of Germany are crater lakes of recently extinct volcanoes. The largest is the Laacher See. Similar lakes are found in the Black Forest, in the Ruwenzori volcanic district of Africa, in several states in the United States, and in the mountains of Chile.

**River-formed lakes.** Many lakes of origins less spectacular than those already mentioned have been formed by simple erosion of the ground by running water. All that is necessary is for a river bedded on an outcrop of soft rock to pass to an outcrop of harder rock, and erosion of the soft rock will outpace that of the hard rock to form a basin where water accumulates behind a bar of hard rock. Such an erosion lake is naturally short-lived, for the rock bar will ultimately be completely worn away by the river.

Rivers form lakes in other ways. Ox-bow or crescent-shaped lakes develop when a river cuts off its bends or meanders. Irregular and erratic meandering deposits made by the river on its delta often create lakes. Lake Pontchartrain on the Mississippi delta, and Zuider Zee on the Rhine delta, have been formed in this way. At the north end of the Gulf of Lower California, the Salton Sea, a huge lake, owes its existence to deposits by the Colorado River cutting off and separating the terminal portion from the main part of the bay. In limestone areas, ponds and lakes form where sink-holes and solution valleys produced by

ground water activity become blocked by rock debris. Most of the lakes in Florida and the Yucatan Peninsula are of this type. Sandbars may form across mouths of bays and estuaries to produce lagoons (shallow lakes occurring near the sea). Thaw lakes appear seasonally in areas of perma-frost when a patch of the frozen soil thaws out in warm weather. Such lakes are commonly found all over the coastal plain areas of north Alaska.

Playa lakes are temporary lakes which form after rains in dry regions. They are often saline and, for obvious reasons, are called salinas. Occasionally playas occur in wind-blown depressions. They are common after rainfall in the semi-arid Great Basin and Range country of Nevada, California, and Sonora. In Black Rock Desert, Nevada, a playa lake seldom more than a few inches deep but covering an area of 450 to 500 square miles appears and remains for part of the winter. After heavy storms it forms a great sheet of liquid mud which later evaporates to cover a broad area with a network of mudcracks.

Meteor craters are a rare phenomenon; few of them contain lakes. Canada's Chubb Lake is one of the few.

Beavers, too, form lakes by damming small streams. In the United States they are now under government protection, and

*Top left*
Wizard Island, rising out of Crater Lake, is an extinct inner volcanic cone.

*Bottom left*
Aerial view of the Everglades Swamp in Florida

*Right*
The Hoover Dam across the Colorado River. Its construction has increased sedimentation in Lake Mead.

are actually 'planted' in certain areas to build lake-forming dams as a conservation measure.

Man constructs artificial lakes by damming for diverse purposes: flood control, migration, water supply, power source. One of the world's largest artificial lakes is Lake Mead in Arizona and Nevada. Produced by damming the Colorado River, it holds 32·47 million square feet of water behind America's largest dam, 726 feet high.

Ponds are of many types and many origins, but essentially they are very small lakes, formed in the same manner or simply wherever the ground surface is irregular and the soil impermeable. Even in areas of sandy soil, percolating waters may deposit an insoluble layer of iron-pan above which ponds may form. Many of the ponds in areas of sandy-soiled woodland and heathland are formed in this way. Those of Surrey in England, and of Sologne in France are particularly good examples.

After the building of the Hoover Dam across the Colorado River, careful studies were made of the sediments brought into Lake Mead. The river built, and is still building, a huge delta at its point of entry into the lake. Eventually the deposit will fill the lake completely, displacing all the water. This will take about 400 years if deposition continues at its present rate. This process affects most lakes, for no matter what their origin, both natural and artificial lakes eventually disappear. Some vanish when their basin rim is breached by erosion and the water flows out. Most, however, are lost by sedimentary filling. Streams, rainwash, weathering and mass wasting constantly add sediment, and in time the lake silts up and becomes a swamp. The normal neighbour of most lakes and their future end, swamps are a feature mid-way between lake and dry land. In areas of strongly seasonal climate, the same basin may range through all three stages each year – lake, swamp and dry land.

Swamps occur, primarily, in three areas: coastal plain regions (e.g. Pontine Marshes); regions of glacial deposition (e.g. Thousand Lakes of Minnesota); and river flood-plain and delta areas (e.g. along the Mississippi River). In moist, temperate regions they are areas of peat-bogs.

There is often much valuable agricultural land to be gained by draining swamps whose high humus content yields richly fertile soils. To this end, the Fen district of East Anglia, the Florida Everglades, the Dismal Swamp of Virginia, and the Pontine Marshes have been extensively drained off through canals leading to the sea.

# Glaciers and Glaciation

Of all areas of the world, mountains are subject to the highest annual snowfall. To appreciate this we must first understand how the amount falling in any given place is calculated, for it is not as simply measured as rainfall. As snow usually drifts to some extent, measuring the depth each day does not give an accurate calculation of the actual fall. The best method is to measure with a nivometer the weight of the fall or the amount of water it yields on melting. The nivometer is a metal container, narrowed at base and top, open above for the collection of the snow and fitted with a tap at the bottom for emptying meltwater. A cylindrical screen around the upper part of the instrument reduces air movement above it. Calcium chloride inside hastens the melting; paraffin oil covers the meltwater to prevent evaporation. As the nivometer also catches any rain which may fall, the total precipitation registered on a rain gauge must be subtracted.

Although there are places at sea level on the west coast of Europe which have an average of only five days snowfall a year, the highlands of central France and western Germany have about fifty days, Canada about a hundred, and the high and exposed summits of the Alps at least 150 days. In Alpine areas of over 3,000 feet, snow generally falls from December to February, and in areas above 5,500 feet from November to April, with an intermission in January.

The highest recorded snowfall is 90 feet. This fell during the winter of 1906–1907 at Tamarack, California, at an altitude of nearly 8,000 feet. The largest average snowfall in the Alps is at 4,300 feet, in the bottom of the Chamonix valley. On the summit of Mont Blanc, where direct recordings are impossible, the snowfall may be five times that recorded at Tour. On high mountain summits, snow seldom falls in the form of flakes; below 32°F. (0°C.) it usually forms as sleet, or as powdery snow composed of crystal fragments which adhere neither to each other nor to the ground.

The snowfall at any given place depends not only on the temperature (determined by the altitude and latitude) but also on atmospheric humidity. It is obvious that even the most intense cold cannot produce snow when water vapour is absent from the air. A wind carrying moisture encounters lower temperatures as it passes up and over a mountain, and is cooled. Cold air being incapable of holding as much moisture as warm air, precipitation in the form of snow or rain occurs frequently on the windward side of mountain ranges. The leeward side receives little or no precipitation, as the wind has already lost its moisture. La Charmette, for example, on the mountain-side of Grand-Chartreuse, receives three times as much snow as Villard-de-Lans, at the same altitude but in a more sheltered position. Again, the Maurienne Alps of Savoy have more snow than the Tarentaise, because the latter are exposed to the lombarde, a spring wind which blows off the Lombardy Plains of Italy. The Pyrenees usually have less snow than the Alps, as their east-west trend does not force the

moist westerly oceanic winds to rise and bring precipitation. In the Himalayas, humid summer monsoons bring more snow to the southern slopes than to the northern slopes. North America generally has more snow than Eurasia, as the latter has a greater longitudinal extent and a less maritime climate.

The term 'permanent snow' is often heard in mountain areas. What is meant by this? No mass of snow lasts for ever; it ends by melting and evaporating, or by being turned into ice and flowing down the mountains as glaciers. Only the snow-cap of a mountain is relatively permanent, perpetually renewed by fresh snow as fast as the old snow melts. For 'permanent snow' to be established, the replacement snow must keep pace with melting snow. The snowline marks the lower limit of permanent snow. Above this the snow falling during the winter equals or exceeds that melted during the summer. The position of the snowline depends also on temperature and atmospheric humidity. In the Alps it lies at approximately 8,550 feet, and in the Pyrenees at 9,870 feet. On the southern flank of the Himalayas it is 16,000 feet, but on the northern flank is not found below 18,000 feet. In the Arctic and Antarctic regions the snowline is at or very near sea level.

If Mont Blanc, half covered by permanent snow at its present latitude, were in equatorial Africa, it would be entirely free of snow; in Spitzbergen or Greenland it would be covered in snow from summit to base all the year.

**Avalanches.** These are the most catastrophic products of snow at high altitudes, especially in late winter and early spring. They are of two kinds: flows of fresh and powdery snow which, having fallen heavily for some days, has not adhered closely to the ground, and starts to slip; movements of a dislodged mass of older, consolidated snow which breaks up as it heads for the bottom of the valley. Size, their unpredictable direction, their speed (anything up to 200 miles an hour), and above all the shock-wave which precedes them make the first type the more dangerous of the two. The shock-wave alone can demolish houses some distance away like the blast from a bomb.

An avalanche, or snowslide as it is called in the American Rockies, can flow for several miles, and size may be considerable. The avalanche at La Blaitière in 1914, in the Mont Blanc area, reached the enormous proportions of 33 million cubic feet.

If the snow is already in an unstable state, an avalanche can be set in motion by the vibration of sound-waves from the slightest noise passing through the ground or the air. A passing train, aircraft overhead, a church bell tolling, gunfire, a single shout, a whistle blowing, or the noise of a football on a slab of rock may be enough to set an avalanche on the move.

The Austrian army learned this to its sorrow during the winter of 1916–17, when its commanding officers were unwise enough to station outposts on the most exposed hillsides. One burst of gunfire was sufficient to loosen the snow on the slopes, and on the 13th of December 1916, thousands of lives were lost in the avalanche.

Avalanches are without doubt one of the worst scourges of mountain regions, and man's defence consists in protection and

The peak of the Matterhorn
is a perfect example of
glacial erosion.

*Top*
The terminal face of the Glacier des Bossons in the Alps, showing glacier convergence.

*Bottom left*
Size, speed and unpredictable direction make an avalanche extremely dangerous.

*Bottom right*
Close-up of snow brought down from the higher slopes in an avalanche.

precaution, since prevention is not at present within his power.

Protection of inhabited areas is afforded by walls, hurdles, or fences set across the probable paths of avalanches to break the force of the descending snow. To protect rail routes, snow-sheds are built with chutes to carry the avalanches safely over the lines to the slopes on the far side.

In Switzerland a Federal Institution for Avalanche Research has been founded. It has a centre on the slopes of the Weissfluhjoch, above Davos. Connected by telephone, radio and teletype with thirty or more observation posts throughout Switzerland and neighbouring countries, its object is the constant study of local conditions: temperature, humidity, snowfall, wind, and sunshine. The forecasts of the Institution have been correct, except under certain exceptional meteorological conditions, and it can generally forecast the dangers which threaten each winter.

## FORMATION OF GLACIER ICE

Whether it falls in flakes or as powder, an accumulation of snow gradually turns into ice. It is principally in the region of permanent snow on the mountain-tops that this phenomenon reaches its fullest development, and becomes the special study of the glaciologist.

Fresh snow is light because it contains a large amount of air, the average density being 0·08. We know that it is made up of hexagonal crystals formed by a skeletal network of long needles of ice. As the snow drifts, the weight causes the crystals to break and melt slightly. They refreeze at night, together with their own meltwater. The snow thus becomes hard and granular, as it

does when a snowball is compressed in the hand. It still contains some bubbles of air, and its density is only six or seven times greater than that of fresh snow (about 0·5). Skiers call it crusted snow, because of its smooth surface. Its formation is hampered by the wind, which evaporates the meltwater as fast as it forms and in consequence reduces the amount of refreezing and welding occurring deeper down in the mass. This reduction in the refreezing of meltwater is the principal cause of the avalanches of snow-slabs which follow several days of the warm, dry foehn and chinook winds.

A rise in temperature, too, especially in the spring, can melt the hardened snow into a mass of separate granular crystals.

Now let us look at a mountain-top hollow in which snow has accumulated year after year. The three reactions just described – consolidation by pressure, melting near the surface, and refreezing lower in the snow mass – all come into play on a larger scale. The grains of ice finally reach an average diameter of nearly two inches and are cemented together by a thin layer of ice. Air bubbles are first reduced in number and then finally disappear altogether. This increases the transparency and density of what is no longer hardened snow but compact ice with a density of 0·9.

The structure of the ice cannot be detected by the naked eye and is apparent only when melting begins and the grains separate as the 'ice-cement' becomes water again. This meltwater is not as pure as that of the grains themselves. Chemical analysis of the latter reveals traces of sodium chloride, the salt found in small quantities in clouds. The ice-granules are pure because they are, as it were, pickled in brine.

Ice floats after freezing at normal temperatures and pressures,

Strong steel fences built to
protect this Swiss railway
line from the threat of
avalanches and rock falls.

Protective galleries against
avalanches in the Spiessgraben
of the Lonza ravine, Switzerland.

as the water then expands and enjoys a lower density. This is the form in which we normally encounter ice, and is called 'Ice I'. That forming at successively lower depths under correspondingly higher pressures and lower temperatures has different characteristics. Numbered from II to VII, the different types have increasing densities, higher melting-point and differing crystal structures. Their densities range from 1·06 to 1·6; melting points are possibly as high as 350°F (176·6°C.) They are difficult to study, and not much is yet known about them, though it is probable that they form a considerable portion of the polar ice-caps.

As snow accumulates in catchment areas it slowly grows in thickness and gradually changes into a granular mass, called névé or firn, and then into ice which shows a layered structure reflecting the individual snowfall seasons. Under the influence of gravity this ice starts to move when it reaches the critical thickness for its slope and climate. The moving ice is known as glacier ice, and the whole moving mass as a glacier.

## CLASSIFICATION OF GLACIERS

There are three major and many minor types of glacier. The major ones are: ice-cap or ice sheet glaciers (called continental glaciers when they cover unusually extensive areas) whose courses are not restricted by topography and whose movement is controlled mainly by the thickness of the ice; alpine or valley glaciers, which are confined within valley walls, where their movement is governed by the shape and slope of the valley; and lastly, piedmont glaciers, which are transitional between the other two, forming as a sheet of ice where two or more alpine glaciers coalesce at the foot of mountain slopes.

## ICE-CAPS AND CONTINENTAL GLACIERS

As the snowline falls progressively as we move from equatorial latitudes to the Poles, it is not surprising that the Arctic and Antarctic regions, when sufficiently exposed to wet winds, are almost completely and perpetually covered with snow and ice. Their surface becomes one immense ice sheet, ice-cap, or inlandsis (a Scandinavian word meaning 'inland ice'). Although thickness may range from several hundred to several thousand feet, gradient is hardly noticeable. With such great depths few mountains project above the surface, and there is very little morainic material. Only the highest peaks, called nunataks (an Eskimo name) protrude here and there from the dazzling white plain of ice.

The two largest ice-caps, those of Antarctica and Greenland, cover about 4,860,000 and 670,000 square miles respectively, with maximum altitudes of 14,000 feet and 10,000 feet. The thickness of the ice has an average of 5,000 feet, with a maximum around 8,000 feet. In Greenland, the curious fact may be observed that at certain points thickness exceeds altitude; that is, the bottom of

the ice-cap and the rocks on which it rests lie well below sea-level. To conclude from this that Greenland is not an island but an archipelago covered with ice would be stretching a point. It seems more likely that it is an island deeply scored with gulfs, and that if, somehow, the ice were to disappear, it would rise slowly to resume its island form.

Throughout Earth's history glaciers have at different times covered extensive areas of the continents. The most recent of these glaciations occurred during the Pleistocene era. At their maximum extent during this period the ice-caps spread over 15 million square miles of the Earth's surface–about 28 per cent of the present land area–covering most of northern North America, extending southward to the Great Lakes in the east, and to the United States-Canadian border in the west. The western mountains were capped by ice and carried numerous alpine glaciers. Similarly, in Eurasia ice-caps covered the British Isles, Scandinavia, and north European Russia, and extend southward into Germany and Poland. The major mountain areas of Eurasia were also ice-capped. Pack-ice existed around the margins of these continental ice-covered regions and probably completely covered the north polar region between Eurasia and North America. The north-eastern part of Asia was largely free of ice because, although temperatures were low, there was not enough snowfall to support extensive ice sheets.

The ancient ice-caps were probably very similar to those of present-day Greenland and Antarctica but on a much grander scale. The thickness of the ice must have ranged up to 10,000 feet at the centres of accumulation, though reliable estimates are difficult to make. Erosional and depositional features were similar to those of present-day ice-caps. In addition, the continental ice sheets indirectly produced many effects: the sea level was lowered as vast amounts of water were taken in the development of the sheets; there were climatic changes of humidity and wind as well as of temperature; later, the sea level was raised again when the ice sheets melted.

The continental glaciers of the Pleistocene advanced, retreated and advanced again. At least four, and probably five, distinct glaciations occurred. They were separated by long interglacial ages. It is still not certain whether the glacial epoch is finally over, or whether we are simply living in another interglacial age.

At the present time it is variously indicated that the rhythm of the world's climates seems to be leading into a long, warm period. A rise in temperatures has been deduced from observations in Russia, Scandanavia, Spitzbergen and Greenland. The Russian expedition which, in 1938–39, followed Nansen's route across Greenland recorded temperatures averaging 10°F. higher than those noted by the Norwegian explorer forty years earlier. In Spitzbergen, the increase in temperature is about 15°F. since the beginning of this century. The polar glaciers are retreating, and the floating ice is withdrawing. In the Barents Sea, the pack-ice has retreated

65 miles over the last twenty years, and the Ross Barrier in the Antarctic has retreated 25 miles in half a century. The biological consequences of this amelioration in temperature are no less obvious. Since 1924 the west coast of Greenland has acquired herring and haddock, species of fish formerly unknown there. Fish which were once encountered only accidentally in Iceland, such as mackerel and tunny, have been caught regularly since 1928. Other species, such as capelin and cod, which used to spawn on the south coast of Iceland, have moved to spawning grounds on the north coast.

## FLOATING ICE

**Icebergs.** Tongue glaciers, often flowing down to the sea, are found on the edges of most ice-cap glaciers. Under the tremendous pressure of the ice behind them their speed may be as much as 100 feet a day, and several miles a year. When they reach the sea, the ice breaks off in gigantic floating blocks, or icebergs, which are carried along by marine currents. Average size is 1,000 to 1,500 feet across, and about 100 feet above surface. Many larger ones have been encountered, particularly tabular icebergs, which measure from 30 to 60 miles long, 5 to 20 miles wide, and 100 to 350 feet above the surface of the water. In 1840, the explorer Dumont d'Urville believed he had discovered a new land, which he called Claire Island, but he was simply sailing along the edge of an enormous block of ice. Today, certain of these larger icebergs are used as temporary air bases and meteorological stations.

The overall size of an iceberg appearing 100 feet above the sea is much greater than might at first be supposed. This can be explained by a mathematical formula. Let $X$ be the width, and $Y$ the total height of the iceberg which, for the sake of simplicity, shall be cylindrical in form. The density of ice is 0·9; that of sea water 1·03. Applying the principle of Archimedes, the following formula can be produced:

$$XY \times 0.9 = X(Y - 100) \times 1.03$$

From this can be calculated an approximate value of 800 feet for $Y$. An iceberg emerging 100 feet from the sea has, therefore, a total height of 800 feet. The unseen 700 feet of its submerged mass constitutes a serious hazard to ships.

Even the smaller icebergs can create unforeseen dangers when carried by currents into the Atlantic shipping-routes. One of the most terrible maritime disasters ever known was the sinking of the liner *Titanic* after a night collision with an iceberg at 41° 46' north, almost the same latitude as Madrid.

Now, however, as the warming-up of Earth's climates proceeds, it is rare for the Greenland icebergs to reach such low latitudes before being broken up by the winds and waves, and melting completely. Their more normal course is to pile up on the coasts of Labrador during their first summer, and then to drift down and melt above Newfoundland Banks during the second. The icebergs of the Antarctic are much larger, and may get as far as the 35th parallel before breaking up. Their debris of ice-blocks is sometimes seen even in the approaches of intertropical seas.

**Pack-ice.** While icebergs are the products of terminal disintegration, or calving, of the polar glaciers, pack-ice is the product of sea water. Icebergs are formed from fresh water; compression of the snow gives it a granular structure when frozen. Pack-ice is formed, theoretically at least, of salt water which has frozen as prismatic crystals, and the first stage in its formation is the appearance of slush, a floating mass of ice crystals on the surface of the sea. Little by little, the crystals increase in size, enclosing salt as they grow. Their ultimate agglomeration forms plates of ice with raised edges. The cold persists and becomes more intense; the pancakes join together to form large slabs of ice, or floes, which may ride up over one another and form hummocks, or may remain separated by channels of open water. The hummocks form sharp, rugged-sided hills of ice which constitute a great danger to aircraft landing on the ice-packs. Any sudden rise in temperature will, of course, lead to the disintegration of these ice-fields. This is demonstrated annually at the break-up of the ice in the spring.

Arctic pack-ice is worthy of particular consideration, as it occupies almost the whole of the space between the North Pole and the North American and Eurasian coasts. Its diameter varies between 1,900 and 2,500 miles. In February it reaches as far as the 60th parallel, with a point advancing to 50° along the coast of Labrador, and an indentation up to 75° in the Barents Sea to the north of Scandinavia. Both anomalies result from the respective influences of the cold Labrador Current, and the warm marine currents which are an extension of the Gulf Stream. In August, the ice is at minimum area. Not only the Bering Strait, but the coasts of Greenland, Spitzbergen, Franz Josef Land, Novaya Zemlya, and a large part of the Siberian coast are free of ice in the summer.

A most important phenomenon in the behaviour of Arctic pack-ice (the phenomenon has not yet been so well studied in the Antarctic) is its displacement or drift under the action of the dominant winds, which seem to be the main influence on its movement. The icebergs for their part, owe movement to marine currents.

The Arctic is generally in the direction of Pacific to Atlantic in the eastern region, and Atlantic to Pacific in the west. A gyratory movement thus exists around the Pole. This is demonstrated by the course of ships frozen into the ice either involuntarily (as the ill-fated *Jeannette* in 1881, and the *Chelyuskin* in 1933) or deliberately (as the *Fram* in 1893 to 1896, and the *Sedov* in 1937 to 1940). All drifted systematically from the Bering Strait towards Spitzbergen. The same movement was observed in the Russian

*Top*
Head-on view of an Arctic iceberg, taken from a low-flying aircraft.

*Bottom*
Antarctic scenery. Pancake ice, forming into floe off Cape Evans, will eventually join to form an unbroken surface.

polar station set up by air twelve miles from the North Pole, on the 21st May, 1937. By the 19th February, 1938, it had drifted so far that it was located near the east coast of Greenland.

Gyratory movement explains, too, why certain points of the coasts of Greenland and Spitzbergen are covered with tree-trunks brought by the pack-ice from the mouths of the great Siberian rivers, and why an American polar station in the western part of the region drifted from 1949 to 1952 in the direction of the Bering Strait before turning towards the Pole.

Floating ice must not be confused with the true ice-islands, the best known of which is Lyakhov Island in the Arctic Ocean, near Taymyr Peninsula. These islands are solid ice, anchored to the sea-bed. Clays and sands found interstratified with the ice contain the remains of the mammoth and musk-ox, pointing to an ancient origin. It is probable that this ice was originally formed on a rocky island now sunk below sea level. Ice-islands are currently beginning to disappear—another result of increasing mildness in the climate.

## ALPINE GLACIERS

When ice is formed in a mountain-top depression it obviously cannot accumulate indefinitely without overflowing. If the depression has no break in its sides, the ice is forced out all round the edge and forms a hanging glacier, or cliff glacier. These glaciers never descend to lower altitudes. If on the other hand, the side of the hollow is cleft, the ice flows through the gap, passing below the level of permanent snow and moving downward until it reaches warmer levels where it finally melts and runs off as water. This is the typical alpine or valley glacier. Unlike the hanging variety, it consists of two separate parts: a gathering-ground in the original hollow, which forms a snowfield full of semi-crystalline ice, or névé; and a tongue of ice forming the actual valley glacier. There is often no sharp line of demarcation between the two areas, the moving glacier ice grading imperceptibly into the motionless ice of the snowfield. There may, however, be a fairly marked change in gradient, the slope of the glacier ice being the steeper. In summer, a pronounced crevasse frequently distinguishes the frontier between glacier and snowfield.

Many alpine glaciers do not originate in hollow catchment basins, but have as their source a large ice-cap or continental glacier. Numerous examples of this type occur, particularly in Greenland.

Mountain glaciers are found on most continents and on many islands. In proportion to its surface area Europe is richer in mountain glaciers than any other part of the world. Although it certainly cannot boast the world's highest summits, its numerous mountain peaks are situated in an area where the atmosphere is always abundantly rich in the moisture so necessary for heavy snowfall. The Alps alone have 1,300 glaciers in a total area of 1,346 square miles. The most extensive and most imposing glaciated area lies in the Swiss Alps where the Aar and Rhône glaciers flow from the Saint-Gotthard, and the Aletsch glacier, which is ten miles long and the largest in the whole of the Alps, flows from the Jungfau glacier. Next in importance are the glaciers of Monte Rosa on the Italian-Swiss border, of Gran Paradiso, of Mont Blanc, and of the slopes of Viso and Pelvoux in the Franco-Italian Alps.

In the United States numerous peaks in the Cascade Mountains, among them Mount Shasta, Mount Rainier and Mount Hood, carry glaciers. Some of those on Mount Hood are almost seven miles long. Valley glaciers occur in the United States only on a few isolated mountain peaks, but to the north, in Canada and Alaska, hundreds exist, some displaying remarkable length. One of the branches of the Hubbard Glacier in Alaska is all of seventy-five miles long.

In Eurasia, the Caucasus, the Pyrenees, and the ranges of Scandinavia also possess glaciers. They are found in Africa, too, near the Equator on Mount Kenya, Ruwenzori and Kilimanjaro. Among islands that have mountain glaciers or ice-caps are Iceland, Spitzbergen, Novaya Zemlya and New Zealand.

Greenland and Antarctica are by far the largest areas of ice and snow in the world, Antarctica having more than all the rest

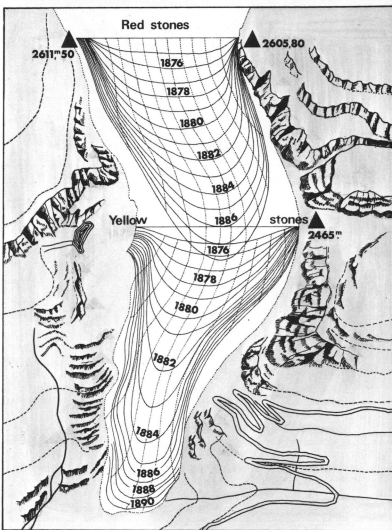

Top
Thorisjokull glacier, Central Iceland. The ice is forming a cliff above a mountain, section about 200 feet.

Bottom
Experiments on the Rhône Glacier, Switzerland, showing the movements of ice. Altitudes are in metres.

of the world combined. In addition to ice-caps, these two areas also have valley glaciers of the polar or high-latitude type, tide-water or tidal glaciers, and many other ice and snow features.

## MOVEMENT OF GLACIER ICE

The inhabitants of the Alps suspected from earliest times that glaciers were not static, that from year to year they advanced or

retreated, and that their appearance changed with the seasons. Yet it took several accidents to arouse the attention of hitherto sceptical scientists in this state of affairs. During his attempts in 1788 to climb Mont Blanc, de Saussure abandoned a ladder in a crevasse on the upper part of the Bossons glacier. In 1832 the ladder was found nearly three miles away. It had moved about 292 feet a year.

In 1824, a guide disappeared in a crevasse of the Maladetta glacier in the Pyrenees. His remains were not discovered until 1931, less than a mile from the spot where the accident occurred. Because the Maladetta is a hanging glacier, rate of movement had been only 40 feet a year.

In 1820, three guides from Chamonix were overtaken by an avalanche and hurled into a crevasse near the summit of Mont Blanc. Their remains appeared in 1861 at the foot of the Bossons glacier, about two miles lower down. The glacier ice had therefore travelled at an average speed of 276 feet a year.

Once these events were confirmed, experimental study of glacier movement began.

The simplest way of measuring consists in driving a row of stakes across a glacier, and taking a bearing from marks on the valley sides. At the end of several weeks, the stakes have moved to form a curve pointing downstream. The speed is thus seen to be greater in the middle than at the edges (sometimes as much as four times greater). Observations of greater accuracy were carried out from 1874 to 1900 on the Rhône glacier to establish the differential movement of glacial ice. Instead of stakes, painted and numbered stones were used, laid out in successive bands, one red, one yellow, and so on. Their displacement on various parts of

the glacier ice was noted monthly for several years. From these and other tests the following conclusions were drawn:

Speed is greater in the superficial part of the ice than in deeper areas. It is also greater in the central area than in the outer portions The resistance offered by the valley walls and floor, and the greater amount of debris carried in lower parts explain this.

Speed increases with increased gradient, and is greater where the valley narrows, and slower where it widens out; it is greater on the concave than on the convex curves. This last can be explained more easily if we take as an example a formation of soldiers wheeling round. The men on the inside of the column must mark time until the outside men are in line again.

To measure the variation of speed at depth, carefully sited stakes were driven down one side of a longitudinal crevasse. The lower stakes were soon seen to move more slowly than those higher up. The same conclusion could have been reached by observing how a transverse crevasse becomes more and more inclined as it progresses downstream, until finally the upper wall collapses on to the lower. The explanation is simple: the bottom of the valley, like the valley walls, acts as a friction brake.

To sum up, glaciers obey the same laws as rivers, though most take a year to travel the distance covered by a river in a few minutes. It takes a century for a piece of ice to move from the head to the foot of the average alpine glacier. Many of the large Alaskan glaciers move more rapidly than those in the Alps. The Muir glacier travels about seven feet a day, and the Child glacier averages almost 30 feet a day in summer. In Greenland phenomenal rates of almost 100 feet per day have been recorded for coastal ice-tongues.

138

*Far left*
*(Above)* Rimaye on the Dôme de Neige des Agneaux in the French Alps. *(Below)* The Aletsch Glacier, Switzerland.

*Left*
A vertical cleft in the Muzelle Glacier in the French Alps.

*Top right*
The glaciers coalesce on Mount Edith Cavell, Jasper Park, Canada.

*Bottom Right*
Seracs on the Rhône Glacier.

**Irregular courses.** How is it that a glacier, apparently rigid, can flow down a valley of irregular shape, pass through constrictions and mould itself to its uneven bed? Many theories have been put forward and many experiments have been made on this subject. A rod of ice spanning two supports bends downward; two blocks of ice pressed against each other fuse together; ice under pressure can be extruded from an aperture; a metal wire weighted at each end and laid across a block of ice works slowly through it and comes to rest leaving the ice in one piece. These demonstrations illustrate two properties of ice: its malleability and its capacity to refreeze.

It is well known that ice, unlike any other substance, is increased in volume when it freezes, and reduced in volume when it melts. If pressure is exerted on a piece of ice, compression reduces the volume and it passes locally and temporarily into a liquid state. When the pressure is withdrawn, the meltwater solidifies. Re-freezing following melting explains the results of the demonstrations just described.

The tremendous pressure exerted on glaciers by the valley walls and floor causes local melting, followed by refreezing. To a certain extent the glacier's advance depends on successive dis-integration and refreezing. Moreover, it must not be overlooked that glacier ice is formed of crystalline granules of fresh water bonded together by frozen salt water. Salt water melts at a slightly lower temperature and under a little less pressure than fresh water. It therefore follows that pressure melts the bond rather than the granules, leaving these loose and free to roll against each other, so that they act rather in the manner of ball-bearings.

The suggestion that the alternate seasonal expansion and contraction of the ice may influence it ignores the fact that the temperature of a glacier invariably remains at about freezing-point to a depth of 100 feet or more. No temperature variation penetrates the ice. At most, the only significant factor here is the expansion of the water which infiltrates into the crevasses and freezes solid.

Finally, it must not be overlooked that a glacier is impregnated with water still in a liquid state, streaming over the surface, circulating in the endless crevasses and cavities, and coursing through it in continuous subglacial torrents. Its millions of fissures are the loci of an incessant water circulation which must contribute to its movement.

**Crevasses.** Subjected to tremendous pressures and dragged in all directions, the ice cannot adapt itself exactly to the form of its bed. As a result, crevasses develop, at first as simple fissures, later as gaping rifts which form transverse, longitudinal, and lateral crevasses.

Transverse crevasses occur at fixed points on the glacier, for they correspond to abrupt changes in the slope of the bed. At such points the ice breaks and forms irregular accumulations of blocks, or *seracs* (named after a white cheese made in the Chamonix valley).

Where glaciers navigate abruptly steep changes of slope or descend cliffs, large masses of ice are detached and dropped, forming ice-falls or ice-cascades. The shattered blocks refreeze at the base of the scarp to become a reconstructed glacier.

Near the glacier head the moving ice is separated from the sluggish or still ice of the snowfield by an especially large gaping crevasse, or series of crevasses. Known in France as the *rimaye*,

The hanging glacier below the Matterhorn (14,560 feet) and the moraine-covered valley glacier descending from the Dent D'Hérens (12,580 feet).

*Right*
*(Top)* Ice front scenery at Enderby Land, Antarctica. In the background is the seaward face of an ice shelf formed by the accumulation of compressed snow over thousands of years. The cliffs were left behind when icebergs *calved,* or broke off. In the foreground the sea has been frozen into ice-floes. *(Bottom)* Close pack-ice which has been compressed, broken and churned up many times. Formed from frozen sea water, it originally resembled the ice-floes in the illustration above.

this is more frequently referred to by its German name, *Bergschrund.* Such a crevasse is constantly being packed with ice from the névé, snow, and avalanches, but all the time the glacial ice is moving away downstream, opening up the gap. In winter the crevasse becomes snow-filled and is frozen, only to open up again in the warmer seasons. Its shape and size vary from one year to another, but its position remains virtually the same.

Longitudinal crevasses, too, have a more or less constant position as they also correspond to irregularities in the bed of the glacier, forming where there are sudden changes of slope or ridges across the valley floor.

Lateral crevasses are quite different. Situated at the edge of the glacier, they move with it and are continually merging and re-forming. They owe their origin to the unequal tension that results from the pull of the central ice and the drag of the marginal ice. As the glacier moves down-valley, tension cracks open out in a fan-shape.

Some crevasses appear suddenly, with the sharp crack of splitting glass or of thunder. A fraction of an inch wide to begin with, they enlarge rapidly as a result of differential movement and melting. Gradually they attain a width of several yards, and depths of 100 feet or more. Lower down the plasticity of the ice and the freezing of infiltrating water tends to close them. In winter they become covered with snow 'bridges', presenting dangerous traps for the unwary climber.

The descent into a crevasse shows the ice to be stratified. As might be expected, it has been formed in annual layers up to three feet thick, each distinguishable by separating layers of darker mud or by differences in the ice-grains.

Although much is known today about the movement of ice, the mechanics of glacier ice movement are still far from fully understood, partly because of the extreme difficulty of practical examination, especially in the deeper and inaccessible portions of the ice.

How thick are glaciers? Measurements have been made by normal mechanical bores and by thermal and seismic soundings. This last method will be described in a later chapter in connection with the exploration of the Earth's crust. In attempts to estimate the thickness of the Mer de Glace for hydro-electric purposes, mechanical bore soundings were supplemented with soundings by electric charges in direct contact with the ice. In this method penetration is achieved by melting, but the difficulties involved in such borings are numerous: refreezing tends to obstruct the bore-hole as fast as it is drilled, stones embedded in the ice suddenly halt operations, fissures allow water to escape and prevent pumping up of waste and so on. Despite the obstacles, the bottom of the Mer de Glace was reached, between 360 and 585 feet deep. Many glaciers are considerably thicker than this. In the Alps, the maximum thickness so far recorded is the 2,570 feet Aletschhorn glacier.

Still greater thicknesses are recorded in continental or ice-cap glaciers, reaching 8,000 feet in Greenland and Antarctica.

## MELTING OF GLACIERS

Under the influence of sun, rain, and warm, dry winds, glacier ice melts on the surface; some even evaporates as water vapour, passing directly from solid to gas.

Proofs of ablation, as the melting is called, are easy to find. First of all, the meltwater can be observed running over the surface of the ice, forming pools and waterfalls before it disappears into crevasses. An abundant supply of water will wear away the upstream side of the crevasse and turn it into a gutter. Later, when the crevasse is closed up again, the channel remains as a sort of pot-hole which the swirling of the water gradually shapes into cylindrical form. This is known as a glacier mill, or *moulin,* because of the noise made by water falling into it. *Moulins* move with the glacier. However, as a crevasse often owes its position to an irregularity on the valley floor, another forms in the same place and, in time, another *moulin.* The process may be repeated again and again, producing a long series of *moulins.*

Another proof of ablation is found in the bevel or dome-shape of the tongue of ice found at the foot of a glacier. The reduced thickness at the edges is caused by surface melting. Melting takes place faster along the edges of the glacier than in the centre–a result of friction and heat-reflection. A glacier is therefore not only convex going downstream (apart from irregularities in its path) but cambered like a bad road surface.

No feature gives the glaciologist more striking evidence of melting than glacial tables, or ice tables. These are slabs of rock, often light in colour, which though once lying on the surface are found perched on pedestals of ice. Sometimes horizontal, the tables are generally inclined to the side most exposed to the Sun's rays, for the rock has played the part of a parasol in protecting the underlying ice from melting. The height of the pedestal indicates the original height of the glacier, until it is itself so reduced by the slow melting of its sides that it can no longer support the weight of the slab, and collapses. When the pedestal melts, the stone slab is sometimes dislodged and slips off, leaving the supporting column of ice projecting above the surface of the glacier. The projections are called ice pyramids. Similarly, thick morainic materials are frequently found on ice walls 100 feet or more above the level of the glacier. In Greenland some ice cliffs topped by moraine rise almost 400 feet high.

Conversely, thin layers of rock of patchy accumulations of dust absorb the Sun's heat, melt the glacier beneath, and sink into it to form pits, or dust mills.

Melting also takes place within the glacier. The temperature of the underlying rocks and friction against the valley bed cause

the formation of subglacial tunnels which meltwater finally reaches by way of the crevasses and glacial mills. Subglacial streams run through these tunnels and are the source of many of the brooks and creeks that emerge at the foot of a glacier. The Mer de Glace gives rise to the Arveyron, a tributary of the Arve, which itself flows into the Rhône. The glaciers of the Saint-Gotthard area feed the Rhône, the Tessin, and the Reuss.

A ridge of rock which is invisible under the ice sometimes dams the subglacial stream to form a pocket of water which may later burst, moving rocks and ice along in the flood waters.

Fed by snows from the lofty snowfield, which melt as soon as they drop below the limit of perpetual snow, a glacier represents the balance between rate of accumulation and rate of wastage. The balance is said to be positive when the accumulation is gaining, and negative when wastage is the greater. There is a point beyond which the rate of wastage, or melting, has the upper had, when all the ice coming from the mountains is transformed into liquid water. In the Alps this glacier front, as it is known, lies between 3,400 feet and 4,225 feet above sea level.

If the glacier front is the point where balance between accretion and ablation ceases to exist, its position is variable, and depends on several factors. Whenever there is considerable accumulation over a period of years, the glacier increases in volume and length. Conversely, every diminution of snowfall in the mountains causes retreat. There are exceptions to this sequence of events. Because the rate of melting may be high enough to equal or even exceed an abnormally heavy accumulation, an ice front may remain stationary or retreat when advance might normally be expected. If wastage is very slow, a glacier will continue to exist and may even advance slightly after snowfield accretion has completely ceased.

During the eighteenth and nineteenth centuries the Swiss glaciers advanced steadily until 1820. Then followed a retreat until 1840, after which there was re-advance lasting till 1860. From that date until recent years most glaciers have steadily retreated. Now French alpine glaciers are again advancing, at the rate of 70 to 150 feet per year. The Glacier des Bossons has advanced in an oscillating fashion, first on one side, then on the other, and then in the middle. Glaciers in most other areas are currently retreating, some of them in Alaska by as much as seven miles in the last twenty years.

If it is generally true that every advance is connected with a cold and wet period, cause and effect are far from closely linked in time. If, for example, there is an extremely high snowfall one year on Mont Blanc, the enormous quantity of ice at the summit added to the Glacier des Bossons would not arrive at its foot for several years. The time lapse would be still greater for the longer Mer de Glace. Thus an area's glaciers may not all show the same variations at the same time. It has even been observed that of two neighbouring glaciers, one advances as the other retreats.

*Left*
The terminal face of the Fox Glacier in the Southern Alps, New Zealand showing terminal and lateral moraines. The glacier rises near Mt Tasman and flows ten miles in a westerly direction.

*Right*
The Mer de Glace seen from Montenvers; the crevasses and trails of mud are concave further upstream.

## WORK OF GLACIERS

Glaciers are one of the greatest erosional forces, moving large quantities of rock-waste in the form of moraine from the higher valley slopes to those down, at the same time wearing away the walls and hollowing out the floor. Like most other agents of erosion, glaciers carry out the threefold process of erosion, transportation and deposition.

The moraines consist of all the material — from rock slabs weighing thousands of tons to fine dust — blown on to the glacier or falling on it. On a simple ice-tongue like the Glacier des Bossons, moraines simply form long lines of debris on each side, called lateral moraines. If two glaciers merge to form a single river of ice, the new, single glacier will support two lateral moraines and a medial moraine (the combination of the left lateral moraine of one glacier and the right lateral moraine of the other). This pattern can be observed on the Mer de Glace. Generally the medical moraine becomes indistinct toward the foot of the glacier.

Sometimes the amount of material carried is so great that moraine covers the whole surface area of the ice. This is seen in the Talefre and Leschaux glaciers whose 'black seam' deposits join the clean ice from the Glacier du Géant, the 'white seam', to flow side by side into the Mer de Glace.

Superglacial or surface moraines, both lateral and medial,

absorb the Sun's warmth, particularly when they are dark in colour. When they form only in thin layers they often melt the ice and sink into it. In this way they become first internal, or englacial, moraine and then subglacial or ground moraine. Englacial and subglacial material is also accumulated by the ice from valley walls and valley floors.

We have seen that subglacial streams run through tunnels beneath the ice. Here, ground moraine mixes with the material eroded from the bed of the glacier by the ice or by the stream, and the stones become rounded and ground down in the middle of a stiff clay, called boulder clay.

Some of the transported morainic material at last reaches the front of the glacier where it is deposited as terminal moraine in the shape of a crescent, or moraine amphitheatre. The convex formation is explained by maximum speed lying in the middle of the glacier. End or terminal moraine is also formed by the snowplough action of the glacier pushing in front of it increasingly large quantities of debris.

While it is morainic material that gathers and scrapes away the bedrock, the great pressures produced by the glaciers' weight make them responsible for a tremendous amount of erosion. A glacier about 1,000 feet thick exerts a pressure of approximately 28 tons on each square foot of the valley floor. Glaciers do not create the valleys they flow in, nor the ridges they override; they

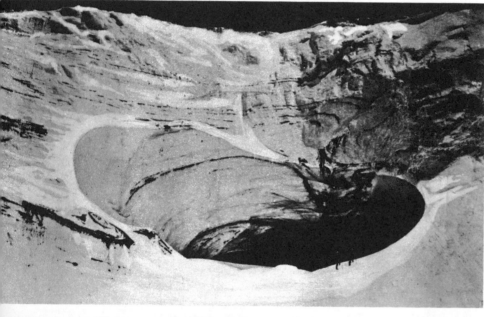

remould them. Alpine glaciers deepen the valleys, sharpen the ridges, and increase the relief and ruggedness of the terrain. Ice-cap or continental glaciers, which cover valleys and hills alike, subdue and round off the land surface into smooth flowing and contours.

The effects of glacial erosion are many and considerable, sweeping away all loose debris in the area, incorporating the debris into the glacier itself, abrading the bedrock with this material, penetrating deep into the cracks or joints in the bedrock, and plucking out and transporting huge blocks of rock.

In recently glaciated areas many features foreign to a stream-eroded region are to be seen. Often found on the floors and walls of glaciated valleys are polished rocks, grooved and striated by pebbles and boulders frozen fast in the moving ice. The trend of the often parallel grooves, sometimes several yards long, indicates the direction of the flow.

The glacier, by abrasion, smooths and rounds irregularities on the upstream side of the valley, the side against which it has flowed, leaving the leeside sharp and angular where rock has been plucked or quarried away rather than abraded. When the smoothed, irregular patches of bedrock are dome-shaped they are known as *roches moutonneés* (named, incidentally, in 1804 by de Saussure, after the curls on the sheepskin wigs of his time).

The abrading pebbles and boulders carried by the glacier are in their turn smoothed, faceted, scratched, and eventually ground into very fine powder, or rock 'flour'. The rock flour, comparable in texture to jeweller's rouge, is carried away by glacial meltwaters, to which it gives a milky colour that can be clearly seen where the waters emerge from the foot of the glacier.

Ice travelling down a former stream valley fills it almost to the brim, and changes its characteristic V form into a broad, steep-walled, U-shaped valley. The stream channel, on the other hand, would occupy only a small portion of the valley. Both stream and glacial channels develop approximately the form of a catanary curve. The major difference in appearance is the proportion of the valley that is occupied by the actual channel.

As the ice moves down the former stream valley it navigates curves and projecting spurs with difficulty, tending to remove obstructions by cutting off the spurs and straightening out the valley as it goes. The scars of truncated spurs are visible as triangular-shaped cliffs, or flat-irons, on the walls of glaciated valleys. The straightness of glaciated valleys is so marked that in Alaska they are often referred to as 'canals'.

The steep-sided, flat-floored fiords of the Scandinavian coasts are submerged glacial valleys, eroded — perhaps below sea level — when the glaciers extended as far as the ocean itself, or maybe at an earlier time when this coastal area lay above sea level.

Glaciers often produce a step-like effect in the longitudinal

*Far left*
*(Top)* Glacial pocket of the Tête Rousse Glacier.
*(Bottom)* A 10-mile stream of ice, the Aletsch Glacier, Switzerland, terminates in the morainic Lake of Märjelen.

*Left*
The Glacier des Bois, the end of the Mer de Glace.

*Top right*
Terminal ice cave under the Chardon Glacier.

*Bottom right*
Hard bands of rock remain resistant to the effects of weathering and erosion which have attacked the surrounding limestone.

profile of a valley, the 'tread' of each step sloping slightly down toward the head of the valley. This is probably the result of the plucking activity of the ice. After glaciation these rocky basins are frequently filled with water, forming a chain of tiny pools, known as paternoster lakes.

Generally in a network of river valleys the tributary streams join the main river valley at the same topographic level. In a system of valleys which have undergone alpine glaciation, the elevation of tributary valleys at their junction with the main valley is much higher than the main stream valley — perhaps as much as 2,000 feet higher. This is because ice erodes in proportion to its thickness; the main valleys, which contain a greater volume of ice, therefore deepen at a far greater rate than the tributary valleys. Tributary valleys whose mouths are high on the slopes of the main valley are called hanging valleys.

Streams flowing in the hanging valleys plunge in a series of cascades or waterfalls to the main stream below. The narrow, misty, high falls along the sides of Yosemite Valley, California, and many Norwegian fiords, are examples of elevation disparity at glaciated valley junctures.

At the head of many glaciated valleys are huge amphitheatre-like depressions with sharply rising walls. Known as cirques, they are among the most spectacular features of the Swiss Alps, the Canadian Selkirks, and the Rocky Mountains. Cirques are cut in all kinds of rock and in all types of rock structure — horizontal, tilted, folded and complex. The steep walls are produced by the plucking and sapping action of the ice at the head of a glacier. The basin-like floor is formed by the deepening action of the ice where it lay thickest. In post-glacial times the floor is often occupied by a tiny lake or tarn.

Arêtes are rocky ridges — knife-sharp with glacial quarrying and frost-wedging — that project between glacier-filled valleys. Where a ring of cirques forms around a high mountain, the peak takes on pyramidal form as a result of headward erosion of each cirque, and is called a horn, or matterhorn after the famous Matterhorn of Switzerland. A col, a sharp-edged saddle or gap, appears where two cirques have cut into a ridge from opposite sides until their head walls met, forming a smooth and elevated hollow.

The features produced by glacial erosion are almost infinite in number and variety. To these must be added the features produced by glacial deposition.

As a result of the high erosive force of a glacier, huge quantities of material are deposited as glacial ice melts away. Glacial till is the name given to debris deposited directly. Till is generally unsorted and unstratified. Glacial drift is that material transported and dropped by the ice and its meltwaters. Glaciofluviatile and glaciolacustrine deposits are those made by glacial meltwater streams and glacial lakes respectively. These deposits are to some extent stratified and sorted.

Glacial erratics are large boulders that have been transported considerable distances and left on bedrock of a different composition. A long series of such erratics is known as a boulder train.

A glacial varve, a glaciolacustrine structure, is an annual deposit in a glacial lake, and consists of a pair of thin sedimentary layers, one darker in colour than the other. Together, they are usually only a fraction of an inch thick. The lighter layer, which is both thicker and coarser, represents the spring and summer deposit, for spring and summer thaws increase the water volume, bringing in coarser material; the darker layer represents autumn and winter deposition, for these periods deposit only fine material. In this way two different layers are formed.

As a glacier melts, its whole load is deposited, and the land is covered with an irregular sheet of debris called ground moraine. Lateral moraines leave ridges close to and paralleling valley walls. Terminal moraines remain at the end of the glacier's course, often blocking valleys and producing lakes by damming. Where there is more than one terminal moraine, only the outermost is the true terminal moraine; the others are recessional moraines, a misnomer really, for they mark moments of extended halt or re-advance of a glacier during a general period of retreat. The terms recession and retreat as applied to glacier behaviour are themselves misleading in that a glacier never withdraws up the valley; it simply melts, and as it melts fastest at its thinnest parts (usually at its foot), the ice-front shifts up-valley, giving the impression, that the whole glacier is actually continuously moving along the slope.

*Left*
A glacial pavement in Yosemite National Park, scratched and polished by a debris-laden glacier. The boulders seen strewn on the bedrock were left when the ice melted.

*Right*
A deeply fluted and scratched boulder in the recently deposited moraine of Athabasca Glacier, in Jasper National Park, Alberta.

Still not fully explained is the origin of drumlins, the elongate, almost tear-drop shaped hills of glacial till that often occur in areas of glacial deposition. They are found in many parts of the world, usually in large groups. Some of the best examples are those situated in Northern Ireland, northern England, New England, and northern New York. Drumlins are 25 to 150 feet high, nearly 500 feet wide, and one-third of a mile to a full mile in length, usually usually aligned with the long axis parallel to the direction of glacial movement, and the wide end facing against the flow. The wider, blunt end of the hill has a steeper slope than the other, which is longer, gently sloping and tapering. Occasionally drumlins are found singly, but usually they form in clusters, often merging into one another. They are thought to be formed by the overloading of the lower portion of the glacier with rock debris, or by the ice overriding heaps of loose debris. Whatever their origin, along with kames, and eskers, they are more commonly associated with continental glaciers than with alpine glaciers.

Kames and eskers are both glaciofluviatile deposits. Kames are rudely conical hills deposited by heavily burdened glacial meltwater streams. Eskers are long, narrow and low sinuous ridges, often stretching for many miles; built up of layered sand and gravel, they are beds of former streams flowing in and under a glacier. Some eskers probably mark the presence of long crevasses which have been filled with rock debris.

Spreading out fan-wise beyond the terminal moraine is the outwash plain, a sheet of clay, sand, and gravel built up by the glacier meltwaters. This material is generally thicker and coarser near the terminal moraine. Small depressions, called kettleholes, may dot the outwash plain and other areas of glacial debris, marking points where huge blocks of ice were left behind as the main body of the glacier melted away.

Areas of recent glacier deposition are hummocky or hilly in appearance. The terms knob-and-kettle or kame-and-kettle are used to describe this kind of topography.

Every feature of glacial erosion and deposition can be found in areas which have undergone glaciation of one form or another. Some features, however, are more characteristic of one particular type of glaciation and are rarely found in association with other types. In regions of alpine glacial erosion, cirques, arêtes, horns and hanging valleys are common, but they are seldom found in areas of continental ice sheets. The same is true of lateral and medial morainal ridges. U-shaped glacial troughs, ground moraine and outwash plains, on the other hand, appear in almost all areas of glaciation. Drumlins, eskers, kames, and knob-and-kettle topography are common features in areas which have undergone continental glaciation but are found only occasionally in alpine glaciated regions.

## CAUSES OF GLACIATION

The glacial theory, developed from observation and comparison of the geology of different areas, claims that in the past huge ice sheets covered regions of the Earth where ice exists no longer, and that today's glaciers once covered areas far beyond their current limits. The ice-caps of Antarctica and Greenland are remnants of vastly bigger continental glaciers of the Pleistocene era.

Early in the eighteenth century the occurrence of glacial features far from glacial regions was explained with theories of giant inundations or of iceberg rafting. The idea that the ice itself brought these deposits was not put forward until the nineteenth century— and then it was not accepted immediately. The most important early contributions to give a sound scientific basis to the glacial theory came from Switzerland, where the actual formation of such features could be observed.

The proof of extensive glaciation lies in the concept upon which the whole of geology is based, that 'the present is a key to the past'. Proof of continental glaciation rests therefore on observation of current glacial activity and on comparison of its effects with older topographic forms found beyond the current limits of land covered with ice.

This leads us to determine what event or series of events brought about glaciation on such a vast scale, for glacial deposits today extend hundreds of miles beyond the present limits of glacial ice. In the United States deposits are found in New York and in a broad belt south of the Great Lakes. Evidence of vast ice sheets is not restricted to recent geological eras, but is found in rocks of long-past epochs, as in the Pre-Cambrian of the Congo. The well-known Dwyka tillite of South Africa is of Carbo-Permian age. It was first described in 1870 and is directly comparable with tillites in India (1847, Orissa, the Central Province, and the Punjab) and with those of South Australia (1859). Extensions of this glacial deposit were found 1888 in parts of Brazil, Uruguay, and Argentina, and over the whole of the Falkland Islands. The wide distribution of this rock indicates the extent to which Gondwanaland was undergoing glaciation while Laurasia was experiencing a tropical climate.

The causes of glaciation are presumed to be largely climatic and are probably based on the amount of solar heat and the type of topography. During periods of mountain building and extensive uplift, climatic contrasts are greater than in relatively stable periods of low relief and widespread continental sea.

Slight variations in the amount of heat received from the Sun would be sufficient to induce glaciation. For example, a relatively small drop in current average annual temperature over large sections of the Earth would reduce summer levels to a point at which they would be insufficient to melt the winter snows. The glaciers would then return. At present an average temperature of 49°F. (9·4°C.) at Geneva halts alpine glaciers at an altitude of 3,400 feet. At a mean temperature of 41°F. (5°C.), only a few degrees lower, they would descend to 1,500 feet and reach the banks of Lake Geneva.

Variations in the amount of solar heat received have been ascribed to different factors: the intensity of sunspots (the fewer the spots, the less the radiation); density of hydrogen atoms in space through which the solar system is travelling (the denser the atoms, the less the radiation); and the relative amount of carbon dioxide and water in the Earth's atmosphere, possibly correlated to the amount of land surface, for it is argued that with a larger land area there would be less water vapour produced, and additional vegetation would tend to use up more carbon dioxide.

Dr. Richard Flint of Yale University, in a recent theory, traces glaciation to a combination of high topography, extensive land surface, and low solar radiation. He assumes greater fluctuations of solar radiation during the past than we experience at present.

Examining these theories more closely, we note that throughout Earth's history glaciation has often followed, at some distance in time, a period of mountain building. But this has not always been so, and it is generally believed that the formation of mountains has no direct relation to glaciation. At most, there is sometimes a certain coincidence between the two phenomena. Cycles of glaciation seem to have occurred about every 250 million years (Pre-Cambrian, Carboniferous, Pleistocene).

The temperature of the Earth follows several superimposed rhythms that depend on the relative position of the Sun, Earth, Moon, and of the solar system in the Milky Way. It is thought that glaciation may well be the manifestation of the periodic return of similar atmospheric conditions. Average temperatures of the Cenozoic, as deduced from a study of floras in Europe, fell steadily from the Eocene, but did not reach optimum levels for glaciation until the Pleistocene.

Other suggested astronomical changes that might affect terrestrial temperatures are: variation in the eccentricity of the Earth's orbit, which would change the length of the seasons; and polar wandering, which would alter the position of the temperature zones (equatorial, temperate and frigid).

Some terrestrial influences that have been suggested are: shifting positions of land and ocean areas, affecting the ocean currents; the effect of mountain building on atmospheric circulation; the blanketing effect of volcanic dust; and fluctuations in the rate of radioactive heat flow from the Earth's interior.

Whatever change or changes caused the Pleistocene climates of the northern hemisphere to vary from warm to frigid, evidence is strong that the temperature changes in the air did not exceed 15°F., and that the tropical oceans fluctuated no more than 10°F. So closely related are the causal factors of climate that a relatively small change in one factor may trigger off profound changes in the whole system.

Perhaps the different glacial ages had different causes. To date there is no conclusive evidence to support or destroy the theory. The puzzle of continental glaciation is still with us, an intriguing and an important one, because in it lies the answer to the question: Are we on our way out of or into a glacial age?

*Left*
Trollfjord in northern Norway.

*Right*
This sheer wall of ice is part of the vast face of the Columbia Glacier, Alaska. It is 300 feet high and 2½ miles wide.

# Volcanoes and Vulcanism

Since the famous eruption of Vesuvius destroyed the towns of Pompeii and Herculaneum in A.D. 79, many violent volcanic eruptions have brought widespread catastrophe. The eruption of Laki in 1783 transformed a tenth of the surface of Iceland into a desolation of lava. In 1902 Mount Pelée annihilated the town of St Pierre in Martinique. Most disastrous of all was the explosion of Krakatoa in the Sunda Strait in 1883, causing a tidal wave which drowned thousands of people.

The immense forces involved in volcanic eruption are such that the Earth vibrates and trembles. There are underground rumblings and deafening explosions. Mountains spilt from top to bottom, flames pouring from the crater. Clouds of gas billow upward. Rocks are hurled into the air. Flows of incandescent lava sweep along the ground, and the sky is dark with a rain of cinders.

A volcano which is in eruption is said to be active; one that has ceased to erupt is dormant; and when volcanic activity is thought to have subsided altogether and further eruptions considered unlikely, then it is said to be extinct.

It is almost impossible to distinguish between an extinct volcano and a dormant one. Volcanoes have been known to lie dormant for centuries without giving any signs of activity. In A.D. 79 no-one suspected the true nature of Vesuvius, whose slopes were covered with gardens and vineyards. Many prosperous towns were spread out round the foot: some were commercial towns but most were holiday resorts for the rich inhabitants of Rome and Naples whose pleasures were unmarred until the 24th of August in that year, when the volcano suddenly revealed itself and engulfed Herculaneum, Pompeii and Stabiae. Almost seventeen centuries later the remains of these cities were uncovered. Roman sentries still standing at their posts, families and their possessions in the apparent safety of their cellar vaults, all had been cast in moulds of volcanic mud.

From then on, Vesuvius erupted every century until the year 1139. A calm period followed, and the volcano became overgrown with trees. In 1631 eruptions started again and have continued until the present day.

The history of Vesuvius shows clearly how a supposedly extinct volcano can suddenly awake from slumber. It shows, too, the longer the period of quiescence, the more violent the subsequent explosions may be.

Let us consider the outward manifestations of a volcanic eruption with its typical succession of explosions, lava flows, and rain of cinders. The chimney of a dormant volcano is often blocked with a plug of solidified lava from an earlier eruption. A lake or glacier may occupy the crater. Smoke escapes from cracks, or fumaroles, at various points in the surrounding area and in the cone itself. A wisp of smoke hangs round the summit, too.

The onset of a sudden eruption is marked by an increase in the vapours and gases: the fumaroles increase in number and activity, and the streamer of smoke from the summit becomes larger and visible from a greater distance. At the same time small earth tremors take place — signs of the movements in the Earth's crust — and faint and intermittent underground rumblings can be heard.

Thus heralded, the eruption begins. There follows a deafening explosion when the plug or crust of lava is blown out of the crater under the enormous pressure built up by steam and gas. A column of vapour shoots up vertically for several miles, then as the force subsides, it is spread out horizontally by the wind. At night the vapour cloud reflects the red heat of the volcano so strongly that it looks like a sheet of flame.

At the present time there are more than 600 active volcanoes and about 10,000 extinct or nearly extinct volcanoes of any size. About 500,000 square miles of the Earth's surface is covered with volcanic material. This represents some thousands of cubic miles in volume.

## ORIGIN OF VOLCANOES

Vesuvius and Etna are too old for man to be able to determine their origin. Within the time of recorded history, however, many new volcanoes have been born. In May 1538, Monte Nuovo ('new hill'), rose out of the plain of Naples, and within three days had reached a height of 400 feet.

One of the most interesting volcanoes to appear unheralded out of level ground was Paricutin, whose development has been watched carefully and filmed. The scene was Mexico, 300 miles from the capital, and it occurred during an afternoon in February 1943. A farmer was ploughing his fields when he was surprised to see something that looked like smoke coming from a furrow. Soon the ground began to tremble and crack. Alarmed, the farmer raced to his village to describe what he had seen. A violent earth tremor cut short his words, destroying part of the village. When the villagers reached the field, cinders, stones, blocks of rock and lava were being showered from a hole in the ground. By the next morning, a volcanic cone 25 feet high encircled the event. A week later the cone was 500 feet high; after two months 970 feet; and at the end of two years it towered 1,500 feet above the plain and was more than three-quarters of a mile in diameter. Over this period the crater continued to pour out gases and clouds of smoke which ascended several miles into the atmosphere, reflecting the red-hot interior of the volcano. A continuous noise was heard, varying in intensity and frequency, resembling at times the roar of waves, at others the crackle of rifle fire or the thunder of cannon. Streams of jumbled and broken lava flowed in every direction, adding more destruction to an area already devasted by flying rocks and volcanic bombs, one of which measured 15 yards in diameter. Flames flickered round the summit, while thunder and lightning added to the terrifying spectacle which brought sighteers and scientists from all over the world.

The Ancients' explanation of volcanic activity was a river of fire, Phlegethon, encircling the Underworld, or the god Vulcan's forges firmly established underneath the mountain Vulcano; they even suggested that 'imprisoned winds' or 'subterranean exhalations' found their way out through openings on the Earth's surface to produce the phenomenon of volcanic eruptions. Much later, when cosmogonic theories had suggested Earth as the daughter of the Sun, chilled and solidified only on the surface, the theory

*Top*
The volcanic island of San Benedicto, with the volcano Boqueron still active.

*Bottom*
The edge of a block lava flow near Cinder Cone, Lassen Volcanic National Park. The flow is only a few hundred years old.

suggesting the existence of a fire in the centre of the Earth was put forward.

Nowadays the central fire hypothesis is discredited. The Earth's crust is known to be solid, and it is only round the outside that a hot and softer layer, called the mantle, is thought to exist. Can we simply replace the conception of a central fire, then, by that of a 'peripheral fire', and regard volcanoes as safety valves? Several facts are against this. To begin with, present-day astronomical and seismological observations indicate that there are no extensive areas of molten rock within 1,800 miles of the Earth's surface. It is improbable, therefore, that the mantle encircling the entire planet should be the direct source of the eruptions manifest on its surface. Volcanoes can be grouped into families, each with its distinctive mode of eruption and each fed, probably, from separate underground reservoirs. Lastly, it can be proved that these reservoirs are situated at depths probably not exceeding 20 miles, not in the mantle but still in the Earth's outer layers. But how is lava melted and what makes it rise?

**Melting of lava.** The temperature of the globe is in part a legacy from the time of Earth's formation and in part the result of the disintegration of radioactive substances, which are almost all contained in the crust of the Earth. The amount of uranium and thorium distributed throughout the first 10 miles of the crust has been estimated at 100 million tons. This calculation has led several geophysicists to suggest that nuclear chain reactions may be produced locally, analogous to the reactions of nuclear piles and atomic bombs. The ensuing melting process would produce an increase in volume which would fracture the overlying rocky crust and provide paths along which the magma (the molten material lying below the solid layer of the Earth's crust) could travel to the surface. As yet we have no proof that such concentrations of radioactive elements exist anywhere in the crust. So there the hypothesis rests.

Another theory postulates that because temperature increases downward in the Earth, convection currents exist in the potentially molten material beneath the crust. The currents, though on a vastly bigger scale, are similar to those that bring scum to the surface in a pan of boiling jam. A description of these currents will appear in the chapter devoted to the structure of mountains (p. 195). Here, it is sufficient to say that the convection currents in question, if they exist at all, must heat the lower layers of the Earth's crust sufficiently to cause local melting and the formation of underground reservoirs of lava.

Rock under the pressure imposed by many miles of overlying material has a higher melting-point than at surface level. We already know that subterranean temperature increases with depth, though not at a constant rate. Basaltic rocks melt at about 2,280°F. at surface level; twenty miles beneath it, their melting-point is raised to approximately 2,600°F. Throughout most of the world, seismic information shows that rocks at this depth are solid, indicating that the temperature does not exceed 2,600°F. But it may come very close to that figure, and if it does, then only a slight increase is required to exceed the melting-point. It is possible that where segments of the Earth's crust are crushed and broken by mountain building, friction may possibly generate enough extra heat to melt the rocks. This could explain why active and recently active volcanoes coincide with the belts of most recently formed mountains, and with active fracturing and faulting. Volcanoes are always associated with great fault zones, though not all fault zones boast volcanoes.

## MAGMA AND LAVA

Lava is the name given to the molten material forced out of the volcano's cone and to the rock that is formed when the molten mass cools and hardens. The term magma applies to the molten material while it still lies below the surface. Lava and magma differ in more than position and degree of solidity, however. Magma contains large amounts of gas and other volatile substances; lava loses most of these volatiles when it erupts, and even before.

We do not know just how magma forms and moves, nor do we know the exact depth of its origin. Studies of existing volcanoes

The floor of the crater of the Ol Donyo Lengai volcano in Tanzania—the last active volcano in East Africa.

provide us with some significant data that suggests the depth is not great — probably some twenty miles below the surface. As liquids rise to heights proportionate to the pressures exerted upon them, the height of each volcano should give an estimate of the depth of its reservoir. Known volcanic cones vary in height from about 2,000 feet to over 20,000 feet (the total height of Mauna Loa, from its base beneath the sea to its crest above sea level, is more than 28,000 feet). Raising the magma to a 2,000-foot summit would require the pressure of a 2-mile depth of rock; to produce the same effect in a 20,000-foot crater, a 20-mile rock overlay would be needed. As long as the supply of magma holds out, the volcano grows to the maximum height to which surrounding rock pressures will force the molten material. Once this height is reached the volcano tends to remain stable. What, then, from this stage on, makes lava rise and overflow from the crater? What makes a volcano erupt with explosive force from time to time?

The answer to both questions is water, in the form of steam. As a magma approaches surface level so there is a corresponding reduction in pressure, and the gases present begin to escape. If the volcano's crater is blocked with lava, the escaping and expanding gases may accumulate until their pressure is strong enough to expel the crusted lava or plug with explosive force. A violent eruption will blow out large quantities of material, large and small rock fragments, molten lava, dust, steam and gases. The smoke cloud hanging over an eruptive volcano consists of volcanic dust and gases.

The amount of water released during an eruption is amazing. At the height of its activity, Paricutin produced 16,000 tons of water daily, as well as 100,000 tons of lava. The problem is to find the source of the water or steam. Does it come from inside or outside Earth's surface? A theory of exterior origin has the support of several authorities. Noting that most volcanoes, active and extinct, are situated near the sea, these geophysicists maintain that oceanic water seeps into the depths of the Earth through fissures and finally reaches reservoirs of magma. If the water came from the interior of the globe, they claim, the quantity emitted would be sufficient to double the volume of the oceans in a couple of million years. But, as far as we know, seas and oceans have not changed appreciably in volume in more than a thousand million years.

Direct observations shows clearly the effect of exterior water on volcanic eruption. Snow melting in a volcano's crater, or heavy rain over the summit area during a period of activity is sufficient to trigger off renewed eruptions. This process was

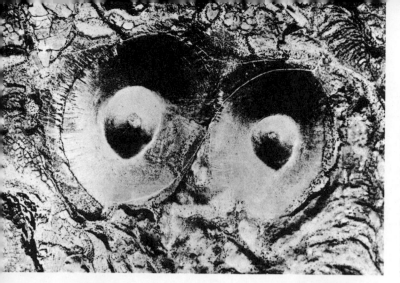

observed in 1915 at Lassen Peak, California, and in Vesuvius in 1926.

Those who uphold the theory of the interior origin of water object to this interpretation. Many volcanoes, both extinct and active, lie several hundred miles from the sea and cannot receive enough water through fissures to support activity. It is suggested instead that magmatic water comes from the melting of deeply-lying rocks which are shifted into a warmer zone or undergo some alteration as a result of earth movements, or are affected by the convection currents already mentioned. Chemical analysis has shown that a kilogram of granite heated until it is redhot liberates 10 grams of water which, at the temperature of the granite, is converted into about 48 quarts of steam. At this rate, a cubic mile of rock would be sufficient to produce the steam supply of a good-sized volcano for several years. When melted, granite also liberates hydrogen, carbon monoxide, carbon dioxide, methane, hydrogen sulphide, and nitrogen. These same gases are given off during volcanic eruptions.

Still other authorities think the magma acquires water from an additional source, from ground water and surface run-off, insufficient quantity to produce eruptions. It still remains for the presence of the large quantities of water in volcanic eruption to be explained to everyone's satisfaction.

## STRUCTURE OF VOLCANOES

Structurally, a volcano must be considered in two separate parts: the outer part (the only visible area) and the internal part (seen only in deeply eroded volcanoes).

A volcano takes the form of a cone or dome, the angle of its slopes being determined by the nature of the materials thrown out. Classification is made according to structure and shape. Where alkaline lavas predominate, the volcano is broad, and the slopes gentle (seldom over 10° near the summit, and only about 2° at the base). This type is known as a shield volcano or lava dome, and its eruptive activity is comparatively mild. Mauna Loa, on the island of Hawaii, is the classic example of a shield volcano. It rises almost 14,000 feet above sea level, with a base diameter of 70 miles. Near the base, slope gradient averages about 2°, steepening slowly upward to 10° at 15,000 feet where it flattens out again.

When rocks and cinders thrown up by explosions form the bulk of the cone, the angle of the slopes is steeper than in a shield volcano, generally about 30° to 40°, the normal angle of repose of coarse materials. These formations are called pyroclastic, cinder or debris cones. Although usually of a violent nature, they are relatively small in size. Paricutin is a pyroclastic cone.

Finally, if lava and debris alternate, a composite cone or strato-volcano results. This type of cone consists of a series of alternating steep and gentle slopes, generally steep near the crest, with an average of 30°, and grading downward to about 5° near the foot. Most large volcanoes fall in this category. Mayon, on Luzon island in the Philippines, is an ideal example. Composite, or intermediate, volcanoes are characterised by their violent explosions, which are generally accompanied by lava flows.

Volcanoes are not necessarily permanently of one type or another; they may change their nature. Mount Etna developed as a shield volcano during the first part of its existence and changed later to a stratocone.

Subsidiary or satellite cones may appear on the slopes of the parent volcano, looking like so many enormous carbuncles. Mount Etna carries several hundred such cones. When their conduits join that of the main volcano they are called parasitic cones. Shastina on the side of Mount Shasta, in the Cascades, is a relatively large parasitic cone, now extinct.

At the summit of its cone, a volcano may open skyward in a funnel-shaped depression called the crater. This may be single, double, intact, or breached. If it has been caused by subsidence or an explosion so violent that the upper part of the volcano is destroyed, it may be large and almost cylindrical, and in this state it is called a caldera (after La Caldera, in the Canary Islands, which is a pit more than three miles wide with surrounding cliffs 3,000 feet high).

Lovely Crater Lake topping the ruins of Mount Mazama in southern Oregon lies in such a depression. The caldera formed

*Top*
The two cones of the Valley of Volcanoes, Peru.

*Centre*
A 3 m.p.h. cascade of lava from a volcano situated in the Congo.

*Bottom*
Subsidiary cone, Piton Bory, Réunion Island.

when Mazama erupted and almost disappeared about 9,000 years ago. The symmetrical composite cone was originally about 12,000 feet high, capped by glaciers whose traces still survive on the remaining slopes. The eruption left a circular caldera over five miles wide and up to 4,000 feet deep. Crater Lake, formed by rain and melting snow, now fills the caldera to a depth of 2,000 feet and is encircled by cliffs 500 feet to 2,000 feet high. Built within the caldera by later eruptions are a small cone and an associated lava flow known as Wizard Island. It is not known for certain just how much of the caldera was formed by subsidence and how much was blown out by the explosive eruption.

The summit of Vesuvius today is a vast caldera more than two miles across, produced by the severe explosion in A.D. 79. This caldera has been breached, and only the north-west part of the rim, Monte Somma, remains. A hollow, the Atrio del Cavallo, separates Monte Somma from the lower, present-day crater. Mount Etna also has an ancient caldera, the Val del Bove.

No fixed size limit distinguishes a crater from a caldera. The caldera of Kilauea is two-and-a-half miles at its maximum width, and that of Mauna Loa is about three-and-a-half miles. Both were formed through collapse when lavas withdrew from the chimneys of the volcanoes. The vent at the top of Mount Katmai, a crater, is over two miles wide and in some places up to 3,500 feet deep. Perhaps the largest caldera known is in the ancient volcanic area in the northern New Mexico. Here the Jemez Mountains contain a caldera, the Valle Grande, which is eighteen miles long and fifteen miles wide. It is accompanied by many smaller depressions — yet some are still three miles in diameter.

Not all volcano cones end in a crater. Those with extremely viscous lava may have plug domes. The dome of Lassen Peak is of this type; so too are many dome-shaped hills of volanic rock found in certain areas of central France.

Larger, non-circular depressions also appear in volcanic areas. Here, tremendous amounts of lava have welled out of cracks or fissures, not vents, withdrawing so much material that the surface has sunk. The subsidence is often accompanied by fracturing. Volcano-tectonic depressions of this kind are particularly common in Sumatra. One depression has especially impressive dimensions; it is 60 miles long and 20 miles wide.

Lava or magma often fills cracks in volcanoes and in the ground around them. The enclosing rocks are softer and more easily destroyed by erosion. When this happens, the lavas are left standing out on the landscape. These dykes, as they are called, may stretch for miles. They can be traced, too, on barren cliffs, for the hardened igneous rock is generally of a different colour from the flanking ground. Dykes are usually arranged more or less radially around the volcano's conduit. Shiprock in New Mexico is an eroded volcanic neck with irregularly radial dykes extending outward. Other dykes are found near the Spanish Peaks, in Colorado.

It is relatively simple to examine the outer structure of an active volcanic cone or dome, but the chimney or conduit in which the lava rises is generally hidden from view. Many extinct volcanoes, however, have been so heavily eroded by rain, streams or ice that the outside layers have been removed, exposing the hardened lava in the chimneys. These solidified plugs are known as volcanic necks. Castle Head, on the shores of Derwentwater in the English Lake District, is a typical example.

A volcanic conduit may descend a dozen miles or so below surface and then divide into branches to feed two or more volcanoes some distance apart. Adjacent volcanoes may come from the same reservoirs, or each may draw from a separate source.

As for the subterranean reservoir of lava, nothing can be observed of its form or structure.

## VOLCANIC EJECTIONS

Gaseous, liquid and solid materials are expelled during a volcanic eruption. The fragmental material, as we have already noted, is known as pyroclastic debris. Large angular masses are called blocks; lava emitted in a fluid state and hardening in the course of its rise and fall is known as a volcanic bomb.

Volcanic blocks are hurled out in the solid state, sometimes breaking as they descend. They may be several inches long, or

*Above*
Paricutin, two months after appearing in 1943.

*Below*
Vertical fissures of Piton Bory, Reunion Island.

*Top*
A 'bread-crust' bomb.

*Bottom*
Agathla Peak, the neck or plug
of lava of an ancient volcano
in Arizona.

*Top*
A 'cow-dung' bomb.

*Bottom*
The dyke or filling of an ancient
volcanic crevasse in the
Kerguelen Islands.

156

they may measure several yards; and they have been known to weigh anything up to 30 tons.

Volcanic bombs, too, vary in size from an inch to several yards in diameter. If the lava is very fluid, the bombs may flatten on impact with the ground (cow-dung bombs), or they may be whirled around in the air until they are twisted at each end with spiral surface markings. If, on the other hand, the lava is viscous, cooling is rapid, and a cracked crust forms on the outside while the inside is still hot and spongy with bubbles of gas. The final formation here resembles a loaf of bread and is, in fact, known as a bread-crust bomb.

Smaller, intermediate-sized pieces of volcanic debris are called cinders, or lapilli ('small stones'). Although they are usually little bigger than nuts, lapilli falling in showers have considerable destructive effect. On the 7th April, 1906, during one of the most important eruptions of Vesuvius in modern times, a hail of lapilli blown by the wind fell on the two nearby towns, Ottajano and San Giuseppe.

'The windows soon broke under the impact of these projectiles,' wrote Lacroix. 'The lapilli accumulated on the roofs, which collapsed into the upper storey of the houses and buried any unfortunate inhabitants who could not or would not take flight, or who, as at San Giuseppe, had taken refuge in a church. The 200 fatalities were all caused by the collapse of buildings.'

Volcanic cinders and ash average anything from a hundredth of an inch to an inch in diameter. Volcanic dust consists of minute particles approximately one-hundredth of an inch in diameter. Being light and durable, dust and ash may be carried for hundreds or even thousands of miles by the wind. Ash from Vesuvius has fallen as far away as Istanbul, while the ash and dust of Krakatoa's 1883 explosion spread throughout the atmosphere of the whole world, and lasted until 1885.

Large-scale volcanic eruptions, which throw a great deal of dust into the atmosphere, can cut down the amount of radiation which the Earth's surface receives from the Sun. Because of this, an unusually cold winter may follow a year of great volcanic activity. The winter of 1783—4, one of the coldest on record in Europe and North America, followed violent eruptions by Asama (Japan) and Laki (Iceland) in 1783. Similarly, 1815, 'the year without a summer', was notable for its long twilights and unusual sunsets — caused by volcanic dust ejected into the stratosphere when Tambora, in Indonesia, erupted continuously for about ten days in the spring of that year. For three days it spread almost absolute darkness over an area extending nearly 300 miles from its crater.

## LAVA FLOWS AND FISSURE ERUPTIONS

One of the most important features of a volcanic eruption is the emission of a more or less fluid lava. This may overflow the crater rim and run down as lava flow. More often, flows are from secondary cones or from fissures in the sides of large volcanoes or at the base of small ones, following the line of the steepest slopes and dividing when they encounter obstacles. Rivers of molten rock, they may extend for several miles, with a volume of millions of cubic yards. Their speed depends on a number of conditions: fluidity, temperature, the quantity of lava, and the gradient. Lavas of the more fluid type, such as the Hawaiian flows, may move as fast as twelve miles per hour. Flows often travel considerable distances — up to 60 miles perhaps — before coming to a halt.

The impressions of the vulcanologist Haroun Tazieff, while examining a lava flow from Etna, are worth recalling here:

'. . . we crossed the heaps of twisted lava with difficulty. Dazzled by the glare of the liquid fire and stifled by the sour wafts of hydrogen sulphide which the wind blew into our faces, we advanced slowly towards the river of lava. Although thirty feet away, we could already perceive the surprising speed of the river of fire. Trembling and coughing, we slowly approached. At twenty paces from the lava it was possible to make more accurate calculations. The molten basalt was descending the valley at more than 12 miles an hour. It was clear that it would lose some of this speed on the way, since two days before we had been observing, unhurried, the slow progress of one of the lava flows at its

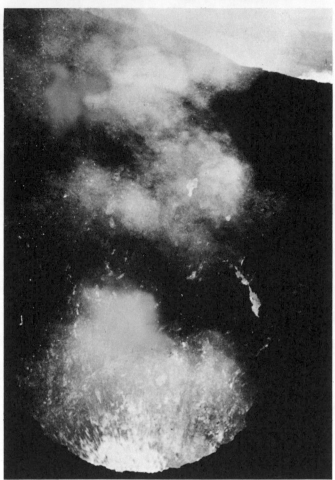

*Top*
A twisted bomb broken open to show the cone round which the fluid lava solidified during its passage through the air.

*Above*
Ash and lapilli being ejected on eruption of Kituro, in the Congo.

*Below*
Dormant crater of Vesuvius.

*Left*
Mount Shishaldin in Alaska, one of twenty active volcanoes in the fog-shrouded peninsula.

*Right*
The crater of an active volcano showing the swirling molten rock or *lava*. Part of the original cone can be seen in the centre.

lowest point, near Rinazzo, at a speed of 30, 50, or at most 100 yards an hour. The lava formed an enormous moving wall, 15 yards high and 200 yards wide, which was steadily and relentlessly pushing blocks of embers and sinister avalanches of solid fire into the vineyards. All were moving slowly forward along the ground in one enormous creeping, viscous mass. Up here at the source, however, the speed of the lava was terrifying. The scorching of our hands and faces did not succeed in driving us back from this spectacle, but the gases finally overcame our resistance. Almost choking and with eyes streaming, we staggered back through the rough basalt, anxiously seeking fresh air.'

The temperature of lava flows is calculated by electric and calorimetric methods. For example, plates of meals with known melting-points are driven into the lava. Here again results vary, and temperatures of 1,000°C., 1,500°C., and 2,000°C. have been given by different observers. The lava which flowed in 1759 from the Jorullo volcano, in central Mexico, was still hot enough beneath the surface to ignite a cigar twenty years later.

It might be supposed that the high temperatures of the lava would give off an enormous amount of heat. This is not so, however, and it is quite usual for a flow to pass through forest or town without causing a fire. One flow from Paricutin even piled up against oaks and cotton-woods without destroying them.

In 1906, lava from Vesuvius passed through the town of Boscotrecase, Italy. A geologist who visited it soon afterward wrote 'houses have been overturned, displaced or buried. Of those overturned, only a few remain, for the ruins have been buried

or carried away; parts of others have been cut clean away, as with a punch, leaving the rest still standing. In some engulfed houses the flow was fluid enough to penetrate all the doors, windows and gratings facing the volcano and fill the rooms. There was one house where it had actually started to mount the stairs, stopping midway between two floors. As for lower-lying houses, some have been entirely buried; others are still discernible by the tops of their roofs — the only part to project.'

How can we explain this anomaly of high lava temperature and absence of fire and flames? To begin with, lava consits of a vitreous mass which is a poor conductor of heat. It also cools quickly at the surface, becoming covered with a crust which in some measure prevents further heat radiation from inside the mass. Thus a lava flow has, as it were, a constantly forming insulating case around its molten interior, so that the front of the flow is preceded by a protecting crust.

The crust is, of course, an obstacle to the release of the gases and steam contained in the lava. Larger and larger bubbles gather under it, until they burst out periodically. As in the parent volcano, small bombs, stones and cinders are shot out under the force of the explosion. Coupled with the movement of the molten material, this often cracks the crust, and the crusted blocks, which are porous and lighter than molten lava, float on the surface of the flow. On cooling, the flow displays extremely rough and irregular surfaces.

Many lava flows develop columnar joints. These are relatively straight, smooth-surfaced fractures caused by the contraction which accompanies cooling. Contraction appears to take place

around more or less regularly spaced centres, causing the rock to break into remarkably regular polygonal columns. These generally hexagonal columns form at right angles to the major cooling surfaces, and stand perpendicular to the top and bottom of the flow. The classic example of columnar jointing in lava flow is the Giant's Causeway on the northern coast of Antrim, Northern Ireland. It is a promontory of columnar basalt, divided by whin-stone dykes into the Little Causeway, The Middle Causeway, and the Larger or Great Causeway. The rock outcrops as irregular polygons, 15 to 20 inches in diameter; the majority of these are hexagons, and there is said to be only one triangular column in the entire exposure. The Great Causeway is in places nearly 40 feet in breadth and is highest at its narrowest part; it extends outward into a platform, and for nearly 300 feet is above water. Neighbouring cliffs, particularly in the bay to the east, exhibit similar columns. The most remarkable is the Pleaskin, the upper pillars of which are 60 feet high; beneath is a mass of coarse black amygdaloidal lava of the same thickness, underlain by a second range of basaltic pillars approximately 40 to 50 feet in height.

The wide extension of lava flows of this particular composition is shown by the occurrence of similar columnar basalt at Fingal's Cave on the Isle of Staffa, western Scotland.

The greatest lava flows, however, generally make their way up to the surface of the Earth not through the volcanoes themselves but along fissures. Vast outpourings, called fissure flows, have emerged from them, not from shields or cones. Sometimes these flows cover hundreds of square miles and are hundreds of feet thick. In north-western United States the Columbia-Snake River plateaux were formed in this manner, and this area of 225,000 square miles is covered with basaltic flows which have an average thickness of 500 feet (over 4,000 feet thick in many places). The Deccan Plateau of India is slightly smaller, covering an area of 200,000 square miles, but the average depth of its basalt beds is probably 2,000 feet. Although fissure flows may contain other types of lava, those of basalt are most extensive. Because they are extremely fluid, they move faster and consequently farther before hardening. Basalt flows of Hawaii have been known to move at ten to twelve miles an hour. Such plateau basalts exist also in Iceland, Africa and South America.

## FAMOUS VOLCANOES AND ERUPTIONS

The differences in volcanoes' activities and shapes stem directly from differences in their magmas, especially in temperature and in the proportions of silica and gases.

Magmas high in silica are very viscous and tend to resist movement until blown out of the vents under accumulated gas pressure. Volcanoes with this type of magma are generally violent, and their steep-sided cones are constructed mainly of fragmental materials. At the other extreme, basic or basaltic magmas, which are low in

*Left*
An active volcano, Mt Ngaurohoc, in South Island, New Zealand. At an altitude of 7,515 feet its snow-covered slopes contrast sharply with the sub-tropical vegetation in the foreground.

*Right*
Part of a ropy lava (pahoehoe) flow in the Craters of the Moon National Monument, Idaho. The surface has a satiny lustre.

silica (less than 50 per cent), usually flow freely and produce gently sloped shield volcanoes and a quiet type of activity characterised by extensive lava flows. If highly charged with gases, however, basaltic magmas can become violently explosive and build up cinder cones.

Most of the well-known volcanoes have magmas which lie between these two extremes and contain only average amounts of silica. They build what we have called composite cones. Magmas of this type flow freely at high temperatures but harden rapidly with cooling. Temperature and gas content determine whether explosion or lava flow accompanies this type of eruption.

The terms used to describe certain volcanic charateristics are often taken from the names of volcanoes known to possess such characteristics. So we have: Hawaiian, Strombolian, Vulcanian, Vesuvian and Pelean (listed in order of increasing violence).

**Hawaiian volcanoes.** The Hawaiian group of islands, rising steeply out of the Pacific at a point where it is 15,000 feet deep, is entirely volcanic. Their volcanoes are typical shield volcanoes, with broad domes or plateaux and gently convex upward profiles. The extremely fluid basaltic lavas and numerous flows issuing from vents on the lower slopes are responsible for the softness of outline. The crust of the Earth under the islands is broken by fractures through which the lava has welled up. Where the fissures intersect the flow has been greatest, and at these points the volcanoes have grown.

The main island, Hawaii, the only one still to shelter active volcanoes, covers an area of 7,600 square miles at sea level. It rises 13,680 feet above the sea, and from sea-floor to summit is over 28,000 feet high. Hawaii was built by five volcanoes: Kohala, Hualalai, Mauna Kea, Mauna Loa, and Kilauea. Kohala and Mauna Kea are now extinct or dormant. Mauna Loa and the adjacent Kilauea are in almost constant activity; both have calders three miles wide, with near-vertical walls surrounding lakes of rippling lava. Fountains of lava gush up in the lakes, sometimes shooting several hundred feet high. Their colour varies from pink to dark red according to temperature on ejection. Sometimes the fountains subside, and a shimmering film forms first floating islands and finally a crust which remains intact until the next eruption. Sometimes the lake overflows its bounds and spreads over the caldera floor, slowly raising its level. Occasionally the lava drains downward and the caldera floor collapses. In time, though, the magma rises again and the pattern is repeated. Attempts have been made to prove that there is a regularity to this cycle, but evidence is far from conclusive.

Data recorded at the Hawaiian Volcano Observatory show that Mauna Loa and Kilauea swell during time of heightened activity. Just before eruption their height above sea level increases, and the volcanoes tilt. After activity subsides, the mountains settle down again.

Hawaiian lavas solidify in long, rope-like lengths, perforated throughout with little cavities formed by expanding bubbles of gas and steam. These relatively smooth-surfaced lavas are called pahoehoe. In places where the release of the gas has been less regular, the whole surface is rough and pitted, like the cinders and slag of a blast furnace. This is a-a, or scoriaceous lava. Both pahoehoe and a-a are terms used to describe certain lava types throughout the world.

Volcanoes similar in form and activity to the Hawaiian shield volcanoes are found in Iceland and Samoa.

An extremely fluid lava offers little resistance to the release of gases and steam. Violent explosions are therefore rare during a Hawaiian eruption, and the inhabitants are often unaware of their occurrence until night falls and their reflection can be seen in the sky. The absence of explosions means the absence of violent ejections. There are no bombs, blocks or ashes. The only materials thrown up are drops of lava which stretch to form long threads as they fly through the air. The filiform products, called 'Pele's hair' in honour of a goddess supposed by the Hawaiians to inhabit the interior of the volcano, are carried by the wind and accumulate in a hardened web-like network of fibres on the surface of the Earth.

**Stromboli.** This volcano is situated in the Lipari islands, not far north of Sicily. Its crater is perpetually full of seething, incan-

descent, semi-fluid lava. Though not as fluid as that of Hawaiian volcanoes, it still produces large flows, which follow the contour of the land into the sea, down a single track on the western side of the island. Steam and gas are not easily released, and small eruptions are therefore almost continuous. The glow of Stromboli is visible from a great distance at night. In ancient times it acted as a beacon for the early sailors. Every ten minutes or so masses of incandescent lava are hurled into the air. Some of it falls back into the crater, some forms cow-dung bombs or spindle bombs — the two solid forms common to soft lavas. Rocks, too, are hurled into the air, but as the explosive force is not sufficient to pulverise them, no ashes are emitted. The gases that are released are colourless but incandescent. The cone is formed mainly of debris. On the northern and southern sides of the mountain — both sheltered from the explosions — wheat, vines and olives are grown. Few active volcanoes are as gentle and regular as Stromboli, which behaves today much as it has done since the days of ancient Rome. Probably only Komba and Batu-Tara, in Indonesia, are comparable.

**Vulcano.** This volcano (which gave its name to the rest) is a close neighbour of Stromboli, lying about thirty miles away. Temperamental and unpredictable, its sudden and violent rages belie its moderate height of 1,250 feet. This vigorous mode of behaviour, which geologists call vulcanian, can be explained by the more viscous nature of its lava. Hand in hand with this characteristic go small lava flows that are quickly cooled on the surface, and a chaotic and scoriaceous crust forming rapidly in the crater and creating an obstacle to the escape of gases. We have already seen that explosion is the inevitable outcome of such conditions.

During the violent eruption which began in 1888 and lasted for two years, millions of tons of lava were reduced to a fine powder. From black clouds pierced by flashes of lightning, bread-crust bombs (typical of this type of eruption) showered down upon the Earth. Even when Vulcano is going through a period of calm, the fumaroles that dot its surface hardly augur well for the future.

**Vesuvius.** Vesuvius, though not the largest, is the most famous of the Italian volcanoes, with a cone rising 3,850 feet above the bay of Naples. The first recorded eruption, in A.D. 79, was extremely violent. The towns lying closest, Stabiae and Pompeii, were buried under ashes, while Herculaneum and others were engulfed by a flow of mud. Part of the volcano was probably blown out in this explosion, as the only part of the former cone remaining is a large semi-circular ridge, Monte Somma.

Terrified by the catastrophe that they had witnessed, the survivors made no effort to restore their towns. The ruins were rediscovered by chance in the eighteenth century, and a start was made in excavating them. Little by little, streets, theatres, public buildings, villas, and shops were uncovered. In the houses there was every trace of life suddenly interrupted and immobilised. Carbonised loaves of bread were found, egg-shells in egg-cups, bottles with their contents reduced to dust. The most impressive exhibits now in the Museum of Naples are the plaster casts taken from cavities in the volcanic rock which once contained the bodies of the petrified victims. Several bodies, including that of a dog, lie in positions showing that death was caused by asphyxiation.

Unlike this early eruption, which was rather similar to that of Krakatoa, the eruption of 1754 was partly strombolian. An eye-witness described 'rivers of lava leaping sporadically in luminous jets like lightning, with red-hot stones hurled out. The reflection formed a brilliant light on the clouds, resembling the aurora borealis girdled by a thick of smoke.'

In 1906, Vesuvius had another major eruption. In the opening phase, flows of lava spread from the volcano; their direction and extent were not at all hard to discover. One flow reached Boscotrecase; another stopped at the gates of Torre Annunziata. Several days later, an explosion blew the summit clean off the cone. An enormous, umbrella-shaped cloud, eight miles high, rained lapilli and ashes on an area to the north-east. The towns of Ottajano and San Giuseppe were partly destroyed, and one night's casualties alone totalled 200 dead and 70 injured.

Another eruption in 1929 was successively vulcanian and strombolian. This time the lava flowed eastward and destroyed

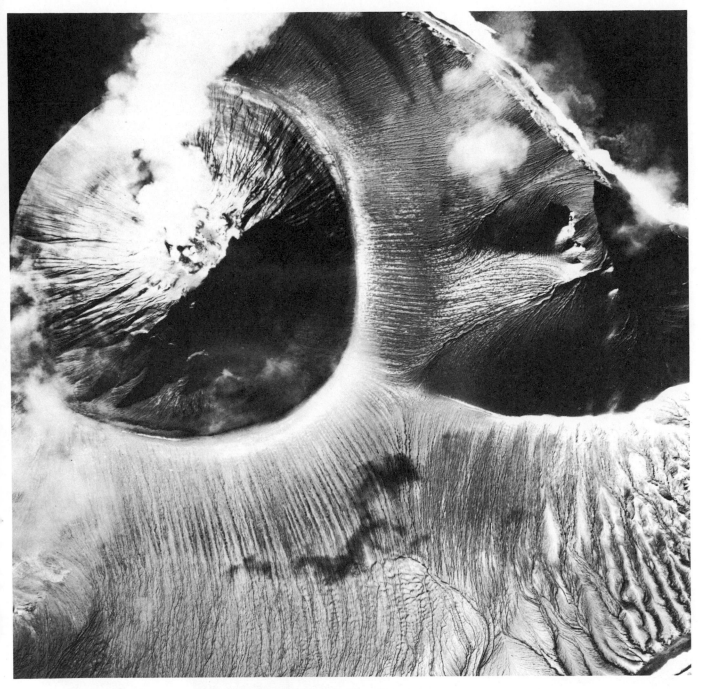

The double crater and liquid basalt flows of an island volcano off the coast of Mexico known as Boqueron.

more towns, among them Lagano, Avini and Campitelli. A further eruption, in 1935, was purely strombolian. During the 1944 eruption first lava and then clouds of ash were ejected.

**Etna.** The island of Sicily is dominated by the imposing mass of Etna, 10,758 feet high, one of the most powerful and variable of active volcanoes. It covers an area of 460 square miles and has a crater about 1,500 feet deep. Since 475 B.C. there have been over 400 eruptions. In 1169 it buried part of Catania, killing almost 15,000 people; in the eruption of 1669 another 20,000 died. In spite of the hazard, Etna's slopes are still populated. Its eruptions are strombolian, vulcanian, or pelean, according to the temperature and degree of fluidity of the lava. The immense central crater very rarely pours out lava, most of it issuing from the many subsidiary cones that pepper its slopes. Thus the steep-sided higher portions of Etna display greater quantities of fragmentals, while the lower and gentler slopes are built up almost entirely of lava. The subsidiary cones are aligned along great fractures which radiate outward from the volcano's centre. New subsidiary crates are constantly formed along old and new fractures, producing in their turn eruptions at various levels.

**Krakatoa.** Situated in the Sunda Strait between Java and Sumatra was a volcanic island which at one time attracted little attention. This island, known as Krakatoa, Krakatao or Krakatau, was buried under luxurious tropical vegetation, and was uninhabited. Suddenly on the 27th of August 1883 almost the entire island (about 18 cubic miles of material) was hurled into the air — some of it to a height of 17 miles. The explosion was heard as far away as China. A gigantic tsunami, or sea wave, drowned almost 40,000 people and destroyed over 1,000 villages on nearby coasts. Large quantities of pumice fell on the surface of the ocean. Dust and ashes were carried round the world by the winds, reaching as far as Europe. Krakatoa once was an island formed by a single volcanic mountain, but in the distant past it had broken apart and reformed as a series of cones: Rakata, Danan, Perboewatan and Verlatan. These merged again to form the island of Krakatoa. After the explosion of 1883, about all that remained was half of Rakata and some land fragments bordering a submarine depression where soundings showed depths of from 600 to 1,000 feet. However, activity beneath the water shows that Krakatoa is rebuilding again.

**Mount Pelée.** One of the most extreme types of volcanic action was typified by Mount Pelée's eruption in 1902, which destroyed the town of St. Pierre, in Martinique, with its 30,000 inhabitants.

The sequence of events has been reconstructed as follows.

Until 1902 the volcano had never been known to erupt. In April of that year, fumaroles and flows of mud were noticed by the inhabitants. These signs, however, caused little disquiet in the flurry of preparations for municipal elections shortly to take place. On May 5, a torrent of hot mud destroyed a sugar refinery and killed several people. Another warning came on May 7, when earth tremors began, and the volcano started pouring out clouds of ash and raining cinders on the surrounding countryside. Panic spread and the exodus began. At two minutes past eight on the morning of May 8 came zero hour. A rift opened in the side of the volcano. With a roar of thunder an immense cloud of thick black smoke gushed out, carrying stones and blocks of rock down to St Pierre at speeds up to 350 miles an hour. As steam and other gases expanded, the cloud grew larger. In two minutes it had reached the town. Nothing could stand up to its advance. The ash in the cloud consisted of lava blasted to minute fragments, and was hot enough to glow redly and melt any glass with which it came in contact (glass flows at about 700°C.). A statue of the Holy Virgin weighing three tons was overturned and carried several yards. Simultaneously the neighbouring town of Sainte Philomène shared St Pierre's fate. Most of the ships in the harbour capsized or caught fire under the volcanic blast. The sea actually boiled, and large numbers of fish and other marine organisms died. In St Pierre there were only four survivors, three of whom died shortly after rescue. Ironically, the only one to escape unharmed was a murderer under sentence of death. He owed his escape to confinement in a very poorly ventilated underground dungeon.

The rare accounts by survivors from boats in the harbour were carefully pieced together by Alfred Lacroix, a French vulcanologist, who visited the scene soon after the eruption.

An observer aboard one of the unharmed boats described his experiences as follows: 'Suddenly there was a violent detonation which shook both earth and sea. It was a tremendous explosion in the mountain, which seemed to be split open from top to bottom, giving passage to a burst of flame and thick clouds of black smoke. These poured down the slopes of the mountain, descending like a whirlwind and sweeping over every obstacle until they reached low ground, where they fanned out and spread toward the unhappy town which was plunged into darkness. The clouds even reached out to the ships in the harbour. Apart from the initial burst of flame, there was no fire, simply a cloud heavy with cinders and pumice, having a tremendously high temperature and able to cover the distance between the volcano and the town, in one-and-a-half minutes. The dark mass caused waves which capsized and overturned many ships. These clouds were extraordinarily dense, masking the town from us, enveloping and penetrating everything...'

These phenomena are vastly different from the characteristics of Hawaiian eruption, and denote a particularly viscous lava. Under the blocked chimney of Mount Pelée, gas and steam had accumulated and finally burst through the surrounding walls, part of which was reduced to cinders and produced *nuée ardente,* or fiery cloud. The plug of solidified lava blocking the chimney caused the explosions to shoot out obliquely or almost horizontally, and probably accounted for the high speed with which the clouds travelled along the ground.

In most volcanic eruptions the ash settles fairly rapidly or is carried upward and dispersed in the atmosphere, but the Pelean cloud is heavily loaded with hot particles of rock and, being much denser than air, tends to hug the ground and move like a flowing liquid. *Nuées ardentes* are not unique to Mount Pelée's eruptions. They are known to have accompanied the eruptions of volcanoes in Alaska, California, Mexico, and many other areas. At Katmai (Alaska), in 1912, an ash and sand flow of *nuée ardente* origin produced a sandy tuff 100 feet thick and covered an area of 50 square miles. Similar deposits are found around the flanks of Mount Mazoma in southern Oregon.

As for the lava plug of Mount Pelée, sometimes it is pushed up and projected little by little from the crater, in the form of a dome or column. Shortly after the eruption in 1902 the plug began to rise. By October it had reached 2,600 feet in diameter and 1,300 feet in height. After that, internal forces, concentrated on one

*Left*
Lava flow of c. 3,000 B.C., and view to cinder cone, Norudurar Dalur, W. Iceland.

*Top right*
Iceland is built up almost wholly of volcanic rock. The surface of Lake Mývatn shown here is dotted with volcanoes and has many craters.

*Bottom left*
Japanese volcano of Mount Aso. Although fairly quiescent, the volcano erupts violently about once a year, spewing forth smoke and ash.

*Bottom right*
Mount Fuji and the waterfall at Shiraito.

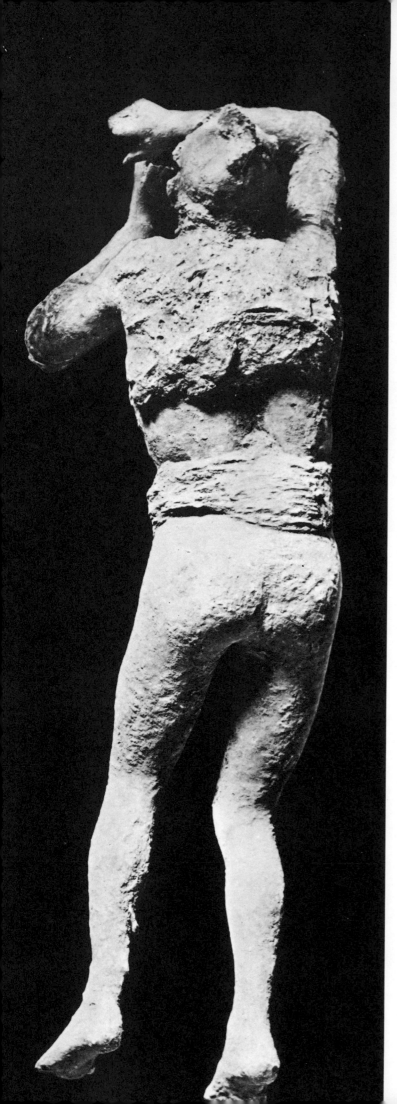

point, pushed up a needle or 'pine' of rock about 1,000 feet high within seven months. Within the following year it slowly crumbled away under erosion and the expansion of internal gas. There was activity again from 1929 to 1932, with hot clouds forming and many small spines growing out of the crater, but they, too, crumbled rapidly.

**Lassen Peak.** Situated in north-eastern California, at the southernmost part of the Cascade Range, Lassen Peak is the only volcano in the continental United States (exclusive of Alaska) to be active in the last century. This 10,400-foot snowy mountain erupted in 1914 and in 1915. It was believed extinct until the 30th of May 1914, when a cloud of condensed steam and ash rose from its summit, and rock fragments were thrown out of a hole which was opened in its snow-covered crater. For most of the following twelve months the activity continued. More than 150 separate eruptions were recorded. The clouds rose to over 10,000 feet in height, and rocks were ejected. Lava rose in the new vent, overflowed, and moved down the snowy slopes. The thawing effect on the snow produced mudflows which carried boulders weighing as much as twenty tons to the valleys below. A hot blast of gases that felled trees three miles away culminated the active period. The force of the explosion was not, however, great enough to dislodge the mountain's central plug and, as in Mount Pelee, the gases were expended laterally from beneath the plug. Since that time — except for a small eruption in 1925 — Lassen Peak has been quiet, and at present emits only a small and generally invisible cloud of gases.

The other famous snow-covered peaks of the Cascades, Mount Rainier, Mount Baker, Mount Hood, and Mount Shasta, are volcanoes that have not been active for centuries and are classified, as was Mount Lassen, as extinct.

**Mount Katmai.** This volcano lies in a relatively uninhabited area, where the Alaskan peninsula joins the mainland. When Katmai erupted in June 1912, the explosion was heard in Juneau, 750 miles away, and in Dawson and Fairbanks beyond the Alaskan Range. Dust fell for three days. About twenty miles from the mountain the deposit was three to four feet thick; 100 miles away, on the island of Kodiak, it was ten to twelve inches thick, and there were traces even 1,500 miles away.

A huge crater about two-and-a-half miles wide and up to 3,700 feet deep was formed. Almost half the floor of the depression was shortly occupied by a lake of warm water. Novarupta, one of Katmai's parasitic cones, extruded a plug 200 feet high, and in a nearby valley hundreds of fumaroles sprang into existence — 'The Valley of Ten Thousand Smokes'. Today only a handful of the many hundreds of fumaroles still survive.

The eruption of Katmai was very similar to that of Mount Pelée in 1902, but on a much bigger scale. It ranks among the half-dozen greatest eruptions of historic times. Yet, because of the sparseness of the population not one life was lost, and it was, in fact, several days before the exploding volcano was identified as Katmai.

## SUBMARINE VOLCANOES

Submarine volcanoes are simply volcanoes which develop on the sea-floor. In principle, they fall into the categories already outlined. Indeed, Vulcano, Stromboli and the Hawaiian volcanoes are very large submarine volcanoes which have ultimately risen above sea level. Few under-water volcanoes, however, construct cones high enough to emerge above the waves. Those which do appear above the water level during an eruption are generally small, and disappear quickly when their unconsolidated pyroclastic debris is attacked and easily removed by the sea.

Several examples of submarine eruptions have been observed. On the 18th of July 1831, the volcanic island of Julia appeared between Sicily and Tunisia. Due to marine erosion it disappeared by the end of December, to reappear briefly in 1863. Three volcanic islands rose off the coast of Iceland on the 15th November, 29th and 30th December, 1963.

The islands of St Paul and Amsterdam in the southern part of the Indian Ocean, and the Santorian islands in the Aegean, are ancient volcanoes partly demolished by eruption. Three islands

*Far left*
The cast of a victim in the eruption of Vesuvius in A.D. 79. The position of his arms in front of his mouth indicates that he died from asphyxiation.

*Top*
The birth of a volcano off the coast of Japan.

*Centre*
In September 1961 earth tremors began on Tristan da Cunha, a lonely island on the Southern Mid-Atlantic Ridge. In October the ground began to crack and the islanders were evacuated by the Royal Navy. Two years later an expedition returned and judged the island safe and in 1963 all but 14 of the islanders returned to Tristan.

*Bottom*
One of the Santorin Islands, Greece; part of the debris of a volcano explosion 2,000 years ago.

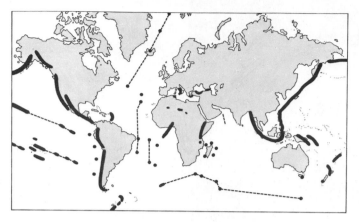

which make up the Santorin archipelago — Thera, Therasia and Aspronisi — represent the edges of the ancient crater which exploded about 2000 B.C. During historic times three submarine volcanoes have appeared between these islands: Palea-Kameni in 97 B.C., Mikra-Kameni in 1573, and Nea-Kameni in 1707. Nearer the present day, fresh submarine eruptions added new lava to these cones in the years 1866, 1925, 1928, and 1929.

Similar eruptions occurred some years ago in the area from which part of Krakatoa was removed by the explosion of 1883. A new island formed in 1927 and disappeared shortly afterwards. It rose into view again during another eruption in 1930 and, most recently, in 1950. The natives called this island Anak-Krakatoa, 'the child of Krakatoa'.

In recent years, violent submarine eruptions have been observed and photographed in the south central Pacific Ocean and in the Aleutian Islands. Cones are often constructed many hundred feet above the ocean level, but with the waning of volcanic activity most of them are quickly destroyed by the waves.

## DISTRIBUTION OF VOLCANOES

Volcanoes have been active throughout the recorded geological past, for their products are found in rocks of all ages, abundantly in some eras and in small quantities in others.

Active volcanoes of the present day, about 600 in number, are not distributed haphazardly over the surface of the globe. Most of them are grouped within belts coinciding with fractures or fracture zones. The belts occur in the roughly parallel areas of the most recently formed mountain ranges. One, often called the 'circle of fire', girdles the Pacific Ocean. A second extends westward from Baluchistan, across Persia, Asia Minor, and the Mediterranean Sea to the Azores and Canary Islands; from there it turns toward the West Indies. In addition to these two major belts there are several smaller areas where volcanic action is concentrated, such as those in the Atlantic Ocean, Cape Verde Islands, St Helena Island, Madagascar, and eastern Africa. A few volcanoes occur independently.

In addition to the principal volcanoes whose frequent and violent eruptions have already been described, the world's active or recently active are listed below.

**Circumpacific Region.** Starting with Alaska, and moving clockwise, the following volcanoes are found:

Alaska: 20, including Katmai and Shishaldin.

Canada: 5, including Mount Wrangell.

United States (Cascade Range): 8, including Mount Rainier and Lassen Peak.

Mexico: 10, including Colima, Jorullo (appeared in 1759), Orizaba, Popocatepetl, Tuxtla, and Paricutin (appeared in 1943).

Central America: 26, including Santa Maria, Fuego, Italco, San Miguel, and Coseguina.

South America (Andean Cordillera): 46, including Purace, Cotopaxi and Sangay.

Antarctica: 2, including Mount Erebus.

New Zealand: 6, including Tarawera and Ngauruhoe.

Polynesia: 23, including Niuafu and Amargura in the Tonga Islands, and Savo in the Solomon Islands.

New Guinea: 30, including Ritter, Ghaie and Raluan.

Malacca and Celebes: 30.

Philippines: 20, including Babuyan Claro, Taal, Mayon, Camiguin del Sur, and Magasu.

Japan: 40, including Asama, Bandaisan, Aso, Fujiyama and Sakurashima.

Kuriles: 18.

Kamchatka: 10, including Kluchevskaya Sopka and Avachinskaya Sopka.

**Mediterranean Axis.** Moving from west to east, and starting from the Antilles we find:

Antilles: 9, including Mount Pelée and La Soufrière in St Vincent.

Azores: 5.

Canaries: 3, the Peak of Tenerife, Lanzarote and Palma.

Italy: 15, including Vesuvius, Stromboli, Vulcano and Etna.

*Top*
World distribution of volcanic activity.

*Bottom*
Fissure lava platform, Grjotagja, Myvatn, Northern Iceland.

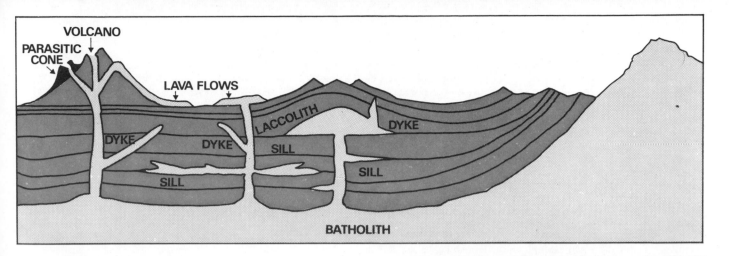

*Top*
The major intrusive and extrusive
igneous forms.

*Bottom*
Mammoth hot springs,
Yellowstone National Park.

Aegean: 3, including the island of Santorin.

The Arabian peninsular and Asia Minor: 6.

East Indies: 79, including Krakatoa, Tangkuban-Prahoe, Papan-daian, Galung-Gung, Merapi, Kelud, Tengger-Bromo, Semeru, Lemongan, Ringgit, Raung, Tomboro, Avu, Tonkoko, Una-Una, Tala, Gamma-Kunowa, Makjan, Api-Banda, Serua, Komba, Tambora.

**Oceanic and African fractures.** The Pacific and Atlantic Oceans, Africa and the Indian Ocean offer the following distribution:

Hawaiian: 5, including Mauna Loa and Kilauea.

Galapagos: 3.

Juan Fernandez: 1.

Iceland: 27, including Askja, Eldeyar, Hekla, Katla, Laki Vatnajokull, and Grimsvotn.

Jan Mayen Land: 1.

Cape Verde Islands: 1, Fogo.

Central Africa: 5, including Nyamlagira and Niragongo.

East Africa: 19, including Teleki, Oldonyo-Engai and Kilimanjaro.

Comoro Islands: 1, Kartala.

Réunion Island: 1, Piton de la Fournaise.

Kerguelen, St Paul and Amsterdam Islands: 3.

## INTRUSIVE VULCANISM

Both volcanoes and fissure flows are surface manifestations of igneous activity, most of which takes place within the Earth's crust.

Its processes and results may remain hidden under a thick layer of rock, or may in time be exposed by erosion. Geologists therefore divide igneous activity, or vulcanism, into two categories: extrusive, which produces volcanoes and lava flows; and intrusive, which produces a number of different types of plutons (the name given to rock masses which have formed through subsurface cooling and solidification of magma).

When intruded into the Earth's crust, the molten materials cool at a slower rate and from rocks quite different in type from those which originate from the surface cooling of lavas. The surface-formed rocks are generally crystalline, and the crystals, except those near the margins of the plutons, are fairly large. Because of pressure, visible gas pockets are absent, and so are porous rocks such as pumice and scoria. Volcanic glasses, such as obsidian, which are formed by extremely rapid cooling are also absent from intrusive rock masses. The intrusive nature of igneous rocks is also revealed by the fact that the rocks above and below them show the effects of heat caused by contact with the molten magma.

If the rocks intruded by the magma are layered, a pluton with boundaries parallel to the layering is described as concordant; if the pluton cuts across the layered rocks, then it is called discordant. Many intrusive bodies, particularly the sheet-like concordant forms, develop columnar joints on cooling, similar to those present in lava flows.

**Dykes.** These are tabular igneous bodies which have been intruded into cracks and fissures in the Earth's crust. They range in size from tiny things a few inches wide to extensive bodies a hundred or more feet across, stretching for many miles and cutting across rocks of all types. Dykes may occur singly or in groups, often in association with volcanoes. They are usually more resistant than the surrounding rocks, and after long erosion remain as ridges raised above the general terrain. Numerous radially arranged dyke ridges occur around Spanish Peaks, Colorado, and a conspicuous formation accompanies Shiprock, an old volcanic neck in New Mexico. In South Africa there is a dyke three to four miles wide that runs across the land for three hundred miles.

**Sills.** Tabular sheets of intrusive material which lie essentially parallel to the layering of the enclosing rocks are called sills. The intrusive body was forced in along the layering surface, where it literally lifted the overlying rocks. Sills vary in length from inches to miles. The best known example in the British Isles is the Great Whin Sill of the north of England. Its westerly extremity can be seen in the steep scarp slopes of the Pennines where it is particularly well exposed in the ice-gouged cleft of High Cup Nick. Farther east, it forms the lip of waterfalls such as High Force in Teesdale; part of Hadrian's Wall is built on its outcrop in Northumberland and finally it appears on the Northumbrian coast at Bamburgh and on the off-shore island of Lindisfarne.

On a smaller scale is the Portrush Sill of Co. Antrim, Northern Ireland. Historically, however, it is a very important locality to the geologist because it was here that the igneous origin of 'basaltic' rocks was finally demonstrated and the controversy between the Neptunists and Plutonists ended.

Columnar jointing develops more often in sills than dykes. The Palisades Sill, 800 feet thick, which forms a cliff along the west side of the Hudson River in New York, owes the steepness of its cliff face to columnar jointing. It owes its name to the same jointing, which resembled the early settlers' palisades.

**Laccoliths.** Magma forcing its way between layers of rock is sometimes so viscous that, instead of spreading out sheet-wise, it accumulates in a large mass, called a laccolith, which arches up the rock layers above. A laccolith may appear as a roughly mushroom-shaped body, or it may be lenticular, oval, hemispherical or irregular. The Henry, La Sal, and Abajo Mountains of Utah are laccolithic in origin.

**Batholiths.** These are the largest of the intrusive igneous bodies, exceeding forty square miles in area and spreading through the Earth's crust to unknown levels. Batholiths may be concordant or discordant in type, and irregular or compact in form. They are generally elongate and make up the cores of many mountain ranges. The Sierra batholith, exposed through long erosion, extends for over 400 miles and forms the bulk of the Sierra Nevada Mountains in California. The core of the Pacific Coast Ranges, extending from Canada into north-western United States, is an even bigger batholith. It averages 80 to over 100 miles wide and totals 1,000 miles in length, covering a total area of 73,000 square miles.

## SECONDARY VOLCANIC PHENOMENA

**Fumaroles.** All lavas rising from the depths of the Earth contain gases, especially steam, which they give off into the atmosphere. The emission of gas does not, however, cease with the explosions, but continues throughout the intervals separating eruptions, forming fumaroles. These gas vents occur on the volcano itself, in the lava flows, or even in the ground surface a considerable distance away from the volcano.

Fumaroles can be classified into five types, according to the gases they release: chloridic, hydrochloric, ammoniacal, sulphuric and carbonic.

CHLORIDE FUMAROLES. Because of their high temperature (from 500°C. to 800°C.), these fumaroles contain no steam, and are sometimes called dry fumaroles. Their principal constituents are the vapours of potassium and sodium chloride. By sublimation (the direct transition from gaseous to solid state) the whitish gases deposit minerals of economic value, such as salts of iron and lead.

HYDROCHLORIC FUMAROLES. These acid fumaroles, as they are more usually called, have temperatures ranging from 200°C. to 500°C. and send forth steam as well as hydrochloric acid and sulphur dioxide in gaseous form. Both sting the eyes and throat of the observer. When dissolved by the rain, these gases can acidify neighbouring rivers, as in the Rio Vinagre, in South America. They also have a strong corrosive action on limestones.

AMMONIACAL FUMAROLES. When the temperature drops below 200°C. ammonia appears. This combines with the hydrochloric acid gas to form white clouds of ammonium chloride, or sal ammoniac, which is deposited as a powder.

SULPHURIC FUMAROLES. After the eye-stinging acid fumaroles and the suffocating alkaline fumaroles, there are fumaroles with a temperature of 100°C. (boiling point) which smell of rotten eggs. Their most important constituents are steam and hydrogen sulphide ($H_2S$). The $H_2S$ gas oxidises on contact with air and, depending on the conditions of the moment, produces sulphur or sulphuric acid.

A solfatara is a fumarole which gives off sulphur gases. Some solfataras are extinct or nearly extinct volcanoes. The volcano La Solfatara, which lies west of Naples, is one of the best known vents yielding sulphur compounds. Another is Vulcano, which supplies about ten tons of sulphur a year. Solfataras occur in many areas, notably in Java, Mexico and Japan.

CARBONIC FUMAROLES. At the end of an eruption and for centuries afterwards, cold, invisible vapours may rise from the ground, chiefly composed of carbon dioxide ($CO_2$). Some of these gases are used in the manufacture of lemonade amd mineral waters. When the carbon dioxide is given off near depressions in the ground or in caves, it accumulates there, for it is heavier than air. The Grotta del Cano near Naples, the Grotte du Chien at Royat in central France, the Valley of Death in Java, and Death Gulch in the Yellowstone district of the United States, are all filled with carbon dioxide.

STEAM FUMAROLES. In areas with a seasonal rainfall many fumaroles are hot springs or geysers during the wet season, emitting gases only during the dry season. Special types, giving off intermittent jets of steam mixed with hydrogen sulphide and carbon dioxide, are called geysers or blowers, *soufflards* in France, and *soffioni* in Italy. The Steamboat Geyser in Yellowstone National Park in the United States has regular pulsations, and jets of steam rise 30 feet or more in the air. In Tuscany, the condensation of steam from *soffioni* fills ponds whose waters are evaporated commercially to obtain boric acid, a kind of gypsum known as Volterra alabaster. This region, as well as Iceland, California, and notably New Zealand are turning more and more to this underground heat as a source of energy for harnessing.

THERMO-MINERAL SPRINGS. Volcanic emanations are not always harmful to human life. Many are of great benefit. Among these are the mineral springs, waters with higher than average mineral content. It is not always possible for the geologist to distinguish whether their waters are of definite volcanic or magmatic origin, which he calls juvenile water, or whether they are waters derived from precipitation, technically known as a meteoric water. Detailed chemical and physical study is necessary to identify them. Meteoric water can penetrate to a considerable depth, undergoing a rise in temperature and mineralisation as it comes into contact with various rocks. True magmatic water is a product of steam condensation coming from magmas.

Virtually all the water in hot springs is meteoric in origin. Only small amounts are of magmatic source, and juvenile water is found principally in a few localities of active vulcanism. But even the presence of an active volcano does not guarantee the magmatic origin of mineral springs, nor does its absence preclude a magmatic source. Each group of hot springs must be examined individually to determine its true source.

Springs of juvenile waters occur near Mount Katmai in Alaska, and in southern Idaho. In the latter, the hotter springs are smaller, suggesting that even here the larger springs with lower temperatures

Geysers along Firehole River,
Yellowstone National Park,
United States.

have been diluted with meteoric water. No volcanic activity has been recorded in the Yellowstone Park area within historic times, yet many of the volcanic rocks of the region appear to be of recent origin. It is therefore probable that the considerable heat in the hot springs and geysers is supplied by a zone of still hot rock lying deep underground, and some of the hot spring water may even be magmatic in origin.

Meteoric water seeping downward through cracks and fissures in bedrock penetrates to depths of several thousand feet, where it comes in contact with heated rocks. In Yellowstone Park hot springs draw their heat from depths of 3,400 to 8,000 feet. Pressure from the weight of the water above pushes the lower water upward again through other joints and seams which widen slowly as solution and other chemical processes take place. At last the waters reach the surface again, appearing as hot springs, pools and geysers.

Where hot spring water contains all the elements present in the rocks through which it has passed, meteoric origin is indicated. Waters of magmatic source contain some of the rarer elements, such as arsenic and boron. Fluctuations in rainfall reflected in the hot springs also suggest meteoric origin, for water levels are higher in the rainy season and lower in the dry periods.

Indications of a magmatic source include absence of seasonal variation in level or flow, constant high temperature, a high degree of mineralisation, the presence of rarer minerals, and radioactivity.

Absolutely pure water is unkown except in the laboratory, for natural waters always contain some mineral matter. Hot spring waters, because of their temperatures and resultant greater chemical activity, have a higher mineral content than warm or cold waters.

Thermo-mineral springs — the term is more precise than the commonly used 'mineral' springs — have been used from very early times in the treatment of various maladies. Religious cults grew up round them, first dedicated to the gods of the Gauls, later to the Roman gods, and finally to saints of the Christian church.

Thermal springs with a mean annual temperature exceeding that of the area in which they occur are almost world-wide in distribution. No continent, it seems, is without them. The possible exception is Australia, and even there, it is not so long since they did exist, for their deposits are still to be found in New South Wales.

Warm or hot springs are found in both cold and hot regions: in the frozen fields of Siberia, in the islands of Alaska, the length of the Andes, the rift valley of tropical Africa and many other areas.

Thermo-mineral springs can be classified according to their chemical and physical properties, or according to their professed therapeutic effects. The first distinction can be made between warm and cold waters, the dividing line being fixed at 65°F (18°C.). Further distinction can be made between alkaline and acid water, between water with and water without dissolved gases. A more precise classification is based on the dominant chemical substances:

Bicarbonate springs have calcium bicarbonate as their principal mineral content. The alkaline waters of the famous springs at Vichy in central France are in this category, as are the springs of Mont-Dore, Puy-de-Dôme.

Chloride springs contain sodium chloride, alone or in association with sodium bicarbonate. These waters have a distinctly salty flavour. Many of these are exploited in France. At Salins-de-Béarn, the water in some of the springs contains seven times as much salt as the sea, and patients undergoing immersion in the baths are held down so they do not float to the surface.

Sulphate springs contain sodium sulphate or calcium sulphate. Those containing the latter are more numerous. Sulphur springs are those in which sulphur occurs in the form of sodium or calcium sulphide, and their waters have a characteristic smell of rotten eggs as hydrogen sulphide is released into the air. Several Pyrenean springs contain sodium sulphide, while calcium sulphide occurs in the Alps. Some springs produce a mixture of the two, as at Eaux-Bonnes and Eaux-Chaudes in the Pyrenees. These sulphurous waters contain a fauna of sulphur-producing bacteria which may contribute in some measure to their healing properties. Sometimes thermo-mineral springs bubbling up through soft soil form deposits of warm mud. At the sulphur springs of Piešťany, in Czechoslovakia, the mud is applied to patients with gout or rheumatism.

A section showing the conditions believed to be responsible for geyser eruptions, based on Bunsen's theory.

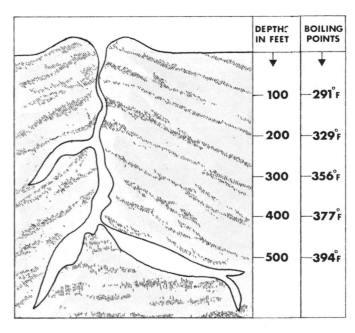

| | DEPTHS IN FEET | BOILING POINTS |
|---|---|---|
| | 100 | 291°F |
| | 200 | 329°F |
| | 300 | 356°F |
| | 400 | 377°F |
| | 500 | 394°F |

餅搗地獄
79

Ferruginous springs contain bicarbonates and salts of iron in solution. These are often described as chalybeate waters. Bath, in Somerset (where the water is at a temperature of 120°F. — 49°C.) is one; others include Forges-les-Eaux in northern France, Orezza in Corsica, and Spa in Belgium. Spa has, indeed, given its name to all health resorts with thermo-mineral springs.

Arsenical springs are those in which arsenic, in the form of iron and sodium arseniate, is associated with one of the above-mentioned salts: with sodium bicarbonate at Mont-Dore, for example, and with sodium chloride and sodium bicarbonate at La Bourboule and St Nectoire in the Puy-de-Dôme area.

Oligometallic springs have poor mineral salt content. These are considered to be beneficial to sufferers from kidney troubles. The waters of Evian on the south shore of the Lake of Geneva, Chaudes-Aigues in the Central Massif of France, and Plombières in the Vosges are oligometallic.

## GEYSERS

The word geyser, meaning spouter or gusher, comes from the Icelandic *geysir*. It is a hot spring from which water and steam are expelled vigorously and intermittently. Geysers are rarer than normal hot springs but, like the latter, are not confined to any particular climatic zone. Excellent geyser formations are found in Iceland, United States, New Zealand, Mexico, the Azores, South America and Japan. In Iceland, geysers and hot springs cover an area of 5,000 square miles. The Great Geyser of this region is particularly well known, and though diminishing in flow, it sends up a daily jet of water nearly 200 feet high and 10 feet in diameter that lasts for twenty minutes. The Gryla Geyser, which plays every two hours, produces a bouncing jet of some 30 feet high.

The Yellowstone Park area of the United States has 200 geysers, besides 3,000 or more hot springs, steam vents, and paint-pots. Here, tiny geysers erupt every few minutes while giant geysers send out enough water to supply a small town. Some erupt according to a regular schedule, others irregularly at intervals varying from one hour to several weeks. The eruptions vary in height from an inch to hundreds of feet, and may last a few seconds or several hours. For many years Old Faithful sent a jet 120 feet into the air

every 65 minutes. But its activity is recently becoming irregular.

The largest geyser on record was Waimangu, in New Zealand, which appeared in 1901 and was active for about two years. At the peak of its activity it spouted a column of water 1,500 feet high, and once threw a boulder weighing 150 pounds a quarter of a mile. Its waters were always dark and muddy, for it was not active long enough to clean and coat its tube with geyserite.

The cause of geyser eruption is not known with certainty. Most geologists agree, however, that outpourings are the result of rapid conversion of extremely hot water into steam. This theory demands a constriction or bend in the geyser's conduit, or an enlargement or ramification of the tube. Given one of these conditions, free circulation of the water as it will be restricted, and steam will be trapped in the cavities. Under normal atmospheric pressure (14·7 lb per square inch) water boils at 212°F. (100°C.) and with increased pressure the boiling-point rises. Under a one-hundred-foot column of water, boiling does not occur till 291°F. (144°C.). Thus, water held by irregularities in the lower portion of the geyser tube may attain temperatures far above the normal boiling-point. As heating continues, a little of this water is converted into steam, despite the high pressures. As water expands greatly when it vaporises, this produces overspill at the top of the geyser, releasing the pressure on the column of water below and thereby lowering the boiling-point. The water below flashes into steam with an almost explosive eruption.

**Geyser and hot spring deposits.** Waters of minerals springs, hot springs and geysers contain a number of salts, most commonly carbonates, chlorides and sulphates. In waters far from volcanic areas chlorides and alkalines are much less predominant. These mineralised waters often build thick and extensive deposits around spring mouths. Several factors combine to produce rapid deposition: evaporation of water at this point, cooling, reduction in pressure, escape of gases. All reduce the water's solvent powers until it deposits some of the material already held in solution. Algae growing in the waters also extract some of the mineral content and deposit it.

Tufa is the general term used to cover all spring deposits, most of which consist in the main of calcium carbonate. When they occur in a pure or almost pure state, they are called travertine.

*Far left*
*(Top)* Hot springs in the Aso National Park.
*(Bottom)* A dramatic moment in the eruption of Mount Hekla.

*Left*
The Petrified Well at Knaresborough, Yorkshire, is heavily charged with dissolved limestone. Any object hung in the dripping water becomes petrified in the course of a few years.

*Right*
Explosive crater of Kerith, now partially filled with water to form a crater lake. S.W. Iceland.

Heavy travertine deposits have been made by the hot springs in the Yellowstone Park area of the United States.

When the carbonate forms in crystalline layers and takes a good polish it is known as Mexican onyx. Siliceous hot spring deposits are called sinter: calcareous sinter or siliceous sinter, depending on composition. Geyserite is siliceous sinter deposited by a geyser. Continuous deposition may produce mounds, cones, and terraces. The most famous are the terraces of the Mammoth Hot Springs in Yellowstone Park whose waters acquire their calcium bicarbonate content as they pass through the limestone bedrock of the neighbourhood. A beautiful series of white and pink deposits, the White Terraces, once existed at Rotomaham in the thermal district of the North Island of New Zealand, but these were destroyed by the catastrophic eruption of Tarawera in 1886. Smaller deposits exist at Karlovy Vary in Czechoslovakia, Tivoli in Italy, Hammam-Meskoutine in Algeria, and in other localities.

**'Petrifying' springs.** Several centuries ago the 'petrification' of objects in certain hot springs was begun as a commercial enterprise in France and England. The term petrification is, strictly speaking, incorrect. The objects immersed in the waters are not turned to stone but encrusted with calcite. All manner of things, pottery statuettes, curios, crucifixes, birds'-nests, baskets of flowers, and fruit were placed in the springs. Over periods ranging from several months to several years the articles were covered with a shining white or tinted stony coat. A variant was to run the calcareous water drop by drop into moulds to form plaques with a clear-cut relief and smooth finish, resembling miniatures cut in ivory.

We should note that certain petrifying springs are indeed a manifestation of vulcanism, often far from actual volcanic centres (as for example, in the Auvergne district of France), but we should remember too that even more of them consist solely of meteoric water, and that they could have been described equally fittingly in the chapter dealing with underground water.

Allied to hot springs and fumaroles are boiling springs, paint-pots, ink-bowls and mud-pots. In the wet season many fumaroles become hot or boiling springs and their gases agitate the waters as they issue forth. Some of these bubbling waters contain rock particles and fragments, fragments of oxidised iron and other substances which colour the water; hence the names ink-bowl or paint-pot. Still others are coloured by the presence of algae, simple forms of vegetable life which usually coat the walls of springs.

## THE BENEFITS OF VOLCANOES

It is usual to think of volcanoes only as sources of catastrophe. The imagination is gripped by eruptions of Vesuvius in A.D. 79, Krakatoa in 1883, and Mount Pelée in 1902, when entire towns were wiped out and thousands of people killed. And yet man has returned time and again to the slopes of the havoc-wreckers, lured back by the remarkable fertility of volcanic soils. The soils which develop from the decomposition of the lavas, cinders and ashes are exceptionally rich in potash, lime and phosphates. In fact, volcanic ash falls short of being the complete natural fertilizer only through its lack of nitrogen. Many districts of the world with a high agricultural population owe the richness of their land to volcanic material. Java, Hawaii and Sicily are paradises of vegetation, thanks to their volcanoes.

Volcanic lavas can also be used as building materials in the construction of houses and metalled roads. Volcanic ash, or pozzolana (after Pozzuoli, where it principally occurs), is also mixed with cement and made into briquettes and blocks for building purposes. Unmodified lapilli are used in some dry areas for mulching the soil, while in the Lipari islands of Italy pumice is worked. Sulphur, boric acid and carbon dioxide are produced by fumaroles. Many of the metallic ores found in rocks are fumarolic in origin, and will be described in the chapter dealing with igneous rocks.

In South Africa diamonds are mined from volcanic chimneys. Thermo-mineral springs, with their therapeutic powers and attraction to the tourist, have been described already. Even the heat of the springs, geysers, and *soffioni* can be utilised as a profitable source of energy.

Economic importance apart, volcanoes have acquired a religious and artistic significance down the ages. Around the beautifully symmetrical Mount Ararat, whose snowy summit towers nearly 17,000 feet, man has interwoven tales closely connected with his own history. In Mexico, the Aztecs offered sacrifices to Popocatepetl to placate the warmth of the devils who lived in their 'Smoking Mountain'. Fujiyama, whose graceful curve places it among the world's most beautiful volcanoes, is inseparable from the art of Japan.

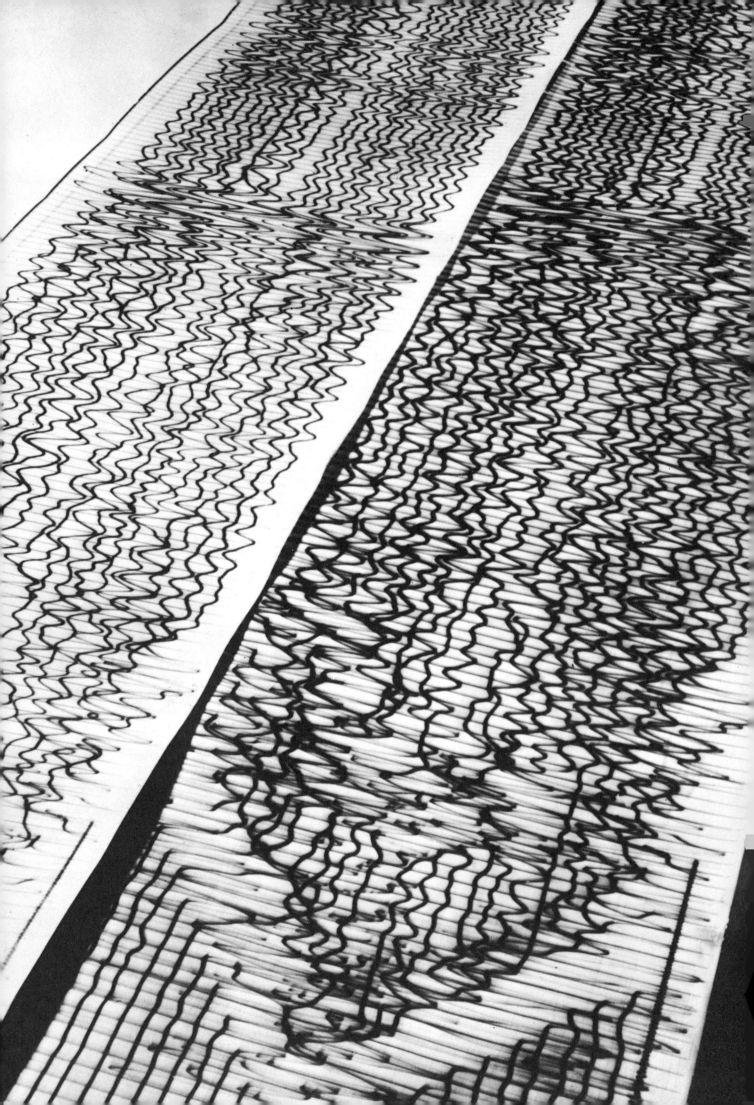

# Earth Movements

We live on a planet whose continents and oceanic basins are in a state of constant change. We have seen that volcanic eruptions can alter the face of extensive areas and destroy whole populations. To their effect on Earth's surface we must add that of earthquakes bringing the sudden subsidence of coasts and wide-scale havoc. Generally, the two phenomena are common to the same areas of the world. Italy, Japan and Chile are the principal regions.

An earthquake is a tremor or undulating movement in the rocky crust of the Earth, set in motion by jarring or a short abrupt movement of the rocks. The study of earthquakes and their attendant phenomena is called seismology (Greek *seio* — to shake); the instruments used for recording such movements are seismographs.

**Seismic phenomena.** The principal seismic phenomena are sound waves, ground vibrations, and 'tidal' waves or tsunamis in the sea.

Witnesses agree in their descriptions of the deafening uproar which accompanies violent earth tremors. They compare it to the rumbling of thunder, the roaring of the wind, the noise of a heavily loaded truck, of gunfire, explosions and violent detonations. Added to the accompanying noise of crashing buildings and surface landslides, these noises from the depths of the Earth may be heard hundreds of miles away from the point of origin.

The worst earthquakes of the last three centuries were those of Callao in 1640 and 1746; Chile, 1751; Lisbon, 1755; Quito, 1879; New Madrid, 1811; Andalusia, 1884; Charleston, 1886; Japan, 1891; Assam, 1897; Alaska, 1899; Calabria, 1904; San Francisco, 1906; Messina, 1908; Kansu, 1920; Japan, 1923 and 1925; New Zealand and Nicaragua, 1931; India, 1934; Anatolia and Chile, 1939; Rumania, 1940; Assam, 1950; Greece, 1953; Algeria, 1954; Agadir, 1960; and Chile, 1960. Japan alone has suffered 230 earthquakes during historic times, and the earthquake of 1923 whose violent motion lasted for only 30 seconds cost her 143,000 dead, 104,000 injured, 576,000 houses completely destroyed, and 126,000 partly demolished. This is by far the heaviest toll taken by an earthquake in our time.

**Intensity of the shocks.** Earthquakes are classified according to the international Mercalli scale of intensity, so that even without instruments severity can be evaluated from observations made by people on the spot without technical knowledge or equipment.

## ABRIDGED MERCALLI INTENSITY SCALE

1. Detected by seismographs and experienced observers at rest.
2. Feeble; noticed only by sensitive persons at rest, especially on the upper floors of buildings. Suspended objects may sway.
3. Slight; felt by many individuals, generally indoors and especially on upper floors of buildings. Standing motor-cars may tremble slightly. Vibration like the passing of a heavy truck. Strong enough to estimate duration and direction.
4. Moderate; felt indoors by many and outdoors by a few—even by people not at rest. Some sleepers awakened. Movable objects, dishes, windows, doors, and even some furniture disturbed. Floors and walls creak. Standing motor-cars rock noticeably. Sensation like heavy truck striking a building.
5. Generally felt by everyone except heavy sleepers. Shifting of furniture, including beds. Ringing of bells. Unstable objects overturned.
6. Shocks violent enough to wake all sleepers. Trees, shrubs, poles, and tall objects sway. Clocks stop. Chandeliers oscillate. General ringing of bells.
7. Severe enough to cause general alarm; people run outdoors. Movable by the people driving cars. Damage negligible in buildings of good construction and design; slight to moderate in normal structure; considerable in badly designed or poorly built structures.
8. Walls fissured. Chimneys, factory stacks, columns, and monuments fall. Heavy furniture overturned. Car-drivers disturbed. Sand and mud ejected from the ground in small quantities. Damage slight in specially designed structures; considerable (sometimes partial collapse) in ordinary substantial structures; heavy in poorly built structures.
9. Buildings shifted off foundations. Damage considerable even in specially designed structures; heavy, with partial collapse, in substantial structures. Noticeable cracks in ground. Underground pipes broken.
10. Some well-built wooden structures destroyed; most masonry and frame structures with foundations destroyed. Ground badly cracked. Railway tracks bent. Landslides considerable in river banks and on steep slopes. Sand and mud shifted. Some flooding from lakes and rivers.
11. Few, if any, masonry structures remain standing. Bridges and embankments destroyed. Railway lines and all underground pipes out of service, earth slumps, landslips and flooding.
12. Catastrophic; damage total. Objects thrown into the air, waves seen on ground surfaces. Lines of sight and level distorted.

**Nature of shocks.** Professor Meunier of Louvain Academy of Arts provides us with excellent descriptions of the shocks he felt in the Nice earthquake of 1887:

'My bed suddenly started to move, first from foot to head, then transversely from my right foot to my left shoulder, and I felt at least fifteen rapid shocks like angry blows delivered alternately from these two directions. Not until this moment did I realise the cause of the disturbance. Then I heard the noises from the street, dogs barking, heavy objects falling, and the bamboos in the garden rustling against the windows, even though there was no wind. The weather remained perfectly clear throughout, temperature and pressure high, and the sea absolutely calm.

To this account may be added that of M. Perrotin, Director of the Observatory at Nice: 'I was awakened before the onset of the shock, and was thus able to observe all the attendant phenomena.

Seismograph records made with a special type of camera attached to twenty-four geophones.

Slight to begin with, they increased with surprising rapidity. At the start I tried to get out of bed, but I could not stand upright: the floor was vibrating from east to west in an extraordinary manner. These oscillations were accompanied by violent shakings, soon over but greater in amplitude.'

These accounts show seismic phenomena to be of some complexity. Each point on the ground is simultaneously moved in three planes and follows a complex trajectory. Vertical and horizontal shock-waves, which combine to give the rotatory or undulatory movement, can be distinguished.

The vertical shocks shake the ground up and down. Mild vertical shocks sometimes produce odd effects. A sugar-bowl, for example, may empty its contents without moving at all. At their maximum intensity, the shocks hurl large objects forcefully into the air. On sandy ground heaps of sand are tossed in the air, and jets of water are often thrown up from ponds.

Horizontal shock-waves can cause rotary movements in objects fixed to the ground at points not directly beneath their centre of gravity. Thus columns, statues and obelisks may be turned or moved on their pedestals.

Most earthquakes produce undulatory movements set up by seismic waves coming from the depths of the Earth in much the same way that ripples are set up in water when a stone is tossed into it. Large surface waves, a foot high and following each other at a distance of about 50 feet, have been recorded. Where waves cross and come into contact with one another, the whole surface of the ground may be covered with temporary hillocks that make it look like a storm-tossed sea.

**Faults and fissures.** A fault differs from a simple fissure in that it is accompanied by movement of the rocks on either side of a fracture. The movement may be vertical, lateral, or in any direction along the surface of the fracture, when such fault movements are a major cause of earthquakes. This type of movement was seen in Japan in 1891, when there was movement along an immense fault running from one side of the island of Hondo to the other. It was 100 miles long, with one side raised 20 feet, and a displacement of 12 feet along the plane of the fracture. No obstacle was able to withstand the movement or turn it from its path. Roads that it crossed were cleft, their height was altered, and they were warped and folded laterally.

The longest fault actually observed in motion was the San Andreas fault, in California, in 1906. It can be followed for over 500 miles along the shores of the Pacific Ocean, cutting its way through the coastal mountain chain. Surveys carried out at a later date by the American geodesy service revealed an alteration in level of up to 3 feet; on either side of the fault horizontal sliding measured 20 feet in places. This fault-line is ancient, but movements still occur along it periodically, producing earthquakes. The San Andreas is a 'living' fault; it is the misfortune of the city of San Francisco to lie across its path.

During the great earthquake of Japan in 1923 the coast was uplifted from 3 to 6 feet. Beds of oysters were found afterwards above sea level, and the bottom of Sagami Bay had been raised by 300 to 800 feet in some places, and in others had subsided more than 700 feet. The uplifting and subsiding was due mainly to the shifting of loose sediments on the floor of the bay.

**Seismic sea waves.** Coastal earthquakes, like submarine eruptions, often cause enormous waves, often erroneously called 'tidal waves', but known to the seismologists as tsunamis. In a tsunami, the sea first of all withdraws from the coastline for several miles, and then returns in immense waves 100 feet or more high and 100 miles or more long, sweeping along at speeds that are sometimes in excess of 400 m.p.h. Often the damage caused by flooding from these waves is worse than the damage caused by the preceding earthquake; anything along the coast that has escaped the earthquake shocks is swept away by the wall of water; ships are carried miles inland, and often thousands of people inland are drowned.

Particularly severe tsunamis occurred at Istanbul (then Constantinople) in 1510, in Peru in 1746 and 1877, in Lisbon in 1755 (60,000 dead), in Chile in 1868, in Java in 1883 on the occasion of the explosion of the volcano Krakatoa (35,000 dead), in Japan in 1896, at Messina in 1908, and in Baluchistan in 1945.

*Left*
Earthquake damage at Diano Marino, Chile.

*Right*
A fissure opened during the Orléansville earthquake, Algeria, in 1954.

*Far right*
Rescue operations in the Orléansville rubble.

A rubble-strewn street of Agadir, Morocco, after the 1960 earthquake which killed thousands and left many more homeless.

# SEISMOGRAPHS AND SEISMOGRAMS

As direct observations of earthquakes can hardly be accepted as wholly objective accounts of events, eye-witness reports are compared and supplemented with the the data supplied by a self-recording apparatus known as a seismograph.

**Principle of seismographs.** Seismographs are founded on the principle of inertia, that is, that the amount of force needed to move a given mass is directly proportional to the size of the mass. For example, in an accelerating bus, a heavy person is less easily thrown off balance than a lighter person. Let us consider another example. When a heavy object is placed on a sheet of paper resting on a table, the paper can be removed with a sharp pull without disturbing the heavy object. A seismograph works in a similar manner. Reduced to its essential parts, it consists of a heavy weight suspended above or in front of a cylinder on which it racks its movements with a needle. During an earthquake, the weight and the needle remain fixed, while the cylinder moves and thereby records its own movements.

In actual fact, the heavy weight of a seismograph can never be absolutely immovable. Even if it resists the first shock, it will probably vibrate with successive waves. Several types of shock-absorber, working by compressed air, liquid or electromagnetism, are used to minimise such effects. However, if the apparatus is to preserve its sensitivity and accuracy, it must not be too well protected. Careful calculation of the maximum shock absorption factor is essential for correct functioning. The optimum condition is for the pendulum to return without oscillation or with only negligible oscillation when moved out of its position of equilibrium. The amount of displacement can then be calculated by interpretation of the pattern, or seismogram, traced on the recording cylinder.

For recording distant earthquakes and minor earth tremors the apparatus must be equipped with a magnifying system, which sometimes consists of a system of levers attaching the needle to the pendulum. In some types of apparatus, however, magnification is obtained by a small mirror continuous with the pendulum, which reflects a ray of light on to a band of photographic paper unrolling before it.

Whatever their type, seismographs are extremely sensitive mechanisms which will record earthquake waves and waves produced by surface phenomena (avalanches, tempests, sea-waves) or by human activities (vibration of machinery, passage of vehicles, explosions). Interpretation of their recordings is therefore complex. The subsidiary recordings must first of all be eliminated by a competent scientist before the earthquake waves can be studied.

Besides studying major seismic phenomena, certain seismological observatories study microseisms; very slight quivers, irregular in motion and occurrence, that may last for hours or days. They have been ascribed to the pounding of surf on the coast, to cyclonic storms over water, and to monsoons. From microseismic records made at specially equipped stations it has been possible to locate a hurricane at sea. Just how a hurricane propagates these waves is not known; land storms do not seem to produce them. Currently much work is being done in this field, in the hope that microseisms will aid in the detailed charting and perhaps even in the forecasting of hurricanes.

Seismographs are today in use in every country, and a permanent International Commission of Seismology centralises and interprets the results of observations made all over the world.

**Seismograms.** The interpretation of seismograms is a complex study that reveals a great deal about the structure of the Earth. The record of a distant earthquake, especially an earthquake taking place on the other side of the world, is composed of a series of graphical undulations corresponding to the series of waves arriving successively at the recording apparatus. Some of the waves have travelled in a straight line through the Earth, while others have followed the curved surface of its crust. Those waves which have penetrated deeply into the interior of the Earth are reflected or refracted on the lithsophere and mantle, and at the junction of the mantle with the Earth's core. Seismic waves are longitudinal or transversal, that is, they vibrate either in the direction of propagation

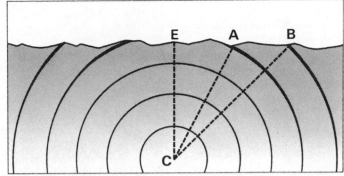

Propagation of seismic waves. The waves spread in concentric spheres from the seismic centre (C) until they reach the surface at points E (epicentre), A, B, etc.

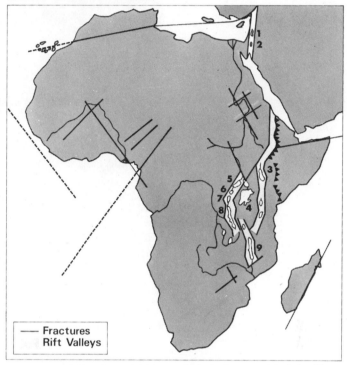

— Fractures
Rift Valleys

Fractures and rift valleys of Africa and Asia Minor. 1. Lake Tiberias; 2. Dead Sea; 3. Lake Rudolf; 4. Lake Victoria; 5. Lake Albert; 6. Lake Edward; 7. Lake Kivu; 8. Lake Tanganyika; 9. Lake Malawi.

or at right angles to this direction. The speed differs in these two forms of seismic waves.

A seismogram, even a very simple one, can be divided into four parts, corresponding to four successive series of waves:

*Primary waves*, called *P* waves. These are longitudinal waves which travel in a straight line through the interior of the globe at an average speed of 4 miles per second.

*Secondary waves*, called *S* waves. These are transverse waves which follow the same path as *P* waves at a lower speed.

*Principal waves*, or long waves, known as *L* waves. These reach the observation point by the longest possible route, that is, round the surface of the Earth, at a speed of 2 or 3 miles per second.

*Posthumous waves*, or waves of feeble amplitude. These are faint tremors that occur in the dying stages of the earthquake.

There are still other series of waves that complicate the seismogram further. Primary waves may be reflected two or three times from the surface of the crust before reaching the observatory (*PP* and *PPP* waves respectively). The same thing may happen to the secondary waves (*SS* and *SSS* waves). A transverse wave can be transformed into a longitudinal wave (*SP* wave), or vice versa (*PS* wave). Similarly, *PPS* and *PSS* waves can be formed, and so on. These few examples will show the complexity of the task the seismologist undertakes. His aim is, not simply to find the point of origin of an earthquake, but also to discover something about both the internal constitution and general structure of the Earth.

THE PRESENT

*Top*
Scene after a Tokyo earthquake.
The buildings were seriously
damaged by fire but survived
the upheaval.

*Centre*
Earthquakes in the Ionian Islands.

*Bottom*
Block diagram of the submarine
canyon of Cap Breton.

## UNDERGROUND CENTRES
## AND PROPAGATION OF EARTHQUAKES

**Focus.** The first point for the seismologists to fix is the depth of the earthquake's underground centre, *focus,* or *centrum.* This he does by comparing several seismographs.

It was once supposed that earthquakes never originated below a depth of 40 miles, that is, below the limits of the lithosphere or Earth's crust. The discovery of centres 200, 300 and even 435 miles deep brought a revision of these ideas. An earthquake can no longer be considered as a simple 'shiver on the skin of our planet'. It is a disturbance that is often deep seated, with its origin in the mantle. It must be noted that this layer, despite its high temperature, preserves enough of its properties of solidity and elasticity to fracture in some places and to flex like steel in others. It has been aptly described as 'solid on a small scale, fluid on a large scale'. That is, it breaks under a sharp impact but, like a plastic body, can resist continuous pressure.

Foci of earthquakes are classified according to depth: shallow, down to 40 miles; intermediate, 40 to 200 miles; deep, below 200 miles. Approximately 85 per cent of the earthquakes of the last fifty years have had shallow foci, 12 per cent intermediate foci, and 3 per cent deep foci.

**Causes of earthquakes.** The major cause of earthquakes is faulting: either movement along an old fault or the development of a new one. Volcanic activity at depth and surface eruptions produce earthquakes that are small or of moderate strength. Still less intense are earthquakes caused by landslides and subsidence in caverns and mines. These last, however, are more likely to be an effect of earthquakes than a cause.

**Isoseismic zones.** When an earthquake originates in the depths of the Earth, seismic waves are propagated as far as the surface. The first surface point to be reached above the focus is the epicentre, best considered as an epicentral area of maximum devastation. Round the epicentre lie the isoseismic zones: the first zone is of partial destruction of buildings (Intensity 9); beyond that is a second zone where walls crack but do not collapse (Intensity 8), and so on, with progressive weakening with distance from the epicentre.

An isoseismic zone could be a perfect circle only if the Earth were homogeneous as far as the focus. In fact, rocks of varying density and elasticity transmit waves at different speeds: for example, the speed through granite is 4,000 yards a second; through sandstone, 1,200 yards; through marble, 600 yards; and through sand, 300 yards. The relief of a region will often favour the propagation of waves in certain directions but hinder it in others. The result is that isoseismic zones are irregular in shape.

Another fact revealed by seismology is that the deeper the origin of an earthquake, the greater its area. The Ischia earthquake of 1883 with a centre of origin less than a mile below the surface did not spread beyond its own island confines in the bay of Naples. In 1886 the Charleston earthquake in South Carolina was felt in Boston, Milwaukee and New Orleans, halfway across the United States. Here the focus was 18 miles deep. Aftershocks of this earthquake were recorded for over a year. In very deep earthquakes, however, (those with a focus between 200 and 450 miles down) extent is not proportional to depth. After travelling such distances the seismic waves are greatly reduced in power when they reach the surface and are, as a rule, restricted in intensity and extent.

The rocks of the Earth are not the only medium to transmit seismic waves. The sea is an easily penetrated and homogeneous medium, but the waves pass through it more slowly because of its lesser elasticity. The shock-waves leaving Chile on the 13th of August 1868 took 12 hours to reach Hawaii, and 19 hours to reach New Zealand. Those released by the explosion of Krakatoa were felt in Europe the following day.

## EARTHQUAKES AND EARTH FRACTURES

From what is known of the origin of earthquakes, it is not surprising that they occur mainly in regions of recent folding and faulting, and also, though rarely and with less intensity, in the regions of ancient

Aerial view of fissures in the
Earth's surface after
a violent seismic shock.

folding and faulting. It is immediately obvious that their distribution coincides almost exactly with that of volcanoes. Yet neither is cause nor effect of the other; both arise from the same general conditions prevailing in a particular area.

The main earthquake zones include the circum-Pacific belt (west coast margins of the Americas and the island archipelagos of Asia) and the Alpine-Mediterranean-trans-Asiatic belt (north Africa, Spain, Italy, Greece, Turkey, Iran, northern India and Burma). These two belts approximately coincide with the main volcanic belts and with the most recently formed mountain ranges. In addition, there are minor earthquake belts along the submarine ridges of the Atlantic, Indian and Arctic Oceans, and in the rift zones of eastern Africa and east-central Siberia. The circum-Pacific zone accounts for over 80 per cent of the total energy released by earthquakes. The second zone, the Alpine-Mediterranean-trans-Atlantic, accounts for about 15 per cent. This leaves approximately 5 per cent elsewhere in the world, most of it in submarine ridge areas.

Not all earthquakes take place within these belts. Some — those at New Madrid and Charleston, for example — occur far from areas of recent fracturing, mountain building or volcanic action. Occasional minor 'quakes occur in areas of recent continental glaciation, perhaps as a result of unloading by melting ice sheets. A few earthquakes are felt in areas of delta building — apparently an effect of loading the Earth's crust by sedimentation.

In attempts to forecast earthquakes, the work of the seismologist is reduced to the task of recording the number of shocks received daily in a given area. As long as the number is increasing continuously, a major earthquake may be on its way, and warnings should be issued. In Japan, for example, the seismographs record 7,500 shocks a year, about twenty a day, even in calm periods. On 17th of November 1930, the daily recording rose suddenly to 100. The increase continued, and on the 24th of that month 700 shocks were recorded. The following day a major earthquake occurred. Increase in frequency of minor shocks does not, however, herald a subsequent major earthquake.

As yet, it is not possible to predict accurately the occurrence of earthquakes. A few facts indicate that the full moon, high tides, sunspots, abrupt changes in air pressure, and a series of smaller shocks may be associated with them, or may trigger them off. At the present time, for instance, a strain is building up along the San Andreas fault, but it cannot be determined when breaking-point will be reached. Because it is not possible to forecast or control earthquakes the only solution is to build structures to withstand the shock-waves, avoiding as far as possible locations where the phenomenon has been most frequent and severe in occurrence.

The trail left by a meteorite on
17th November, 1955. It was
observed to enter the atmosphere
at an altitude of more than 80
miles over the west coast of
France; it disintegrated over
central France.

# Exploring Earth's Crust and Interior

In our study of volcanoes and earthquakes, we have already made passing reference to the internal structure of the Earth. We shall now examine the subject more carefully, exploring first the surface, and then the depths.

The search for economic deposits, oil, metals, coal and underground water, has led to the development of a new science, geophysics, and a new technique, geophysical prospecting, whose methods are as highly specialised as they are spectacular.

**Gravimetric method.** Gravity is the force of the Earth's attraction which pulls objects groundward when they are dropped from a height. It is the force responsible for keeping us on the surface of our planet. It accounts for the 'weight' of an object though that weight is not uniform over the whole of Earth's surface, for it increases with latitude and decreases with altitude. Weight may also vary with topography. When minor causes of variation are eliminated, weight remains fundamentally due to the nature and structure of the rocks beneath. The presence of thick masses of certain rocks increases weight; the presence of other rocks tends to diminish it.

Instruments used in gravity surveys include the pendulum, the torsion balance, and the gravimeter (gravity meter). Today only the last two are commonly employed in exploratory geology. An instrument of great sensitivity, the gravimeter detects the slightest variations in gravitational attraction, in other words, the variations in weight. The gravimeter is moved through a series of stations, and the readings with corrections are used to plot a gravimetric map which supplements the information given on a normal geologic map.

The intensities measured by the gravimeter are dependent mainly on the spatial arrangement of rocks of different specific gravities. The gravity response to the rocks beneath the surface is affected by the depth, the size and the shape of the rock mass, and by the nature of its density contact with adjacent rocks. This type of geophysical surveying is used to best advantage in areas where the topography is gentle and the difference in relief is slight, where rock density differences are large, and where the rocks lie within 200 feet of the surface.

**Electrical methods.** If the speed and cheapness of gravimetry is useful for making a preliminary survey, it is not always sufficiently accurate for detailed survey. Here, recent developments in electrical methods of prospecting may be substituted, and either the natural electrical currents of the Earth are employed or artificial currents are induced.

Continuously passing through the Earth are electric currents called telluric currents. Their precise origin is unknown. All that is certain is that they are connected in some way with sunspots, protuberances and eruptions of the Sun, as well as with fluctuations of the Earth's magnetic field, polar aurora, and other natural electrical phenomena. The intensity and direction of these currents being variable, the only way to apply them in prospecting is to take simultaneous readings from several stations. Generally, the readings are taken from a base station situated in the centre of the zone to be studied and also from mobile units, which plot a network of observations over the whole area. At the base station and at each of the mobile units, the variations and differences of potential are measured between the ends of two cables, several hundred yards long, which are laid out at right angles to each other and are in contact with the ground through metal rods or terminal electrodes.

Let us take as an example the electrical exploration of a region in which the ground-rock is formed by a covering of sedimentary rocks (good conductors of electricity) over a base of crystalline rocks (poor conductors). Where the resistant base lies near the surface, the currents have less room to flow freely and the intensity of the electrical field builds up at this point. From intensity recorded, then, it is possible to calculate the depth of the crystalline rock. With a large number of readings taken over the sea under study, and with the necessary calculations, the geologist is able to plot the undulations and shape of the deeper layer as clearly as if the overlying rock were transparent.

The method of prospecting by electrical soundings, now widely used in the petroleum industry, was developed by the Schlumberger brothers, members of the Societe de Prospection Electrique. The recording apparatus, too, was invented by members of this body. An electrical current is artificially produced in the ground, and its propagation and the fall in its potential are recorded at various points. From variations in electrical resistance, the nature and arrangement of unseen layers of rock can be deduced. The procedure consists in passing an electric current into rock formations through a number of electrodes, two of which are used for current flow and the remainder for measuring the resulting variations. The greater the distance between the two main electrodes, the farther the current penetrates into the rock layers. The recording apparatus used is called a potentiometer. For deep measurements, the general name 'sonde' is given to the operations.

**Magnetic method.** The third method of geophysical prospecting is based on terrestrial magnetism, and a number of different devices are employed in its application to magnetic survey work: the dip needle, the Hotchkiss Superclip, and the magnetometer. All these instruments measure particular variations in the Earth's magnetic field. The most sensitive and most widely used is the magnetometer; it is even used in making aerial surveys, when ground instruments measure the vertical or the horizontal components of the field, and air-borne instruments the total field. This method is effective in outlining areas in which magnetic intensities are much greater or much less than the general regional magnetic background. It is an invaluable way of surveying inaccessible land areas and extensive areas of water. Certain regions are particularly suited to this method of surveying because of the presence in the subsoil of a layer of sandstone rich in iron oxides.

The same method is eminently suited to the detection of metalliferous deposits, particularly of magnetite and pyrrhotite.

There are, however, certain disadvantages in the magnetic method. First, magnetic storms make readings subject to error. Secondly, the magnetic balance used for these observations is of such sensitivity that it detects the presence of any nearby metal object and its readings are correspondingly influenced. One

prospector recounts how his balance was long subject to abnormal displacements until he discovered that a neighbouring well, to all appearances a harmless thing, contained a load of scrap-iron dumped there by the local farmers.

**Seismic method.** Of all the modern methods of prospecting, the most impressive is that of seismic sounding, which proceeds by man-made explosions in the ground. The ensuing vibrations are measured as they travel through the layers of rock. The waves are refracted from some layers, that is to say, their direction of travel is changed just as the path of a ray of light changes as it passes from one transparent medium to another. From other layers the waves are reflected. There is thus seismic refraction and seismic reflection, either of which may be utilised. The recording apparatus in seismic methods is a seismograph, or geophone, a smaller version of the instrument used for recording earthquake movements.

The application of seismic technique demands highly qualified personnel and complex machinery. First, there is a mobile drilling-rig. Though smaller, it works like an oil-boring derrick. Once the hole is drilled (anything from ten to a hundred feet or more in depth), a shot-firing truck carrying a tank of water replaces the drilling-rig. A charge of dynamite is lowered into the hole, and the hole filled with water. Geophones are laid out round the borehole, synchronised so that their response to the shock-waves is recorded at precisely the same instant. At a given signal the charge is detonated electrically. A column of earth and water is thrown up, and the shock-waves transmitted through the ground are reflected and refracted from the various layers of rock. From the data recorded by the seismometers the nature and arrangements of the underlying rocks can be determined.

In the United States, seismic surveying is an important aspect of industry, using annually several thousand tons of explosives, 1,000 miles of film for seismogram (the 'pattern' of waves traced by the seismometer) recording, and drilling a million holes the total depth of earth drilled amounted to over 60 million feet.

**Coring.** Excellent though the results from gravimetric, electric, magnetic and seismic methods of prospecting are, they must generally be augmented by a programme of drilling before the oil or mineral deposits can be worked. This may sometimes take years. Holes must be drilled down to the level of the deposit, and cores of rock extracted during the boring. The coring tool is a crown drill set with small industrial diamonds which cut into the rock as the drill rotates. As the drill sinks down, a core of rock rises into the pipes of the drill-rods. The cores are then examined in the laboratory by geologists and chemists. This procedure is very costly for, even discounting the price of the diamonds, a great deal of time and labour is involved. Work is extremely slow, since every time a core is raised, the whole series of drill pipes must be withdrawn from the hole. Even more serious is the fact that, despite all precautions, certain friable rocks — such as poorly cemented sandstones — crumble or disappear completely in the core. It is even possible to drill through a coal seam without being aware of it.

**Electromagnetic, radioactive and sonic methods.** In recent years these methods of surveying have been more generally applied. The first utilises a loop of wire to induce secondary currents in any conductive substance around it. The electromagnetic fields created by the secondary currents are then carefully examined. The instruments used in this method have been adapted for foot-by-foot examination of drill holes. They have also been successfully used in aerial surveys.

Radioactivity methods and devices have developed as a direct result of the search for uranium during recent years. They are now used extensively in the petroleum industry. The instruments used measure the disintegration taking place in radioactive elements present in the rocks. Familiar among the devices used is the Geiger counter, which measures gamma ray intensities. Less well known, but very widely used, is the even more sensitive scintillometer, measuring both gamma and beta ray intensities. As it registers

even minute differences in radioactive intensity, it is used in detailed evaluations of rock properties and also in aerial surveys.

The most recently developed of the various subsurface exploration techniques is sonic logging, which involves recording the time a sound wave takes to travel through a specific length of rock formation. The speed of sound in subsurface formations depends on their elastic properties, their porosity, their fluid content, and their pressure. Differences in speed are, therefore, directly related to subsurface conditions. The sonic log, which is detailed and records with a high degree of accuracy minute changes in the rocks, is proving excellent for correlation of data. For example, electric logging (resistivity and other types), electromagnetic or induction logging, radiation and sonic logging devices have all been adapted for use in oil drilling procedures. Their use in this field in recent years has reduced the amount of coring needed, because the transmitting and receiving instruments are small and can be speedily lowered and raised in a drill hole, and the log — generally photographic — is recorded rapidly on a truck-mounted receiver near the hole.

A new technique of using X-ray techniques to study rock formation is providing important data on the structural framework of rock. Sandstones, for example, which appear to be completely 'structureless' in outcrops and hand specimens, show various types of delicate layers in X-ray pictures. These studies provide information on the origin of rocks, and give clues to their economic importance as well.

## THE DEPTHS OF THE EARTH

By the application of the many methods of geophysical prospecting the Earth's structure can be examined from the surface right down to the bottom of oil drillings several thousand feet deep. But these depths are fractional compared to Earth's radius of approximately 4,000 miles. The geophysicist finds himself, as it were, in the position of a doctor who can examine only the skin of his patient. There

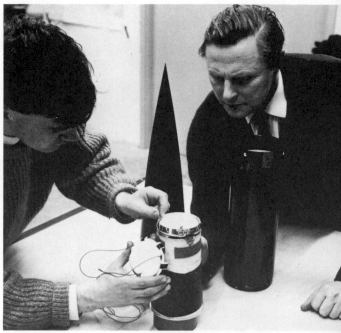

*Top*
Galitzin Pendulum for vertical seismic component.

*Bottom*
Electric boring methods. The shot-firing truck fills the hole.

*Top*
Preparation of a rocket sonde prior to being placed in nose cone. Rockets carry radio sondes, which measure temperature, humidity and high-level wind speeds and directions. The sonde is returned to the earth by means of a parachute.

*Bottom*
Final explosion in electric boring method.

are various ways of obtaining information about the deeper constitution of the Earth, but ultimately they produce only unverifiable hypotheses.

**Density of the Earth.** If the Earth could first be weighed we would have some assessment of its mass and density. The proposition is absurd, but in 1798 an English physicist, Cavendish, did calculate Earth's mass and density, basing his method on the principle of attraction, or universal gravitation. That is to say, he worked on the principle that two bodies are drawn together in the direct ratio of their mass and inverse ratio to the square of their distance.

His apparatus was a torsion balance composed of two fixed spheres of lead which attracted two smaller spheres of the same metal. From the twisting induced in the silver wire acting as support, the force of attraction of the larger spheres was calculated and, by relating this to their weight, the force of attraction, mass and density of the Earth was deduced. Despite later improvements in techniques, Cavendish's deductions remain almost without modification, and the density of the Earth is taken as 5·5 grams per cubic centimetre.

Most surface rocks, such as granite, limestone, clay and sand, weigh only 2 to 3 grams per cubic centimetre. What are we to conclude from the difference between the average density of the Earth and the density of its crust? Evidently deep down there is material with a density of more than 5·5 forming a heavy kernel.

**Terrestrial magnetism.** Let us approach the problem from another angle. We know that the Sun possesses an enormous electric charge and while spinning on its axis sets up a magnetic field which extends well beyond the orbits of the farthest planet in the solar system. While rotating in this magnetic field Earth becomes electrified. But to explain terrestrial magnetism Earth must contain sufficient metal, particularly iron and nickel, to act as a magnet. Where are these heavy metals with an average density that must equal 15, if not in the centre? The heavy core, necessary to explain the density of the Earth, can only be of iron and nickel.

**Temperature of the Earth.** Another major phenomenon is Earth's temperature. At the surface, the temperature of the planet varies with the seasons. On the other hand, at shallow depths of 50 feet or more the temperature remains constant at about 55°F. (11°C.). This is the temperature recorded in many cellars and caves, 20 feet to 200 feet deep. The Paris Observatory has recorded the temperature of its 74-feet-deep cellars since 1718; the variation observed has not been more than 1° centigrade.

Normally, temperature underground increases with depth. The rate of increase is known as the geothermal gradient. Below an upper zone (50 to 400 feet) where temperature is affected by changes in the atmospheric temperature and by circulation of

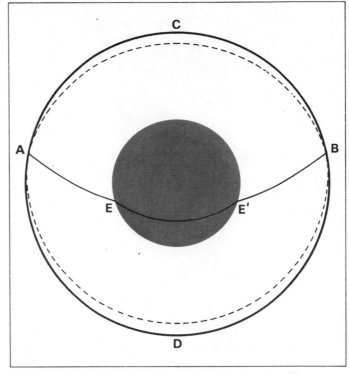

Transmission of earthquake from A to B by the crust (ACB, ABD) and by the Earth's core.

ground water, the gradient is more or less constant, with an average increase of 1°F. per 60 feet, or 150°F. per mile.

The rate of increase varies, of course, with the nature of the surrounding rocks. Volcanoes and thermal springs, too, produce local heat increases in rock. Sometimes radioactive rocks suddenly raise the temperature above the expected limits.

All observations made to date of temperature in artesian wells, mines, borings and tunnels lead to the same result: temperature rises steadily away from the surface of the Earth.

Temperatures in oil wells are measured with a recording thermometer in a device called a temperature bomb. The whole apparatus may be lowered into the hole or the recorder may remain at the surface while only the thermometer is lowered. At the bottom of a 6,000-foot well the temperature reads about 100°F. more than average surface temperature. The geothermal gradient is doubtless steeper near the surface than at depth. Temperature increases may be as high as 1°F. in 30 feet near the surface or, as in the gold mines of the Transvaal, as low as 1°F. per 250 feet. In California,

Principle of isostasy. The crust is greater under the mountains than under the sea. Weight is decreased at A, since the density of the crust is less than that of the substratum.

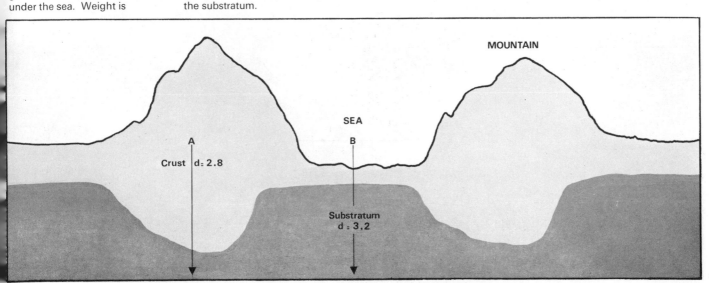

the bottom of a 16,000-foot well showed a temperature of 204°C.

Scientific speculations based on geothermal gradient regarding the temperature at the centre of the Earth had to be given up when the existence of radioactive substances was discovered. These give off heat, but their total quantity and their distribution in the Earth is still unknown. Probably they are confined to the outer layers; if they extended to a depth of ten miles in the quantity in which they occur in granite rocks, for example, they would produce more heat than that shown in the geothermal gradient. Such temperatures are unlikely because they would raise the level within the solid crust above the melting range of all rocks.

Most scientific estimates for the temperature of the Earth's centre range between 2,000°C. (3,632°F.) and 4,000°C. (7,232°F.), although a few quote figures ranging from 6,000°C. (about the same as the Sun's surface) to several hundred thousand degrees.

Pressures existing at the Earth's centre are also a matter for divided opinion, but there is much smaller schism here, for the densities of the rocks are known and pressures at depth can be calculated. The pressures at Earth's centre are calculated to be approximately 3 million atmospheres, or more than 20,000 tons per square inch.

What, then, produces Earth's heat? Here, too, there are several schools of thought. Some believe that the heat is the initial heat remaining from the time of the Earth's molten origin. Following this theory, calculation shows that for a thousand million years Earth should have been completely isothermic, that is, at the same average temperature (11°C. or 55°F.) from the centre to the surface. There would, in addition, be an exact thermal equilibrium between the amount of solar radiation to be given out at any particular time.

This hypothesis, however, does not take into account the radioactive elements with which the rocks of the Earth are impregnated. Uranium and thorium (the principal of these elements) and potassium emit by disintegration a continuous and practically inexhaustible flow of atomic energy. A single gram of radium, a by-product of the disintegration of uranium, produces 0·037 calories per second, that is, about 31 thousand million calories in the course of its life. The potential of nuclear energy is so considerable that its manifestation is viewed with some alarm lest radioactive elements exist in every part of the Earth's mass. Happily, many geophysicists consider they are present only in the Earth's outer zones and that the interior is almost free of them.

A theory that at once suggests itself on considering Earth's temperature is that volcanoes may be fed from the very centre of the globe and are proof of great internal heat. However, two facts refute this theory: the temperature of lava is only 600°C. to 1,500°C. (1,112°F. to 2,732°F.), and seismic evidence indicates that volcanoes, sources are not more than 25 miles down. Vulcanism is, then, a very superficial phenomenon—no more than blisters breaking on the surface of the skin.

**Earthquakes.** Records of earth tremors and earthquakes are another source of information about the interior of the globe. We have already described the seismograph which notes the smallest earth tremor. The same shock can be recorded after direct passage (primary or *P* waves) more or less through the centre of the Earth, or after following the Earth's crust (principal or *L* waves). The direct trajectory is not, however, straight but interrupted or curved as a result of the refraction produced by layers of rocks of different density and physical condition. This can alter the method of propagation of earthquakes and indirectly give information on the constitution of the layers in the interior of the Earth.

The primary earthquake waves transmitted by the Earth travel at rates which range from 3·4 to 8·5 miles per second; waves passing into the deeper layers travel at about 4 miles per second. From this it can be deduced that the deeper parts of the globe have greater rigidity than the surface layers, and that their elasticity surpasses that of steel. It is a conclusion that lends support to the theory of the existence of a metallic core situated in the centre of the globe.

The seismograph shows, too, that there are in the Earth's interior several surfaces of discontinuity which produce abrupt changes in the propagation of shock waves. The most important of these surfaces occur at depths of 6 – 9 miles, 18 – 25 miles (Mohorovicic discontinuity), and at 1,800 miles (Dahm discontinuity). Less important ones, producing less abrupt changes, are found at 250 miles and 435 miles. Later we shall see what such surfaces reveal about the structure of the globe.

**Isostasy.** To understand the fundamental phenomenon of the alternating subsidence and uplift of the Earth's crust, let us first take a concrete and recent example, that of Scandinavia. Weighed down by ice during the glacial period, this part of Europe sank several hundred feet. Since the end of this period, it has risen and continues to rise today at the rate of 3 feet per century. What is true of Scandinavia also holds true for other regions. The Earth's crust is made up of segments which rise or fall in relation to· one another. A lowering movement in each segment temporarily weighted is compensated by an opposite movement in an adjacent section.

'This condition of equilibrium of figure to which gravitation tends to reduce a planetary body, irrespective of whether it be homogeneous or not' (Dutton, 1888) we call by the name of isostasy (Greek *isos* — equal, and *stasis* — stability). This conditions the regular astronomical movements of the Earth and suggests, too, that its rigid crust rests on a plastic layer into which it can sink.

If the principal of isostasy is correct, the rigid crust should be deeper in the mountain areas (because of their thickness) than in crustal areas elsewhere. In other words, the crust must be much thicker under the mountains than under the sea, as the diagram on page 187 illustrates.

Can this be proved? Theoretically, the weight is equal at all points sharing the same altitude and the same latitude, that is to say, at all points lying the same distance from the centre of the Earth. But this supposes that the crust is homogeneous in constitution. If, on the contary, the crust is of lower density beneath the mountains and the continents, and of higher density beneath the sea, the oceanic segment must be at a deeper level than the continental sections. Although the density differences are small this is in fact so. In the diagram, density is less at point *A* than at point *B* (*see* p. 187). Gravimetry thus confirms isostasy.

## METEORITES

Throughout history man has been fascinated by 'shooting' stars. They were once thought to be gifts from the gods, and even the smallest were collected and carefully preserved in the temples. The temple of Diana at Ephesus, for example, contained a stone supposed to have been thrown by the goddess; while the famous 'Black Stone' enclosed in the Ka'ba in Mecca is still an object of pilgrimage for all Islam.

Early witnesses of these falling objects did not hesitate to embellish their accounts at the expense of true fact, and attributed meteorites to extraordinary causes. It was not until a fall at Benares in India, in 1798, that scientists began to attribute the phenomenon to a natural cause. In April 1803 the first scientific study of a 'hail of stones' was made at Laigle in France. The Academy of Sciences of Paris sent an investigation committee under the leadership of the physicist Biot. He reported:

'On Tuesday, 25th April, toward one o'clock on a fine afternoon, an extremely bright flaming sphere moving rapidly through the air was sighted. Several moments later a violent explosion lasting five or six minutes was heard over an area about thirty leagues in diameter. The explosion consisted of three or four cannon-like reports, followed by a sort of discharge resembling first a fusillade and then a deafening roll of drums . . .

'This noise came from a little cloud, rectangular in form, its longest side lying in an east-west direction. The cloud appeared stationary throughout the duration of the phenomenon. Only the vapours composing it spread out on all sides from the effect of the successive explosions. This cloud occurred about half a league north-north-west of the town of Laigle. It was high in the sky, for it was simultaneously observed overhead by the inhabitants of

190,000 square miles of the United States and Mexico photographed from an altitude of     500,000 feet. The black patch (top left) is the Gulf of California.

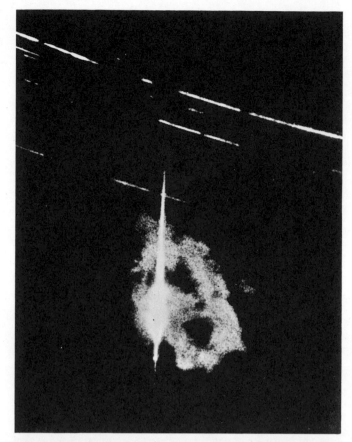

*Top left*
Meteorites exploding. The white
splash represents the point of
explosion, the trail of light,
the projection of gases.

*Bottom left*
Nebula and Ntar trajectory.

*Right*
(Top) Iron and nickel
Giron olivine meteorite about
12″ long from Wyoming.

(Centre) A Meteorite from
Saskatchewan, showing torn
and partly fused surface.
(Bottom) Stone meteorite,
Texas.

villages situated more than a league from one another. Throughout the area over which the cloud hovered were heard whistlings like those of stones hurled from a catapult, and at the same time a shower of solid masses exactly like those called meteorites was observed. The biggest of those retrieved weighed 8½ kilograms; the smallest I saw weighed only 7 or 8 grams. The total number that fell can be estimated at two or three thousand.'

Since 1803 many falls of meteorites have been observed. Some have been spectacular. In January 1868, in Poland, more than 100,000 small meteorites fell in an area of cultivated land covering no more than 15 square miles. On the 30th of June 1908, a fiery meteor estimated to weigh more than 40,000 tons crossed the sky of Siberia and fell in the region of Lake Baikal. The noise of the explosion was heard 600 miles away. Its debris pitted the ground with hundreds of craters similar to shell-holes, measuring 1 to 50 yards across. The countryside was devasted for forty miles around, and more than eighty million trees were blown by the blast of the explosion. Calculations established that had the meteorite fallen 4 hours 47 minutes earlier, the city of St Petersburg would have been at the centre of the crater; the destruction would then have been as complete as that of Hiroshima by the atomic bomb. In 1946 an enormous meteor fell in East Africa, and another in Siberia in 1947.

That cosmic bombardments have occurred unknown to man is proved by meteorites found in different parts of the world. There is a 14½-ton block found in Oregon and now the property of the Planetarium of the American Museum of Natural History, New York; three meteorites found in Mexico, ranging from 11 to 17 tons and formed almost solely of iron and nickel; a block of 60 tons from south-west Africa; two meteorites (one 3 tons and one 34 tons) brought back by Admiral Perry from Cape York, Greenland, and now displayed in New York's Planetarium; and a block of one million tons reported from Adrar in Mauretania.

In the United States, Meteor Crater (formerly Coon Butte) in Arizona is a favourite haunt for tourists. The crater is generally believed to have been formed by an enormous iron meteorite. The depression measures 4,500 feet across and 600 feet deep, its rim rising 130 to 160 feet above the surrounding desert. The meteorite causing such a depression must have weighed thousands of tons. It apparently broke into several parts during its fall.

At the bottom of the crater lie fossiliferous lake deposits, 70–90 feet thick, dating from about the penultimate glacial substage. Ancient (Permian) sandstones and limestones surround the crater, inclining outward from it in all directions. Fragments and blocks of rock (mainly limestone) are scattered over the rim, in the crater itself, and for about six square miles around. Some of the blocks weigh as much as 4,000 tons. Thousands of pieces of meteoric iron have been found in the area, though very few within the actual crater. Some of the pieces are quite heavy, weighing up to 100 lb. So far something like twenty tons of meteoric iron has been collected.

Several theories, besides that of meteoric origin, have been put forward to explain the crater. One suggests volcanic origin — but there are no volcanic rocks in the immediate vicinity. Another suggests origin through solution of limestone, salt and gypsum layers which lie at depth, followed by the fracturing and collapse of a low structural mound above these subsurface solution cavities. The circle of the crater is, however, almost too perfect for such a theory. Moreover, both this and the theory of volcanic origin have to admit at least a shower of small meteorites to explain the presence of the meteoric iron.

Several hundred thousand dollars have been spent on exploring the crater in hope of finding a large body of metal. Drilling revealed crushed, metamorphosed and fused sandstone down to a depth of 620 feet; beyond lay the first undisturbed sandstone. No large body of metal was found. This cast doubt on the meteoric origin of the crater until it was realised that all the evidence indicated that on contact with Earth's surface the meteorite had not buried itself but exploded.

There is a still larger meteoric crater, about 11,500 feet in diameter, in the far north of Canada, at Ungava. Chubb Lake, a magnificently blue lake, as circular as if marked out with a pair of compasses, occupies the depression excavated in granite gneiss. The rim of the crater stands 300–500 feet above the neighbouring

Meteor Crater, Arizona, viewed from the air. The depression measures 4,500 feet across and is 600 feet deep.

plain, and the lake is 1,300 feet deep in places. It is estimated that the crater was formed fifteen thousand to thirty thousand years ago. Around it are three sets of fractures which are attributed to the explosion of the meteorite, but meteoric origin has not been proved beyond doubt. There is, however, no evidence of solution or volcanic activity in the area.

Craters six and eight miles wide have been reported from regions as far apart as the Ivory Coast and the Arctic land areas. Probable meteorite craters currently on record include those in Odessa, Texas; Compo de Cielo, Argentina; Henbury, Australia; Wabar, Arabia; and a small group in Siberia.

**Origin of meteorites.** What exactly are meteorites? Are they fragments of a hypothetical planet, Vulcan, which it has been suggested once lay between Mars and Jupiter? Vulcan, this theory holds, was about as large as the Moon, and its explosion would have created a considerable number of blocks of all sizes. Many of these — the biggest — still continue to revolve round the Sun. They are the asteroids, planetoids, or minor planets, such as Ceres, Pallas and Vesta—estimated at 485, 300 and 235 miles in diameter respectively. More than 1,500 asteroids are known; there may be many more still to be discovered.

Meteors are solid objects heated to incandescence by friction as they pass from space into the Earth's atmosphere. Several millions — most not bigger than grains of sand — fall daily and vaporise before reaching the surface. Larger meteors may travel so rapidly through the atmosphere that only their outer surface is vaporised, leaving a solid core. If their volume is great enough

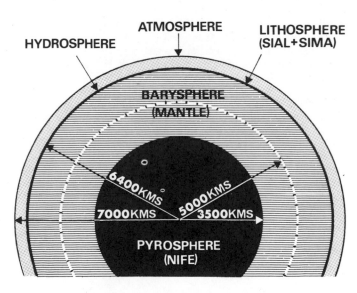

191

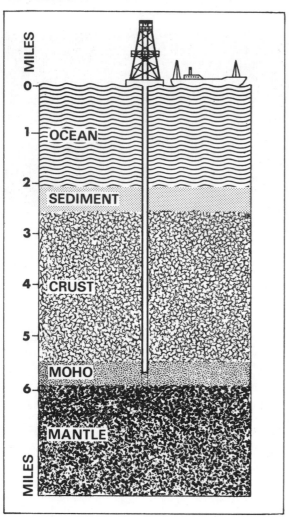

The sketch shows how scientists to the Earth's mantle.
would drill through the 'Moho'

these larger meteors may plunge into the ground, exploding and producing craters. Most of the metal content vaporises in the explosion. In a slower fall, a meteor buries itself in the ground to become a meteorite. Meteors generally travel in groups or swarms; some of the swarms appear from the same direction at regular intervals and are part of the solar system, and have an orbit around the Sun. Others follow no recognisable pattern and probably come from outside the solar system. Tiny meteors, or cosmic dust, which fall in such great numbers daily, give rise to those spectacles of the night sky, the shooting stars. It is certainly a splendid effect to be produced by so slight a cause. The height at which they pass and become sufficiently heated to be visible is estimated at about 60 to 100 miles. Surfaces of fresh snow have sometimes been observed to be covered with minute drops of melted iron after a group of shooting stars has crossed the sky.

Meteorites move at an estimated velocity of 25 miles a second. The Earth moves round its orbit at a speed of 18·5 miles a second. If the meteors are moving in the same direction, the speed of encounter as they overtake the Earth is about 6·5 miles a second. On the other hand, if the meteors are moving in an opposing direction, the velocities are added to one another, and the speed of encounter is then 43·5 miles a second. In any case, the friction against the Earth's atmosphere is sufficient to raise the surface temperature of the meteors to several thousand degrees and make them incandescent. Colour varies with the surface temperature of the meteor. Sometimes it is white, sometimes red, yellow, green or blue. An iron meteorite as small as the head of a pin can shine with a light equal to that of a star of the first order.

**Composition of meteorites.** From knowing the origin of meteorites it is only a short step to knowing what they are made of.

In theory, the hypothetical planet Vulcan was solid when it exploded, and consisted of a solid core surrounded by concentric

layers of crust. It would not have differed much from the other heavy planets of the solar system. Studying meteorites is rather like studying fragments of this deceased planet, and enables us to draw parallels between it and the most inaccessible parts of our own.

Meteorites can be classified according to structure and chemical composition.

Numerically largest are iron meteorites. They can be compared to nickel-steel, chrome-steel, and cobalt-steel. Though extremely hard and rigid, they are brittle. They form the bulk of the meteorites which explode in our atmosphere. The nickel-iron belonged perhaps to the core of the planet Vulcan, and gives a possible indication of the constitution of the interior of the Earth.

Perhaps a similar origin can be attributed to the carbon meteorites. These are, however, rare; only ten are known at the present day. The carbon is not free but combined with nitrogen, hydrogen and oxygen. Sometimes solid or liquid hydrocarbons similar to these in petroleum are found but they have probably been formed not by the alternation of organic matter but by the action of superheated steam on metallic carbides.

Stony meteorites form a third category. These are composed mainly of iron and magnesium silicates, which turns thoughts at once to the composition of the Earth's crust. Others, less common than the previous group, have a vitreous structure, resembling large drops of glass that have melted and then solidified. They are presumably the result of the complete fusion of other stony meteorites. The vertical fall from the sky in a liquid state is sufficient to explain their pear-shaped form.

Many meteorites contain radioactive elements and, as a result, their absolute age can be determined by radioactive methods. The average result of these determinations is 4,600 million years, the same age as the Earth, significantly indicating that meteorites are part of the cosmic material from which the Earth and the solar system were formed.

**Tektites.** These are found in many parts of the world, especially Australia. They are translucent stones the size of marbles, coloured black or dark green. Their origin is a mystery, but the presence of nickel-iron spherules of cosmic origin in the Philippine tektites shows that meteoric impact is involved in their formation. It is also known that tektites must have originated somewhere in the Earth-Moon system because they lack any appreciable measure of the aluminium-26 isotope which is found in all debris exposed for long periods to the cosmic rays of space. They have been melted twice, first to form the entire glassy body, followed by a partial remelting of the surface due to passage through the upper atmosphere.

It is not known, however, whether they are bits of the Earth splashed off by a cosmic impact, which then re-entered the atmosphere, or lunar debris following a meteoric impact which arrived either independently or attached to some parent body. Evidence, and objections, exist for all three theories; possibly the samples of the Moon's surface brought back by satellites will help solve the mystery.

## LAYERS OF THE EARTH'S CRUST

The two preceding sections bring us nearer a consideration of the internal structure of the globe. We have seen that there is a heavy metallic core at the centre of the Earth, then a viscous overlying layer on which floats a rigid crust of a varying thickness. Let us try to bring all these facts together into one linking theory.

In 1875, the Australian geologist Suess concluded that Earth consists of a collection of concentric layers which become denser toward the centre. No subsequent discovery, either in astronomy, geophysics, geochemistry or in geology, has basically altered his conclusions.

Starting with the outside, the successive layers are the atmosphere, the hydrosphere, the lithosphere, and the barysphere. Let us ignore the atmosphere and the hydrosphere which we have studied in previous chapters, and examine instead the lithosphere and the barysphere—the mineral layers of the globe (see diagram on page 191).

The lithosphere is the solid crust on which we live and whose thickness is variable from 20 miles to 40 miles. The first thing to be noted is how thin it is compared to the radius of the globe — hardly a hundredth part of the whole. Sedimentary rocks (limestone, clay, shales, sandstone) make up part of the lithosphere; igneous rocks (granite, gneiss, basalt) make up the main part. Two of the principal elements contained in the lithosphere are silicon *(Si)* and aluminium *(Al)*. This explains the name sial which is given by geologists to the material which forms part of the Earth's crust. The crust or outermost portion of the Earth consists of two zones or layers, which vary in composition and thickness below the continents and ocean basins. In the continental crust, the outer or sialic layer (often called the granitic layer because of its main constituent) is about 10 to 15 miles thick. This measurement is based on the study of primary and secondary earthquake waves. The lower layer, or *sima* layer (named after the proportionately larger amounts of silicon and magnesium it contains), is about the same thickness as the sialic layer. It is also known as the basaltic substratum, because it corresponds in density and earthquake wave velocities to basalt. Seismic studies near earthquake centres show that in continental mountain areas the granite zone thickens under crystalline mountains. This leads to the conclusion that many mountains have deep-seated granite extensions or roots. The thickness under the Alps is about 40 miles, whereas the average for plains and hills is only 10 to 25 miles.

Under the oceans the evidence from studies of primary, secondary and long wave studies is less complete. In the Pacific basin the sialic layer appears to be replaced by sima, and the crustal thickness is about 20 to 30 miles. The Atlantic and Indian ocean basins are still in debate because of the lack of reliable information on the speeds of seismic waves travelling through these regions. So far the data indicate that the sialic layer is missing here, too.

The zone beneath the dual-layered crust is called the barysphere or *mantle*; it extends downward to about 1,800 miles. This segment seems also to have two distinct regions or layers. Our knowledge of the mantle is again based on the behaviour of primary and secondary waves recorded between 700 and 7,000 miles. The abrupt changes in speed of seismic waves as they pass through this region show that the nature of the material has altered. We have no certain knowledge as to the constituents, but it is probable that this zone contains a much higher concentration of ferromagnesium minerals than the crust. We do know that it is solid, as it allows the transmission of secondary waves, which do not travel through liquids. Indeed, there is increased speed through this zone, indicating an increase in rigidity. An abrupt increase of speeds occurs at a depth of about 300 miles, indicating yet a further change; beyond this the waves act uniformly down to the 1,800-mile lower limit of the mantle. There is much discussion on the consistency of the mantle, as it must be sufficiently fluid to allow the sinking and displacement of the divisions of the crust, yet sufficiently solid to crack and give rise to deep earthquakes. Some scientists suggest that the upper part of the mantle is glassy and may undergo slow flowage as it adapts itself to changing conditions on the surface. We can usefully compare it with ice or sealing wax. Each breaks under a sharp blow, but flows under firm, slow pressure.

The mantle is possibly in a curious state of thermal diseqilibrium because the upper layers, apparently, are less hot than the lower, and therefore heavier. Convection currents should result from this state: warm currents rising from below, and cold currents descending to the depths, with horizontal currents joining the two to form a convection cycle. The speed of these theoretical currents has been calculated at half-an-inch to four inches a year. The downward pull is at any rate considerable on the lower side of the Earth's crust, and many of the main internal phenomena — vulcanism, earthquakes, continental drift, and formation of mountain chains — may be attributed to it.

The last major zone of the Earth's interior is the core, or pyrosphere, extending downward from 1,800 miles to the centre at a depth of 3,950 miles. It forms the largest part of Earth's mass. Analysis of earthquake records reveals that the core also has two sections: an outer core, 1,360 miles thick, and an inner core with a radius of 790 miles. Secondary waves, which travel only through solids, are eliminated at the 1,800-mile depth, and primary waves which travel through the outer core are first reduced in speed and then quicken pace again at a depth of 3,160 miles. This indicates

The first stage of operation 'Mohole', which took place in 1961—62 off the west coast of Mexico.

that the inner core is solid, though there is much uncertainty about its constitution. It must be heavy (with a density of about 15) to compensate for the low density (less than 3) of the Earth's outer layers. The rocks in the mantle, if similar but under great pressure, would still average only about 5·7. Therefore the specific gravity of the core should be about 15 to give the calculated average density of 5·5 to the globe. It is therefore supposed that the core is composed of iron with about 8 per cent nickel and some cobalt. These are roughly the same proportions found in some meteorites. The pressure it supports is enormous. Is it solid or liquid, or a combination of both? And have the words solid and liquid any meaning in this context? The metals may all be above their critical temperature, and in that event they can only be gaseous. The disintegration of matter suggests that the metals' molecules and atoms may be dissociated, that the atomic nuclei themselves may be broken, and that there may be an accumulation of electrons, neutrons and positrons which offers no evidence of its existence.

This takes us far from the simple concept of a nickel-iron core, to which Suess gave the name 'nife', a word made up on the same principle as sial and sima.

Lastly, some scientists consider the core as the remains of solar matter in which hydrogen was dominant but compressed to the point of taking on metallic consistency.

How much further knowledge of the Earth's interior we will gain is uncertain. The United States hopes to take a step forward with their project 'Mohole', an attempt to drill a hole right through the crust down to the Mohorovicic discontinuity ten miles down which forms the upper surface of the mantle. In the first phase of the project samples of basalt rock were brought up from 1,000 feet below the ocean bed and in a subsequent exploratory experiment from a platform at sea resting on two remote-controlled submarines which were submerged to achieve greater stability. Two promising sites are being surveyed, in the Atlantic and the Pacific, where the crust is thinner than that which lies underneath the continents.

# Diastrophism: Mountain Formation

Diastrophism is the term used to describe any movement of a section of the Earth's crust with respect to any other segment. There are two main types of such movement: orogenic and epeirogenic. Orogenic, or mountain-building, movements produce folded rock structures and great thrust fractures along which there is considerable lateral movement. Epeirogenic movements are mainly vertical shifts which produce domes, basins, and steeply dipping fractures along which large blocks of the Earth's crust rise or sink.

A period of mountain-building activity is called an orogeny; a time of very extensive diastrophism, a revolution.

## OBSERVATIONS IN MOUNTAIN AREAS

Anyone passing through a mountain area is at once aware that some of the rocky cliffs are bare of snow and vegetation. These cliffs often show bending of the rock layers, some displaying oblique beds, others vertical beds and folds of all kinds. The same arrangement is found deep inside the mountains when tunnels are constructed. In many mountain areas rocks have been folded and refolded upon one another, and fractured into the bargain. There are few more striking contrasts than, say, that of the Alps, an intensely folded region, and a quarry near London or Paris, where the beds of rock succeed one another more or less horizontally. Let us emphasise that it is folding rather than altitude which distinguishes a mountain formation, for mountain peaks may be eroded away over the years, leaving only their structural roots as evidence.

Can the folds so clearly visible in a recently formed mountain be detected in the heart of ancient mountains, which have been planed down by erosion until they are now reduced to a gently undulating land surface, called a peneplain? If we examine a gorge, a quarry, a road, or railway-cut in one of these peneplains, we shall find the same anticlinal and synclinal folds of rock as in the Alps. The only difference is that these folds have been cut off at ground level. They are the roots of ancient folds which may once have risen thousands of feet.

Different types of rock are found in mountain areas: in the Alps, there is the granite of Mont Blanc, the mica schist of the Aiguilles ranges, the crystalline schists of St Gotthard, the marble of the Jungfrau, the dolomite of the Tyrolean Alps; there is the folded sedimentary rock of the Appalachians; the lavas and tuffs of the Cascades. Some of these rocks have an igneous origin, and some a sedimentary origin. The igneous types (granite is one) have risen from the depths of the Earth in molten masses and generally form only a very small area of the mountains. The sedimentary types, limestone, marl, shale and sandstone among them, have been deposited mainly in water. A third type — the metamorphic rocks such as gneiss, mica schist and marble — is formed by transformation of either igneous or sedimentary rocks under heat pressure. Sedimentary and metamorphic types are the more abundant at the surface of Earth, and we may therefore conclude that the rocks which compose most mountains have been

Mount Cristallo in the
Italian Dolomites.

formed under water before undergoing changes after deposition.

Were they deposited in fresh or salt water? In shallow or deep? The answers are furnished by fossils, for in the limestones and shales of some mountains ammonites, belemnites, oysters and nummulites occur in profusion, while some rock masses are formed wholly of coral reefs. The fossils are usually those of marine animals, and moreover, those which lived between the surface and a depth of about 600 feet. Sediments of the kind deposited in the abyssal depths of the ocean are unknown, but land deposits, though less abundant than marine deposits, are not infrequent and may contain freshwater shells. The conclusion is that most of the rocks of mountains were deposited along the margins of the oceans, or in ancient seas whose depth rarely exceeded 600 feet.

If this thickness of the rocks is measured perpendicularly to their stratification, a mass of sediments is revealed which reaches approximately six miles in the Alps, seven in the Rockies, and eight in the Appalachians. The implication is that several miles of sediment accumulated in a sea which was only 600 feet deep. There is clearly conflict in the conclusion.

## WHERE MOUNTAINS ORIGINATE

The problem is to reconcile the two facts: deposition in a shallow sea, and the great thickness of the sediments.

Two American geologists, Hall and Dana, found an answer by postulating special areas in the Earth's crust, deep trenches in which the bottom was continuously subsiding during sedimentation. These enormous trenches Dana (1873) named geosynclines.

In order to examine a concrete example, let us imagine the present-day Alpine areas during the Mesozoic period, or the Caledonian region in the Palaeozoic. Instead of high mountains we would have before us seas several thousand miles long and several hundred miles wide. These were the Alpine or Caledonian geosynclines. The Alpine geosyncline shores were formed in France by the Massif Central and the Vosges, in Germany by the Black Forest, and in Italy by a land mass which has now subsided beneath the modern Piemonte. These shores were the forelands of the geosyncline. The Caledonian geosyncline extended from Scandinavia to Britain. To the south-east, it was bordered by a vast lowland area, the Baltic Shield. To the north-west, it is likely that the foreland formed part of the Canadian Shield. To the east lay an irregular and often mountainous land area, or a chain of volcanic island arcs. Terrigenous sediments from the adjoining land areas were brought down into the geosyncline in the usual way.

A phenomenon then occurred which was to distinguish geosynclines from the other seas of the globe. The geosynclinal trough subsided and deepened. Because the movement was slow, the deepening was continuously compensated by sedimentation. The depth of water generally remained at 600 feet or less. Corals and other marine organisms were abundant. Sometimes sedimentation progressed more rapidly, and the seas regressed; the geosynclinal area then became a broad sedimentary plain across which streams flowed and built their flood-plains and alluvial fans. Swamps and ponds formed. Land-living animals and plants grew.

*Top*
Glacier, National Park,
Montana, U.S.A.

*Centre*
The Black Forest, an ancient
mountain area.

*Bottom*
Folds of an ancient mountain
chain levelled off by erosion,
below the village of Firmy,
Aveyron, France.

Continental sediments were deposited and erosion occurred. This explains why there are some gaps in the succession of rocks, and why there are intercalations of gypsous and rock salt; lake, dune and river deposits; lignite, plant fossils, and other coal, as well as footprints and bones of land-dwelling animals.

The oldest sediments, occupying the lower part of the geosyncline, were gradually pushed down several miles below the surface and placed in a favourable situation for transformation into rock, with desiccation, temperature and pressure all contributing to the change.

Many geosynclines were not simple troughs, but were divided into multiple geosynclines, or polygeosynclines, by long, narrow ridges, often called geanticlines or axes. Their sedimentary histories are more complex, particularly as individual troughs did not always subside at the same time or rate. The geosynclinal areas, with their great thicknesses of unconsolidated or semiconsolidated sediments, were areas of weakness in the Earth's crust and slowly, through the action of compressive forces in the crust, these areas yielded. The rocks bent, gradually rising, folding and fracturing to form mountains. But even as they grew, the agents of erosion (streams, glaciers, wind) began downcutting and levelling long before the mountains had fully developed. Mountains, like all Earth's phenomena, begin to age as soon as they are born.

The theory of geosynclines, first expounded by Hall and Dana, was elaborated by the French geologist Emile Haug. His classic publication *Treatise of Geology* (1912) extended application of the theory to all the mountain chains of the globe and to every geological epoch. The Earth's surface, he stated, contains at any given moment both geosynclinal areas and continental areas. In the first, subsidence brings a deep accumulation of sediments of similar composition; in continental areas, which are more or less rigid, there is a lesser thickness of varied sediments. In practice, the distinction is not as marked as this, and Haug misinterpreted somewhat the meaning of geosyncline. A non-mountainous region can offer sediments just as thick as a chain of mountains. In southern England and the Paris basin, where there is no formation of mountains, borings show thicknesses of Tertiary and Mesozoic sediments up to several thousand feet. Probably every sedimentary basin sinks slightly under the weight of its contents and thus resembles a geosyncline. The thickness of sedimentary continuity which characterises both features reveals the phenomenon of subsidence. A geosyncline is simply a large sedimentary basin which may ultimately evolve into a mountain chain.

## SIMPLE FOLDS AND THRUSTS

Mountains, then, are regions of contorted rocks. Before trying to analyse the causes of their formation, it is necessary to study the folds and other features of the rocks of which they are composed. This architectural study of mountains is known as tectonics. When rock layers are under pressures which exceed the elastic limit of the rocks, they will yield by bending, folding or fracturing. The fractures appear as a series of folds with alternating crests and troughs. The arch or crest of an upfold is called an anticline; it is flanked on either side by downfolds or troughs, called synclines.

A simple fold whose limbs converge or diverge at the same angle is a symmetrical fold. A fold whose limbs are unequally inclined is an asymmetrical fold. Few folds are perfectly symmetrical. An overturned fold is an asymmetrical fold in which one limb is bent so that it lies beneath the other.

A fold whose limbs are horizontal is a recumbent fold. Here, as in other overturned folds, correct order of strata is reserved in one of the limbs: the oldest lies above the youngest, that is, the overturned limb above the normal limb.

The overturned limb of asymmetrical and overturned folds is usually stretched. Stretching may continue to breaking-point, when the anticline becomes fractured and separated from the adjacent syncline, sliding along the plane of fracture or faulting, or thrust-plane, to become an overthrust fold.

Single folds rarely constitute a mountain. Generally they occur side by side, like a succession of waves. In the Jura, for example, the folds are vertical or slightly overturned, and roughly parallel to each other. This simple folding is known as Jura-type folding. It occurs again in the Appalachian mountains. In the

*Top*
Beds folded into a right angle at St Jean de Luz

*Bottom*
Vertical folds which have been subject to considerable weathering in the Alps.

197

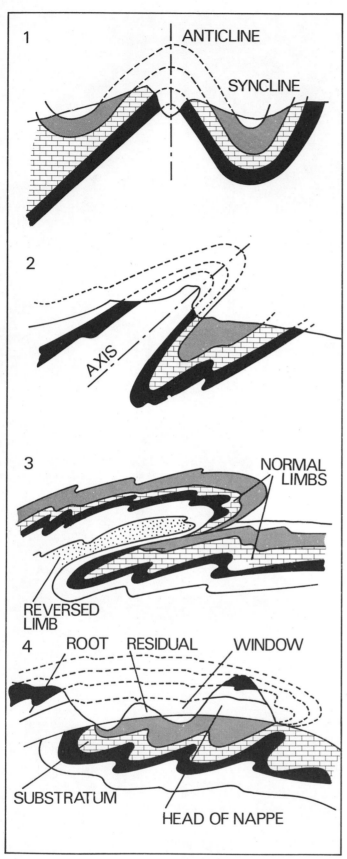

Alps, some folds are inclined towards France and some towards Italy. In Mont Joly, in the Mont Blanc area, the recumbent folds are piled up, one above another. These are overthrust folds which have slid over one another like the slates on a roof.

The most extraordinary arrangement of beds is that first brought to light in the Franco-Belgian coalfield, then in the Alps, and finally in many mountain areas of the world — the nappes, decken, or thrust-sheets first described by Marcel Bertrand (1847 — 1907).

While he was still a young engineer, Bertrand was struck by a curious geological anomaly in a small hill near Toulon, where Triassic rocks lay above Cretaceous rocks, instead of vice versa. Sedimentary rocks must always be laid in chronological order. What had happened here? The Triassic formations had certainly been laid down before those of Cretaceous times, yet now they overlaid them. Bertrand suggested that a section of the Triassic rocks had been detached, fractured or faulted, and thrust over until it rested over more recently formed rocks. Since this discovery, many others have followed. The combined efforts of many geologists have since revealed the Alps as a vast pile of superimposed nappes. Similar thrust-faulted areas occur in the northern Rocky Mountains of United States and Canada.

We have seen that in an overfold the overturned limb may stretch until it breaks into a thrust-fault, while the normal limb slides along the fault plane. This phenomenon may extend over distances of 50 to 200 miles, and it may transport enormous masses of rock, called nappes or thrust-sheets. These are recognisable by their position above a substratum of completely different rock. Palaeozoic rocks can be found in conjunction with Tertiary, without the intervening Mesozoic strata that would normally separate them; ancient rocks can lie above those more recent.

Some nappes form mountain areas, either on their own or by heaping up in series or piles. Sometimes they are continuous and can be followed uninterrupted over long distances. Sometimes erosion has carved openings or gaps in them, so that the underlying rocks are exposed in so-called windows or *fensters*. In other places, they have been reduced to small fragments scattered over a broad area. Such an erosional remnant is known as a nappe outlier, or a klippe.

It is obvious that a nappe must grind down the thrust-plane considerably as it slides over it, and at its base there is generally a layer of crushed rock, fault breccia, or gouge. Thus, when orogenic movements take place, folds of many different shapes, and large and small thrust-faults are produced.

## VERTICAL MOVEMENTS

All the orogenic movements just mentioned are caused by horizontal or compressive movements in the Earth's crust. To these may be added epeirogenic, or vertical, movements such as subsidence and uplift, doming and block-faulting. Their relative importance varies according to the mountain region concerned.

The simplest structure produced by vertical movement is the block-fault, a fracture or crack in the rocks along which, generally, there is movement bringing a change in level between the two sides. One side may rise or sink while the other remains static; both sides may move in almost opposite directions.

The diagram given here shows that a fault can be vertical *(A)* or oblique *(B, C)*. When erosion has rounded the surface of the rocks, and soil covers them, a fault may become nearly invisible on the surface *(D)*, and any break of slope it may have formed may disappear completely *(E)*. A natural section (a cliff, for example) or an artificial section (a cutting or mine-shaft perhaps) shows that the geological strata do not correspond exactly on both sides of the fault. A segment of the Earth's crust left standing between two faults is known to geologists as a *horst (F)*. The Hercynian mountains of Europe, such as the Thüringerwald and Harz, are examples of *horsts*.

Similarly, part of the Earth's crust can subside between two faults and become a graben *(E)*. The Rhine Valley and Death Valley are grabens, and the biggest of all is formed by the series of straight-sided troughs that run from the Dead Sea down to the lakes in the Great Rift Valley of Africa. Mountains in which the major structure is a series of high-angle faults are block mountains.

*Above*
Folding systems.
*(Top to bottom)*
1. Vertical fold.
2. Oblique fold.
3. Recumbent fold with reversed limb stretched.
4. Nappe.
   Broken lines indicate parts removed by erosion.

*Top right*
Acutely folded Upper Culm measures in the cliffs, Cornwall.

*Bottom right*
Layers of a limestone beach bent by igneous intrusions which forced their way in molten state between the limestone.

*Left*
The Policeman's Helmet, an anticline at Septmoncel in the Jura.

*Top right*
Steeply inclined Silurian strata at Bank End near Grizebeck, on the western side of the Lake District dome. The strata were laid down by sediment under water millions of years ago, and later pushed into their present position by powerful earth movements.

*Bottom right*
Diagram showing a horst and graben formation.

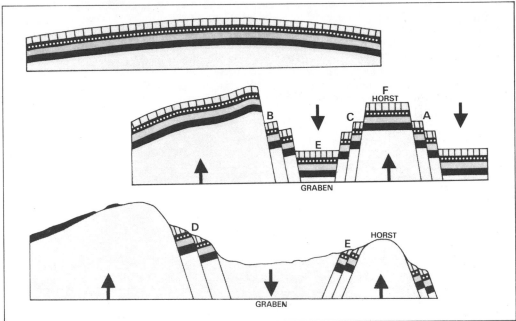

## SIMPLE FOLDING OF THE JURA

The simplest mountain chain in the world is the Jura, a huge crescent-shaped area extending from Chambéry to Zürich, bordering the Alps and contiguous with the Massif Central, the Vosges and the Black Forest. A traveller following the Rhône and Valserine valleys cannot fail to notice the limestone cliffs that extend from Grand-Colombier to Colomby. For forty miles he follows the line of the highest summits of the southern Jura. Seen from the air, it is a cluster of parallel ridges: monoclinal, anticlinal and synclinal. The same topography is found in the western Jura, and the Appalachian mountains.

In the central Jura, on the contrary, the folds become lower and lower from south-east to north-west, and give way to faulted plateaux only slightly undulating. For the traveller going from Switzerland toward Dijon, the line of separation is crossed in the Pontarlier area.

Leaving the Swiss and French Jura for the Swabian and Franconian Jura, the layers of rock here are absolutely horizontal. No folding has taken place; only plateaux resembling those of the tabular Jura are found.

There are thus two Juras: the tabular and the folded. Inhabitants of the area know this well, calling the first area the 'plateau' and reserving the term 'mountain' for the folded Jura. The side-by-side existence of anticline and syncline in the folded Jura influences rivers, vegetation, distribution of population, and topography.

The Appalachian Mountains of eastern United States are very similar in structural type to the Jura. In Palaeozoic times there was a huge geosyncline in this region, extending southward from Nova Scotia to Alabama. Over the course of millions of years it subsided slowly and filled with over 40,000 feet of sediments. Adjacent to this trough, to the east, lay a series of volcanic island arcs, a hilly linear land area; to the west lay a broad lowland, the main body of the continent. Toward the close of the Palaeozoic era, compressive forces crushed the great geosyncline, folding and fracturing the sediments and producing a series of north-east to south-west anticlines and synclines, and long, low-angled thrust-faults. The Earth's crust was shortened 200 miles or more. Erosion began to cut into these mountains as soon as they began to form, producing a rugged topography. These early Appalachians were probably as high as the more recently formed Rockies.

Since their origin the Appalachian Mountains have been levelled by erosion two or three times. After each levelling they have been rejuvenated by uplift of the region. The topography we see in the region today is of relatively recent origin (late Cenozoic) but the structures which form the body of the mountains are over 200 million years old.

In the east, the folds are closer and tighter, while the mountain ridges are more rugged; westward both folds and mountains become gentler and broader, until on the western margin the folds die away and the rocks lie horizontally, forming the Allegheny (or Appalachian) Plateau. A few large rivers such as the Susquehanna cross the mountain area, their tributaries excavating large valleys in the softer rock belts. Here and there sink-holes occur in the limestone strips. Most of the prominent ridges are formed of resistant sandstones or conglomerates and are monoclinal in structure. Few are synclinal or anticlinal.

## THE COMPLEXITY OF THE ALPS

While the Jura can be considered a perfect example of simple mountains, the Alps offer maximum complexity both in history and in structure; to understand them the successive stages of their formation must be followed, at least in outline.

First let us consider the area in Carboniferous times, some 250 million years ago. At this period the Alps did not exist; their territory was occupied by part of the Hercynian chain, the mountains which then formed the greater part of western Europe. Remains of these earlier mountains still survive as five blocks of hills in the Alps: Mont-Blanc, Pelvoux, Mercantour, Esterel and Maurée.

From the end of the Palaeozoic period, after erosion and subsidence had done their work, the whole area of the Alps was low and swampy ground. The region then deepened and was invaded by the sea, forming the Alpine geosyncline, which consisted of two marine trenches (the Dauphiné geosyncline on the French side; the Piemonte geosyncline on the Italian side) separated by a median ridge or geanticline (the Briançon geanticline which coincided roughly with the present French-Italian frontier). These three features of the Alpine geosyncline suffered variations in extent and depth during Mesozoic and Tertiary times, the most

constant being the Piemonte geosyncline. Sedimentation in the area naturally reflected the vicissitudes. At times limestones were deposited, at others shales or sandstone. We shall refer again later to these varied formations. Here we shall study only their part in the formation of the Alps. Compressive forces folded the sediments into a series of waves, pressed against one another and broken and thrust northward or eastward in the form of nappes. The ancient Hercynian blocks rose, and the nappes became moulded to them or wedged between them where they were unable to ride over their crest. This happened in the Mont Blanc block. The structure was further complicated by the formation of new folds on the outside of the Alpine arc, today forming the Pre-Alps.

The Caledonian mountain ranges, which extended from Scandinavia to Scotland, show both Jura and Alpine characteristics. The south-east front is well preserved in Scandinavia where outward thrusting toward the foreland of the Baltic Shield is conspicuous and involves movement of up to eighty miles. This front is poorly defined in Britain and is largely hidden by later sediments; where the older rocks are exposed, as in Shropshire, the folding is open and broadly undulating. The north-west front is cut off by the Atlantic except in the north-west Highlands of Scotland where thrusts (for example, Glencoul) splay out over the north-west foreland, now represented only by a narrow strip of land. This relic may indicate the extreme limit of the Canadian Shield.

The Urals, the Andes, the Alps, the Highlands of Scotland, and the Northern Rockies are other examples of mountains formed through folding accompanied by thrust-faulting.

Mountains are found also in areas where crustal movement is

mainly vertical rather than lateral. The resulting structures are steeply dipping fractures and broad, gentle warps which produce domes, basins and block mountains. In block mountains, where faulting has taken place on a large scale, the individual blocks shift differentially along the fractures, and the movement of any one block may be up or down, or first one way and then the other. The geologic history of such a region may be very complex. The mountains of the Great Basin or Basin Range province of south-western United States are of this type. This region has been fractured by numerous approximately north-south faultings which have a series of mountain blocks and structural basins.

**Fault-Scarp Mountains.** The Vosges and Black Forest Mountains which form the walls of the Rhine graben are fault-scarp formations similar to the Sierra Nevada. In these, one side of the mountains is formed by a great fault fracture, while the other is merely inclined or tilted. The faulted and tilted block of the Sierra Nevada is about 400 miles long and 75 miles wide. On its faulted eastern side, slopes are abrupt and steep; on the inclined western side they are gentler.

Dome mountains are a product of vertical movements generally resulting from deep-seated igneous intrusions such as batholiths or laccoliths. The intrusions push up and arch overlying sediments, which then dip or tilt away from the central area. Erosion attacks the crest more vigorously until in time the underlying intrusive mass is exposed.

The Black Hills of South Dakota are a classic example of an exposed igneous core of crystalline rocks in dome formation.

*Left*
Oblique fault in cliffs.

*Right*
Col des Aravis, a valley in the Savoy Alps. Mont Blanc is seen in the far background.

Before the central area was extensively eroded, the crest may have reached 6,000 feet above the adjacent plain. Today the granite core forms such mountain peaks as Harney. Into the surrounding inclined sediments have been carved a series of ridges and valleys which almost completely encircle the granite area.

## CAUSES OF OROGENESIS

So far we have described only the structures involved in the formation of mountains: a geosyncline giving rise to mountains through sedimentation and compression; horizontal movements responsible for folding and thrust-faulting; vertical movements producing block-faulting, doming, subsidence and uplift. All this generally takes place under a thick cover of superficial sediments, at depths, pressures and temperatures at which many hard rocks become plastic. Although the movements are so slow that geologists measure annual speeds on a scale of inches, there is no difficulty in recognising the existence of orogenic phenomena. Explaining why they occur is another and more difficult matter.

Geologists for long believed that mountain building was confined to certain epochs in the geologic past and did not take place outside these periods. In fact, the geologic time scale still in use today was based on the concept of periodicity in diastrophism. The evidence of recent research shows, however, that throughout geologic time there has always been crustal activity going on somewhere, at times more pronounced and more extensive, at others less vigorous, but ever present. Any explanation of orogeny must account for this fact as well as for the immense energy required to fold and fault

rocks and thus shorten thick sections of the Earth's crust. What are the forces of energy involved? And how are they produced? The answers still lie within the realm of theory.

**Terrestrial contraction.** This was the earliest theory to be put forward and was much favoured during the last century. Taking as a premise Laplace's hypothesis of the original fluidity of the globe, de Beaumont stated that it cooled and contracted in the course of time. The cooled and rigid crust therefore became too big for the shrinking interior, and was subjected to horizontal squeezing which folded the rocks as it adjusted itself to the diminishing interior. Mountain formation was compared to the formation of wrinkles on a withered apple. As the geometric solid possessing the greatest surface for the least volume is the tetrahedron, it was suggested that the Earth tended to form a tetrahedric shape, or something near it.

Unfortunately several facts are opposed to the tetrahedral hypothesis. To begin with, if we can imagine the folds that form the Jura restored to their former size and spread out horizontally again, it would seem that they must have occupied an area 12 miles wide instead of their present-day 10 miles. The folds of the Alps, by the same argument, would have occupied a width of 650 miles instead of their 325 miles. To produce this amount of contraction during cooling, the Earth must have had a temperature diminution of 2,400°C. (4,352°F.) during the Tertiary alone. At this rate, the temperature of the Earth in the earlier Palaeozoic would have been too high for the existence of a solid crust. The tetrahedral theory, then, does not allow for enough shrinkage, nor does it account for localisation of mountains. On the other hand, there is nothing to prove that the Earth is cooling down. The breaking

down of radioactive substances is sufficient to compensate for its loss of heat. Some geophysicists even suggest that the Earth is becoming hotter and expanding.

**Continental drift.** Wegener's theory of continental drift gave another slant to the ideas about the mechanism of mountain building. The Earth's crust is fractured in many regions and is thought by some to float on a viscous magma disturbed by convection currents. It is easy to suppose from this that the continents are only separated fragments being moved around the surface of the mantle like the icebergs on the surface of the sea. Firmly upheld by some authorities, and regarded as utter nonsense by others, the theory still remains doubtful. Although most geologists do not subscribe to Wegener's theory, most admit that there has been some kind of relative continental shift of position through geologic time.

**Wegener's theory.** Wegener began by supposing that to begin with all the continents were united. To this single continent he gave the name Pangea (Greek *pan* — all, *ge* — earth). Pangea was concentrated almost entirely in the present-day southern hemisphere. The position of the Poles was not thought to have coincided with present positions; the South Pole was located about 20° S and 20° E, and the North Pole somewhere around 20° N and 160° E. This gathering of the continental masses and inclination of the polar axis would certainly explain admirably why traces of a Carboniferous glaciation (on the evidence of striated and polished rocks) are found today in southern Africa, Madagascar, India, Australia and Patagonia, the regions which were concentrated round the South Pole at that time. The same explanation would hold good for coal-producing Carboniferous forests, which were localised in

*Left*
Lac du Sautet, Savoy, looking
towards the Devoluy massif.

*Right*
Screes tumbling into Wastwater
in the English Lakes.

an equatorial zone, that is, in present-day North America and Europe. As for the North Pole, according to Wegener it lay in the open sea; no trace remains of what must have been an enormous icepack.

The theory is that Pangea was divided at the beginning of the Mesozoic period, the continental segments moving apart. These continents may be compared to the ice-floes which follow the break-up of an ice sheet. What happens to floes? Carried by marine currents, they form a slight bow-wave and eddies; pushed against one another, they end by overriding each other. Such is the behaviour of Wegener's continental masses, riding on a hypothetical subcrustal zone of weak and heavy material. With a melting icepack, Pangea broke into pieces which began tó move apart under the influence of convection currents inside the Earth, centrifugal force, and tidal pull. The principal thrust was towards the Equator, with a general displacement from east to west. The 'continental rafts' were even pushed over one another: the Asiatic continent over India, Africa (including Italy and Spain) over Europe, thus forming the giant Himalayan and Alpine chains of mountains.

The Americas, dragged westward, pushed up a ridge of sediments with their outside edge, forming the Rockies and the Andean cordillera. Australia, moving eastward, formed the Great Dividing Range. It was not until sometime after this that a fresh drifting of Europe towards the north formed the present-day Mediterranean. The Balkans, Hungary, Italy and Spain would then be fragments of the African continent, remaining attached to Europe after this new episode.

Without pursuing sweeping assertions further, let us say only that the second argument put forward for the drifting theory is the curious shape of the west and east coast of the Atlantic. If America is cut out of a map, it can be fitted fairly neatly into the Europe-African block; the projection formed by Brazil fits into the Gulf of Guinea; the Antilles are the counterpart of the Mediterranean; Labrador and Greenland fit into Scandinavia, and so on. It is not only the coasts which correspond, but also the ancient chains of mountains, those of Norway, Scotland and Canada in the north and, further south, the Armorican mountains of Europe with the Appalachians of North America. The gneissic plateaux of Africa and Brazil form a perfect continuation of one another, and the geological affinities of South Africa and Patagonia are indisputable. The Old and the New Worlds are like pieces of a jig-saw puzzle which can be fitted together.

Elsewhere coincidence of shore-lines is equally striking. In the Red Sea, the Asian and African shores are similar in shape, and the conformity extends to the geological formations. Everything seems to point to a recent separation of Africa from Arabia.

To these geological arguments, Wegener and several other authors add the evidence of palaeontology. Similar animals and plants exist today in widely separated parts of the world. Wegener's theory of continental drift provides a simple explanation for the wide distribution. In particular, South America, Africa, Madagascar, India and Australia may have formed during the Palaeozoic and Mesozoic eras a vast continent to which has been given the name of Gondwanaland (from the name of an Indian province). On this hypothetical continent lived the ancestors of the plants and animals which today are so widely scattered. One of the most ardent supporters of this conception was the entomologist Jeannel, whose evidence and arguments in favour of this theory are explained in detail in his book *The Origin of Terrestrial Faunas* (1942).

**The evidence against Wegener's theory.** Objections are numerous and of considerable importance:

The connection between the American and Euro-African continents is not quite perfect. The west and east coastlines of the Atlantic cannot be fitted together without suggesting that the continents have undergone considerable deformation and torsion. If the drifting theory allows such deformation to be produced during movements of the continental masses, it must meet the objection that there is also proof of non-deformation.

If the reunification of the continents into a single mass, Pangea, is an admirable explanation of the distribution of Carboniferous glaciers and forest, it is contradicted by the evidence of more ancient glaciations, notably those of the Cambrian period. The reunification theory would then have to admit that the continents, once dispersed, had come together again, and again been separated. This seems to be going a little too far in juggling with the continental masses.

Wegener's theories demand, too, a considerable displacement of the Poles during the course of geological time. Today, this movement is very restricted. Each continuously describes the same circle, not more than 50 feet in diameter. These minor variations are far less than those necessary to Wegener's theory. Here again, it must be admitted that movements in the past may have been very different from those of the present day, and that it is debatable

Towering dolomitic crag in one
of the highest parts of the
Tyrolean Alps.

The Triassic forms a large part
of the dolomitic Alps in
Italy.

*Above*
Moraine Lake Valley of the Ten
Peaks, Banff National Park.

*Right*
Stair Hole, to the west of
Lulworth Cove, Dorset. The

Dorset coast is famous for its
variety of geological formations.
At one time a thrust fault

pushed up a natural sea wall,
leaving behind this depression,
which is open to the flood tide.

how far past events can be accurately reconstructed from those of the present day. Few geologists would care to enter the lists decisively in favour of any of the theories.

If the continents are floating on the surface of the mantle, it ought to be possible to measure variations in the relative distances between any two points at the present day. At first it was thought that a positive result had been found in the comparisons of longitude made between 1823 and 1933. It seemed that Greenland had moved nearly a mile away from Europe during the century; that Paris and Washington increased their distance by fifteen feet a year; that San Diego and Shanghai had come six feet nearer in the same time, and so on. It was found later, however, that the evidence supplied came from early records and gave inaccurate measurements of the longitude of the various points studied. An International Time Bureau has since been set up, with its headquarters at the Paris Observatory where it tests the accuracy of chronometers and receives time signals from every observatory in the world. It is now a simple matter to work out simultaneously the distances between the observatories. The result is correct to within a few inches. Since these accurate measurements have been recorded, no trace of movements has been noted.

Continental drift implies not only a continuous alteration of the continental masses but also a continuous alteration of the oceans. In Wegener's hypothetical Pangea, the single continent was associated with Panthalassa, a single ocean, today divided into the Atlantic, Pacific, etc. This conception is totally opposed to our knowledge of the permanence of these great oceans. It is almost certain, for example, that the Pacific Ocean has always existed in its present location.

Concerning the distribution of living things, it would seem that resemblances between certain geographically distant floras and faunas have been exaggerated, and the effects of normal migration underestimated; such factors as transport by wind, birds, floating wood, ships, man, and the previous existence of land-bridges such as those of the Bering Straits and Antarctica have been ignored or underestimated. Taking these factors into consideration during a study of the origin of the fauna of Madagascar, the French zoologist Professor J. Millot opposed the 'Gondwana' myth and refuted Wegener's theories with all force.

Wegener's idea of continental drift was a splendid hypothesis, backed up by evidence and arguments that appear convincing but which cannot altogether hold in face of results produced by more recent research. On the other hand, the hypothesis cannot be wholly explained away by other theories.

Whether continental drift exists or not, how do geosynclines form and deepen, and how finally are they compressed until a chain of mountains rises? What forces exist to produce such spectacular results? We are brought back to convection currents, already invoked to explain vulcanism, earthquakes and the structure of the globe.

**Convection currents.** We have already compared the outer layers of the Earth to the contents of a pan of boiling jam. The bottom being hotter than the surface, the jam rises in the centre, moves to the edge, carrying the scum with it, dives down again at the rim of the pan, becomes heated at the bottom, rises again, and so on. The movement is cyclic, and the pan is the scene of a cycle of convection. Earth's mantle, too, may contain convection currents. The only difference between it and the pan, setting aside proportion, is that the mantle's currents are much slower (a few inches a year at the most) and a number of currents spread out side by side from the centre of the Earth, circulate and spread to the extremities.

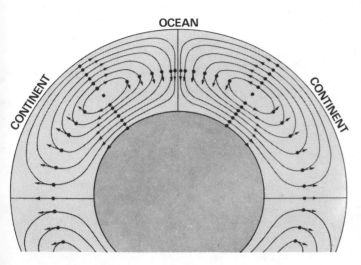

Convection cycles operating
within the pyrosphere.

Let us consider two neighbouring cycles. Warm ascending currents or cold descending currents will exist at their contact. In the first, there will be a tendency for the crust to rise. In the second, on the other hand, there will be a downward suction tendency in the crust, and the formation of a geosyncline. The geosyncline's walls are subjected to a steady tearing-away which manifests itself in faults, giving rise to earthquakes and volcanoes. Thus evidence suggests that orogenesis, earthquakes and vulcanism could have their roots in the same cause.

It may be that convection currents were once more rapid and widespread, and that they have become slower and more localised as the Earth gradually cooled down. This would explain why the most ancient chains of mountains are wider than the most recent.

According to the calculations of Griggs, each convection cycle passes successively through a phase of acceleration, a phase of stabilisation, a phase of slowing-down, and a phase of stoppage, lasting in all some millions of years. This periodicity fits in wonderfully with what we know of the successive formations at long intervals of the various mountain chains.

Griggs goes still further in calculating the size of the convection cycles, or the convection 'cells'. Their width being roughly equal to three times their height, it can be said that the cycle is three times the thickness of the mantle, the width of the Pacific, or the distance which separates the American chains from the mountain chains in formation of the Asiatic coast (the island arcs).

More recently still, the geologist R. Perrin has put forward a new theory to explain mountain folding. Mountains are known to contain a great deal of metamorphic rock, that is, former sediments or igneous rocks that have become cleaved and crystallised — crystalline schist, mica schist and gneiss. To date, folding was believed to be responsible for this. Perrin envisaged it, on the contrary, as the cause of the uplift of mountains. His hypothesis is based on the discovery of chemical reactions which can take place in the solid state. Materials that have been sufficiently pulverised to come into extremely close contact can react with one another and combine at temperatures very much lower than their melting-points. Thus, at temperatures well below 1,000°C., silicates can be formed by bringing silica, lime and magnesia into contact, or even kaolin and calcium carbonate. These reactions in solids are accompanied by a considerable increase in volume. Why, supposes Perrin, should similar reactions not occur in geosynclines? Great weight of overburden brings the minerals at depth into extremely close contact. Why should their sedimentary contents not be able to uplift and fold themselves under the sole influence of the chemical reactions taking place in their depths? The hypothesis is certainly attractive, and must be given serious consideration despite its radical departure from current beliefs.

Still more recently (Heezen, 1960) a theory has been put forward based on the idea that the Earth's interior is expanding and causing the crust to crack along major rift lines such as the Mid-Atlantic ridge and the Afro-Asia Minor rift. Material from beneath moves slowly upward in these rift zones, dividing segments of the Earth's crust and producing mountains directly as well as by 'squeezing' action as the crustal segments move. As always in geology and geophysics, the phenomenon of diastrophism is the result of many interlocking forces, of which only too many are still unknown.

*Left*
*(Top)* Famous limestone terraces, about 36 feet high at Hamman Peak, Algeria, showing the boiling waterfalls of ferruginous and sulphureous springs.
*(Bottom)* Wave rock, believed to be the result of wind erosion. This spectacular granite formation stands near Hyden, 240 miles east of Perth, Australia.

# Earth in the Service of Man

## Minerals and Mineral Ores

Even the deepest drill holes and the most profound mine shafts probe but a thin slice of the Earth's crust. What lies still deeper is unknown and can be the subject only of hypotheses and theories. If, however, we restrict ourselves to considering the thin slice that is accessible for study, we find that the materials of the Earth's crust consist of minerals and rocks.

Minerals are naturally occurring inorganic compounds possessing a characteristic internal structure (atomic) and physical and chemical properties which are fixed or vary but slightly. The main groups of rock-forming minerals are silica in its various forms, silicates, oxides, carbonates, sulphides and sulphates.

Rocks are assemblages in varying proportions of different minerals. Granite, for example, is a rock composed mainly of the minerals quartz, feldspar and mica. Whatever their composition Earth's rocks fall into one of three categories: igneous, sedimentary or metamorphic. Igneous rocks are those which have risen from the depths of the Earth as hot, molten masses; sedimentary rocks are those formed by the breakdown of pre-existing rocks and deposited by wind, water or ice. Metamorphic rocks are the result of transformation in igneous and sedimentary forms subjected to high temperatures and pressures.

### PHYSICAL PROPERTIES OF MINERALS

A detailed study of the chemical, physical and optical properties of minerals would be beyond the scope of this book. There is, however, one property with a scientific interest that surpasses all others; that property is the crystalline state and, often, the crystalline form.

The immediate fascination of mineralogical collections is the geometrical forms of the majority of the exhibits. Here are the yellow, octahedral crystals of sulphur, transparent cubic crystals of rock salt, hexagonal crystals of quartz or of amethyst. Grown under ideal conditions, each mineral has a fixed characteristic crystalline form, but because perfect developmental environments are rare in nature, perfect crystals are equally rare.

Research has produced a theory that all crystalline forms can be grouped into one of six or seven crystal systems, each comprising a basic, or normal group of forms and all forms derived from them. The main crystal systems are the isometric, tetragonal, hexagonal (sub-divided by many mineralogists into two systems: the hexagonal and the trigonal, or rhombic), orthorhombic, monoclinic and triclinic.

Let us take, for example, the cube, the basic form of the isometric system. This polyhedron has six faces, eight corners (coigns), and twelve edges, all in perfect symmetry. Suppose that the corners become truncated at an angle which is the same with respect to the adjacent faces. The result is a crystal with the six primary faces of the cube, plus eight faces which have replaced the corners of the cube, that is, fourteen faces in all. This is a combination of cube and octahedron. In other words, the basic cube crystal has been modified by the octahedron form, which is

a simple eight-faced crystal resembling two pyramids joined at their bases. If the truncations grow until they join up with one another, eating away the cube faces, an octahedron is formed.

Returning to the basic cube, if instead of the corners, all the edges are truncated at equal angles to the adjacent faces, a crystal with these twelve truncation faces and the six primary faces of the cube results, giving a cubododecahedron. A dodecahedron is a crystal bounded by twelve rhomb-shaped faces.

The term truncation is used when the modifying crystal faces make equal angles with adjacent similar faces, otherwise the terms 'cut' or 'modified' are used.

Instead of the cube, we might take the rhombohedron or any of the various prisms. In each case the symmetrically disposed truncations would lead to an infinite number of derived forms. When two crystals grow together the 'twin' which results exhibits a secondary symmetry no less perfect in form.

If symmetry dominates the mineral world, animals and plant life are equally subject to it. There is the radial symmetry of flowers, of corals, of starfish, and of sea-urchins; there is the bilateral symmetry of worms, of crustacea and of vertebrates. Man, too, displays a certain symmetry with his two ears, two eyes, two lungs, two kidneys, and so on. Few beings escape the universal law of symmetry.

Cleavage is another property of minerals that is controlled, like their crystal form, by their internal structure. Some minerals split in certain constant directions or planes called cleavages. The result is to reduce the mineral finally to the basic form of the system to which it belongs.

Many other physical properties of minerals are used in identification. One of the most valuable is hardness, defined as a mineral's resistance to scratching, and measured against an arbitrary numerical scale consisting of ten minerals, numbered from one to ten in increasing degree of hardness: 1, Talc; 2, Gypsum; 3, Calcite; 4, Fluorite; 5, Apatite; 6, Orthoclase; 7, Quartz; 8, Topaz; 9, Corundum; 10, Diamond. Few minerals have a hardness exactly matching any one of these; most lie somewhere between two of the standards. For example, a mineral which scratches apatite but not orthoclase is said to have a hardness of 5·5.

Another important identifying physical property, especially in metallic minerals, is streak. This refers to the colour of the powdered mineral, which may or may not be that of the solid mineral. For example, hematite, an iron-bearing mineral, occurs in a number of colour forms — grey, black, green and brown — but no matter what its colour, its streak is always a rich red-brown. Streak for any particular mineral is a fairly constant property and a useful means of identification, even for an untrained amateur. The technique is, simply, to rub the mineral under inspection on a piece of unused porcelain. Gold and 'fools gold' (pyrite), are easily distinguished in this way; pyrite produces a black or dark streak, and gold a streak identical with the visible colour of the solid.

Colour can be used in a general way for identification purposes. The weakness of the test is that not only are many minerals the same colour, but they are also colour variable. In fact, colour

Geologists examining rock samples.

constancy is a rare mineral characteristic; most occur in several colour forms. This is especially true of colourless or white minerals in which even very minor impurities will produce colour. For example, quartz, colourless when pure, occurs in an almost infinite colour range in its impure forms.

Other useful identifying physical properties include lustre, manner of fracture, specific gravity, magnetism, iridescence and fluorescence. This last property is one of the most striking, and certain minerals when exposed to ultra-violet light display brilliant colour effects: fluorite gives violet, calcite gives scarlet, and so on. Fluorescence is so marked a feature of certain valuable minerals that portable 'black light' lamps are used in detecting them.

## PRINCIPAL MINERALS

The common minerals, which form the largest part of the rocks, are silica, the silicates, the carbonates, and the iron oxides.

**Silica.** Gelatinous silica, that is, very fine particles of silica in suspension in water (colloidal suspension), is present to a greater or lesser extent in water. It is silica's method of travel. All colloidal solutions finally coagulate, and gelatinous silica is no exception. The constituent particles join together into flakes which increase in size to form first a compact but not very solid mass, and finally hydrated silica or opal, an opaque or transparent, colourless or variously coloured substance.

While remaining hydrated, opal can acquire a crystalline form and is then known as chalcedony. This form of silica is generally found as fibrous, kidney-shaped masses which are resolved under the microscope into a felt of fine needles. In the Chalcedony Park in the United States there are the vegetable fibres of trees which have been completely replaced by a succession of different forms of silica. In France, in the coal basin of Autun, the same phenomenon has occurred, and the structures of plants of the Carboniferous period are preserved intact.

Related to opal, chalcedony and flint are the ornamental stones such as jasper, agate and onyx. Chert is compact, dense, very finely crystalline, siliceous material which occurs as layers or nodules in other sedimentary rocks, generally limestones. Flint is a dark-coloured variety of chert and is, essentially, silica with some water. Agate is the term applied to colour-banded varieties. The exact origin of these minerals is difficult to determine. They appear to have come into existence in a number of ways; the terminology, classification, origin and exact nature of this group of siliceous minerals or rocks are still far from clear.

The last and most important form of silica is quartz. In this, the silica is anhydrous and perfectly crystallised either as bi-pyramidal prisms, or as bi-pyramids belonging to the hexagonal system.

Quartz is harder than steel and glass but not as hard as diamond. It is affected only by hydrofluoric acid, and fuses only in an electric furnace. Generally transparent, like crystal (hyaline quartz or rock crystal), it is sometimes translucent or nearly opaque and occurs in colourless and coloured forms; amethyst, milky quartz, and smoky quartz are some of the coloured varieties. Silica-bearing waters deposit quartz in rock fissures and cavities as veins, or geodes. Many igneous rocks and metamorphic rocks contain quartz, and when they are broken down by erosional agents the unalterable quartz forms sands which on cementation become sandstones. Quartz, alone or together with opal, chalcedony or other silica minerals, is to be found in most rocks.

**Silicates.** The formula of silica ($SiO_2$) is very similar to that of carbon dioxide ($CO_2$), producing silicates just as carbon dioxide produces carbonates.

Silicates (combinations of silica with different oxides), are extremely important constituents of the Earth's crust, forming the greatest part of crystalline rocks and found, partially decomposed, in stratified rocks.

The principal silicate families are the feldspars, the amphiboles, the pyroxenes, the olivines and the micas.

The feldspars are anhydrous silicates of aluminium combined with potassium, sodium, or calcium, depending on the species, and they form almost one-half of the Earth's rocks. Pure potash feldspar ($KAlSi_3O_8$) is called orthoclase, pure soda feldspar ($NaAlSi_3O_8$) is albite, pure lime feldspar is anorthite, and there

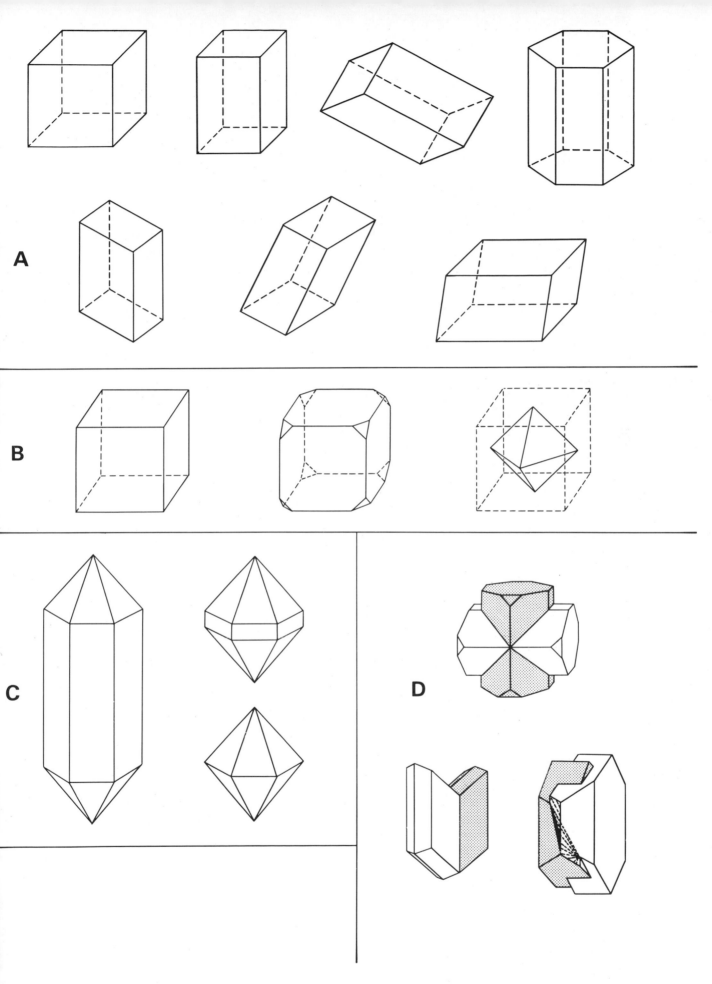

Elements of crystallography: *(Top to Bottom)*
A. Primitive forms of the seven crystal systems.

B. Passage from the cube to cubododecahendron and to the octahedron by the process of truncation.

C. Crystalline forms of quartz.
D. Twins of gypsum (spear head), orthoclase (Carlsbad) and staurolite.

Calcite flowers in the Grotto
of Deramats, Vercors France.

one direction. This property allows them to be split into thinner and thinner laminae (flakes) for various commercial purposes. White mica, or muscovite, is used for windows, lanterns, ovens and stoves because it is infusible and transparent. Once used extensively in Russia for windows and in portholes of warships, it became known as 'Muscovy glass'. Isinglass is another name for it. Because it is a poor conductor its uses today are mainly in electrical equipment — in vacuum tubes, coils and condensers; and because it withstands high temperatures it is utilised as an insulator in all types of heating appliances, toasters, irons, coffee pots and so on. In commerce and in the chemical laboratory it has extensive applications because it resists the action of acids and caustic solutions. In ground and powdered form, muscovite is used in the manufacture of paints, inks and insulating board. The United States produces large quantities of ground mica. The main source area of sheet mica is India; a poorer quality comes from South America.

Different varieties of mica have special uses: phlogopite is employed in high temperature equipment as it withstands heat up to 1,000°C; lepidolite and zinnwaldite, both micas, are minor sources of lithium; and roscoelite yields vanadium. Dark mica, such as biotite, is used principally in the electrical industry as an insulator.

**Carbonates.** Streams and subterranean waters always contain a fairly large amount of calcium bicarbonate, Ca $(HCO_3)_2$, the form in which calcium carbonate is transported.

In contact with the air, the bicarbonate gives off carbon dioxide and splits up according to the following formula:

$$Ca(HCO_3)_2 = CaCO_3 + CO_2 + H_2O$$

Calcium carbonate ($CaCO_3$), being insoluble in water, forms a powdery and, later, a compact precipitate.

Conversely, calcium carbonate can be dissolved in water rich in $CO_2$:

$$CaCO_3 + CO_2 + H_2O = Ca(HCO_3)_2.$$

The two formulas given are reciprocal to one another and express what is called a reversible reaction.

In rocks calcium carbonate may form crystals visible to the unaided eye (crystalline) or recognised only under the microscope (cryptocrystalline). In crystalline form it occurs as calcite or aragonite.

Calcite is a whitish or transparent mineral, little harder than a finger nail and effervescing under the action of cold, weak hydrochloric acid. It is one of the most widespread minerals and, like quartz, is found in veins and geodes. It constitutes the sparkling matter of stalactites, stalagmites and crystalline marbles. Often it occurs in concentric layers and constitutes the rounded grains, or oolites, of many thick limestone beds. Along with aragonite, calcite is the normal shell-forming material. Its most interesting and rarest variety, Iceland spar, occurs as beautiful, transparent rhombohedra. These crystals are doubly-refracting and any object viewed through a piece of Iceland spar gives a double image. The crystalline forms may be rhombohedral with many truncations or may occur as bi-pyramidal prisms known also as 'dog-tooth spar'. In all these cases the crystals cleave into rhombohedral shape, the fundamental crystal form of calcite.

The carbonates of calcium ($CaHO_3$) and magnesium ($MgCO_3$) often crystallise together, and there is also a double carbonate, called dolomite after the French geologist, Dolomieu. This mineral is harder than calcite and effervesces only with strong acids. Calcium, magnesium and calcimagnesian carbonates in the form of limestone, chalk and dolomite are widespread throughout the world. In industry, these rocks are used as building and decorative stone, and in the manufacture of cement, plaster, fertilizers and pharmaceutical products.

## PRECIOUS STONES

Too hard to be drilled and requiring, in consequence, a metal setting, precious stones or gems did not figure extensively in human history until the Bronze Age. Decorative stones, however, came

are several lime-soda feldspars, or plagioclases. All these substances are white, pale grey or lightly coloured, and are almost as hard as or harder than steel. They crystallise in more or less elongated or flattened prisms; frequently, several crystals grow together as 'twins'. One of the best known twins (Carlsbad) is formed by the interpenetration of two orthoclase crystals.

Unlike the feldspars, the amphiboles and the pyroxenes always contain calcium, magnesium and iron. This last element imparts a dark colour to the mineral. Crystallographically amphiboles occur in rocks as prisms with an octahedral cross-section. Asbestos, used in the manufacture of fireproof fabric, is a variety of amphibole with a fibrous structure. Crysotile, or fibrous serpentine, is another mineral which is commonly called asbestos. The ornamental stone, jade, also owes its toughness to the intertwining of very fine fibres.

While most of the silicates so far named contain aluminium and calcium, these elements do not exist in the olivines, which are solely ferromagnesian silicates. Olivine, so called because of its yellow-green colour, is one of the main constituents of basalt, in which it can be recognised by its colour and granular form. The name 'olivine' refers to a series of minerals which consists of admixtures in various proportions of fayalite ($Fe_2SiO_4$) and forsterite ($Mg_2SiO_4$); fayalite and forsterite are called the 'end members' of the series.

Finally, the silicate series is completed by talc, kaolinite and the micas which are chiefly hydrated members of the family and may or may not be ferromagnesian. The name mica (Latin 'to shine') refers to their glistening, almost pearly appearance. Micas containing only aluminium and potassium are called white micas; those which contain, in addition, iron and magnesium, are the dark or black micas. Both may be found as hexagonal plates or in flakes, because they possess a cleavage which is perfect in

*Top*
Drilling holes for explosive to blast the diamond-bearing rocks in a South African diamond mine.

*Bottom*
At the same mine diamond workers put gravel through a trommel at a prospecting trench.

into use long before man utilised metals, probably as far back as 22,000 B.C., and it is known that the semi-precious stone turquoise was used between 7,000 and 3,400 B.C. Succeeding India and Persia the New World became a supplier from the beginning of the sixteenth century, the period when lapidaries, goldsmiths and jewellers vied with each other in attractive presentation of precious stones. Though the courts of Europe competed in their turn for possession of the world's biggest and most beautiful stone, they never approached the magnificence of the oriental courts.

The chief precious stones are diamond, emerald, ruby and sapphire; those of secondary importance (semi-precious stones) are alexandrite, topaz, aquamarine, garnet, amethyst, jasper, onyx, agate, cornelian, opal, jade, turquoise. Most of these stones occur as constituents of igneous rocks but they are frequently found as secondary deposits in alluvium resulting from the break-down of such rocks. The principal producers are South Africa, India, Ceylon, Burma, Indonesia, Brazil (Minas Gerais), Colombia and Peru; the biggest markets are found in London, Amsterdam, Antwerp, Paris and New York, and the chief cutting centres are Rotterdam and Antwerp.

Apart from diamond, which is pure carbon, all other precious stones are compounds, either oxides of silicon (amethyst) or aluminium (ruby and sapphire) or silicates (emerald and topaz), or even phosphates (turquoise). The primary quality of gem stones is their beauty. Although standards of beauty have changed through the ages and according to the dictates of fashion, most gem stones prized centuries ago are still sought after today. Important properties include brilliance or lustre, unusual and attractive colour, special optical effects ('stars' in sapphires and rubies, or a play of colours in opals), the ability to take a high polish or delicate carving, durability and rarity. Scarcity value sets the price. Emeralds, for example, are scarcer than sapphires and therefore command a higher price on markets strictly controlled to maintain a constant high value.

Colour was once the prime quality of gem stones, and a purple amethyst was considered as precious as a blue sapphire, but with the development of faceting the difference in brilliance became apparent. The sapphire, having a higher index of refraction, bends light more and therefore glitters more. If the gem bends the violet wave-length more than the red, it will give off flashes of light. This characteristic, called dispersion, is highly desirable in gem stones.

Durability in a gem stone implies hardness, enabling it to be worn without showing signs of wear. To be considered durable a gem stone must have a hardness greater than 6 in the numerical scale.

The various properties that are so valued can be produced by the lapidary with an appropriate cut — cabochon, rose, brilliant, step, or mixed cut.

Cutting, the craft of the lapidary, demands extraordinary patience and skill. The rough stone must first of all be cleaned of its stony skin, then split along its cleavage planes by sharp blows with a 'hammer' or cut with bronze discs rotating at great speed. The cutting proper is then carried out on grindstones covered with diamond powder. Finally comes polishing and setting in a gold or platinum mount.

Precious stones are weighed in 'carats'. The metric carat was standardised in 1932 at 200 milligrams and so is a little smaller than the earlier standard carat, the English carat, which is 3 grains of one-eighth of a penny-weight.

## Diamond.

A crystalline form of pure carbon, the diamond belongs to the cubic or isometric system and is generally found as an octahedron. From a chemical point of view it is the simplest precious stone, the hardest, and the least subject to chemical attack. Colourless diamonds are highly valued, while tinted varieties (yellow, grey, brown) are not generally so desirable. However, the very rare, deeply coloured blue, green or red diamonds are the most treasured of all gem stones. There is no doubt that diamond is of deep and igneous origin if one is to judge from the temperature of 5,900°F. and pressure of 1·5 million pounds which the General Electric Company of the United States had to employ recently to effect synthesis of some fractions of a grain of the mineral.

Before 1870, diamonds were obtained only from the alluvial deposits (gravels) in river beds in South Africa, Brazil, India and Borneo. In that year, at Dutoitspan in South Africa, a rich diamond-bearing yellow clay, which bore no resemblance to a river deposit, was found in a hollow. On digging into the yellow clay, it was seen to be underlain by a diamond-bearing rock, called 'yellow-ground' by the miners. This in turn was found to pass downward into 'blue-ground' also rich in diamonds. The lateral spread of the 'ground' was discovered to be limited but its depth is measurable in thousands of feet. Such occurrences were named 'diamond pipes' and are similar in shape and dimensions to the throats of volcanoes but there is no evidence that the sides of the pipes were heated. They are attributed to explosions within the Earth which drilled holes through the crust. Blue-ground is a fragmented rock containing minerals such as corundum, mica, chrome-diopside, garnet, diamond, and rare specimens of a rock called eclogite. Eclogite is a mixture of garnet, olivine, chrome-diopside and iron oxide, and is believed to have consolidated at great depths below surface. Industrial diamonds (bort, carbonado) which make up four-fifths of the total world production come from the Congo and Brazil. 'Bort' is the name used in commerce for splinters, rough fragments and imperfect crystals of diamond. 'Carbonado', or black diamond, is a variety of crystalline carbon related to diamond but which shows no crystal form; it is found only in Brazil. These varieties are used in industry as abrasives.

The most famous diamonds are the Cullinan which originally weighed 3,024 carats (21·32 ounces), the 'Woyi' (770 carats), and the 'Jonker' (726 carats). A sizeable fragment of the Cullinan now adorns the British sceptre.

## Emerald.

This deep green stone, rival of the diamond, and one of the more valuable of gem stones is a silicate of aluminium and beryllium, and an ornamental variety of beryl. It crystallises in the hexagonal system and usually contains impurities which, far from adversely affecting its value, enhance its sparkle and produce what is called the 'jardin' of the stone. Large flawless specimens are unknown except in fiction and fable. Even slightly flawed stones of moderate size are extremely rare and very costly. The most valued emeralds are deep grass green and have a velvety appearance. The finest stones come from the Muzo or Chivor Mines in Columbia, Peru, and the Transvaal Mountains; the Urals also produce quality specimens.

Other beryl minerals of gem stone value include heliodor, a yellow form from south-west Africa; morganite, a pink variety from California; and goshenite, a colourless type.

Aquamarine, coming from Brazil and many other areas, is a pale bluish-green cousin of the emerald. Alexandrite, a silicate of beryllium without aluminium, is less transparent than emerald and possesses a change of hue which makes it grey-green in daylight and wine red in artificial light. The colour of many of the beryl varieties can be improved by heat treatment.

## Ruby and Sapphire.

These precious stones are clear coloured varieties of corundum ($Al_2O_3$) tB which oxides of chromium, iron or titanium impart a red or blue colour. They have a hardness of 9 and no cleavage, and are therefore very durable. The finest rubies are pigeon's blood red, a deep purplish red, and most of them come from near Mogok, Burma. The finest sapphires are Kashmir blue or cornflower blue. The best gems are those that are cut parallel to what was once the end of the crystal, for here the colour is at its most intense.

Fibrous internal structure may be present in both rubies and sapphires, producing cloudy star stones which have recently been very popular. The famous soft grey Star of India sapphire and the purplish De Long Star ruby (both on display at the American Museum of Natural History in New York City) are superb examples of these unusual stones. Sapphires come from Kashmir and Queensland. The last produces some remarkably deep blue, almost black, gems. Ceylon, too, has been a source for fine

Workers in Sierra Leone washing the diamond-bearing gravels with sieves and large flat pans in their search for alluvial diamonds along the banks of a stream.

General view of the Premier
mine, Transvaal.

rubies and also for sapphires for well over two thousand years.

Rubies and sapphires can now be synthesised so perfectly that it is impossible for the layman and difficult even for the expert to distinguish synthetic from natural stones.

**Synthetic gems.** Synthetic jewels have long been made from a special glass called 'paste'. Most are easy to detect, but a few gem stones — spinel (titania), alexandrite, emerald, diamond, sapphire, ruby and the star stones — can be synthesised very faithfully.

Unless tested carefully the expensive synthetic stones can fool even the most experienced. The jewels produced by the 'flame-fusion' process using an oxyhydrogen flame are excellent in quality and very difficult to distinguish from natural stones. Before World War II most synthetic gems were produced in Switzerland, in France and in Germany, but England and United States are now leaders in this field. Synthetic rubies and sapphires have

come to replace the natural stones as jewel bearings in watches and delicate instruments for obvious reasons of economy.

## ORES

It is known that Neolithic man used copper during the late Stone Age. Along with gold, it appears to have been the first metal utilised. Then followed bronze, an alloy (an intimate admixture in solid solution) of copper and tin, and finally iron, which dates back to more than one thousand years before Christ. Up to the eighteenth century only lead, zinc and silver had swelled the list of metals worked. Later, as the techniques of metallurgy were developed, metals of low specific gravity such as magnesium and aluminium were made available to industry. It is now possible to obtain by a process of alloying special steels such as nickel manganese, chromium, cobalt, molybdenum, tungsten and vanadium. Some metals are utilised for special purposes: antimony

The process of diamond cutting
and forming the facets.

for hardening printing type; vanadium as a catalyst in the manufacture of sulphuric acid; titanium or, better still, titanium white to replace white lead or zinc white for painting; cadmium for galvanising iron; caesium in photoelectric cells; cerium in lighter flints. The rare and consequently precious metals are, of course, gold and platinum, but there are others such as iridium, palladium and rhodium which are even rarer and which are alloyed with platinum in commercial usage. Finally there are the radioactive metals — radium and uranium.

**Ore Deposits.** Ores are combinations of metals and non-metal elements such as oxides, sulphides, sulphates, carbonates and silicates. For example, cassiterite (tin oxide) and galena (lead sulphide) are both ores. An ore deposit is an accumulation of minerals or native metals which can be mined at a profit. It is not easy to define the word 'metal' but it can be best described as follows: a substance which possesses many of the following

properties — solid at ordinary temperatures; opaque; when polished, a good reflector of light; a good or fairly good conductor of heat and electricity; when melted and allowed to cool, it solidifies into a compact mass of crystals.

Native metals are those which are very nearly pure or are allied with other metals; for example, there is native copper, native gold, and electrum — a mixture of gold and silver.

When the Earth was a liquid globe at high temperatures, it seems likely that the metals were arranged from the surface to the centre in the order of their atomic weights. Much later, after the Earth had cooled and a crust had congealed at the periphery, the light metals were quite naturally found to make up part of this crust. The heavy metals, on the other hand, have probably been able to reach the surface only by means of swirling eddies in the heart of the still liquid mass and by eruptions enabling this mass to penetrate into the crust. In other words, the heavy metals, which without that process would be inaccessible to us, have

risen to the surface with intrusive magmas and lavas. By heavy metals we mean all except aluminium, magnesium, potassium sodium and calcium.

As the majority of metals have an igneous origin, let us consider a hot sticky magma rising from the depths of the Earth. It moves upward with entrapped water-vapour and other volatiles, of which some are metallic. The volatile materials escape from the magma along fissures in the Earth's crust and may be deposited as veins, perhaps several miles long and half an inch to several hundred yards thick. A vein, which is the filling of a fissure in the rocks, usually comprises four parts: the wall or enclosing rock, the wall coating (generally argillaceous), the economically unimportant material, or 'gangue', which is usually calcareous or siliceous, and finally the ore which occupies the central cavity. Several veins which fall close together and allow mining as a unit are called a lode deposit.

Not all metalliferous deposits occur in veins. An ore may be deposited in the cavities of a porous rock after being transported by running waters, or it may be deposited in layers in lakes or in the sea. Here we shall study only the main ores.

**Iron ores.** Iron is at once the commonest metal and the most useful in its various forms (pure iron, cast iron, steel). Iron salts are found in most rocks: the monosulphide (troilite, $FeS$) imparts a grey colour to potter's clay; the bisulphide (pyrite, $FeS_2$) forms golden-yellow cubic crystals in slates; magnetite ($Fe_3O_4$) also known as magnetic iron ore, limonite ($2Fe_2O_3\ 3H_2O$), and glauconite (iron silicate) colour the rocks black, red, and yellow or green respectively.

The economic potentialities of these ores, from a metallurgical standpoint, are restricted because account must be taken of the conditions of the deposit and of the richness of the minerals in metallic iron. The only salts exploited commercially are magnetite (72 per cent iron), hematite (70 per cent), limonite (60 per cent) and iron carbonate or siderite (30 per cent).

The chief iron-producing countries are the United States, the U.S.S.R., France and Sweden. Areas rich in iron ore include the Lake Superior region of the United States (hematite), the magnetite mountain of the Urals, the masses or mountains of magnetite in Sweden (Kiruna), and the 'minette' of Lorraine. This last deposit is oolitic hematite, that is, it is composed of rounded grains rather like fish roe. Each grain is made up of a small central nucleus with concentric surrounding layers of hematite and limonite. The ore itself is disposed in beds within marly limestones. It is the most extensive deposit in France, furnishing nine-tenths of the total production of the country.

Britain, too, is well endowed with iron ores, but they are usually low in iron content. The Jurassic contains several horizons of sedimentary iron ores, of which the most important is the Northampton Ironstone extending from Grantham to Northampton with an estimated reserve of about 1,000 million tons of ore. The ore is oolitic, with oolites of chamosite (iron silicate) or siderite (iron carbonate) set in a muddy or sideritic matrix. It is also somewhat siliceous.

The Frodingham Ironstone, near Scunthorpe in Lincolnshire, consists of limonite oolites in a matrix of chamosite and siderite mixed with a considerable proportion of shells and shell fragments. This reduces the percentage of iron to 20 but makes the ore self-fluxing.

The early iron and steel industry was based on the Clay ironstones — nodules of siderite in shales. These contain 30 per cent iron and are associated with the Coal Measures but they are discontinuous in their distribution. They constitute a vast reserve if the other bodies become worked out.

Finally there are the hematite ore bodies of Cumberland, Lancashire and Glamorgan. They occur in the Carboniferous limestone, and the remains of corals, crinoids and other shells preserved as hematite in a similar matrix reveal them as replacement bodies. This ore is high grade, up to 60 per cent iron, but known reserves are not great.

An important consideration in iron smelting is the need for large quantities of coke for treating the ore. There is thus a need to establish a link by waterway, preferably, or by rail between

the iron-mining regions and the coal-mining areas. Examples of coal-steel linkage include the Lorraine and the Ruhr; Lake Superior iron and Pennsylvania coal; iron from the Urals and coal from Siberia.

In the United States the low-grade oolitic hematite ores of Alabama are mined extensively because of a fortunate arrangement of circumstances: the presence of coal in the same locality, the self-fluxing character of the ores, and the location of the ore deposit in an area where transport and labour is relatively cheap. These oolitic hematites are scattered over a wide belt from Pennsylvania to Alabama; the available reserves amount to about 1·9 billion tons. The only important mines, however, are in Alabama where the ores are thickest. The major United States iron production comes from the Lake Superior region in Minnesota, Michigan and Wisconsin. The ore is mainly hematite and occurs in folded sedimentary formations that outcrop in narrow belts known as ranges. There are seven ranges, the largest of which is the Mesabi Range. Total production of the Lake Superior district averages 90 million tons annually, representing about 80 to 85 per cent of total United States iron production.

**Nickel ores.** Nickel is a metal which often occurs with iron and is used pure in small amounts for plating. About 95 per cent of nickel is used in steel alloys. The main suppliers are Canada (about 75 per cent of the world's production), U.S.S.R. (much of U.S.S.R. production comes from the Russian-annexed Finnish Petsamo area), New Caledonia, Cuba and Union of South Africa. The ore from New Caledonia is an apple-green silicate of nickel, garnierite — the richest and most valued ore. The United States produces a small quantity of nickel as a by-product in copper mining.

**Copper ores.** Copper is the first metal to have been used by man because it commonly occurs in the native state, that is, pure, and is soft enough to be worked immediately. Today it is used for electric cabling and telephone lines, electrical and cooking equipment, and ornamental purposes.

Copper generally occurs in association with other metals and in almost every ore deposit. Nearly every country with ores of commercial value produces some copper. More than 150 copper minerals exist but only about eight occur as deposits of commercial value. In addition to native copper, the principal ores are the sulphides bornite ($Cu_5FeS_4$), chalcocite ($CuS$), and chalcopyrite ($CuFeS_2$); the oxide, cuprite ($CuO_2$); and the carbonate, malachite ($Cu_2(OH)_2CO_3$).

The chief deposits of native copper are those located on the Keweenaw Peninsula of Lake Superior (United States). Early Indians in the area collected the copper for arrow and spear points, ornaments and tools. The original deposits have been mined for over eighty years and are now nearly exhausted after producing more than four million tons of copper. In the same region a new copper deposit of low-grade sulphide ore is now being mined extensively.

The largest and most productive copper mine in United States is located in the Bingham area in Utah. This mine is also an important source of other metals. It ranks third for silver, second for gold, second for molybdenum, and yields large amounts of lead and zinc. The income from the by-products (gold, silver, etc.) covers the cost of mining and processing; the proceeds from the copper are profit. The open-pit, contour-terraced mine is spectacular. It covers 933 acres and within the pit there are differences in elevation of 2,000 feet. Fullsize railroad cars run on more than 170 miles of standard gauge railroad track.

The third largest copper producer in the United States is Butte, Montana, a region which first developed as a gold and silver district, but since about 1892 has made copper its major source of income. This is one of the world's greatest copper deposits and is an outstanding example of vein lodes. In addition to copper, zinc, lead, manganese, silver and gold are produced. The mine is about 8 square miles in area, with over 1,000 miles of underground workings.

The world's main producers of the various copper ores are at present the U.S.A. (about 30 per cent), Chile, Northern Rhodesia,

the U.S.S.R., and Canada. England was once in the front rank but as the ores of the Cornish deposits were worked out, her importance waned along with most of the countries of Europe.

**Tin ores.** The main use of tin is in the manufacture of commercial tin and tin alloys needed in plating other metals. It is also used in combination with other metals for the manufacture of solder (tin-lead) bronze (copper-tin), Babbitt metal (lead, tin, antimony) and type metals. The tin ore minerals are the oxide, cassiterite ($SnO_2$) and the sulphide, stannite ($Cu_2FeSnS_4$).

The major areas of tin production lie in south-east Asia, with China, Malaya, Indonesia, Thailand and Burma yielding almost 63 per cent of the world's production. The deposits are mainly placer (alluvial) which are dredged or mined by hydraulic methods. Cassiterite is the main ore mineral; there is also some stannite.

Bolivia is the world's second largest producers. Because of inadequate means of transport and inaccessible location, only the highest grade deposits (both cassiterite and stannite) are mined here, and Bolivia has vast low-grade ore reserves. By-products include copper, silver, bismuth, lead, zinc, molybdenum and tungsten.

**Lead and Zinc ores.** Lead and zinc, although very different in their chemical and physical properties, generally occur together in nature. With few exceptions lead districts produce zinc, and vice versa. Next to copper, lead and zinc are the most essential non-ferrous industrial metals. About one-third of the lead produced is used in pigments, while the remainder goes into storage batteries, electrical coverings, tetraethyl lead and solder, pewter and alloys. Galvanising accounts for about 40 per cent of the zinc produced. Other applications include castings, rolled zinc, pigments and alloys such as brass.

The principal lead ore is galena ($PbS$). It frequently occurs with silver — argentiferous galena is an important source of silver. Zinc is extracted mainly from sphalertie ($ZnS$, zinc blende) and to a lesser extent from hemimorphite [$Zn_2SiO_3(OH)_2$] and smithsonite ($ZnCO_3$). Gold and copper may be present in the lead and zinc ores, and bismuth, antimony and cadmium are common by-products in zinc mining. The major producers of lead and zinc are United States, Australia, Mexico, Canada and U.S.S.R.

The Tri-state (Oklahoma, Kansas, Missouri), or Joplin District, in the United States produces large quantities of zinc and lead from deposits in cavernous limestone. No other metals are produced in this district. The lovely Coeur d'Alene Lake District, in Idaho, is best known for producing more silver than any other area in the United States, but its largest yield is lead-zinc ore. Gold and copper are by-products, and the area is mined to depths of 5,300 feet.

Kimberley, British Columbia, has the world's largest lead-zinc mine, employing over 1,200 men in mine and mills; it produces silver as well. Mexico has one of the oldest lead and zinc mines still in operation. It opened in the middle 1500s.

**Aluminium (Aluminum) ores.** Up to 1888 aluminium, or aluminum as it is called in the United States, had been manufactured only in small quantities and by a single chemical process. Since that date, because of the development of electrolytic methods of separation, and because of its extraordinary lightness and malleability, aluminium has sprung to the forefront and in quantity of output now rivals steel and cast iron. It is used in aircraft construction in the building industry, and in the manufacture of toys, boats, automobile parts and kitchenware.

Its chief ore is bauxite (hydrated aluminium silicate), generally coloured red or brown by iron oxide. Bauxite is not a single, pure mineral but a clay-like mixture of several, all of which are hydrous oxides of aluminium. It often contains pea-sized, concentrically layered pellets called pisolites, formed by intense weathering in hot humid climates. In processing, first pure alumina ($Al_3O_2$) is extracted chemically. This is then treated electrolytically in carbon crucibles in the presence of cryolite (fluoride of aluminium and sodium). This has the effect of lowering the temperature of treatment to about 1,800°F. The aluminium collects at the bottom of the crucible and is finally recast into ingots.

Bauxite was first discovered in 1822 at Les Baux in France. Today it is mined in the U.S.A. (near Little Rock, Arkansas) Surinam, Jamaica, British Guiana, Hungary and France. Production is beginning also in Venezuela, Australia and Ghana. France has excellent deposits in Languedoc and Provence; the deposits are sedimentary and are located in pockets on the surface of Tertiary terrains. They are gernerally worked by open-cast methods. The bauxite is then transported to coal centres for chemical treatment and to hydro-electric centres for electrolytic treatment.

**Magnesium ores.** Magnesium, the third lightest metal after aluminium, is widely used in aircraft construction and in the manufacture of flares, incendiary bombs and flash-bulbs. It is extracted from a number of ores, from magnesite ($MgCO_3$), from dolomite [$CaMg(CO_3)_2$] and from carnallite, a double chloride of magnesium and potassium. However, most magnesium production is from natural brines and sea water by electrolysis after preliminary treatment with lime and hydrochloric acid; there are therefore unlimited quantities available to almost all nations. The leading producers are U.S.A., Germany, Japan and the United Kingdom.

**Mercury ores.** Mercury or 'quicksilver' is a metal that is liquid at ordinary temperatures. Because of its quick-moving, elusive globules it was named after the fleetfooted Greek god Mercury. The only important ore of mercury is the sulphide, cinnabar (HgS). It usually occurs in porous rocks, to which it imparts a vermillion colour. The ore is calcined in furnaces, and the liquid metal is extracted, distilled, and put into bottles. Deposits include those of Almaden (Spain) which have been worked for 3,000 years, Idria (Italy), New Idria (California), Nevada, Oregon, Idaho, Alaska, Mexico, Yugoslavia and Russia.

**Silver ores.** Yesterday a precious metal, today an industrial one, silver has acquired for itself a special place in certain technical processes (silver plating and soldering) and in the manufacture of photographic films and paper. About 70 per cent of the total silver production is used for monetary purposes — as coin or bullion. Much is used for jewellery and tableware. Silver has ores of its own but is extracted mainly as a by-product from those of other metals, lead, zinc, copper and gold. It is also found as a native mineral. Mexico has long been the leader in silver production; mines started in 1530 are still being worked. Other countries, U.S.A., Canada, Peru, U.S.S.R., Australia, Bolivia, Germany, Japan, Yugoslavia and the Congo, also produce silver.

**Precious metals.** Gold and platinum, by reason of their resistance to chemical attack, are considered the two principal precious metals. Gold, in particular, has become the standard for currency, mostly as bullion. It is also used as gold leaf, as a plating on ornamental objects, and has extensive applications in jewellery. It is generally alloyed to increase its strength. White gold is an alloy of gold with silver, nickel or platinum; green gold contains silver and cadmium; red gold contains copper; purple gold contains aluminium.

Native gold is the most important ore. It may occur as microscopic specks, thin wires, masses of dust, or nuggets, and often contains silver as an impurity. In a few areas sylvanite [$(AuAg)Te_2$] and calaverite ($AuTe_2$) are important ore minerals of silver and gold. Gold also occurs as an impurity in ores of many other metals and can be found in most geological types of ore deposits. In minute amounts, it is a very common metal. Ocean waters probably contain a great deal of gold (estimated at 10 million tons) but extraction is uneconomic.

The most easily workable deposits are gold-bearing alluvials deposited by rivers which have for a long time eroded and leached pre-existing rocks. These alluvial deposits are called placers. Where the placers are buried under a variable thickness of unproductive alluvium they must be worked by pits (shafts) and drifts; where they are found at the surface, the gold can be recovered directly by panning or dredging. The great Californian Gold Rush in 1849 was precipitated by the discovery of placer gold.

A gold mine at Kalgoorlie, Australia. The heaps rising behind the buildings are debris from the mine.

The processing of alluvial deposits differs according to their contents and their richness in gold. Often washing alone is sufficient to separate the nuggets, and early prospectors, or 'gold panners', washed their pans under a stream of water or shook them in the wind to get rid of the impurities. If the parent rock is hard and compact and if the gold is finely dispersed, the rock must be crushed, ground to a powder, and then mixed with water and potassium cyanide to dissolve the gold; zinc powder is added to the solution which precipitates the gold as a black slime; the slime is melted and run off into ingots for subsequent refining. Sometimes five tons of rock have to be treated to obtain a piece of gold the size of a walnut.

In the United States the Homestake Mine in South Dakota, the most important mine in the country, has yielded over 500 million U.S. dollars in gold. The surface plant of this mine handles 4,000 tons of ore daily. The amount of gold recovered per ton is about 0·5 ounce. The famous Mother Lode deposit in the Sierra Nevada of California is a series of closely spaced gold-bearing veins occurring in a zone about one mile wide and 120 miles long. This deposit has produced 250 million dollars in gold. Much of the original deposit was eroded and the gold re-deposited as placers.

The most important gold region in the world is the Rand gold district in Transvaal, Union of South Africa. It yields about one-third of the world's production and has produced a total of over ten billion dollars worth of gold. The gold occurs in a curved belt about 90 miles long in which about 40 mines (some of the world's deepest) work at levels of 3,000 to 6,000 feet beneath the surface. Some will eventually be worked to depths of 9,000 and 10,000 feet.

The main gold-producing regions of the world are in the Union of South Africa, Ghana, Southern Rhodesia, the Congo, the U.S.S.R., Canada, U.S.A., Australia, Columbia and the Philippines.

Platinum is more useful commercially and rarer than gold. Its industrial uses, as a catalyst, for laboratory dishes, for dental plates, and for dies used in the extrusion of glass and rayon, make it more and more sought after. Platinum generally occurs with other rare and chemically similar metals such as iridium, rhodium, osmium, palladium and ruthenium. The market value of some of these other rare metals is as high as platinum, and exceeds 100 U.S. dollars per troy ounce. The most common ore is native platinum but there are several others, mainly natural alloys with other metals.

The major producer of platinum and other rare metals is the Union of South Africa. Canada follows, with the U.S.S.R. in third place. United States and Columbia produce small amounts.

## THE IGNEOUS ROCKS

An igneous rock is any rock which has formed from molten magma or lava. It is also the primary which formed as our molten Earth cooled, and the ancestor of all other rocks. Igneous rocks make up about 85 per cent or more of the Earth's crust.

The most common and universally known igneous rock is granite, found in Brittany, in Devon, New England, California, Alaska and in many ancient and recent mountain regions. Granite is often grey and is composed of grains (whence its name), some white and others black. With the aid of a hand-lens, it is not difficult to recognise that the grains are crystalline minerals. Among the white minerals are transparent ones (quartz) and opaque ones (feldspars). The black crystals include amphibole and the micas, which can be split into laminae with the point of a penknife.

The minerals constituting granite have crystallised from an igneous mass, that is, in the course of the cooling of a molten magma which was able to rise from the depths of the Earth and to move toward the surface. In other words, granite is at once a crystalline, igneous and intrusive rock.

The constituent mineral grains of granite are visible to the naked eye and are of almost equal dimensions. Such a perfect degree of crystallisation could only have been achieved slowly and at considerable depth within the Earth's crust. Granite is therefore a plutonic igneous rock. Massifs made up of granite, which today are visible at the surface of the Earth, have been exposed by the erosion of several superincumbent layers of rock.

Let us now move on from granite to basalt, the most widespread type of lava (volcanic igneous rock). When examined with the naked eye, hand-lens or microscope, this rock reveals a different aspect from granite. In most basalts, large crystals are absent altogether or few in number. Those that are present are embedded in a homogeneous looking mass, called the matrix or groundmass, which is made up of minute crystals called microlites (Greek *mikros* — small, and *lithos* — stone). Because basalt crystals are generally present in a wide range of sizes, crystallisation cannot have taken place at the same time. The larger crystals must have developed slowly during the ascent of the magma, and the small crystals rapidly while the lava was being poured out on to the surface of the Earth. In short, basalt is frequently a surface-formed igneous rock that is commonly found in lava flows.

To sum up we shall say that igneous rocks are those which have risen from beneath the Earth's crust as hot viscous masses: some, such as granite, consolidated at depth; others, the volcanic rocks, completed their consolidation on the surface.

Cooling at different rates, a single magma may produce many types of rock. The outer surface will cool more quickly and be finer textured. The interior will cool slowly. Different minerals crystallise out at certain temperatures, so that some will do so fairly early in the cooling process, and if they are heavier than the liquid magma they will tend to settle slowly downward. Thus a large igneous mass will have zones of different minerals and zones of different textures.

**Classification.** Igneous rocks are classified on the basis of texture and composition. Texture refers to the size, shape and arrangement or pattern of the mineral grains or crystals of which the rock is composed. The following terms are commonly used:

*Pegmatitic:* particle size must be over half an inch in length.

*Granitoid, granular or coarse-grained:* particle size averages $\frac{1}{4}$ in. in diameter but may range from just within the vision limits of the naked eye up to half an inch.

*Aphanitic, felsitic or fine-textured:* individual mineral grains are present but are not visible to the unaided eye.

*Glassy texture:* no mineral grains present, the rock is a smooth, homogeneous mass.

*Porphyritic:* two or more distinct grain sizes present. The finer grains make up the groundmass or matrix; the larger grains are known as phenocrysts.

Textures are mainly controlled by the rate of cooling. Slow cooling takes place when the igneous masses are large or deeply located, or both. Rapid cooling occurs when the igneous masses are small and at or near the surface. Slow cooling produces coarse textures; rapid cooling, fine textures; and extremely rapid cooling, glassy textures. Porphyritic textures are produced by two or more cooling stages, first a slow stage in which the phenocrysts form, then a fast one, producing the groundmass.

Chemical composition, particularly the amount of silica and volatiles, is also an important and controlling factor in the development of textures. Magmas which are high in volatiles (fluids and gases) or low silica, or both, are more liquid, forming larger mineral grains and coarser textures.

Within each of the textural groupings of igneous rocks, the subdivision is on a basis of mineral composition. For example: a granular rock composed mostly of quartz and feldspars with some mica is called a granite; a fine-textured rock with this same composition is called a rhyolite. The different texture groups grade one into the other and so do the composition groups.

## COARSE-TEXTURED IGNEOUS ROCKS

Coarse-textured igneous rocks are extremely widespread in both ancient and present-day mountain chains. As examples we can

Some colourful minerals:
*(Top left to bottom right)*
1. Sicilian sulphur;
2. Malachite;
3. Cheyssilite from Bisbec, Arizona;
4. Cinnabar;
5. Agate;
6. Smithsonite (New Mexico).
7. Fluorite;
8. Orthoclase and quartz;
9. Cerussite;
10. Grandiorite.

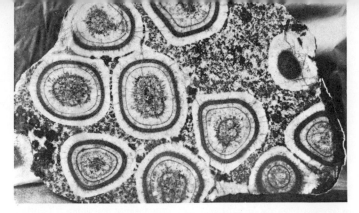

cite the English Shap granite with big pink orthoclase crystals, and the Dartmoor granite with white potash feldspars six inches long, and in France the pink granite of Tregastel and Cape Frehel, the porphyritic granite of Cherbourg, and the pegmatite of Limousin.

Granite is a hard and resistant rock, capable of taking a polish, and has a specific gravity of 2·7. It is used as a building stone for jetties, breakwaters, bridges, paving stones and curb stones, and as an ornamental stone on building facings, churches and tombstones.

Less resistant, muscovite-biotite-granite disintegrates and cannot be used in the same way as granite. Its easy decomposition by atmospheric agents produces kaolin (a clay mineral) which is used in the porcelain industry. Cornish china-clay, particularly that extracted near Saint Austell, is kaolin produced by ascending hydrothermal solutions, not by descending atmospheric solutions.

Granite pegmatite also decomposes easily and provides workable crystals of quartz and mica. In places pegmatites are a source of precious stones (emeralds, topaz, tourmaline).

In addition to granite and its near relations, the coarse-grained group includes rocks deficient in or devoid of quartz and richer in amphibole, pyroxene and olivine. These are ferromagnesian minerals which impart a dark, sometimes completely black, colour to the rocks. Such rocks are the syenites which have much the same composition as granite but not as much quartz and more amphibole; diorites in which the feldspars are darker and amphibole is associated with pyroxene; gabbros in which the feldspars are dark and pyroxene is associated with olivine. Admittedly these differences can scarcely be appreciated by the layman and they often necessitate the use of a hand-lens and, above all, an eye very practised in distinguishing one mineral from another.

Commercially, the rocks just described have much the same properties and the same uses as granite, and are often sold under the name of granite, which has an appeal not possessed by the terms syenite or diorite. Most commercial granite in the United States comes from the Appalachian Mountain belt or from Minnesota, Wisconsin, or South Dakota. There is much granite in the western United States mountains but very little is quarried.

Because the cost of cutting, shaping and polishing granite is high, other rocks are sometimes used as ornamental stones. Minette (mica syenite) and kersantite (mica diorite) — both rich in dark mica — are both used; so is orbicular diorite, a beautiful ornamental rock consisting of rounded masses or rosettes composed of alternately dark and light layers of minerals. Widely distributed in Corsica, this rock is also known as corsite and as napoleonite.

Peridotite, which is made up almost entirely of olivine, is related to gabbro. The decomposition of some of the olivine grains in periodtite lends to it reddish or blackish flecks which, sometimes enhanced by the presence of garnets, produce a very handsome ornamental stone, often used for jewellery and other ornamental objects.

## FINE-TEXTURED AND GLASSY ROCKS

The fine-textured and glassy igneous rocks occur mainly in smaller intrusive igneous bodies and in lava flows.

Typical of the fine-textured group of igneous rocks is basalt. Many present-day volcanoes, such as those in Hawaii and Réunion Island, emit lava of basaltic composition. In the past a part of Auvergne was overwhelmed by huge sheets, the existing parts of which now constitute the 'plateau basalt' so widespread in the Mont Dore, Cantal, Aubrac and Cézallier. In India one basalt sheet is known to cover an area of about 115,000 square miles. Fragments of a once extensive sheet covering a large area in the North Atlantic can be found in Antrim, Skye, the Faeroes and Iceland; in north-western United States, the Columbia Plateau consists of lava beds thousands of feet thick.

Basalts are often split into long prismatic columns running from one side of the flow to the other. The prisms generally have 4, 5 or 6 or, more rarely, 8, 9 or 10 sides. They were formed by tensional stresses in contraction shrinkage during the cooling of the lava flow, and are arranged perpendicular to its surface. A front view of a group of columns looks like a colonnade or an

organ; from above it looks like a causeway. Among the most famous of these 'organs' are those of Murat and Saint-Flour and Esplay in France, and those of Fingal's cave on the Island of Staffa in Scotland. The most famous causeway of all is the Giant's Causeway on the north coast of Ireland which exposes its 40,000 prisms at wave level. The Devil's Post Pile in California is yet another fine example of columnar jointing.

Basalts are black or dark grey—or occasionally reddish or greenish in colour — heavy, compact and very resistant rocks. Locally they are used as a building stone and for road-fill, flooring and facing tunnels. The 'basalt in half-mourning' of the Mont Dore, so called because of the contrast between the large white feldspar phenocrysts and the black groundmass, is a porphyry.

The main constituents of basalt are pyroxene, plagioclase feldspars and olivine.

Where amphibole takes the place of olivine in basalt another rock, andesite, is formed — very common in the cordillera of the Andes. In France it forms a large part of the Cantal, the Mont Dore and the chain of the Puys. It is also the lava of several present-day volcanoes, including Mount Pelee and Krakatoa. In Great Britain it is common in Snowdonia and in the Lake District (Helvellyn and Scafell). Andesite is a fine-grained rock, often grey, green or reddish in colour, rough in appearance, fairly hard and resistant to weathering. It is transitional in composition between rhyolite and basalt. Many buildings and monuments in Clermont Ferrand have been constructed with andesite, which is worked at Volvic near Riom. The Volvic stone takes a good polish.

The red and green porphyries of antiquity are ancient andesites which, once they have acquired a patina, make very beautiful ornamental stones. The red porphyry of Upper Egypt was very widely worked by the Ancients. The plinth of Napoleon's tomb in the Invalides in Paris is of green porphyry.

Another porphyry, the blue porphyry of Esterel, has been worked since early times between Agay and Saint Raphael, and many monuments in Arles, Orange and even in Rome owe some of their beauty to it.

Glassy igneous rocks are produced when lavas or magmas cool so quickly that the minerals have no chance to crystallise. Massive or non-porous glass is known as obsidian. Porous glass with fine, closely spaced openings is pumice; that with few and larger pores in scoria. The openings in porous glasses are caused by expanding gases which produce a frothy texture in the upper portions of lava flows.

Obsidian is a glass which is generally black but sometimes grey or brown in colour. It has a smooth curved fracture and a high lustre.

Pumics are generally light in colour, buff or pale grey. They are abundant on the Lipari Islands, where they are worked for a variety of domestic and industrial purposes. The dome of St Sophia in Constantinople (Istanbul) is made of pumaceous rock.

## ORIGIN OF SEDIMENTARY ROCKS

Sedimentary rocks originate at or near the Earth's surface under much lower temperatures and pressures than igneous rocks.

Most of them are deposited as sediments in the calm parts of rivers, lakes and seas. They underlie about 80 per cent of the entire Earth's surface, form a large portion of the visible crust, determine many of the surface features, and contain coal, oil, gas and many other substances of economic value.

Almost all the sediments producing these rocks are formed by the destruction of the continents by erosional agents. They include terrigenous or detrital sediments (gravel, sand and clay) from the disintegration and decomposition of other rocks, and products of chemical solution and precipitation, such as rock salt, gypsum, and some limestones. Some are formed by the activity of living organisms secreting calcite or silica. To this category of sediments, described as organic, belong the coral reefs, diatomaceous and globigerina oozes, and coals.

Because sedimentary rocks are deposited by many different agents of erosion they differ considerably in physical and mineralogical properties. However, all have the common characteristic of layering, bedding, or stratification. There are only a few sedimentary rocks which lack this distinguishing feature.

Stratification stems from variations in depositional conditions. Differences in the volume and velocity of a river, for example, produce sedimentary layers of varying texture, with coarser materials carried during times of high water and rapid velocity. Differences in colour or type of material also cause stratification. Normal bedding is horizontal or nearly horizontal in the original sediments. There is also the possibility of cross-bedding, graded bedding, and lensing of beds.

Mud cracks, ripple marks, rill marks, wave marks, raindrop impressions and fossils are minor features which reflect conditions of the depositional environment and which have accumulated through time in sedimentary rocks.

Sediments are converted into rock, or lithified, mainly by two processes: cementation and compaction. Either of these processes can convert sediment into rock but generally both partake in consolidation. Desiccation and crystallisation are also often involved in lithification if conditions are suitable.

The basalt columns of Giant's Causeway, Northern Ireland.

COMPACTION. Under the weight of a heavy overburden of other sediments the particles in the lower sediments move closer and closer together. Interstitial water is driven out and the sediments lose volume and are compacted. Compaction is most effective in fine-grained materials, and some rocks (claystone and shale) are formed by this process alone. A clay under a 1,000-foot layer of other sediments loses about 20 per cent of its volume; under 5,000 feet it is about 55 per cent. In coarser materials — sand is one — there is little loss of volume but the grains may be forced close enough to be held together by molecular attraction. In this way a very friable sandstone may be formed.

DESICCATION: The process of lithification is speeded up when the water is squeezed out of the sediments or when they are exposed to the air.

CEMENTATION: In addition to suffering desiccation and compaction most sedimentary rocks are cemented together. The cementing material is carried in solution by water — either the original water in the sediments or ground water entering the sediments after the initial water has been removed. Changes in environmental conditions cause deposition of this material from solution into the pore spaces between the rock particles, thus binding them together into solid rock. Sand and gravel for example, become sandstone and conglomerate. Silica and calcium carbonate are the commonest cementing materials.

CRYSTALLISATION: This may occur in directly precipitated deposits or later, after a rock is partly or completely formed.

Generally, sedimentary rocks are divided into two groups: clastic (or detrital) and non-clastic (or chemical). Clastic and detrital sedimentary rocks are composed of broken rocks or mineral fragments. The non-clastics include all other sedimentary rocks, mainly chemical or organic precipitates. However, there are many classifications of sedimentary rocks, all based on very similar principles, and differing only in detail. None is the perfect or universal system, because sedimentary rocks are too diverse in origins and character to be fitted into a relatively simple usable pattern. Some classifications are based on the agent of erosion (rivers, seas, ice, wind) which deposited the sediments. Others are based on the mineral composition of the rock, or on the size of its particles.

## CLASTIC ROCKS

**Conglomerates and Breccias.** These two rocks are formed by the cementation of gravels, pebbles, cobbles and boulders. If the component particles are rounded, the rock is called a conglomerate; if they are angular, the rock is a breccia. The cementing substance is generally a mixture of sand and/or clay particles, with silica or calcite as the binder.

Conglomerates and breccias are used for ornamental purposes, building stones and ballast.

**Sandstones.** Sandstones are clastic rocks with a grain size lying between 1/16 mm. and 2 mm. Generally the sand particles consist of quartz but may be composed of any other mineral. The commonest cementing materials are silica, calcite or iron oxide. Sandstones may be pure white or multicoloured, depending on the mineral grains or cement present.

Pure quartz sandstones are used in the manufacture of optical lenses, glassware, windows, building materials and insulators.

*Arkose* is a variety of coarse sandstone, usually reddish in colour and containing much the same minerals as a granite: quartz, feldspars and mica. Arkose is usually produced by the rapid erosion of granites, although it may also form more slowly under extremely dry or extremely cold conditions which will slow down or prevent decomposition of the feldspar and mica minerals.

*Greywacke* is another variety of sandstone. Usually grey, blue-grey or brownish-grey, it originates under conditions similar to those producing arkose. The parent rock is, however, not a granite but a gabbro, basalt or some other basic igneous rock.

*Top*
A section of volcanic rock showing, from top to bottom: a solid basalt dyke; a vesicular dyke formed by bubbles of gas; carboniferous limestone.

*Centre*
Cavernous millstone.

*Bottom*
Pudding stone a (conglomerate).

231

**Shale.** This is the general name given to most argillaceous rocks. Shales are very fine-textured (grain size is less than 1/16 mm.) and thin-bedded, and show frequent changes in the fineness of their particles. They are composed mainly of clay or mud but may also contain variable amounts of other substances. Shales containing large amounts of organic matter are called carbonaceous shales, and are usually black, perhaps grading or interbedding with coal. Those which contain calcium carbonate are calcarous or marly shales; those with sand are arenaceous shales; those with iron are ferrugenous shales. Argillaceous rocks which lack the thin-bedded characteristic of true shale are siltstone or claystone.

Shales are detrital rocks lithified by compaction and desiccation only; there is no cementation.

The clay minerals and the rocks which they form are a very complex group. They form part of the residual soil and sometimes of the transported soil, as well as forming sedimentary rocks such as shale. Other varieties of clay-formed rocks include: claystone, similar to shale but lacking the thin-layered bedding; siltstone, a slightly sandy claystone; bentonite, a clay derived from volcanic rocks; marl, a limey claystone. All are formed by deep weathering of many different rocks, impure limestones, igneous rocks, glacial till and others. Commercial clays are obtained from several varieties of shale and from loess.

**Kaolin and other Clays.** The purest argillaceous rock or mineral is kaolin, which is usually found in weathered portions of feldspathic rocks (granulite, pegmatite and granite).

Plastic clay (bentonite) is usually grey (due to iron monsulphide) but sometimes it is tinted or even mottled with bright colours (iron oxides). It is soft and plastic, remains unchanged at ordinary temperatures, and absorbs and retains large quantities of water. Pugged with water, it forms an easily worked paste that can be moulded between the fingers. When saturation point is reached it becomes impermeable and any superfluous water will run off.

Plastic clay is the modelling clay of sculptors. It is also used, as are other less pure clays, in making all manner of things: bricks, tiles, pipes, pottery. 'Potter's clay', plastic clay mixed with water and sand, is pugged in special machines to produce a workable paste which can be moulded or passed through a mill and stamped out to the required shape before being dried in an air tunnel. Then follows baking in a furnace where the heat hardens the clay with a slight decrease in volume. At the same time the decomposition and oxidation of the iron monosulphide into hematite turns the colour from grey to red. Gas or oil may be used to fire the kilns instead of coal.

In addition to plastic, there is an infinite variety of clays. Varieties very rich in iron oxide include red ochre or bloodstone, yellow ochre, ombre, and Sienna all used in painting. Red ochre can be obtained by calcining yellow ochre — the limonite becomes hematite on dehydration.

**Marls.** This is the name given to rocks which are mixtures of clay and lime. They are often grey or buff in colour. They may be classified either with shales or with limestones because they grade one into the other. Marls are very widespread rocks and are important because of their many applications.

When added to siliceous soils (process of marling) they enrich the land with the lime and clay necessary to crop growing. When burned they provide lime and cements. They are also used as mortar for bonding stones together and for the rough casting of houses.

Hydraulic cement or natural cement is derived from marls containing between 15 per cent and 25 per cent clay. Their high clay content makes them suitable for use under water, where they set as perfectly as in the open air. An obvious application of this property is in bridge building, wherever the structure comes

*Left*
Excavation by shale plane of carbonaceous shales of Lower Oxford Clay formation.

*Right*
A china clay mine near St. Austell, in Cornwall.

into contact with water. It is also used for jetties, breakwaters, and lighthouse foundations.

Cements are obtained by heating to high temperatures marls containing more than 25 per cent clay. Some set in a few minutes and are used for blocking leaks; others set more slowly. Portland cement, invented in 1824 by Joseph Aspdin, an English bricklayer, was so named because when set it looks like Portland stone. It is produced in Britain by burning a mixture of chalk and clay, but can also be manufactured by burning impure limestones.

There are many kinds of cement, designed to serve various needs, but the basic materials are the same — lime, alumina, silica and iron oxide. The lime may be obtained from limestones, marls, shells or blast furnace slag. The alumina may come from clay, coal-ash slag or igneous rocks, and the silica from clay, shale, coal-ash slag, sand or sandstone. The iron oxide may come from certain limestones, clays and shales, or from pyrite, slags, cinders, and mill scale.

The constituents are crushed and mixed in proper proportions with water and burned in rotary kilns. The product of this process is hard pellets which are ground into fine powder — cement.

Concrete for house foundations and road making is a mixture of cement, pebbles, sand and water. The resulting substance is conglomerate. Reinforced concrete is composed of cement, gravel, sand and water intimately mixed and poured round a metal framework.

Recently a new technique has been developed: the manufacture of pre-stressed concrete, enabling sections prepared in the factory to be fitted together later as piles, stressed tubing, railway sleepers, and key-stones of bridges. The sections are built upon a framework of steel rods, particularly elastic and resistant to rupture. During the casting and setting of the concrete, the rods are subjected to tension. Afterwards they are released, compressing the whole structure and strengthening it.

## NON-CLASTIC ROCKS

**Calcareous rocks.** These are composed essentially of calcium carbonate ($CaCO_3$) in many forms and with many impurities.

Calcareous rocks can be scratched with a steel nail or with a knife. They are white when pure, and grey, black, blue, red or buff when impure. Grey limestones are the commonest. The majority (chalk and shelly limestones among them) are both detrital and organic. Others, such as the oolitic limestones, tufas and travertines, are precipitated from solution.

Limestones are used, according to their properties, as building stones (freestone or quarry stone) or as raw material for the manufacture of lime. Heated in masonry furnaces they give off carbon dioxide and are converted into quicklime according to the formula:

$$CaCO_3 = CaO + CO_2$$

The quicklime is mixed with sand and water to give mortar. The quicklime combines with the water, giving off a great deal of heat, to become slaked lime according to the formula:

$$CaO + H_2O = Ca(OH)_2$$

When set between the stones of a wall as mortar, the mixture is quick to harden because the slaked lime, in contact with the carbon dioxide content of the air, is reconverted into calcium carbonate:

$$Ca(OH)_2 \quad CO_2 = CaCO_3 \quad H_2O$$

Slaked lime in very dilute solution constitutes whitewash.

Lime water, obtained by filtering whitewash, is a reagent used in chemistry.

**Chalk.** Chalk is a white — sometimes grey or greenish — limestone so friable that it crumbles and turns to powder when rubbed

between the fingers. It is extremely porous and consequently permeable. Chalky soils readily absorb rain water, and because of their great dryness are naturally infertile.

Chalk is composed mainly of microscopic particles of $CaCO_3$, minute tests (or shells) of Foraminifera, and assorted organic debris in which the palaeontologist is able to identify sponge spicules, shell and carapace fragments, etc. The various deposits are dated geologically by their rich fossil content, whole and fragmentary.

Several types of concretions may form within chalk: flint nodules drawing their silica from sponge spicules and the skeletons of radiolaria; nodules of marcasite, the iron sulphide which stems from the putrefaction of organic matter originally contained in chalk; nodules of calcium phosphate, also of organic origin; and grains of glauconite or iron silicate which impart a greenish colour.

Study of its fossils indicates that chalk is a deposit made in a fairly shallow sea (a few hundred feet deep). It is not comparable with the globigerina ooze being laid down at great depth in present-day oceans.

Chalk occurs over a very large part of England, including Salisbury Plain, the Chiltern Hills and the North Downs. In France, it is found over a large area but is usually covered by more recent sedimentary rocks. Chalk forms the cliffs of Dover, of Beachy Head, of the 'Needles', of Dieppe, of Fécamp and of Etretat, where regular, parallel bands of flint nodules can be seen in it.

White or pure chalk is easily distinguished from glauconite or green chalk, from marly chalk coloured grey or tan by clay, and from rough chalk or 'tuffeau' which contains sand, mica and phosphatic chalk. Most of the chalk beds of the western United States are of the buff marly variety.

Too friable and too readily broken down in solution or by frost to be used for permanent buildings, chalk is ideal for interior decoration where it has the advantage of being easy to work. In powder form, as Spanish white or Meudon white, it is used for cleaning window panes and silver plate.

**Shelly limestones.** Chalk is a rock in which shell fragments predominate and has been studied separately because of its geological importance and its rather special properties. The other shelly limestones are named according to their predominant fossils.

Miliolid limestone contains innumerable tests of Foraminifera with a maximum diameter of 1/25 inch. Nummulitic limestone is made up of coin-shaped Foraminifera. It is abundant in the Mediterranean regions, where, as in the pyramids of Egypt, it has been used as an ornamental stone. Cerithium limestone is characterised by the presence of numerous conical shells of gastropods. Often there remains only a cast, resembling the thread of a screw, the rest of the shell having been dissolved by percolating waters. Because of its porous texture this limestone can be used only as an ornamental stone. Limestones of Mississippian age are quarried extensively in the central United States for building stone.

The limestone of Derbyshire and Wales is crinoidal and contains the remains of sea lilies. In Burgundy and Lorraine there is an abundance of crinoidal limestone which owes its beauty to the presence of numerous glistening plates of the heads of crinoids, to dissociated pieces of crinoid stems and to a mass of sea-urchins' spines cemented by calcium carbonate.

Where the shells and shell debris are not cemented, the result is a shell sand or *falun*, like that found in the 'Crags' of East Anglia and in Touraine. Accumulations of uncemented or poorly cemented shells and shell fragments, called coquina, are forming today along the coast of Florida and elsewhere.

**Reef limestones.** In contrast to the limestone accumulation of dead organic debris just described, these 'constructed' limestones are produced by the activity of living organisms. There are ancient or fossil coral reefs more or less similar to those which flourish nowadays in warm waters. These are called bioherms, reef limestones, or coralline limestones; fine specimens are found on the outer edge of the Paris basin. Reef limestones are also found in the Carboniferous limestones of Lancashire and Yorkshire and also in the Permian beds found throughout most of Texas.

*Top*
Coarse grit with shale fragments; Triassic from Budleigh Salterton, Devon.

*Centre*
Nummulitic limestone.

*Bottom*
Oolitic limestone.

*Top*
Crinoidal limestone.

*Centre*
Wet clayey soil.

*Bottom*
Part of an ancient dried out mud bank. When the water returned it filled the cracks with sand which remained while the mud hardened gradually into stone.

**Oolitic limestones.** These limestones are accumulations of small, rounded grains (oolites) resembling hardened fish roe. Each oolite is a concretion of several hundred very thin layers of calcite deposited round a nucleus. The oolites are cemented together by calcite to form an attractive rock that is frequently used as a building stone.

Most oolithic limestones are of marine origin and were deposited in the neighbourhood of coral reefs. For their development, they need strongly agitated water rich in dissolved carbonate. Each particle in suspension becomes a nucleus round which successive layers of carbonate are deposited.

Portland stone is an oolitic limestone widely used as a building material in London and the south of England. Bath stone, Osmington oolite from the Dorset coast, and Caninia Oolite from the Carboniferous are other British examples. In France oolitic limestones are very common in Lorraine, Burgundy, Calvados and Poitou. Caen stone has been used to build most of the cathedrals of Normandy. In the United States Indiana Bedford oolite is quarried extensively as an ornamental building stone.

**Travertines and Tufas.** Travertines are fine-grained limestones produced by the precipitation of calcium carbonate in fresh waters particularly rich in bicarbonate. Examples are found at Tivoli in Italy, Sézanne and Champigny in France, and Matlock in England.

Calcareous tufas are soft, light, cavernous rocks, often formed by deposition on mosses and algae. The great avidity of these plants for carbon dioxide expedites the precipitation of calcium carbonate. Both travertines and tufas are deposited today by many hot springs and geysers. In the United States such deposits occur in Yellowstone Park and most of the other hot spring areas.

**Dolostone** or **Dolomite.** Even though they consist mainly of magnesium carbonate, the dolomites—mixtures of calcite ($CaCO_3)_3$ and dolomite ($CaMg(CO_3)_2$) in varying proportion—are classified with the limestones. Dolomitic limestones are more resistant to weathering than pure limestones. When some parts of the rock are more dolomitic than others, differential weathering produces fine ruiniform limestones such as those of Montpellier-le-Vieux. The classic dolomite region is the Dolomitic Alps, that part of the Alpine chain located on the frontier between the Tyrol and Venezia. Its picturesqueness depends above all on the contrast between the high rocky walls of dolomite cut by erosion and the gentle slopes of marl on which they rest.

Dolomitic limestone occurs in the north-east of England, around Sunderland, and is used as a building stone; parts of the Houses of Parliament at Westminster are built of dolomite.

**Lithographic stones.** Lithographic limestone, or Bavarian stone, is an impermeable, hard and fine-grained clayey limestone which will take a polish. It finds an application in painting, in stone engraving (lithogravure) and in writing on stone (lithography). The technique is, briefly, this. The artist makes a design with a stick of grease-paint on a flat surface. When the design is finished, acid is applied to the surface of the stone. The action of the acid makes all save the artist's grease-paint design resist an ink impression.

**Evaporites.** Under this heading are brought together rocks of dissimilar chemical composition but possessing one common feature: all are precipitates formed by evaporation of waters in littoral lagoons, or in arid climates. Evaporites appear to have been precipitated from sea water according to a definite sequence. The more highly soluble minerals remain in solution longer. The less soluble ones are precipitated out as evaporation progresses. Anhydrite and gypsum come out of solution early, and halite later, as evaporation continues.

The principal evaporite rocks are gypsum, sea salt, rock salt, potassium salts and phosphates.

**Gypsum of Plaster stone.** Gypsum is a soft, generally white or light-coloured mineral which can be scratched with the finger nail. Usually the crystals are very small and form a rock somewhat

235

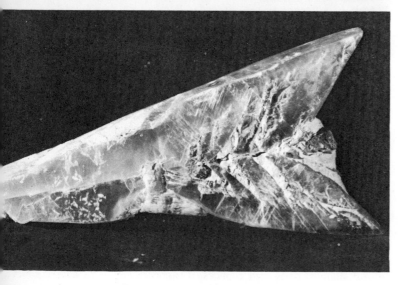

like sugar. This is saccharoidal gypsum. Sometimes the crystals are a little larger and are called 'larks feet'. Still larger, but more uncommon, are spear-head crystals, twins of two individuals joined together along a plane.

The main property of gypsum is its capacity to be transformed into plaster by loss of water on heating. Heated to 350°F., for example, it loses three-quarters of its water content and acquires the formula $2CaSO_4 H_2O$ (plaster of Paris). Reduced to powder and mixed with water, this gives a sort of gruel which sets quickly and gradually reforms again as tiny interlocking, needle-like crystals of gypsum.

The large beds of gypsum in the Midlands (Nottingham, Staffordshire and Derbyshire) where it is quarried for conversion into plaster of Paris. In the United States the most extensive gypsum and anhydrite deposits are found in Texas and New Mexico.

Gypsum is also used to slow down the rate of setting of Portland cement; its sulphate contents make it a fertilizer; it is used as a filler in paper manufacture, in paints, as a nutrient in yeast growing, and as plaster in building.

**Salt.** Sodium chloride is the commonest salt in sea water, and it is four or five times more abundant than all the other salts together. The water of the Atlantic Ocean contains on average 28 parts sodium chloride in 1,000 parts of water; that of the Mediterranean 32 parts in 1,000. It has been calculated that the total mass of sodium chloride in the sea reaches the colossal figure of thirty thousand million million ($3 \times 10^{16}$) tons. The mind is incapable of appreciating such a quantity. However, on average it requires the evaporation of 10,000 feet of sea water to deposit a 100-foot-thick layer of salt. The mineral salt, called halite, is characterised by its easy solubility in water, its distinctive taste, its cubic cleavage and its crystal form. It is generally colourless but is often tinted grey, brown, yellow, red and blue by the presence of impurities.

Commonly called kitchen salt, or simply salt, sodium chloride is one of Earth's most useful substances. Human blood is made up of 90% water, 0.9% salt, and small quantities of proteins and other substances. Under normal conditions this salt content is continually being expended and must be renewed. In short, human development is linked to the availability of supplies of salt. It is one of the reasons why the shores of the Mediterranean have been the cradle of civilisation. The Romans, in fact, paid their mercenaries partly in this commodity, and from this practice arose the custom, common to many countries, of calling payment for services *salary* (Latin *sal* — salt).

Salt is recovered from four different kinds of deposit: seawater and underground brines; playas and saline lakes; bedded salt deposits; and salt domes. In the United States all types are worked commercially.

**Sea salt.** Sea salt is extracted from the sea in salt pans or salt marshes, which may be looked upon as agricultural workings. In fact, the work in them is cyclic and subject to the rhythm of the seasons. They depend strictly on atmospheric agents, and productivity is assessed on the yield per acre. Salt is produced by solar evaporation in almost all maritime countries, including France, Spain, Italy, India, the United States and also in the U.S.S.R.

**Atlantic salt marshes in France.** There are considerable differences between the salines of the Atlantic coasts and the salines of the Mediterranean coasts. On the ocean coasts, sea water is stored at the high tides of full and new moon in basins where it deposits its mud and calcium sulphate content. The water is next passed into evaporators where it is concentrated. Finally it is run into the saline proper, which is a sort of shallow basin with a flat bottom divided into numerous compartments. Becoming more and more concentrated under the action of the sun and the wind, the water finally arrives in the central compartments where the salt is deposited. Extraction takes place from May to September, after the salines have been repaired after the damage caused during the winter. The salt begins to appear by about mid-June, and

*Top*
This gneiss was once layers of sand and shale, crumpled during mountain building periods.

*Centre*
Lithographic limestone.

*Bottom*
Spear-head gypsum.

*Top right*
Drilling operations in the rock salt mines at Meadow Bank, Cheshire.

*Bottom right*
Transporting rough rock salt along the main motorway of Meadow Bank mine.

because of the uncertainty of the climate it must be collected each evening before the slightest rain has a chance to dissolve it.

The salines of the west of France, which enjoyed peak prosperity in the fifteenth and sixteenth centuries, have steadily declined. A hundred years ago they had an annual production of more than 200,000 tons; today they produce on average 45,000 tons.

**Salines of the Mediterranean.** The organisation of the salines of the Mediterranean is quite different. In France the salt marshes have, for the most part, a direct link with railways or port installations. Those of Aigues-Mortes and Salin-de-Giraud annually produce 100,000 and 200,000 tons of salt respectively, two to five times more than the Atlantic coast salines. The method of exploiting the sea is totally different there.

The absence of tides requires, first of all, that the sea water should be pumped into the reservoirs, whence it is led into a series of basins. These latter correspond to the evaporators of the oceanic coast, but with this difference — they have a surface area of several hundred acres. In them the sea water deposits its mud and its calcium sulphate and is concentrated to about the point at which it is saturated with sodium chloride. At the end of this first phase of concentration the sea water is pumped into new and shallower salines of smaller area (about $2\frac{1}{2}$ acres). Here the bottom has previously been levelled and rolled to facilitate final collection of the salt. Evaporation comes at the end of August when the salt forms a crust several inches thick.

The crude salt is then taken to the washing plants to be freed of impurities. This is effected by mechanical stirring in a pit filled with its mother liquor, water saturated with salt and barely able to dissolve any more. On leaving the washing plant, the salt is carried by conveyor to perforated steel plates where it drains. A belt conveyor then leads it away to be tipped by a loader on to a pile which may finally contain several thousands of tons of salt.

The salt remains in heaps for several years before being taken by narrow-gauge railway to the grinding room and then bagged. Rain falling on the heaps dissolves only the superficial layers, which soon become covered over with a protective crust.

**Rock salt.** Rock salt is the salt formed by the evaporation of ancient seas, lagoons, salt lakes and shotts.

If it is difficult to understand how extensive deposits of this kind could have formed over the centuries, a solution is offered by present events in a gulf of the Caspian Sea, the gulf of Kara-Bugaz. This gulf, which is about 95 miles long and 80 miles wide, has an area of 7,000 square miles and a depth not exceeding 34 feet. It is linked with the Caspian only by a narrow channel, across which there is a sand bar allowing only surface water to enter. The rate of evaporation in the hot dry climate is so rapid that a current of water flows over the bar into the gulf and replaces the losses caused by evaporation. Thus the waters of the gulf have become five times as saline as ocean water, although the addition of fresh water from large rivers makes the water of the Caspian only one-third as saline as ocean water. Because of the rapid evaporation, the surface water of the gulf becomes denser and tends to sink. To date, gypsum and sodium sulphate are deposited. Salt is not yet sufficiently concentrated.

The most important salt horizons in Europe are those of Stassfurt (Germany), Salzburg (Austria) and Wieliczka (Poland). Rock salt is exploited in Britain in Cheshire (Northwich, Nantwich), Walney Island and Billingham. In France, it is worked in Lorraine, in the Jura and in the south-west. The deposits of Lorraine are the most important since they occupy an area of 600 square miles with an average thickness of 150 feet.

In the United States some wells produce natural brine, some of which is sea water trapped in sedimentary rocks, but most of which is ground water which has dissolved bedded salt rocks.

Bedded salt deposits are quite common, especially in rocks of Silurian and Permian age. Usually layers of gypsum, anhydrite and sometimes potash minerals occur with them. All were formed in the evaporation of ancient seas. Large deposits occur in the United States. Silurian salt deposits are found in Michigan, Ohio and New York, with a total thickness of 300 feet in New York, increasing westward to about 800 feet in Michigan. The

Permian salt deposits of the United States (Oklahoma. Texas, Kansas and New Mexico) covering about 100,000 square miles, on average 200 feet thick, are the world's largest known deposits.

Extensive salt deposits may also form in arid interior basins which have no outlet. Streams flowing into these basins may form lakes, such as Great Salt Lake, Utah. In other basins, where the water supply is low or intermittent, the basins contain lakes only after rain. The stream waters carry sediment and mineral matter into the basins and as the water evaporates, salts and sediments accumulate. The salts of these deposits vary according to the rocks being eroded in the area. In such regions the salts are quarried or mined when the playa lakes are dry, or are extracted by solar evaporation in the permanent lake basins. A little of the United States salt production is of this kind.

**Salt domes.** Salt domes occur in the United States Gulf Coast, Germany, Iran, Rumania and Spain. Those of the Gulf Coast are the most famous because of the tremendous petroleum production associated with them. These domes are great plug-like salt masses, one or more miles in diameter, which have moved upward and up-arched or pierced the overlying sediments. The domes have been deeply drilled in the search for oil but their bases have yet to be reached. Seismic sounding shows that they extend to depths of 17,000 to 20,000 feet. Most of the domes are fairly deeply buried but a few have been exposed at the surface. The manner of their origin is still in the realm of theory. The most popular view is that they come from thick, deep-lying salt beds. Salt is lighter than most sediments and under a heavy weight will flow like the ice of a glacier. So, when the salt beds are buried under many feet of sediments, there is an upward flow wherever there is a point or zone of weakness.

In Louisiana and Texas salt is mined from some of these domes. It is very pure and the reserves are tremendous. The United States has led the world in salt production in recent years, producing some 23 million metric tons annually. Other important salt producers are Russia (over 7 million tons), China (over 6 million tons), West Germany, India and France (each about 3 million tons). Total world production is almost 80 million metric tons.

Rock salt is extracted in two ways: either by mining or by the Frasche process. In surface and subsurface mines, conditions differ from those of coal-mines, being marked by the cleanness of the material, the absence of firedamp, and the ease of using electricity and explosives. In the Frasche process of salt production, holes are drilled to the salt bed. Stream or hot water is pumped into the bed through some of them to form artificially produced brine which is pumped out through others. The brine is then purified and evaporated to recover the salt.

Salt is sold in a number of forms; as blocks of rock salt for 'lick stones' for cattle, or broken up and reduced to 'powdered salt' for industrial purposes. It reaches the consumer in various grain sizes. Its main uses are in flavouring food, tanning leather, preserving wood, destroying weeds, manufacturing chlorine, hydrochloric acid, caustic soda, dyes, paints, lacquers and cements, and also in pharmaceutical preparations.

## POTASH SALTS

The potash salts, as they are commonly called, are really mixtures in variable proportions of the chlorides of potassium ($KCl$) and magnesium ($MgCl_2$) and of sulphates of the same metals ($K_2SO_4$ and $MgSO_4$). They are also called deliquescent salts, because they have high powers of absorption of atmospheric water-vapour. They are generally the source of potash fertilizers. Generally they are accompanied by gypsum and rock salt. This indicates that they were formed from evaporation of ancient lakes and seas.

In Europe the biggest deposits of potash salts are those of Stassfurt and Alsace, dating back to the end of the Palaeozoic era. At Stassfurt the beds are 800 feet thick. Working is by shafts and tunnels. Although the deposits are known to be over 5,000 feet deep, the beds are mined only at 1,000 and 2,000 feet where they have been raised by domes and anticlinal folds.

Lake Searles in the Mojave Desert of California is a playa lake, which sometimes contains water but is generally covered

These women in Angola are pouring their baskets of salt into barges for transporting to salt storage piles.

with a crust of soluble minerals. The superficial upper crust, 110 feet thick, rests on a bed of mud below which lies a second layer of salts, and so on. Until the development of the New Mexico deposits this area led in United States potash production. Today, Lake Searles produces several valuable salts (sodium, bromine and lithium). Originally the Searles deposit was exploited for borax (today a by-product). The main United States borax production is currently in Kern County, California. The Kern deposits are extensive and unusually pure. It is quite possible that the Dead Sea, in Palestine, will meet the same fate and will gradually be converted into a deposit of salts. Its mineral resources are already being exploited.

Potash deposits at 3,000 to 5,000 feet have been discovered in Eskdale, North Yorkshire. Their presence had long been considered a possibility but it was not until the D'Arcy Exploration Company put down a boring for oil in 1938 — 9 that the deposits were proved. Minerals include polyhalite, sylvite and carnallite.

Japan produces small amounts of potash salts, mostly from seaweed and as a by-product of cement and salt.

**Magnesium and Bromine.** Sea water is also the source of two very important substances — magnesium and bromine.

The extraction of magnesium from sea water and natural brines had its origins in 1916 at the height of World War I, and since then has developed and been perfected. Magnesium is also extracted from dolomite, magnesite and brucite ($Mg(OH)_2$), but even today magnesium of marine origin is less costly than that derived from any other source. One thousand tons of sea water produces on average one ton of metal. The United States, West

Germany, Japan and the United Kingdom are the main producers.

Sea water is the principal source of bromine. The extraction process is simple and consists only in driving bromine from its salts by bubbling chlorine through the water. The demand for bromine is increasing in pharmacy, in photography, in synthesis of dyes, in the paint industry, as a non-flamable solvent for celluloid, resins and gums, as an anaesthetic and sedative, and as ethyline dibromide (tetraethyl anti-knock). A modern plant located on the Dead Sea at Sodom, Israel, has recently been constructed with an expected production of 1,250 pounds annually.

**Phosphates.** The mineral apatite, $Ca_5(FCl)(PO_4)_3$, common in igneous rocks, is probably the prime source of all phosphate. When rocks containing apatite decompose, the phosphate is made available for plants in the soil or is carried in solution to the sea. Marine organisms use it in the formation of shells and bones, and some bacteria contain it in their tissues. Eventually, on death, the phosphatic material accumulates with the sediments on the sea floor. To this may also be added some phosphates precipitated directly from the sea water.

Small amounts of phosphates are present in many sedimentary rocks but higher concentrations occur only in a few beds. Recent guano deposits are the only other source.

In the United States, Florida is the leading phosphate producer. In this area phosphates occur at the surface in beds of sand-to-pebble-size particles. Phosphate-rich rocks of Pennsylvanian age occur in Wyoming, Idaho, Montana and Utah. The geologic structure of the region is complex and the mining is underground. Copper, lead and silver smelters in Montana produce sulphuric acid as a by-product. The acid is used in the treatment of the phosphates. This close source of acid makes the low-grade deposits reasonable commercial propositions that can complete with the products of southern phosphate mining, despite the distance from major markets. The phosphate reserves of this area are enormous; vanadium and uranium in small quantities occur in the same deposits.

In Europe the phosphatic soil of the Ardennes and the phosphatic chalk of Picardy were once widely exploited. Today the big producers are the United States (about one-quarter of the world's production), North Africa (Morocco and Tunisia), the U.S.S.R., and the islands in the Pacific and Indian oceans. The African deposits enjoy the double advantage of exceptional richness (up to 75 per cent pure phosphate) and proximity to marine transportation. The two clearing ports are Casablanca and Safi.

**Guano.** Guano is a rare deposit of partially leached animal excrement, generally from birds or bats but occasionally from rats, ground sloths and other animals. Certain equatorial islands, principally off the coast of Peru, are frequented by vast flocks of sea-birds who find abundant food in the fish shoals along the coast. The most important of them is the cormorant, followed by the gannet, pelican and albatross. The area becomes covered with the birds' droppings which, in the course of time, bacterial activity converts into guano — a Peruvian word for dung. Guano is a hard, white, lustrous substance, partly crystalline and consisting largely of phosphates and nitrates (both very soluble and valuable as plant food and as sources of fertilizers). Because of their high solubility they are uncommon and occur mainly in arid country or caves. Deposits of guano have been exploited wherever they occur, and today all the known Peruvian superficial deposits have been exhausted. The guano islands have had to be divided into three groups, and the fresh droppings are gathered from only one group every third year, thus allowing time for hardening and bacterial action. The annual production of this recently formed guano is about 150,000 tons.

Other smaller accumulations of guano are also known; there are pigeon deposits in Spain and Palestine, and bat deposits in some caves. In the United States, during the war of 1812, bat guano from Mammoth and other caves was a source of the saltpetre (nitre, potassium nitrate) used in the manufacture of gunpowder.

Ancient caves inhabited by carnivorous animals and bats may become sealed by red water-borne clay containing phosphorites or concretions of calcium phosphate. These so-called 'bone phosphates' are shiny masses, irregular in shape, and resembling porcelain. Their development has been accompanied by decomposition of the guano and any animal carcasses present, and by interaction between the phosphates and the surrounding limestone. The widely known phosphorites of Florida are of such origin, and are worked for phosphates and fertilizers exported to all parts of the world. The phosphorites of Quercy, situated in a limestone region of central France, yielded a famous collection of bones of extinct animals during their exploitation for fertilizer during the last century. Most phosphate rock is acidulated with sulphuric acid to make various soluble phosphates used in dental cements, fireproofing compounds, baking powder, water softeners, matches, and in tracer bullets and other ordnance objects. Small amounts of 'raw' phosphate rock are used for soil treatment.

**Nitrates or Saltpetre.** The damp walls of stables and cowsheds may carry a deposit composed mainly of potassium nitrate (saltpetre). Nitre ($KNO_3$), which appears as a white efflorescence on the soil of hot humid regions, has the same composition. In both cases nitrogenous organic matter undergoes bacterial decomposition to produce ammonia for the formation of the nitrates.

In Peru and Chile important deposits of sodium nitrate are worked as Chile saltpetre.

**Sulphur ores.** Sulphur occurs frequently in the native state, often in the vicinity of active volcanoes. Free sulphur is deposited in fissures in the ground or in ancient volcanic craters by fumaroles. It is also found in association with salt domes and in some sedimentary rocks, most frequently interbedded with gypsum, salt

and limestone or with bituminous deposits. Sicily is particularly rich in such deposits, both open-cast and underground workings. The lumps of impure sulphur which have been mined are stacked in masonry furnaces. Part of the sulphur (15 — 20 per cent) is used as fuel to melt the rest. In spite of the resulting loss, this process is still the most economical in a country devoid of fuel. The melted sulphur is collected into moulds and turned into bright yellow ingots.

Accumulations of combustible hydrogen sulphide gas produced when the sulphur is burned make sulphur working a highly dangerous occupation. In the past impure sulphur was also melted simply by igniting stacks up to a hundred feet in diameter and about fifteen feet high. This was a wasteful method, for 50 per cent of the sulphur was burned and lost to recover the remaining 50 per cent.

Today the Frasche process used in the recovery of salt is also used in sulphur production. Hot water is forced into wells drilled down to the sulphur beds. This melts the sulphur, which sinks to the lower levels of the water. There it is drawn off through other wells where it is piped into huge vats or pools to be cooled and dried. When there is a deposit several hundred feet square and about sixty feet thick, the sulphur is broken up by power shovels and shipped. The sulphur obtained thus is 99 per cent pure, and needs no further processing.

It is also possible, as in Iceland, to pass sulphurous vapours from the ground through water-cooled tubes, where sulphur is then deposited.

Large quantities of sulphur are produced from pyrite ($FeS_2$), usually as a by-product in metal mining. Another source of supply is the water which has been used in the 'scrubbing' of coal-gas.

Before 1904 most of the world's sulphur came from Sicily. In that year the Frasche process was first used in Louisiana; twelve years later the United States took over as the world's leading producer. Today about 80 per cent of the world's sulphur is produced there. Mexico is now the second largest producer. Italy, Japan, Chile, Argentina and France also have some deposits.

Sulphur is used extensively in industry, in the manufacture of sulphuric acid, fertilizers, paints, explosives, artificial textiles, synthetic rubber, paper and dyes.

## METAMORPHIC ROCKS

The same forces which produce folding, faulting and large magmatic intrusions in the Earth's crust can also bring about structural and mineralogical changes in rocks. This process is metamorphism. Changes produced by weathering and erosion are arbitrarily excluded from metamorphism by the majority of geologists in order to prevent unnecessary confusion.

Under metamorphism sedimentary rocks lose many or all of their characteristics. Stratification, fossils, textures and other features become contorted and eventually disappear. Similarly, igneous and even metamorphic rocks themselves change in response to new environmental conditions, developing new structures and new minerals. The agents of metamorphism are heat, chemically active fluids and deforming pressures.

The pressures producing the changes may be static or dynamic. While the static pressure from an overburden of 20,000 to 30,000 feet of rock is insufficient to produce metamorphism in most rocks, it is powerful enough to make salt beds and similar weak deposits flow. A rock overburden of 30,000 feet, as in the lower portion of a geosynclinal section, induces plastic flow in most rocks, and a change in their nature.

Further change results when compressive forces are sufficient to fault or fold rocks. The amount of change or degree of metamorphism is controlled by the degree of squeezing that takes place. The tighter the rocks lie, the more complete is the metamorphism. In the Appalachian Mountains is the United States, the most pronounced changes have occurred in the more intensely folded eastern section and in the older, deeper lying rocks. The upper folded beds show little or no metamorphism. This is also true in other folded mountain areas.

Heat is important in metamorphism because it speeds most

Phosphate piled up to dry
at Khouribgha, Morocco.

chemical actions and thus frees the various rock elements to combine again in different ways. The more heat present the greater the metamorphic changes. Two main sources of heat are active in rock metamorphism — that resulting from friction produced in the deformation of the rocks, and that from magmatic intrusions. The latter is more local in nature and affects only a small area of rock immediately adjacent to the intruding mass.

When magmas move up through the Earth's crust, hot gases and liquids penetrate the adjacent rocks. These chemically active fluids react with the rock minerals, combining with them to form new minerals. The fluids may also remove material from the rocks and in this way too leave the original rock changed.

Metamorphism caused by heat and hot solutions is called contact metamorphism because only a relatively small zone — a few inches to a few hundred feet — is altered, forming an aureole of new minerals in contact with the magmatic intrusion. The effects of contact metamorphism vary according to the size, composition and heat of the invading magma and according to the basic nature and particular characteristics of the intruded rock.

Metamorphism produced by large-scale diastrophic action is known as dynamic or regional metamorphism. Its effects are controlled by the intensity of the deformation and the amount of heat developed.

The minerals found in metamorphic rocks are those commonly encountered in sedimentary and igneous rocks. The feldspars — quartz, biotite, hornblende, calcite, dolomite — may well have been present in the unmetamorphosed rock or they may have been produced during metamorphosis. In addition to these, metamorphic rocks contain many other minerals. The new minerals include, particularly, silicates of alumina such as sillimanite, kyanite and andalusite, together with silicates of iron, calcium and aluminium, such as garnet, staurolite, chlorite and epidote. Some of these minerals are characteristic of the degree of metamorphism which has occurred. Chlorite is typical of slight or low-grade metamorphism, almandite garnet of moderate or middle-grade metamorphism, and sillimanite of severe or high-grade metamorphism. Most other minerals are of little or no significance in revealing the degree of metamorphism.

241

Boiling sulphur pit at Namafjell,
Northern Iceland.

There are two structural or textural groups of metamorphic rocks: foliated and non-foliated. The non-foliated group is characterised by rocks with a massive, dense or granular texture. The foliates, on the other hand, have a texture which may range from fine to coarse and are characterised by a distinct layered, laminated or banded structure. The layered structure is called foliation, and the laminae may be thin or thick, smooth and even, or crumpled and crenulated. The laminae are made up in good part of flaky, tabular or needle-like mineral particles, organised with their long axes parallel. This causes foliation and cleavage.

## FOLIATED METAMORPHIC ROCKS

The commonest of these are the slates, phyllites, shists and gneisses.

**Slate** is a fine-grained, homogeneous, low-grade metamorphic rock. It splits or cleaves readily into thin and thick sheets. The minerals present, though in particles too small to be seen with the naked eye, include mostly mica flakes and some chlorite and quartz. The original layering of the parent shale is generally lost in metamorphism but is sometimes visible as coloured bands running at all angles to the flat cleavage surfaces. Slates occur in a wide variety of colours but the commonest are dark grey, black, green and red. The reds and purples are caused by the presence of iron and manganese, the greys and blacks by carbonaceous material. Crystals of pyrite also occur in shales.

Because of their smooth cleavage, slates are used for roofing shingle, stair treads, table tops, blackboards and trimming stone. In the United States commercial slate is quarried in the Appala-chian Mountains, from Maine to Georgia. Vermont and Pennsyl-vania slates are famous. In Britain slate is found chiefly in North Wales, North Devon, the Lake District, Ballachulish in Scotland, and Kilkenny in Ireland. The main horizons in France are at Fumay, Trelaze and Allassac. The working is usually by shafts and underground galleries and is controlled, as it is in other mines, by the disposition of the beds.

**Phyllite.** There is no sharp line of demarcation between phyllite and slate. The mineral composition is similar except that in a phyllite the mineral grains are slightly larger, and the texture thus coarser and the lustre higher. The foliation plates in a phyllite may be perfectly smooth, as in a slate, or they may be slightly rippled. Phyllite displays a slightly higher degree of metamor-phism than slates, and besides having a coarser texture may contain new metamorphic minerals, such as tournaline or magnesium garnets.

**Schist.** Schist is one of the most abundant of the metamorphic rocks. It is coarser in texture than slate or phyllite and most of its mineral grains can usually be seen with the unaided eye. The foliation of a schist is in discontinuous layers of uneven thickness, and is gnerally crumpled, bent or curved. Schists are named according to the dominant mineral: mica schist, garnet schist, talc schist, and so on. There are many varieties because schists are formed from any igneous, sedimentary or low-grade meta-morphic rock.

**Gneiss.** This rock is a product of high-grade regional metamor-phism. Gneiss is coarse textured and typically banded by layers

Marble quarrying at Carrara, Italy.
The workmen who do the
cutting must be expert climbers.

The oldest slate quarry in
Britain at Delabole,
in Cornwall.

of different minerals. Its foliation is poor because many of its mineral bands have a massive structure. Gneisses may be formed from igneous or sedimentary rocks. The minerals present give a clue to the parent rock. Gneisses of igneous rock origin display layers of quartz and feldspar alternating with biotite and hornblende. Clayey sedimentary rocks produce gneisses in which bands of quartz and feldspar alternate with tabular and fibrous minerals such as chlorite, mica, graphite, kyanite, etc.

## NON-FOLIATED METAMORPHIC ROCKS

This group includes marble, quartzite, hornfels, anthracite and soapstone. Anthracite will be discussed in more detail with other coals. Soapstone is a smooth, light-coloured, greasy textured rock composed mainly of talc.

Hornfels is a low-grade, compact metamorphic rock. It is very fine in texture and drab in colour. Minerals present may include biotite, garnet, and feldspars. Hornfels is the product of metamorphism in a variety of sedimentary rocks, particularly in shale or impure limestones, or of the metamorphism of basic igneous rocks.

**Marble.** Much used as an ornamental building stone, marble is a coarse-grained, crystalline calcareous rock produced by the metamorphism of limestone or dolomite. The principal mineral constituents are therefore calcite and dolomite. Pure marble is white but impurities in small quantities commonly colour it. Reds, pinks, yellows and browns are produced by iron compounds. Grey and black are produced by carbonaceous matter, and green shades by chlorite or serpentine. Commercial 'serpentine marble'

may contain very little calcite and may actually be serpentine, which forms from ultrabasic igneous rocks. Many commercial marbles are nothing more than pretty limestones, unmetamorphosed.

Carrara marble from Italy is found in lenticular masses in a bed of Triassic limestone 3,000 feet thick. The marbles of Pharos, Mount Pentelicus, Mount Humettus and the Acropolis are of Cretaceous age. In France, the Saint Béat marble is also Cretaceous but is of coarser grain than those already named. In the United States Vermont, Alabama, Tennessee and Georgia yield beautiful marbles; Maryland produces *verde antique,* and Arizona onyx marble, not a true marble but a spring deposit.

Architects and monumental masons give the name marble to all types of limestones and breccias that take a polish. These marbles may or may not be crystalline. Some are pure white, while others are coloured in various ways. Many are veined or irregularly stained; these are old, fissured limestones in which the fractures have been filled up with a secondary deposit from percolating waters. Some marbles owe their beauty to fossils which contrast with the general tint of the stone.

**Quartzite.** This is a solid quartz rock produced by the cementing of sand grains with crystalline quartz. Quartzite is a smooth, fine-textured, extremely hard, glassy rock used primarily as road-stone. It is generally light in colour, white when pure and tinted grey, tan, red or purple in its impure forms. Close examination reveals tiny, rounded quartz grains.

Hard and extremely compact, quartzites form ridges which have been thrown into relief by erosion. The Twelve Pins of Connemara, Errigal in Donegal, and the Stiperstones in Shropshire are two outstanding British examples of quartzite hills.

243

Miner drilling at the coal face.

# Mineral Fuels

## WHAT IS A FUEL?

Mineral fuels are divided into three categories: solid, liquid and gaseous. In the first are the coals proper, that is, the mineral coals which together constitute black coal with its principal constituent of carbon. Some mineral coals — anthracite is one — are almost pure carbon. In the second is petroleum, rich in both carbon and hydrogen, and giving a whole range of fuels and lubricants. In the third category fall the natural gases (often occurring with petroleum deposits) and the butane gases on the one hand and, on the other, coal gas (city or artificial gas) and the producer gases made by passing air or water vapour over white-hot coke.

Liquid and gaseous fuels are frequently known as gasolines (or petrols). Their physical state allows them to pass directly into spark-ignition engines.

Fuel, gasoline (or petrol) and ignition are words which call to mind the same phenomenon, that of very rapid oxidation.

The two principal combustible elements common to wood, coal and petroleum are carbon and hydrogen. Of the two, hydrogen is more efficient. Let us try to show why.

The value of a fuel lies primarily in its calorific value, that is, in the number of British Thermal Units liberated by the combustion of one pound of that fuel. (The B.Th.U. is the quantity of heat required to raise the temperature of 1 pound of water 1°F.

Since pure carbon has a calorific value of 14,137 B.Th.U. and hydrogen a value of 61,493 B.Th.U., the higher the proportion of hydrogen a fuel contains, the better it will be. The high hydrogen content of liquid and gaseous fuels (from 10 to 50 per cent by weight) explains why they give out far more heat on combustion than solid fuels, which generally contain less than 6 per cent, but may contain over 11 per cent.

There is a further important consideration. The less oxygen available in the fuel, the more eager hydrogen and carbon are to burn. Therefore, the lower the oxygen content of a fuel, the better it will burn. The ideal fuel would be pure hydrogen. Other factors to be considered in assessing the merits of different fuels are moisture content, and ease of extraction, transportation and utilisation.

Coal and petroleum appear to have had an organic origin. We are certain of coal's plant-derived origin, although the exact mechanics of the process may not be agreed upon. Most scientists agree, too, on an organic origin for petroleum. Evidence seems in favour of this theory, but it has yet to be proved conclusively.

It is thought, then, that the majority of fuels come directly or indirectly from carbohydrates (vegetable substances which are themselves the result of photosynthesis occurring in the green plant). In this sense we owe most of our fuels to the Sun. When we burn wood or alcohol, both extracted from living plants, we are merely recovering recent solar energy. When we burn coal, or city gas we are redistributing ancient solar energy. We are using, so to speak, the sun of bygone ages.

Even though petroleum has almost replaced coal as a fuel in the United States and challenges it elsewhere, coal is probably still the world's leading fuel. It is burned to yield heat for many purposes — to generate electric power in steam turbines, to provide mechanical power in locomotives and ships, to make combustible gases that can be used directly or mixed with petroleum gas.

## PEAT

Any study of solid mineral fuels must begin with peat, which is a transitional stage between living vegetation and coal, and gives us first-hand knowledge of the origin and development of the latter.

We know that some peat has its origin in clear and freshwater bogs, or in meadows with a spongy and loose soil (trembling bogs) caused by the invasion of earlier ponds or lakes by pond weeds, water-lilies, reeds, rushes and, above all, by floating algae and aquatic mosses. Today peat is forming in many small lakes, bogs, swamps and marshes — particularly in the moist, cooler climates where glaciation has occurred recently and where drainage is sluggish. It is also accumulating in some large swamps in warm climates.

Peat formation starts with the death and partial decay of plants bordering pond or swamp. As the vegetation grows, dies and decays year after year, an accumulation of peat forms along the shore-line, and the water-loving plants shift their growth farther out in the lake. In time, first a greyish-green or whitish moss — sphagnum moss — and then trees grow over the peat accumulation. Eventually the lake may be completely grown over.

The most hygrophilous of the mosses are the sphagna and hypna, which absorb moisture from the soil and from atmospheric water vapour. They are constantly growing upward as their bases gradually die and decompose at the bottom of the pool under the action of anaerobic bacteria. A trench cut in a peat-bog exposes the successive stages of peat development: living moss, dead moss, peat moss, laminated peat, solid peat.

Peat is a brownish porous substance, light in weight, greasy to the touch, and in its natural state literally gorged with water. Even after drying it burns badly, giving off a lot of smoke and leaving a copious residue. It is, in short, a mediocre fuel in spite of its fairly high carbon content (50 to 60 per cent) and its calorific value of between 7,200 and 10,800 B.Th.U. It is chiefly used in the form of briquettes, ovoids, or compressed tablets for industrial heating (producer gas). In its countries of origin it is also used in the production of tiles and manure. On distillation it yields coke, tar, ammonia, fuel gases and other products.

The age of some peat can be deduced from the objects found in it: household utensils or tools of recent date, coins bearing royal effigies, Roman pottery, polished or worked flints accompanied by the bones of prehistoric animals. The Irish peat-bogs, for example, have yielded the bones of large deer contemporary with primitive man. Straying into the bogs to graze, they were sucked into them 20,000 or 30,000 years ago. This is, incidentally, the maximum age of peat that can be classified as Quaternary coal.

Peat-bogs are sometimes classified as upland and lowland bogs. Lowland peat-bogs are more or less water-logged swamps, and are known also as infra-aquatic peat-bogs. Their waters are alkaline and rich in mineral matter, and the bottom is often calcareous; the dominant mosses are the hypna. Lowland bogs generally occur in plains or flat-bottomed valleys. The red bogs of Ireland come into the latter category.

Upland peat-bogs, on the other hand, are swampy meadows (trembling bogs) occasionally rising above the surrounding ground, as, for example, the mountain brown bogs of Ireland. The water is enclosed in the mossy lining (primarily sphagnum), and is pure and slightly acid. The bottom is formed by silt, sand, clay or granitic rock. In mountain districts, upland peat-bogs may be found on slopes without any outlet for the trapped water.

Under suitable conditions either of the two types of bog may change into the other. A lowland peat-bog may become sufficiently acidified to create conditions suitable for the formation of an upland peat-bog, with the hypnum gradually giving place to sphagnum.

Large swampy areas where peat is accumulating are found in Sumatra, and in the mangrove areas along the coasts of Florida, Cuba, Ceylon and New Zealand.

Dismal Swamp, on the border of Carolina and Virginia, in the United States, is an enormous swampy forest covering over 780 square miles. The original area, before partial draining was effected, was about 2,200 square miles. The rocks beneath the swamp are stratified marine sands and limes, probably Pliocene in age. The swamp is located on the Atlantic coastal plain some 5 to 20 feet above sea level. A very slight drop in level could produce dry land, a very slight increase inundation by the sea. The surface of the land is gently undulating, and the relief is slight. Some areas are fairly dry while others have extensive stretches of open water, like Lake Drummond. Bald cypress *(Taxodium)*, juniper and black gum characterise the lower levels. Pines, particularly the southern pine, grow in the higher areas. The cypress and gum are adapted to a semi-aquatic life. From their underground shoots and arched roots issuing from the base of the trunk, adventitious shoots extend above ground and carry respiratory stomata. The junipers grow only in the areas exposed during the dry season. The water of the swamp is clear and has a slight flow. On its bottom there is an accumulation of trees, branches, needles, spores, cones akin to pine cones, seeds and other vegetable debris, which gradually turn into peat and eventually into lignite. The accumulation of peat ranges from 1 to 20 feet thick. It is estimated that an area of about 1,500 square miles is now covered by an average depth of seven feet of peat. The total available quantity is about 672 million tons, and the rate of accumulation may be anything from six to twelve inches in ten years. About twenty feet of vegetable matter are needed to form one foot of coal.

The old and obsolete methods of working peat (the only ones, incidentally, suitable for small bogs) contrast with the methods currently employed in the large peat-bogs of the Netherlands, Denmark, North Germany and Russia.

The most primitive methods take advantage of low water (at the end of the summer) to cut the peat into parallelepiped sods which are then placed side by side on the ground until dry enough for stacking.

In large-scale peat-working the bog is first dried out by means of drainage ditches and canals, or by pumping. The purpose of this initial operation is to expose the peat and facilitate the manoeuvring of bucket excavators, grabs and mechanical shovels. In this method the peat is stacked in layers on the ground by mechanical spreaders and transporters. Subsequently sods are cut by other machines.

The surface layer of the peat may also be reduced to a fine dust by milling-machines fitted with teeth revolving at great speed. The dust is gathered in after drying. Another method is to liquefy the peat with jets of water under pressure. The pulp thus obtained is pumped into tanks and sent to spreading and drying grounds. This method is indispensable in deposits containing tree stumps (as in Russia) which make extraction by mechanical means difficult.

Drying is one of the great problems of peat-working. It can be effected naturally, or by heating, compression, and so on. No general rule can be laid down about the operation, nor about any of the various methods of extraction. Everything depends

*Above*
Cracked peat surface, Newfoundland.
*Below*
Peat-working in Ireland.

*Right*
Loosely stacked piles of peat slabs. They take about two months to dry before they are ready for sale.

upon the size of the deposits, their location, atmospheric conditions, the quality of the material and several other relevant factors.

## LIGNITE

Although lignite is essentially a low-rank coal dating from the Tertiary or from the end of the Mesozoic era, a good deal is known about its origins from a wide variety of sources. As its name implies (Latin *lignum* — wood), this coal frequently has a fibrous structure similar to that of charcoal or tree branches. If we compare its chemical composition with that of wood we find a relative gain in carbon and a loss in oxygen. The relationship between lignite and its raw material is thus similar to that between peat and the cellulose of mosses.

The main lignite deposits are found in the United States, Russia, Germany, Czechoslovakia and other European countries. Exploitation is carried out either by opencast mining or by underground-shafts and galleries. In the Rhineland (near Cologne) there are extensive deposits in which the lignite, lying close to the surface and at times reaching a thickness of some 150 feet, is mined in tiers with the aid of giant dredgers mounted on rails or caterpillar tracks. In France the largest deposit occurs in the Fuveau coalfield which covers an area of 390 square miles. It consists of three strata of which the last is some 800 feet thick. Extraction is, of necessity, by means of shafts and galleries. In south-east England lignite has been mined in the Weald.

Whether brown or black, spongy or compact, dull or brilliant, lignite is utilised for industrial purposes after drying and briquetting. Its carbon content (between 60 and 80 per cent) and its calorific value (between 10,800 and 12,600 B.Th.U.) are little greater than those of peat. Like the latter, it gives off a lot of smoke on burning and leaves a heavy residue of ash. The pyrites content often gives rise to an acrid and pungent emission of sulphur dioxide. In short, lignite is a poor fuel. On the other hand, its industrial uses in some regions are of great importance. On distillation it yields producer gas, city gas, coke, synthetic gasoline, bitumen, ammonia, colouring matter, and numerous pharmaceutical products. We shall discuss these derivatives further when dealing with those of coal, to which they are allied.

## COAL

With mineral coal we reach the study of true coal. Apart from some Tertiary (Spitzbergen) and Mesozoic deposits, most deposits date from the Palaeozoic era, particularly from the Carboniferous period. We shall study in a later section the special conditions of climate and vegetation which made this period the great 'carrier of coal' that its name suggests.

A thin section of coal cleaned with a mixture of nitric acid and potassium chlorate and examined under a microscope reveals a mass of vegetable debris: fragments of bark, wood, leaves, vessels, cells, spores, grains of pollen and algae all immersed in a kind of jelly. In short, coal is an extraordinarily fine vegetable pulp which has accumulated over the ages and which has solidified.

The strata of sandstones and shales occurring between coal strata contain enormous quantities of vegetable imprints and constitute a veritable herbarium of the plants which produced the coal.

Compared with its parent cellulose and wood, coal shows a gain in carbon and a corresponding loss in oxygen:

|  | Carbon | Hydrogen | Oxygen |
|---|---|---|---|
| Cellulose | 45% | 6% | 49% |
| Wood | 48% | 6% | 46% |
| Soft Coal | 85% | 5% | 10% |
| Anthracite | 95% | 3% | 2% |

Part of the hydrogen has also been liberated in the form of water and methane. Coal, then, is formed by the same chemical process which produces peat and lignite. The differences between these fuels depend upon the original vegetation, the extent of the maceration, the bacterial species involved and, finally, the toxics which arrested decomposition at a fairly early stage.

While it is accepted that coal was formed from vegetable debris derived from Carboniferous forests, it still remains to be shown how the debris reached the site of its transformation. One theory proposes local subsidence of former forests, another suggests a drift of vegetable alluvia into an arm of the sea or into a lake.

Let us consider the two theories in turn.

**In Situ or Autochthonous theory.** In the swampy forests of the Palaeozoic era, old trees fell in the course of time and decayed where they had stood; leaves, fragments of wood, and bark accumulated in the hollows; spores and grains of pollen were blown away to fall wherever the wind carried them. Under a thin layer of water peat formed. During seasons of flood or high water, mud and sand often interrupted the accumulation of vegetable matter. This was the time when the 'herbaria' of vegetable imprints of today's museums were created. Several millennia passed. If the area lay on a coastal plain, the sea may have transgressed and inundated it, halting the deposition of organic debris. And then the sea would regress again, allowing the forest to take over. Inundation and regression were repeated time after time. For example, where there are some 400 coal seams in the Franco-Belgian coalfield they occur because the course of coal formation was interrupted 400 times. The theory supporting this method of deposition is called the *in situ* theory or autochthonous theory (Greek, *autos* — self, *khthon* — earth). The evidence favouring it is weighty:

The freedom of coal from mineral matter favours accumulation of organic matter in swamps rather than in transported deposits.

Today many large deposits of organic debris are known to be forming in swamps on a scale large enough to produce coal in quantities comparable to those of the past. Dismal Swamp in the United States is one example of a currently developing coalfield.

Tree trunks with their roots embedded in sediments lying beneath the coal are found in coal seams, and sometimes the small roots penetrate pieces of wood buried in the sediments. Old soils sometimes underlie the coal layers. The arrangement of the vegetable debris in the coal is haphazard.

**Allochthonous or transportation theory.** A second theory puts forward the idea that vegetable matter from more or less distant forests was once carried by rivers to some arm of the sea or some lake, and that these plant fragments formed the starting point of the formation of the coal seams. This is the theory of allochthony (Greek *allos* — other, and *khthon* — earth). It is based on the fact that coal, after all, is simply a sedimentary, stratified rock like sand, clay and marl. The vegetable debris found in it generally lies flat or is steeply inclined and, moreover, is graded in order of density. Certain seams contain only branches; others consist exclusively of leaves or bark, or again of spores and grains of pollen. Everything is arranged as if currents of water had done the grading. Vertical trunks are extremely rare. Of the few that are found, some are clearly drift-wood. Parallels are found today in the great tropical rivers in flood, when portions of virgin forest torn from their banks course down the Amazon, the Plate, and the Congo. These materials are deposited on the bottom of seas or lakes in the same way as sediments. Here they are later compressed into strata of coal.

Data on which the allochthonous theory rests includes:

Large quantities of timber are yearly rafted down streams, especially in areas of virgin timber, such as those of Canada.

Peat and brown coal have been found in some present-day deltas.

In the sedimentary series which accompanies the coal beds, each bed as it nears a coal seam becomes finer in texture, indicating normal specific gravity sorting.

Marine fossils are often in beds immediately next to coal layers.

Vertical tree stumps, apparently in place, could occur by settling and gradual burial by sediments and coal-forming matter. Here it is argued that tree stumps often float roots downward.

In a discussion of the Commentry coalfield in France, Fayol (1887) advanced a further argument in favour of allochthony. This coalfield has formed by the filling-in by two deltas of a former lake surrounded by high land. The coal series has interbedded sediments which include cag with 1½-inch boulders. It has been possible to reproduce experimentally the deposition of its fertile (coal) and sterile (sandstone, shale) strata by directing several currents of water carrying vegetable debris, sand and silt into a specially devised tank. Fayol estimates that the Commentry coalfield could have been filled by vegetable alluviation in the space of 17,000 years, whereas its formation through the local subsidence of a series of forests would have required a minimum of 800,000 years.

The autochthonous and allochthonous theories are not mutually exclusive. Evidence exists for each which cannot be explained away by the other. Autochthonous and allochthonous coals can co-exist in the same coalfield or occur independently in different coalfields. While Commentry and Decazeville are certainly allochthonous, Saint-Etienne is probably autochthonous. In the Franco-Belgian coalfield autochthony and allochthony are combined. Every coalfield has its own history which only research can uncover.

One of the world's biggest dredgers at the Fortuna lignite mine near Cologne. Its daily output is 100,000 tons.

# CLASSIFICATION OF COAL

Most of the names used in the classification of coals have existed since the early days of the coal trade. Because of this they are inexact. As the industry grew, and differences in composition, heating value, ash content and volatiles were noted, the need for a more precise system was realised.

No single classification has proved universally satisfactory, for coal, like all sedimentary rocks, is highly variable; a classification adequate for one region where most of the coal conforms to certain standards may not prove acceptable in another. Usually the coal rank classification based on the carbon content, the earliest system, is still used. The names for the types of coal used in this classification usually appear in most of the others.

Coal is an extremely complex and variable mixture. A complete chemical analysis or 'proximate analysis', is most widely used in the industry. This determines percentages of water, volatiles, fixed carbon, ash, sulphur and other items.

Water is most often present in the lower rank coals, ranging from 43 per cent in lignite to a mere 4 per cent in anthracite. When heated to red heat coal gives off its volatiles, which include carbon dioxide, carbon monoxide, hydrogen and hydrocarbons. The volatiles, depending on which are present, may increase or decrease a coal's heating capacity. The presence of hydrogen and hydrocarbons in some lower rank coals produces more heat.

The amount of ash left after burning determines the coal's grade (not synonymous with rank). High ash residue means a low-rank coal. An anthracite with impurities may leave more ash than a lignite. The melting temperature of ash is important because large, porous clinkers which become pasty when hot may clog furnace grates and hamper operations. Sulphur impurities, often as pyrites and marcasite, may be present. These give off sulphur dioxide when the coal is burned, yielding an unpleasant odour and causing the smoke to act as an acid. B.Th.U. values are also included in most proximate analyses.

Fixed carbon figures indicate the amount of carbon present uncombined with other substances. The higher the rank of coal, the higher the fixed carbon percentage. Anthracite may have 96 per cent fixed carbon and lignite as little as 38 per cent. Except for the coals high in favourable volatiles, coals high in fixed carbon generally make the best fuel.

There are three main ranks of coal: lignite, bituminous and anthracite. The term 'rank' is an apt one because coal in its process of development passes from one rank to the next. All bituminous coal was at one time lignite, and anthracite has also passed through a lignite stage. Even though it is used as a fuel, peat is not considered a coal, but rather as the initial or embryonic stage in coal development. In their incipient stages all coals were peat.

**Lignite or brown coal.** These terms are used interchangeably in America even though in their original usage, back in the early 1800s, they probably referred to different varieties of coal. Lignite has a characteristic layered structure and may contain fragments of wood in a pulpy mass. Its calorific value on extraction is 5,500 to 7,000 B.Th.U. When free of moisture and ash, it is 10,000 —12,000 B.Th.U. It is generally very high in volatile matter.

An undersea boring tower working off the coast of Durham. The drill tip reaches 2,500 feet below the sea bed.

*Top*
Universal coal-cutter at work.

*Bottom*
Automatic coal-cutter in operation.

249

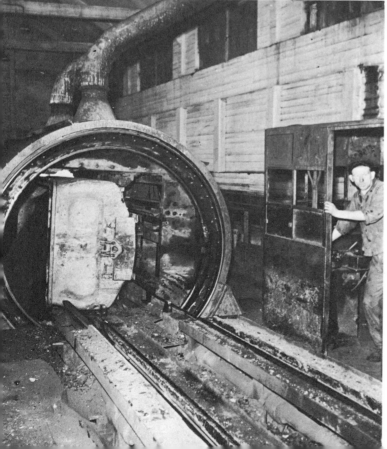

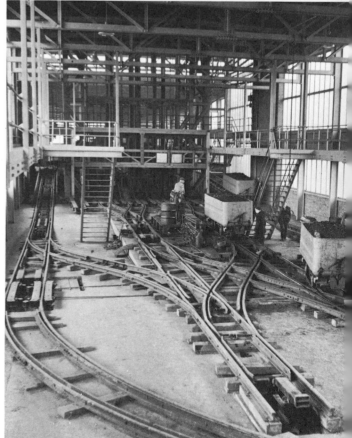

**Sub-bituminous coal.** This is an official U.S. Geological Survey term to cover the rank between brown woody lignite and bituminous coal. It is often called black lignite because its structure resembles that of lignite. The B.Th.U. values of sub-bituminous coals range from 8,000 to 10,000. These coals are often very high in volatiles and moderate in fixed carbon amounts.

**Cannel coal.** Bogheads and cannel coals are composed entirely of plant spores (or caneloid), grains of pollen, microscopic algae, and small fragments of vegetation. One can imagine old bogs on which a layer of algae grew at water level, subsequently sinking to the bottom only to be replaced by another, and so on. At the same time clouds of spores and pollen carried along by the wind descended upon the water and underwent a similar fate. Bogheads contain up to 36,000 algal cells per cubic inch. As a result of the method of their formation and their raw materials, bogheads and cannel coals are rarely found outside certain areas — Scotland is one of them. Cannel coals, originally called candle-coals, are so named because they burn with a bright, smoky flame. Scotland's 'parrot coal', a variety of cannel, produces a pronounced crackling sound while burning.

Cannel coal is a variety of bituminous coal. It is dull and satiny, shows no layering, and breaks with smooth fracture surfaces. It is high in volatile matter and has a moderate fixed carbon content. It ignites readily and is clean to handle. The B.Th.U. value is about 13,770. It is often used domestically.

TORBANITE is a variety of boghead from Scotland which originates from algal accumulation. It is dark brown, dull and lustreless, fracturing irregularly or smoothly. Like other bogheads and cannels it has a high content of volatiles, that include illuminating and lubricating oils, paraffins and illuminating gases.

JET is usually classified as a variety of lignite but is considered by many as wood which has been changed into a form of cannel. It resembles cannel but the woody structure revealed when it is thin-sectioned proves that it cannot have the same origin. Jet ranges from fibrous to compact and takes an excellent polish. It occurs in England, France, Spain, Bohemia, and Asia Minor, where it was known and used in ancient times. It is used in Europe in the manufacture of jewellery and other ornaments.

**Bituminous coal.** This is an inaccurate name because coal lacks true bitumen in any substantial amount, but substitute names have not proved popular. Bituminous coal has a shiny, dark grey to black colour and is denser than lignite. It is usually banded or layered, the layers varying in lustre. It breaks parallel to the bands or sometimes at right angles to them, giving a blocky fracture. It yields a smoky yellow flame and a strong bituminous odour. Its heating properties are greater than lignite, and its B.Th.U. is 12,000 to 15,000. More bituminous coal is consumed than any other variety because it is abundant, has a high fuel value, and can be easily adapted to special uses. There are many varieties, which fall in two main divisions: caking (coking) coal and non-caking (non-coking).

CAKING OR COKING COAL. This type of bituminous coal softens and runs together into a pasty mass as soon as combustion takes place. At higher temperatures it gives off its constituents as bubbles of gas. A hard, grey, cellular mass called coke is left after heating. There are no simple tests for separating coking from non-coking coals. Best bituminous constituents or varieties are vitrain or clarain. Byerite is an intensely black variety which yields a large quantity of gas and tarry oils. It is sometimes included with the boghead coals but it does produce crumbly porous coke.

NON-CAKING OR NON-COKING COAL. This type resembles coking varieties outwardly but differs in the ratio of hydrocarbons to oxygen. It does not generally cling to the sides of an agate mortar when rubbed with the pestle. Non-coking coal burns easily without swelling or softening, and leaves a powdery residue.

English cherry coal is a non-coking variety, so called because of its velvet black colour and fine lustre. It is brittle and breaks easily. Splint coal, slate coal, and 'hard' coal are English names for other non-coking varieties. They are greyer and duller than cherry coal. Durain and fusain are non-coking coals, as are the cannels and bogheads.

Both coking and non-coking bituminous coals are also called fat coals. Coking varieties are fat, short-flame coals; non-coking types are fat, long-flame coals.

Fat, long-flame coals, also called gas coals, have a high proportion of volatile matter (30 — 49 per cent) which makes them unsuitable for heating purposes because they give off abundant black smoke on combustion. On the other hand, they are very suitable for the production of coal gas. The microscope reveals a great abundance of spores, grains of pollen and leaf cuticles. The cement which binds them has its origin in the nitrogenous and hydrocarbonaceous substances accumulated as a result of the chlorophyll function which occurs mainly during the early stages.

*Top*
An Anderton Shearer drill working at the coalface.

*Far left*
The truck tipping point.

*Left*
Car-handling arrangements at the top of the shaft.

*Right*
The modern colliery system.

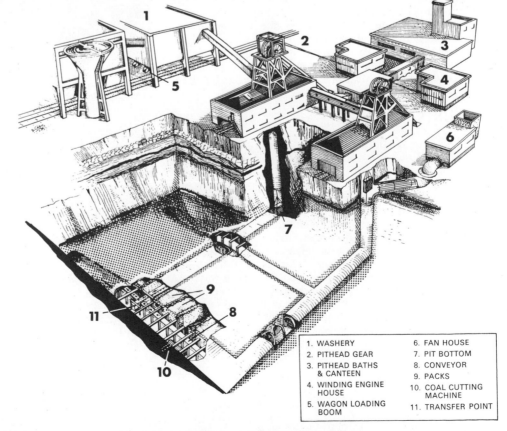

1. WASHERY
2. PITHEAD GEAR
3. PITHEAD BATHS & CANTEEN
4. WINDING ENGINE HOUSE
5. WAGON LOADING BOOM
6. FAN HOUSE
7. PIT BOTTOM
8. CONVEYOR
9. PACKS
10. COAL CUTTING MACHINE
11. TRANSFER POINT

Fat, short-flame coals contain between 18 and 29 per cent volatile matter. They have a higher wood fragment content than the long-flame paste varieties and are used to produce metallurgical coke.

**Semi-bituminous** is also called lean or dry bituminous (as opposed to fat coals), steam coal, or superbituminous. The last name is probably the most appropriate because the coal lies between bituminous and semi-anthracite in rank. In the United States the name low-volatile bituminous is now replacing other terms. Semi-bituminous has a high fixed carbon content — almost as high as semi-anthracite — and a B.Th.U. of 14,000 — 15,000.

**Semi-anthracite.** This term includes a range of coals between semi-bituminous and anthracite. Semi-anthracite is very similar to anthracite but is more friable and less smoothly fractured. It has a higher volatile content and ignites more readily, burning with a yellow flame. Its efficiency is greater than anthracite's because it burns more rapidly. Its B.Th.U. value is 12,450 to 14,200.

**Anthracite.** This is the hardest of coals, with a jet-black colour and a dull-to-brilliant or submetallic lustre. Because of its hardness it has even been cut and polished for ornamental stone. It is clean to handle and burns with a short, pale blue flame. Anthracite does not coke. It breaks with a smooth, curved fracture, sometimes showing small rounded or elliptical forms (a bird's-eye coal).

Because of the absence of soot or dust, and its long-burning quality, anthracite is excellent for domestic use, though it is more difficult to ignite and its B.Th.U. value, on average, is lower than that of semi-bituminous or semi-anthracite. This is an indirect result of its low volatile content. B.Th.U. value may range from 9,200 to 16,000, and generally averages 14,000 — 15,000 for a good grade. The fixed carbon content in anthracite is the highest of all coals, running over 90 per cent at times. Some anthracites, such as those from Rhode Island, may be graphitic in places and their fixed carbon may be half the normal average.

*Peacock coal* is not a separate type of coal but a form of anthracite or bituminous coal. It owes its name to its iridescent colours. It occurs generally only in the upper levels of a mine, where there is much fracturing of the rock and coal. The waters seeping through the cracks deposit a thin film of iron oxide along the cracks. Sometimes the colour film is formed by crude oil or sulphur dioxide but the main cause is iron oxide, produced by the oxidation of pyrite near the surface of the ground.

*Carbonite* is a natural coke produced in areas where magmas have invaded coal seams. It has a dark grey to black colour, a graphitic to submetallic lustre, an even fracture, and a texture ranging from porous to compact. Carbonite may grade into anthracite, which in turn grades into bituminous. It resembles artificial coke but is more compact and contains a higher percentage of volatiles. Most carbonites make excellent fuel. In United States it occurs in New Mexico, Colorado, Utah, and Alaska, and south of the United States border in Mexico.

## COAL-MINING

The exploitation of a coalfield and, to a lesser degree, of a coal-mine is a considerable and costly undertaking requiring much preparatory work by geologists, engineers, and economic experts who attempt to predict the future yield of the enterprise. Then begins the work of prospecting, boring and rock analysis, which will in turn be succeeded by the sinking of the shafts and galleries and by the construction of the necessary underground and surface installations. Several years pass before any effective extraction of coal can begin.

It should be pointed out that the operation is simplified when open-cast methods can be used where there is only slight overburden. The United States, Germany and Russia are fortunate in being able to exploit a large number of their deposits in this manner. In Britain only about 5 per cent of the total coal production comes from open-cast mines.

Open-cast or strip mining involves stripping away the overburden of barren rocks, followed by dredging and hauling of the coal. Today overburden is removed by steam or electric shovels

Open-cast coal mining (mainly bituminous deposits) in Northumberland.

or a special scraper called a dragline. Many of these mechanical monsters have booms over 100 feet long and dipper sticks about 85 feet long. The dippers can hold 30 cubic yards or more. Debris can be moved at the rate of 700,000 cubic yards a month. Loose mantle rock is easily and quickly scraped away; consolidated rock requires preliminary drilling and blasting. Once uncovered, the coal is brushed with steel brooms and then drilled and blasted. Sometimes, when it is friable, it can be removed without blasting.

Strip mining has more than economic advantages. Fewer men are needed to produce a given quantity of coal, and larger trucks and cars can be used for haulage. Timbering is not required. There are no gases and therefore no explosion or fire hazard. Transfer of shovels and other equipment to another area when the field has been worked out or sold involves smaller loss than in an underground operation. Strip mines are quick to put in operation yielding a faster return on initial, investments.

The biggest drawback of this method lies in areas of severe climates, where the work may be stopped or slowed in bad weather.

A second drawback is that it destroys the land. However, experiments show that modern land reclamation methods over a period of time will return part of the affected area to service.

The other major mining method is underground or closed work. Underground mines may use the room-and-pillar of the longwall technique, or one of the numerous variations of these two systems. Mine openings are necessary in all underground operations and from these a shaft or slope is sunk, or a tunnel or drift is driven. The shaft is a vertical opening and when it leads only from one seam to another is called a blind shaft. A slope, as its name implies, is an inclined opening, generally used as a roadway to and from a coal seam. Tunnels and drifts are horizontal passageways; tunnels are usually open at both ends, while drifts generally lead from a seam to the surface.

Shafts vary in size and shape but all are located in the more stable portions of the area and carefully reinforced. In the United States the law demands at least two separate mine openings as a safety requirement. Shaft width may range from five to ten feet,

and length may exceed 50 feet, depending on the number of hoisting compartments and the size of the coal cars. Although dual openings may create an extra fire hazard, they are necessary in most mines, especially those extending far under the sea.

Legal requirements governing the amount of air needed in a mine differ from country to country. In a mine with a considerable amount of gas, a minimum of 200 cubic feet of air a minute should be supplied to each working area.

A number of gases are found in coal-mines. Those present in quantity include carbon dioxide, methane, ethane, hydrogen, carbon monoxide and hydrogen sulphide. Generally, only methane and carbon dioxide occur in hazardous quantities, except during blasting operations.

Methane ($CH_4$), known as marsh gas, forms fire-damp when mixed with air. It is lighter than air and accumulates along the roof of the tunnels, where its tastelessness, odourlessness, colourlessness and non-poisonous nature may allow it to pass unnoticed. When mixed with air it is violently explosive and creates a hazard.

150 ranging shearer digging the top 5 feet section of the seam.

In the early days of mining the methane layer was often burnt off before a day's work. This dangerous practice is now prohibited. Instead, flame safety-lamps with specially designed wicks are used to detect its presence.

Carbon dioxide, also called choke-damp or black-damp, is heavier than air and sinks in tunnel or shaft. Like methane, it is a product of vegetation decay. It is produced by the explosion of coal gas or dust, and by blasting.

Heavy accumulations of methane and carbon dioxide have been known to blow out a whole section of a mine. Such violent reactions may be associated with what miners call 'bumps'. Bumps occur when pillars rupture, or a strong layer of rock in the roof fractures, sending shock-waves through the mine and shaking the pillars with disastrous effects.

Coal dust igniting in the air, too, can produce explosions as violent as those produced by the gases. Fire and explosion cause the worst mine disasters. Once under way, fires are difficult to check and mines have been known to burn for years, involving huge losses of coal. Some fires can be stopped by chemical extinguishers and water. In others where the methods are not effective, the burning area is sealed off to prevent oxygen from feeding the flames. If this fails because air still seeps through the rocks and fans the blaze, the fire may be smothered with carbon dioxide or steam. As a last resort the mine may have to be flooded, if this is possible. Naked lights, defective electrical equipment, and improper handling of explosives are the main causes of mine fires. In the United States, safety campaigns and sprinkling rock dust to make the coal dust inert have produced a major drop in the loss of life and property from explosions.

**Construction.** Each of the shafts is surmounted by a metal structure or pit-head frame, or by a winding-tower supporting a winding-pulley over which the cable passes. From this are suspended the cages (not unlike ordinary elevators) in which the miners are taken up and down. The steel cable is coiled around an enormous drum operated electrically or by steam.

We have said that the shafts give access to horizontal roadways, called gangways, main entries or main headings. In coalfields where the seams lie horizontal or dip gently, these galleries are cut into the coal itself, or very close to it. This simplifies extraction. In coalfields of the Franco-Belgian type or in the Appalachian anthracite belt, however, the seams and veins are inclined, highly folded, and faulted. Entries can merely cross the seams at intervals, as the ground structure permits. At the point where a gangway or entry cuts a vein, a new horizontal road, cross entry, or butt is dug, generally at right angles to the previous one and situated wholly inside the seam. If it leads off the gangway giving access to the return shaft, it constitutes the bottom level of the mine; if

it leads off the gallery giving access to the intake shaft it constitutes the top level. In deep mines intervening levels linked with the shaft are constructed as the work progresses.

Finally, between one entry and the next, ascending gangways called shoots are cut. These follow the lines of the seam's steepest gradient. The coal is mined upward along the dip of the seam almost as far as the gangway above. Chutes convey the coal down the dip to the lower gangway.

Where the vertical distance between bottom and top levels is too great, the intervening space is divided into several stages by cross-cuts — roadways similar to the gangways but not leading into the principal shafts. The cross-cuts are interlinked by inclined places or false vertical shafts called blind shafts.

Along the entry a series of cross-entries lead off, from which the rooms, chambers or breasts are driven along the seams. The end of the chamber where coal is being worked is called the face, or working face.

A coal seam is described in terms of thickness, direction, and inclination or dip. The last varies with the diastrophic history of the seam. An almost horizontal coal bed is a flat seam; a steeply inclined seam is an edge or pitching seam; a sudden change in direction is a bend. Below a seam extends a barren stratum (sandstone, shale) which constitutes the footwall; above it extends another barren stratum which constitutes the roof. As coal is worked at a face it is necessary to support the gradually exposed part of the roof with timbering, piles of rock, metal props, or pillars. Pillars are simply large columnar sections of coal left in place. In the longwall technique all the coal is removed continuously along the face of the seam, and supports of various types are used. This system requires more skilled workers than the room-and-pillar method.

In the initial working by the room-and-pillar method anything from 30 to 65 per cent of the coal may be left to form pillars. When a panel or block of workings has been finished or when the mine has been worked to its boundaries, all the pillars which can safely be removed are worked. Under excellent conditions almost 100 per cent of the coal is recovered. A good average for bituminous is 80 — 85 per cent; for anthracite 75 — 80 per cent represents maximum recovery and 45 per cent minimum recovery. Removal of pillars is difficult and dangerous, and only highly experienced men are employed. Methods vary from slice-by-slice removal to single-blasting clearance. Sequences for removal vary, too, depending on the particular conditions. Temporary timbers hold the roof during the process and are then themselves removed and the mine allowed to cave. The practice of filling in, or stowing, the worked out mine with rocks and earth is more costly. Stowing is common in Europe where population is dense. In the United States, on the other hand, mines are in sparsely

Interior of the 500 tons an hour
Baum washery at Killoch
Colliery, Ayrshire.

inhabited areas where ground surface subsidence and collapse are not damaging to property. Even in settled areas of the United States mines are sometimes caved.

**Working the coal.** Before the introduction of modern techniques, coal was undercut, broken down with a pick or bar, or blasted out. Nowadays primitive equipment has been succeeded by the percussive drill, the plough, the coal-cutter, and other drilling, cutting and hauling machines operated by compressed air or electricity. Percussion tools are widely used in Europe. Except for the pneumatic pick, most of the coal-cutters in the United States are chain-type machines driven by electric motors. Electric headlamps and wire-gauze Davy lamps reduce the risk of fire-damp and dust explosion risk to a minimum. The Davy lamp is also used as a gas detector, the size of the flame indicating the presence of gas.

Two problems normally encountered in mining are the evacuation of seeping water and the removal of coal loosened in the course of work at the face. It is impossible, even if the porous walls are specially concreted, to prevent underground water from infiltrating. Such water is therefore drained off by canals into tanks and then pumped to the surface. The quantity that has to be evacuated varies between 1,750 and 350,000 cubic feet in a day.

Much coal today is cut mechanically by a rotating cylinder with metal teeth that gouge into and along the coalface. The coal is then automatically transported by armoured flexible conveyor down to the bunker. New types of roof supports are now in use consisting of hydraulic rams that are moved forward as the removal of coal requires the machines to bite deeper into the seams. In the most modern mines coal cutter, armoured conveyors and hydraulic roof supports are operated by remote control and nuclear physics has contributed a radioactive device for steering the power loader.

Inside the main roadways or galleries the coal, once drawn by men or by ponies is nowadays hauled on rails by electric, diesel, compressed air and storage battery locomotives. In gassy mines locomotives must meet very rigid specifications because of the fire hazard. Although they are expensive to purchase and maintain, storage battery locomotives are widely used in such mines in America. The coal is transported down chutes to large bunkers and from these skips lift it to the surface. A skip is a kind of large bucket-lift into which the coal is dumped directly from the cars. Objections have been raised, however, that skip hoisting adds to the dust in the mine and increases breakage of coal. Skip shafts are smaller than cage shafts, and have the advantage that fewer men are needed to operate the system.

255

Once it has reached the surface the coal is delivered to conveyors which carry it to the coal-dressing works. It is estimated that an average shaft some 17 feet wide allows the extraction of over 1,000 tons per day from a depth of 1,700 to 2,000 feet.

Mining methods are essentially the same all over the world. In Europe mines tend to be deeper than in the United States; shafts of 3,000 feet are not uncommon. —

These are generally elliptical and reinforced with brick, stone or concrete in contrast to the American rectangular, wood-supported shafts. Better ventilation is usually obtained in these well-lined shafts. The longwall system of mining and the filling of mines are more common in Europe than in the United States. A smaller though increasing proportion of cutting machines is used in Europe because of the coal type, structure, and mine depth, as well as the labour situation.

On arrival at the pit-head, the 'run-of-the-mine', as the mass of unsorted products brought to the surface is called, contains between 50 and 75 per cent coal, against 25 to 50 per cent rock from the roofs and footwalls.

All metallic particles and objects such as shale, rock and dirt are separated out in the Coal Preparation Plant near the mine. In some mines a huge magnet removes metal from the coal passing beneath it. Other machines wash, sort and separate until only coal remains.

The coal is sometimes placed on perforated screening-plates and shaken to-and-fro. The coal first loses its dust, and is then sorted into chunks of increasing size. During screening, larger pieces of rock, generally slate, are removed manually or mechanically. Anthracite is usually broken before screening and washing.

In some mines special cone-shaped washing machines using sand and water separate out smaller rock fragments by specific gravity: coal is much lighter than the rock and is carried along by the water currents. The rheolaveur washing system was developed in Belgium and is used extensively for bituminous coal and some anthracites. In addition, there are many types of sorting, washing and floating techniques used for cleaning. Wet-wash processes are more common in the United States. Dry-wash systems are used more extensively in Europe, especially Great Britain. The advantages of the dry technique lie in better-looking coal which sells at a higher price, and absence of freezing if the coal has to be left in cars during winter. Coals for domestic use are dyed to hold down the dust. This also serves as a trademark.

At the end of production operations the coals are classified into commercial categories: run-of-the-mine, lump, furnace, small egg, stove, nut, etc. In earlier days most United States coal was sold as run-of-the-mine. Advances in cleaning methods have since been made, especially in Europe, and now a greater percentage of the coal is sorted and cleaned. The percentage is higher in Europe than in United States because the European coal is more friable and cleaning is a necessity.

Coal dust, fine and ultrafine particles of coal, even silty coal materials, which were once discarded, are also recovered and sold under size names: buckwheat, rice, barley and ultrafine. These 'fines' are used in briquetting, recarbonising steel, smelting ores, in certain furnaces, and in paint manufacture.

After classification coal is stored until sold, or it may be stored by the buyer. Storage presents a problem and an expense for manufacturer as well as buyer. Peak coal use is in the winter when adequate supplies must be on hand. Difficulties arise in storing large quantities of coal. Other problems include fire hazard, coal breakage, cost of rehandling, and deterioration from weathering. All these raise the price of the coal.

## COAL DISTRIBUTION

While the Chinese are thought to have used coal since very early times, its utilisation in Europe does not go back much further than the ninth century.

It has been suggested that the Greek and Latin words for coal (anthrax and carbo) designated charcoal, which may have been the only 'coal' to be used throughout the whole of Graeco-Roman antiquity. Yet from the writings of Theophrastus, it is implied that the coal mentioned is brown coal, probably from Thrace in

northern Greece or Liguria in north-west Italy. It was called at that time Thracus lapis and gemma Samothracia.

We have little knowledge of coal use in the Middle Ages, and what we have is more often derived from legend than from history. Although it is certain that blacksmiths in coal-bearing countries were using the fuel locally from the ninth and tenth centuries onward, it has been established that mineral coal did not supplant wood as a domestic or industrial fuel (forges, lime-kilns) until the twelfth century. The word coal derives from the Saxon col; in England it was first known as kohle and later cole until about the seventeenth century.

From the twelfth century onward the records grow more numerous. Deeds of sale and purchase, concessions, title-deeds and titles introduce us to the world of coalmines. The substantial capital needed to start up a new mine led to the creation of the first mining companies. Through illustrations we are well informed on their ways of working and handling coal. The buckets were hoisted up by large spoke- or paddle-wheels. As today, there were arrangements for ventilation, and protection against flooding and fire-damp.

Despite the development of the mines, coal was very rarely exported outside its countries of origin before the eighteenth century. Wood remained the principal fuel. However, increased coal utilisation was made inevitable by the depletion of the forests and the expansion of the metallurigcal industry, quickly followed, early in the nineteenth century, by the emergence of the steam-engine and coal gas. In 1801 came Philippe Lebon's first gas lamps; in 1804, Trevithick's Penydarran engine, the first railway locomotive in history; 1807, Fulton's first steamship; 1829, Stephenson's Rocket and the opening of the Liverpool and Manchester railway. The first record of coal discovery in North America was in 1672, in Cape Breton Island in Canada. Nearly a year later coal was found in the United States, but almost another hundred years went by before anthracite was located in Pennsylvania. In 1730 bituminous mining started in Virginia. Anthracite was first shipped in 1776. The total United States anthracite production in 1814 was only about twenty-five tons, and bituminous only a couple of thousand tons. Million-ton annual production was not reached till around 1850. In the next hundred years the industry boomed, and by 1940 United States production was nearly 500 million tons. Total production has declined steadily since the peak years of the early '50s, which were not much greater than the '30s. Anthracite production has declined since the '30s, although bituminous production rose until the early '50s.

The development of coal exploitation followed that of the utilisation of coal and its numerous derivatives. The competition of oil and water-power in the twentieth century need not, it seems, restrict its uses. Coal, waterpower, heat energy from the soil, petroleum, and now solar energy and wind-power, must combine their resources if man is to meet his ever-increasing power needs. Moreover, its countless chemical derivatives make coal a raw material capable of holding its own with petroleum.

The world reserves of coal are at present estimated at 6 million million tons. As the annual rate of production has for some years been stablised at approximately 2,200 million tons, simple calculation shows that current output can be anticipated for another three thousand years.

If we examine the international coal situation it is apparent that almost 40 per cent of the total reserves are in the United States, and many of their coalfields enjoy the additional advantage of horizontal seams and slight overburden. Of the remainder, some 40 per cent are divided between U.S.S.R. (which has tremendous reserves), China, Germany, Great Britain, and Canada. Australia has some of the world's thickest coal seams and a fairly large reserve of high-grade coal, but it is not yet a large factor in the world's coal production. Although every other continent, including Antarctica, has some coal, Africa, Central and South America are almost totally lacking in it.

Coal-bearing rocks have been discovered in the systems of a number of geologic periods. The largest deposits of bituminous and anthracite are found in Mississippian, Pennsylvanian and Permian rocks. Lower grade bituminous and lignite occur in

Mesozoic rocks, particularly Cretaceous sediments. Tertiary rocks, too, produce lignite and low bituminous and locally a little good bituminous.

In the United States there are three major coal provinces: the Appalachian fields, the Interior fields, and the Rocky Mountain fields.

The Appalachian fields, extending from New England to Alabama, are possibly the most important coal deposits in the world. They produce all the anthracite and 70 per cent of the bituminous used in the U.S.A. The coals of this region occur mainly in rocks of Pennsylvanian age, with small amounts found in Mississippian and Permian rocks. The rock layers and the coal beds have all been folded, some very tightly. Towards the western margin of the region the folds become gentler and finally level out. The anthracite occurs in the steeply dipping beds of the eastern portion of this belt. Erosion has probably removed over 90 per cent of the original deposit but over 4·5 billion tons have been extracted and about twice that amount still remains. Mining costs are high in the anthracite area because of the complexity of the structure.

The Interior coal province lies in an east-west belt which cuts across Kentucky, Indiana, Illinois, Iowa, and Missouri and then runs south through Kansas, Oklahoma, Arkansas and Texas. The coal is about the same geologic age as that of the Appalachian fields. The main fields are in Illinois, Indiana and Kentucky—second only to the Appalachian fields as sources of bituminous. The coal of this area is used extensively for domestic purposes and steam but not generally for coke. In some parts competition with the petroleum industry limits production.

The western coalfields occur in scattered patches through the Rocky Mountain area from Montana south to Arizona and New Mexico, and continue northward into Canada. Most of the coal in this region is Cretaceous. Lignite and sub-bituminous are commonest but here and there, where the rocks have undergone intense folding, bituminous and anthracite occur. Most of the coal from these small fields is consumed locally or remains un-exploited because of the small market and available petroleum.

To the main coal provinces must be added a considerable area of Cretaceous lignite on the inner margin of the Gulf Coastal Plain. Small deposits of low-rank Eocene coal are found in California, Washington and Oregon. In Alaska, coal areas are small and scattered and include all ranks of coal from lignite to anthracite. Some of the coal is used locally but intensely folded and fractured beds and the cost of transport make production costs so high that development is restricted.

The coalfields of Britain are found in three main areas. The southern area includes south Wales, the Forest of Dean, Somerset and Kent; the central area covers Yorkshire, Lancashire, the Midlands and parts of north Wales; the northern area is Durham, Northumberland, Cumberland and the lowlands of Scotland. All the coal is of Upper Carboniferous age except for one small mine in the Jurassic of north-east Scotland and a little lignite in Devonshire. The chief anthracite area is in the north-west of the south Wales coalfield.

Post-Carboniferous earth movements gave rise to the various coal basins; some, like those of Scotland and south Wales, have axes which run east-west; others, in England, are north-south. The strata within these basins are slightly inclined or flat, with a few notable exceptions in west Wales, Lancashire, and parts of Scotland. Compared with those of European coalfields, the seams are relatively undisturbed and such faults as do occur are generally small.

Each separate field contains many seams which vary in thickness from 25 feet, as in the Dysart mine of Fife and the thick coal of south Staffordshire, to the 12 to 14 inch seams of Co. Durham. The quality is high, particularly in the lower levels, and the coals generally are bituminous gas or coking varieties.

## STEAM COALS

A country's electricity requirements cannot always be met solely with the resources of its running water. Moreover, flow may be irregular and impossible to adapt to the consumers' needs. It is then necessary to supplement the hydro-electric power stations

Coke ovens being discharged on to a steel wharf.

with steam-generating stations utilising pulverised, poor quality coals or blast-furnace gases. Coals are also used to produce steam in locomotives, ships, and stationary boilers.

The ideal coal for the production of steam is one with a high calorific value, low smoke, clinker production, and long burning capacity. Steam coal should also ignite readily and burn fairly rapidly. All these requirements point to coals with a fairly high volatile content. In addition, the steam coal must allow storage over fairly long periods. A wide range of sizes and ranks is used, selected to suit the specific need. Low-volatile bituminous, or 'smokeless' coal is the ideal. This type combines the highest calorific value of any coal with faster combustion than anthracite. In the United States, West Virginia, Virginia, Maryland and Pennsylvania produce this type; in Britain, it is estimated that steam coal forms approximately 10 per cent of the total output.

The steam-generating stations must of necessity be situated close to railroads or rivers capable of supplying their fuel. It is an added economical advantage to site them near coalfields or heavy electricity consumer areas, close to hydro-electric power stations.

In a steam-generating station the coal is first stored in a large yard, dredged, and mechanically conveyed to the crushing machines which reduce it to a fine powder. Strong air currents then carry it into the burners or combustion chambers heated with fuel-oil to a temperature of about 900°F. The combustion chambers have a capacity of several hundred cubic feet and a heating surface of several thousand square feet. The water circuit pipes are attached to their walls. The water must be absolutely pure, distilled, and so conditioned that its vapour when raised over 900°F. and a pressure of 34 lb. to the square inch cannot damage the turbines it supplies with power. While the turbines drive the alternators and generate the electric current, the water vapour emerging from them passes into condensers where it regains its liquid form on cooling, and is once more ready for re-use. Cooling is effected by circulation or by a shower of river water. The quantity of water needed to operate a large power station is a flow roughly equivalent to that of the Seine, and its coal consumption is between 1,000 and 3,000 tons a day.

Not all steam-generating stations are fuelled by coal. A typical non-coal station in France, at Herserange, is fed with gases from seventeen blast-furnaces from four iron works in the Longwy coalfield. In England and Wales, where over 98 per cent of

*Left*
The terraced landscape of
an open-cast mine at
Decazeville, France.

*Top right*
Bevercotes, the world's most
advanced coal mine, in
Nottingham, is fully
automated.   All operations are
linked to a central control room
at the pithead.

*Bottom right*
A thermal station operating at
Hornaing, France.

the power supply comes from steam generating stations, coal
accounts for 85 per cent of the fuel used in the stations, the balance
being supplied by oil (12 per cent) and uranium (3 per cent).   The
thermal efficiency of modern stations is about 40 per cent, the
upper limit being set by the maximum temperature that present
day materials for boiler and turbine fabrication will withstand.

## COKE AND TOWN GAS

Bituminous coals, those richest in volatile matter, are the ideal
coking and gas coals.   For high-grade coke, the volatile content
of the coal should range from 25 to 30 per cent; the sulphur content
may run higher for second-grade coke.   Coking coals may yield
from 50 per cent to 80 per cent coke.   To be profitable the coal
should produce at least 65 per cent coke.   Cannel coal and high
volatile bituminous coals are used for high quality hard coke.

The operation by which these two constituents of coal are
separated is variously called distillation, carbonisation, pyro-
genation, and coking.   It differs from combustion in that it takes
place in a closed vessel in the absence of air, thus preserving the
full calorific value of the products.

The modest retorts of earlier gasworks have today been
replaced by large vertical or horizontal ovens in which gas, coke
and by-products are produced at high, mean or low pressure,
sometimes continuously, sometimes discontinuously.   In the United
States many types of oven are in use, ranging from the almost
extinct, less efficient beehive oven to the copper retort ovens and
the by-product or chemical recovery ovens which save all the
volatile constituents as well as producing coke and using less
fuel.   As we cannot discuss these different commercial processes
in detail we shall confine ourselves to that of discontinuous
distillation at high pressure in horizontal ovens.

In large plants the horizontal ovens, anything from a dozen
to 600 or more, are arranged in batteries of 50 to 60.   Each oven,
roughly 45 feet by 13 feet high by 7 feet wide, made of steel with
a lining of silica bricks, is hermetically sealed by a door at each
end.   In the partitions or sidewalls separating the furnaces, spaces
or flues are left for boilers which, at a temperature between 2,200°F.
and 2,700°F., burn a mixture of rich gas from the ovens and poor
gas from gas-generators.

Firing an oven is a slow process requiring gradual build-up of
temperature, without undue expansion.   Once fired, the oven

will burn continuously until it requires remaking. Each oven is fed with about 15 tons of pulverised coal by an automatic coal-car. The doors are then hermetically sealed and distillation begins. An inspection-hole allows progress to be followed. The coal becomes pasty and, starting at the walls and proceeding towards the centre, it cakes into a parallelepiped of white-hot coke. The firing operation may last from 12 to 24 hours, depending on the coal used and the products to be obtained. Foundry coke is fired much longer than most types.

When distillation is complete, one door of the oven is opened and the coke is pushed horizontally out of the other by the coke-pusher, a shield operated by a bar and controlled by the coke-ram. The coke tumbles into a coke-car, which immediately carries it under a quenching-head to be showered with cold water. Finally, it is tipped on to a quenching-floor where cooling is completed. It is later conveyed to the separators and then to the storage area.

In the most modern coking-plants water-quenching has been replaced by dry-quenching, a method in which the coke is smothered in iron or masonry chambers. In this way much of the heat of the coke charge is recovered for the various needs of the plant.

Coking may be effected by low- as well as high-temperature processes. Results vary with different types of coal but, in general low-temperature processes produce softer coke and a larger proportion of certain by-products. High temperatures are required for hard cokes.

The coke obtained in horizontal ovens is harder and more suitable for iron-smelting than that produced in vertical ovens. Other factors have a part to play too. In theory, coals very rich in volatile matter should produce a friable and porous coke whereas coals very poor in volatile matter, such as anthracites, should produce a pulverous residue. Theory suggests coals of average composition to obtain a hard and solid coke. In practice, every plant uses a mixture, or 'coke paste', suited to its own requirements. The needs of gas coking-plants differ from the needs of gas generators, and each uses a different 'paste'. The poorest quality coke is utilised in gas generators and in the manufacture of water-gas.

Most coke manufactured is used in blast-furnace operation. Together with a small amount used in foundries, this accounts for about 90 per cent of the total. The remaining 10 per cent is used in the manufacture of water-gas, producer gas, domestic heating, export and miscellaneous industrial purposes. Most of the gas produced in the process is returned to the ovens for under-firing, most of the remainder goes to steel mills or allied plants.

But let us return to the crude gas being given off from the ovens. It is a heavy yellowish gas, with a temperature of some 1800°F. Before it can be utilised it has to be scrubbed and cooled. For this, a whole series of apparatus is used, modelled on that of the early gas-works but benefiting from the resources of modern techniques. The gases are drawn out by powerful exhausts and sent through a series of condensers and scrubbers. First there are the troughs, giant pipes placed above the ovens in which the gas is subjected to a continuous shower of cold ammoniacal water. At this stage it precipitates some of its tar. Then follow primary and secondary condensers in which the rest of the tar is deposited. The condensers are succeeded by scrubbers in which water absorbs the ammonia and the oils are distilled. Iron oxide purifiers remove the malodorous and sometimes toxic products (sulphuretted hydrogen, hydrocyanic acid). Not until these various processes are completed is the gas ready for storage and consumer use.

In round figures, distillation of one ton of soft coal yields 1,550lb of coke, metallurgical and foundry coke as well as fines for briquetting or for boilers, 100lb of tar (used for road construction and waterproofing or broken down further for special products), 10lb of benzol (used in that form or distilled to give gasoline and napthalene), 7lb of ammonia (used in manufacture of fertilizers), 12,000 cubic feet of gas (used for heating or scrubbed for illuminating purposes).

## THE PAST AND FUTURE OF CITY GAS

The discovery of town or city gas was, as it were, 'in the air' by the end of the eighteenth century, when the attention of public authorities was drawn to it by Jean Pierre Minkelers in Holland

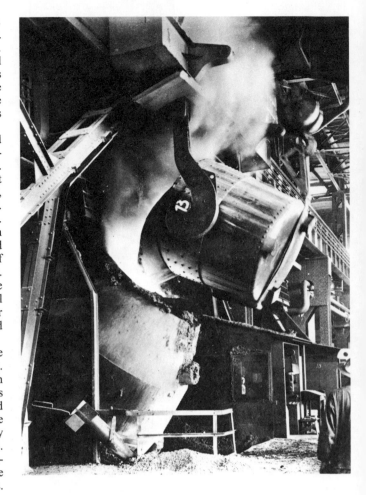

in 1784, William Murdock in England in 1792, and Philippe Lebon in France in 1796. Philippe Lebon intuitively foresaw the future vouchsafed to inflammable gas. Only the special circumstance of war was responsible for the failure of his schemes to develop public and domestic gas lighting in France. Murdock in England set up a small experimental plant in 1795 and lighted a Birmingham factory by gas, receiving a medal from the Royal Society of London for his invention. It was left to Winzler, or Windsor, a businessman rather than a scientist, to found the first gas companies in England, build the first gas-works at Westminster and light successively the streets of London (1813) and the Passage des Panoramas in Paris (1816). From that time on the industry never looked back.

Following its discovery and success in a few European cities, gas lighting in the United States got its real start in Baltimore. After the lighting of Rembrandt Peale's Museum, pipes were laid and the city of Baltimore adopted gas lighting. Other cities followed Baltimore's example shortly after.

Although gas is used today for all domestic purposes, electricity offers a real threat to a commodity which, not without reason, is claimed to be dangerous, malodorous and dirty. These failings are inherent in its chemical composition, for analysis yields the following breakdown: hydrogen, 50 per cent; methane, 25 per cent; carbon monoxide, 10 per cent; acetylene, ethylene, sulphuretted hydrogen, ammonia, carbon disulphide, benzene and tar droplets, 15 per cent. Of these, hydrogen causes explosions and fires, carbon monoxide is toxic, tar forms a blackish deposit on walls, and most of the other constituents give off an unpleasant smell. The balance sheet is thus unfavourable.

The idea of distilling coal seams in situ was fostered by the Russian chemist Mendeleev (1889) and the English chemist Ramsay (1912). In 1935 an experimental station was set up in the Donetz coalfield. Despite the difficulties and the inevitable set-backs, the tests proved so conclusive that industrial development was started in several countries.

One method consists in selecting an inclined panel of coal

650 to 1,000 feet long, and surrounding it at its lower end and laterally by fire-proof corridors to stop fire breaking into other parts of the mine. The lower corridor is the fire chamber where the coal is ignited by remote control. At first the fire is fanned by cold air blown into one of the lateral corridors. The gases derived from the distillation rise to the surface through the other corridor. When the reaction is well under way, the temperature is raised and the proportion of oxygen in the incoming air reduced. Distillation spreads from the bottom to the top of the coal panel.

In a second process, called electro-linking, two boreholes are drilled close to each other. An electric cable terminating in an electrode is lowered into each and a current of 2,000 volts is passed through. In spite of the poor conductivity of coal, this is sufficient to close the circuit. By reason of its resistance to electricity the coal seam grows hot, then white-hot, and finally begins to gasify. After this initial stage of electro-carbonisation it is enough to pass a current of fresh air into one of the boreholes in order to collect from the second hole a combustible gas suitable for many purposes.

There can be no doubt that underground gasification eliminates the miner's work and the manhandling of coal; moreover, its operational costs are relatively low, and it can be used in seams which are too thin or too poor in quality for normal working. But the method produces no coke and its application cannot therefore become general.

**Transportation.** Transporting coal and gas is a major problem in the economics of the industry. Coal may have to be taken long distances from collieries to coking plants whose coke products may travel still further to reach the plants which need it.

Generally, a high percentage of coke plants is controlled by steel and iron companies, and plants are often located near cities, for domestic use of coke and gas. Transportation at one end or the other is usually eliminated by careful siting of coking plants. By-products, however, may have to travel long distances to the consumer. Steel smelters are often located in coal-mining areas so as to be near to coke plants, as is the case in Pennsylvania.

A final problem is that of storing the gases as a precaution against constant fluctuations in consumption. Although water gasometers, and dry tanks have steadily increased their capacities over the years, they can hold only a limited reserve. For this reason it was suggested that in petroleum-bearing areas porous rocks occupying anticlinal vaults could be used as gasometers after their contents had been removed in the course of petroleum exploitation. Tanks of this type are in service in the United States and about a quarter of the gas used in the United States is pumped each summer into 258 storage pools and 10,521 disused wells to be stored until the cold weather. The artificial storage of gas underground is at present illegal in the United Kingdom, but legislation is being drafted which may overcome this obstacle.

## NATURAL GAS

In the United States natural gas has superseded manufactured gas to a large extent, chiefly because improved means of transportation make it possible to bring it from remote areas. Natural gas supplies nearly a third of the total energy requirement of the United States and is the sixth largest industry there in terms of investment. Output in 1962 reached an estimated value of over 2,000 million dollars and sufficient reserves have already been discovered to satisfy twenty years of consumption at the 1963 rate.

Natural gas is used for domestic purposes, and also to perform many processing and manufacturing tasks, and its by-products are used in thousands of domestic products. Research is being conducted into the possibility of converting natural gas into electricity and results so far indicate that an efficiency level of 60 — 80 per cent can be achieved compared to 40 per cent efficiency for conventional methods of producing electricity.

## TAR AND ITS DERIVATIVES

Formerly tar derived from coal gas was used unprocessed for tarring roads and caulking ships. Chemists eventually realised,

261

however, that it contained a large number of exploitable products, and from their discovery a new industry has emerged. Coking plants and gas-works (except those producing high-grade coke) are currently striving to produce the maximum amount of tar, either by using coals particularly rich in volatile matter or by processing coals at a relatively low temperature. A coal which at 1,800°F., for instance, would produce only 100lb of tar, produces 200lb at 900°F.

The tar obtained is later distilled at a constant pressure in tube-stills where the crude product, introduced in small quantities, gradually replaces the distilled product which passes into fractionating columns. These are divided into a series of superposed chambers, or vats, in which the rising vapours successively deposit their constituents in order of increasing volatility.

Let us examine a selection from the most important of the tar derivatives.

**Light or benzene oils.** The most volatile constituents of tar, these oils are composed mainly of benzene ($C_6H_6$), a hydrocarbon known to be the starting-point of the 'aromatic' series. Its formula can be represented as a hexagon with a carbon and a hydrogen atom at each of its apexes. Carbon being a tetravalent body, each of its atoms holds a hydrogen atom and three neighbouring carbon atoms.

All the properties of benzene derive from this formula, in which each of the hydrogen atoms can be replaced by another atom or by a group of monovalent atoms. Thus, starting with benzene, we may obtain chlorobenzene, nitrobenzene, aminobenzene or aniline, toluene, phenol, xylene, etc. Several 'benzene nuclei' can, moreover, combine to form more complex bodies, of which the two most important are naphthalene and anthracene, distinctive components of the middle and heavy oils which will be discussed presently.

From these benzene derivatives — and many others could be named — it can be seen that the uses to which light tar oils can be put are manifold. Pure benzene is used in dry-cleaning and stain removing, paradichlorobenzene as an insecticide, trinitrobenzene and trinitrotoluene (tolite) as explosives, aniline as a dye, the phenyles, benzyles and benzoates in the synthesis of perfumes. Benzene derivatives are employed, too, in the manufacture of plastic materials and synthetic rubber (buna), and numerous pharmaceutical products are also obtained (antipyrin, pyramidon, dulcin, saccharine, luminal, stovaine, cocaine, coramine, paregoric, sulfonamides).

Lastly, there is industrial benzol, formed from a mixture of benzene, toulene and xylene. Used alone or blended with gasoline or with gasoline and alcohol, its high calorific value (greater than that of petroleum spirits) and above all its anti-knock quality are particularly well suited to certain combustion engines. It must be remembered that benzol is extracted from light tar oils and, in still greater quantities, during the final filtering operation in coal gas production.

**Phenolic oils.** The oils distilled from tar immediately after the light oils are the phenolic oils. Their principal constituents are the phenols and cresols. Pure phenol or phenolic acid is well known as a powerful antiseptic and disinfectant. Together with the cresols, it is used in the manufacture of the antiseptic soaps known as cresyls. In addition, it is the starting-point in the synthesis of unexpectedly diverse products: photographic developers (hydroquinone and diaminophenol), explosives (picric acid and melinite), pharmaceutical products (salicyclic acid, guaiacol, menthol and aspirin), colouring matters (picric acid and fluorescein), plastic materials (bakelite), and synthetic fabrics such as nylon and imitation leather.

**Middle or naphthalene oils.** These oils settle in that part of the fractionating column with a temperature of 410 — 460°F. They are yellowish in colour and on cooling liberate their principal constituent, naphthalene. Its uses are as numerous and varied as those of the phenolic oils. Important derivatives are used in the manufacture of dyes, perfumes, plastic materials, pharmaceutical and photographic articles. After the deposit of naphthalene, creosote oil is left behind. This is utilised as a wood preservative.

**Heavy or anthracene oils.** Obtained at the bottom of the fractionating column, at temperatures varying between 460°F. and 680°F., these greenish oils have anthracine as their main constituent. They are used in impregnating cardboard, tar-lined paper and packcloth. Some of their derivatives are used in the manufacture of dyes.

**Pitch.** This is the heaviest tar product left in the retorts. It has a characteristic blackish colour. It is the best binding material for coal-dust in the manufacture of ovoids and briquettes. Mixed with heavy oils, it forms road pitch, competing in this application with petroleum-extracted bitumen.

In the compass of a brief chapter it is not possible to cover all the resources of the coal, gas and tar industries. The achievements of the chemists in these fields are truly prodigious: cars, tarred roads, synthetic rubber, imitation leather, plastic materials, aniline paint, dry-cleaning preparations, pharmaceutical products, film developers, nylon, synthetic perfumes and a range of fabrics.

Purification plant with tower purifiers, Kent.

Crew working on the drilling bit
of an offshore well on Lake
Maracaibo, Venezuela.

# Petroleum

In 1808 a scientific expedition of the Imperial Academy of Sciences of St Petersburg declared on its return from Baku that 'petroleum is a mineral of no usefulness'. Today, over a century and a half later, 1,487 million tons are being utilised annually for the most diverse purposes.

Petroleum is no recent discovery. With its derivatives it has been put to many and varied uses for a very long time. The soil of the Middle East has always been known to be impregnated with oil, and in Chaldea, Egypt and China the distillation of petroleum was familiar several thousands of years ago. It was used in lighting, in the treatment of many ailments, and also for purposes of war. Marco Polo writes more than once of petroleum in the account of his voyage across Asia.

The word 'petroleum' did not exist until the Renaissance Rock-oil, mineral oil, or naphtha, were the names given to the mineral before that time. 'Petroleum', declares a sixteenth-century book, 'is an oil which oozes and trickles out of the rocks on the estate of the Duke of Ferrara, near a town called Modena'. The engraving accompanying the text shows a gushing spring and people filling vessels from it. The same book goes on to explain the uses to which the petroleum was put. 'First, this oil purges and cleanses any ulceration and heals all old wounds ... it brings balm to those hard of hearing ... and, whosoever has been scalded either by the heat of boiling water or pitch or has been chilled by frost, let him rub himself with petroleum twice a day ... whosoever has been struck, disfigured or thrown, so that the bruises and other effects are visible, let him rub them with petroleum twice a day and he will be cured ... whosoever has injured himself internally or has caught a chill, let him take half a shell of a filbert or hazel-nut and place in it a little petroleum, and let him drink this with a goblet of warm beer ... he or she who has been bitten by a mad dog or has been stung by a poisonous animal, let him or her rub the wound or bite with petroleum ...'

Petroleum was, in short, a panacea for all ills, for we must also credit it with curing coughs, bronchitis, pulmonary congestion, cramp, gout, rheumatism and eye-strain. Modena petroleum was sold as far away as France, where it competed with the national oils.

Rock-oil, mineral oil, or the 'liquid ore' as it was also called, had a well-established reputation, then, as a therapeutic product in the Middle Ages and in the succeeding centuries. How did it subsequently come to be utilised as a motor-fuel? And how, after its origin in early European and Asian history, did it become an essentially American product before assuming its current cosmopolitan character?

Let us go back to the eighteenth century, when the history of American petroleum began with the first contact between redskin and paleface. 'Black oil' collected from the surface of the marshes was traded for glass beads and alcohol. The colonists used the petroleum to grease the axles of their wagons, to heal the wounds of their horses, and to treat their own rheumatism and injuries. Towards 1800 drysalters began selling the 'elixir of long life' which the Whites soon managed to procure for themselves in large quantities. Even the shallowest well sunk into the

Pennsylvanian soil would fill rapidly with salt water covered with an oily and evil-smelling film. To throw back the water and retain the oil was child's play. A salt merchant from Pittsburgh hit upon the idea of making a business of it, and the elixir was sold in its natural state, 'unchanged and undiluted', in elegant bottles whose labels praised the unrivalled virtues of the contents.

In 1840 a chemist from Yale University distilled the contents of such a bottle. The extract he obtained was, in his own words, 'light and inflammable, and shed a brilliant light'. He had rediscovered the lamp-oil that had been known to the peoples of antiquity. The discovery came at an opportune moment. Whale oil which had previously been used for lighting was becoming scarce. Oil lamps were developed forthwith, notwithstanding their noisome smell and the danger of explosions.

By 1859 existing sources were no longer sufficient to meet the increasing demands of lighting and therapeutic applications. At this juncture a youthful New York lawyer, George H. Bissell, decided to exploit the mineral systematically. In association with James M. Townsend, a banker from New Haven, and Benjamin Silliman, a chemistry and geology professor at Yale University, he founded the Pennsylvania Oil Company, the ancestor of the giant trusts which today span the world with their networks. From a farmer at Titusville they rented 125 acres of ground, and one of the shareholders of the Company, Edwin L. Drake, undertook to sink the first well.

It must be said that Drake, an ex-railroad conductor who was in no way prepared for his task, owed his success only to unbelievable luck. By the greatest of chances he happened to pierce the ground just at the right place; at a depth of a mere $69\frac{1}{2}$ feet he struck a large pool. On the advice of an old well-digger he did not dig at once with a pick but, instead, sank an iron pipe into the ground. A small steam-engine was moved in, a wooden scaffolding constructed, a chain forged, and a vertical ram installed. Drilling started in June 1859. On the 27th of August oil began to appear. The following week, it flowed out at the rate of ten barrels a day.

Thereupon there was a general stampede for the new El Dorado, and literally thousands of wells were drilled in Pennsylvania in the next few months, and an ever-increasing stream of speculators invaded the region, which was quickly parcelled out into countless plots. Each plot-owner immediately assigned all his remaining savings to the purchase or hire of some makeshift equipment. There was nothing scientific about the speculators' approach. They simply dug at random. While some grew rich, others were ruined. Distillers set themselves up alongside the wells and distilled the crude oil in primitive stills, extracting from it the commercially valuable lamp-oil. Gas and gasoline was rejected as useless. Haphazard boring was accompanied by a waste of products. Yet in spite of everything production rose. In the State of Pennsylvania alone it jumped from 2,000 barrels in 1860 to three million in 1862. Deforested, drilled, oil-polluted, infected with the stench of sulphur and gasoline, shrouded in fumes, what had a few years earlier been a rural countryside had become an 'ante-room to hell', in the words of a reporter of the period. The

search soon extended into the neighbouring states. By the end of the century oil was being produced in Ohio, West Virginia, Kansas, Texas, Oklahoma, Colorado, Wyoming, California, and overseas in Russia, in the Dutch East Indies and Poland.

In New York the financiers speculated. Petroleum was quoted on the Stock Exchange and with each insignificant report about the sinking of new wells or the abandoning of old ones suffered fantastic gains and losses within the space of a few hours. In 1862 production greatly exceeded demand, and prices collapsed. The value of a barrel (U.S. barrel equals 42 gallons) dropped from twenty dollars in 1859 to ten cents. Clearly, the petroleum industry needed completely reorganising. John D. Rockefeller was the man who undertook the task. Leaving the risks of prospecting and extracting to the producers, this financier of genius set up a first company, the Standard Oil Company, to centralise and standardise operations of storage, refining and transport. The distilled products were to have a uniform quality which would bring stabilisation of prices. Agreements were made with railway and navigation companies to transport the oil even to the most distant countries. Where necessary, pipelines were built, despite the often very active resistance of the road- and rail-carriers.

As a result of agreements signed voluntarily or under intimidation, Rockefeller controlled practically the entire oil industry of the United States by 1878. Federal Government concern at this 'State within the State' resulted in anti-trust proceedings being instituted against Standard Oil. In 1911 it was finally forced to liquidate and surrender its autonomy to the thirty-three subsidiary companies of which it was composed.

The outstanding dates in the development of petroleum products are:

1860: lamp-oil begins to be used on an industrial scale;

1900: development of the internal combustion engine; utilisation of gasoline;

1915: manufacture of chemical products with a petroleum base;

1930: first use of liquefied propane and butane;

1933: development of the diesel engine; utilisation of gas-oil;

1950: development of the jet engine; renewed utilisation of paraffin.

Because of the diversity of petroleum's products, the high heat value of its fuels, and its economical transportability, man has made petroleum, in the form of natural gases, gasoline, gas-oil, fuel-oil, bitumen, etc., a major factor in the world's economy. Quantitatively it is the most important of all its raw materials.

## WHAT IS PETROLEUM?

Apart from the experts, few are familiar with crude oil, a substance which emerges from the ground as a thick, viscous, and sometimes evil-smelling, brown or dark green liquid. Certain kinds, like the deposits of Venezuela, contain primarily gasolines; others, like those of Texas, are richer in fuel-oil or in mazut suitable for furnaces; others again, like those of Mexico, yield principally bitumen for asphalt. Specific gravity varies from type to type and provides an indication of quality. Other criteria include pour- and cold-points, viscosity, optical properties, odour, flash- and burning-points, colour, and coefficient of expansion.

Chemically, crude oil is a naturally occurring liquid, a mixture of compounds of carbon and hydrogen. The proportions of the hydrocarbons vary considerably. Some of them occur as gases, others as solids, and both are in solution in liquid hydrocarbons. Since rock-oil or crude oil is a mixture, it does not have a definite chemical composition nor fixed physical properties. Despite this, the elementary composition of crude oils is surprisingly constant, ranging from 83 to 87 per cent carbon, and from 11 to 14 per cent hydrogen.

The amount of hydrocarbons in a crude oil is a measure of its purity — generally all but 0·3 to 3·0 per cent by weight consist of hydrocarbons. The main hydrocarbons in oils are paraffins, naphthenes, or aromatic groups or complexes of them. The aromatics are the least common; most oils have a paraffin, naphthene or mixed paraffin-naphthene base. In paraffin oil there is a higher hydrogen content in relation to carbon. In naphthene oil

Top
The first oil well at Titusville.

Bottom
Oil wells in Kilgore, Texas.

266

the situation is reversed. Naphthenic oils are heavier and proportionately higher in viscous but volatile lubricating oils. Naphthenic oils are often referred to as asphaltic base because distillation produces solid or semi-solid asphaltic residues. Lower densities and residues of petrolatum or paraffin wax characterise a paraffinic oil.

The hydrocarbons are divided into a number of chemical series, each with somewhat different chemical and physical properties. The four series which make up the bulk of crude oil are the methane (normal paraffin) series, isoparaffin (branched-chain paraffins) series, the cyclo-paraffin (naphthene) series, and the benzene (aromatic) series. Crude oils are often labelled according to the series predominating in the oil.

Let us take the simplest example, that of the methane, or paraffin, series. This is a saturated, straight-chain, homologous series, advancing by a $CH_2$ increment from its simplest member methane ($CH_4$) to complex molecules with more than 60 carbon atoms.

A carbon atom is a kind of four-armed creature (four valencies) which can retain four hydrogen atoms, because each hydrogen atom has only a single valency. The resultant hydrocarbon is methane, the chemical formula of which can be written as $CH_4$ or $CH_3 - H$. This is the simplest of all the hydrocarbons, and the most stable. It is the main constituent of natural gas and forms in swamps, as 'marsh gas', from decaying vegetable matter.

The formula $CH_3 - H$ draws attention to the group $CH_3$, which the chemists call methyl and which plays an important part in the composition of the hydrocarbons. The methyl group, be it noted, has only one valency and thus replaces any one hydrogen atom of a pre-existing hydrocarbon. Methane ($CH_3 - H$) is followed by ethane ($CH_3 - CH_3$), propane ($CH_3 - CH_2CH_3$), butane ($CH_3 - CH_2 - CH_2 - HH_3$), pentane ($CH_3 - CH_2 - CH_2 - CH_2 - CH_3$), and so on. The formulae of these hydrocarbons are in open chain. They are normally written in their condensed form as $CH_4$, $C_2H_6$, $C_3H_8$, $C_4H_{10}$, and $C_5H_{12}$, etc. The general formula for the hydrocarbons of the methane series is $C_nH_{2n+2}$.

Most crude oils are formed chiefly of these hydrocarbons — gaseous hydrocarbons ranging from $CH_4$ to $C_4H_{10}$, liquid hydrocarbons ranging from $C_5H_{12}$ to $C_{15}H_{32}$, solid hydrocarbons ranging from $C_{16}H_{34}$ to $C_{35}H_{72}$.

However, an oil containing only these thirty-five hydrocarbons would have to be regarded as very simple. In practice, they are joined by others (isomers) having the same composition but differing in molecular structure and therefore differing in their properties. To make this clearer, let us return to the structural formula of butane ($CH_3 - CH_2 - CH_2 - CH_3$). On this linear chain a lateral one can be grafted by replacing a hydrogen atom of the group $CH_2$ with a methyl group. Thus we have, for example,

$CH_3 - CH_2 - CH - CH_3$
|
$CH_3$

This new hydrocarbon is methylbutane. Its shorter formula is $C_5H_{12}$. exactly the same as that of pentane. Methlbutane and pentane are thus isomers. There are three pentanes, each with the formula $C_5H_{12}$ each containing 83·33 per cent carbon and 16·76 per cent hydrogen, and each with a molecular weight of 72·15 but a different boiling-point.

It can be shown by the theory of probabilities that the number of isomers increases with the $n$ number of carbon atoms: 3 isomers when $n=5$, 5 isomers when $n=6$, 9 isomers when $n=7$, ... 62 billion isomers when $n=40$. As we see, astronomical figures are quickly reached. In all probability only a very few of these compounds actually occur in measurable amounts in crude oil. However, the tremendous complexity of the hydrocarbon family is impressive. The total number of individual hydrocarbon compounds that may be present in oil is not yet known.

A further complication arises from the fact that a crude oil is never composed of a single hydrocarbon series. While the methane of the paraffin series predominates in oils of Pennsylvania and Rumania, these oils contain also hydrocarbons of the naphthene ($C_nH_{2n}$) and the aromatic ($C_nH_{2n-6}$) series. In another oil, that of Baku, the naphthene series predominates, but not to the total exclusion of the other series.

It should be noted that the hydrocarbons of the naphthene and aromatic series differ from those of the methane or paraffin series in that their carbon atoms do not carry all the hydrogen atoms to which they are entitled. They are unsaturated hydrocarbons, in contrast to paraffinic hydrocarbons which are completely saturated. It follows that while the latter are stable, the former are not and tend continuously to acquire new hydrogen atoms. The diagram on p. 268 depicts saturated carbons and unsaturated carbons.

Bearing in mind all the complications, it is not surprising that the refineries modestly estimate the number of hydrocarbons figuring in the composition of ordinary oils at 8,000. Other substances present include in varying quantities sulphur, oxygen, nitrogen and metals.

Some quantity of sulphur (0·1 — 5·5 per cent by weight) occurs in most crude oils and in each of the fractions that form the oil. It is rarely present as elemental sulphur, but generally occurs in compounds such as hydrogen sulphide ($H_2S$) and the organic sulphur compounds. Sulphur and sulphur compounds cause corrosion, bad odour and poor combustion. Until modern refining methods were developed the presence of sulphur in a crude oil made it much less desirable and less valuable. Now that the unwanted substance can be eliminated the sulphur-bearing crude oils are almost equal in price to the non-sulphur crudes. In fact, in some gas pools with a high sulphur content, sulphur may even be extracted at a profit. Generally crudes of a high specific gravity contain more sulphur than others. Heavy Mexican crudes carrying from 3 to 5 per cent sulphur are among the world's highest sulphur bearers. Crudes with less than 0·5 per cent sulphur, such as the Pennsylvania oil, are low-sulphur crudes; those with more are high-sulphur crudes. Almost half of the United States oil production is low-sulphur.

Almost all crude oils contain small quantities of nitrogen, ranging from less than 0·05 to a maximum of 0·08 per cent in some California crudes.

The amount of oxygen in crude oil ranges from 0·1 to 4 per cent by weight, generally averaging less than 2 per cent. It may occur as free oxygen or in various compounds. Like sulphur, oxygen occurs in larger amounts in the heavier crudes.

In addition to sulphur, nitrogen and oxygen, crude oil usually contains numerous miscellaneous substances, both organic and inorganic. Visible under the microscope are fragments of petrified wood, spores, algae, insect scales, tiny shells and shell fragments, and particles of lignite and coal. Metals found in crude oil ash are many, including vanadium, nickel, cobalt, copper, zinc, silver, lead, tin, iron, molybdenum, chromium, manganese, arsenic and uranium. Sodium chloride, too, is present in most crude oils; when the quantity exceeds 15 to 25 pounds per 1,000 barrels of crude oil, desalting is necessary because, like sulphur, salt is corrosive. Some of the salt is in the form of crystals in the oil, and some is dissolved in the water which is normally produced along with the crude oil.

Petroleum and natural gas underground and in their natural state are under greater pressures and higher temperatures than they are at the surface. As the crude oil travels upward in a well, from the reservoir to the surface, it encounters changes in pressure and temperature which induce changes in its original composition. Even with advanced laboratory techniques it is possible to determine the original composition only approximately. Even the make-up of crude oil at the surface is only partly known. In some twenty-five years of study to isolate the individual hydrocarbons that can be separated out of crude oil between 1·04°F. and 356°F., only 92 have been recovered out of the 500 or so occurring in this temperature range.

Passing from the chemical to the physical aspect, we find that crude oil is, properly speaking, a liquid containing gases in solution, and microscopic particles of solid hydrocarbons in suspension. Because these substances have neither the same boiling-point (the temperature at which they pass into the gaseous state) nor the same freezing-point (the temperature at which they pass into the solid state), it is possible to separate them under the effect of temperature increases. On this are based some of the methods of distillation of petroleum in the refineries. First the gases (methane to butane) are released, then the liquids in order

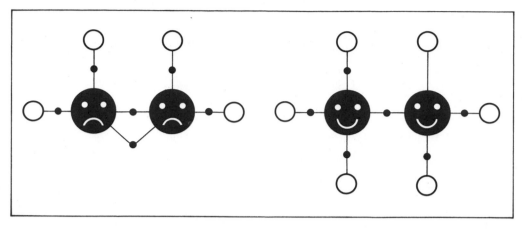

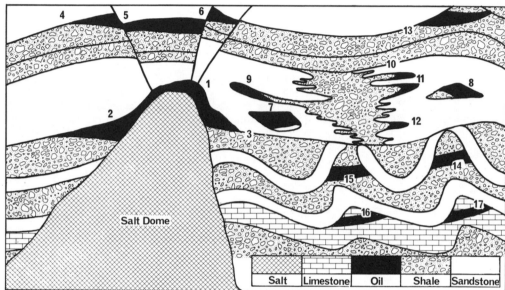

*Top left*
Diagram illustrating the difference between saturated paraffinic hydrocarbons (smiling) and unsaturated hydrocarbons (shown with tears).

*Bottom left*
The position of oil traps:
1 cap rock oil; 2 and 3 flank accumulations; 4, 5, 6 fort traps; 7 and 8 sand lens trap; 9, 10, 11 and 12 stratigraphic traps; 13, 14, 15 and 16 structural anticlinal or dome traps.

*Right*
An oil prospector in the foothills of Brooks Range, Alaskan Arctic.

of growing complexity—pentane at 97°F., hexane at 156°F., heptane at 208°F., octane at 257°F. The solid and semi-solid hydrocarbons form the residue of the distillation.

Because petroleum is a mixture, its physical properties vary considerably, depending upon the type and relative proportions of the hydrocarbons and impurities present. More than a dozen physical properties of petroleum are commonly determined. Some of these are density, viscosity, optical activity, refractive index, colour, fluorescence, odour, pour- and cold-points, flash- and burning-points, coefficient of expansion, surface and inter-facial tension, capillarity, absorption.

Of these properties, that most often referred to is the density or 'gravity'. The density of a substance is the weight of a given volume — say, pounds per cubic foot. Specific gravity is another way of expressing the same thing without specifying a unit of measure. It is the ratio of the weight of a given volume of any given substance and the weight of an equal volume of pure water at a particular temperature and pressure. In the United States the practice is to give the specific gravity of oil at 60°F. and one atmosphere of pressure, and because the price of crude oil is generally based on 'gravity' these measurements are important. With most liquids one or two arbitrary scales are used: degrees Baume or degrees A.P.I. (American Petroleum Institute). The first is a European scale, the latter an American. By the A.P.I. scale, water has a value of ten, heavier substances have numbers lower than ten and lighter substances have higher numbers. Crude oils range in density from types that are heavier than water (some California crude with an A.P.I. of 5° — 7°), through oils of 10° A.P.I. (Venezuela) up to 57° A.P.I. and even higher. The bulk of crude oils average 27° to 35° A.P.I.

The density is determined simply by putting a hydrometer with the A.P.I. scale calibrated upon the stem into crude oil. Temperature and pressure readings are recorded at the same time, so that the particular reading can be converted to the standard (in U.S. 60°F. and 1 atmosphere).

Viscosity or resistance to flow (the opposite of fluidity) is measured by the time taken for a given amount of oil to flow through a unit opening at a given temperature. Crude oils vary greatly in viscosity, light oils and natural gas being very mobile or low in viscosity. Others may be highly viscous, in fact, plastic. Viscosity of oil depends mainly on the amount of gas dissolved in it and on the temperature. The higher the temperature and the more gas, the less viscous the petroleum. This is an important characteristic because at ordinary temperatures some crudes are so viscous that they must be heated to be pumped. This can make production and transportation very expensive.

The colour of petroleum varies from almost colourless to red, green, and even opaque black. Generally lighter colours occur in the low specific gravity crudes. Also, all oils show some degree of fluorescence, particularly the aromatic oils. Colours range from yellow through green to blue when the oil is viewed under ultra-violet light. This property is used to spot oil shows in cores, cuttings and drilling mud. Fluorescence allows the detection of the minutest quantities (1 part to 100,000 parts) by the unaided eyes; with laboratory techniques, one part in hundreds of millions can be noted.

Some petroleums have a gasoline-like odour, described as agreeable or sweet. Those with quantities of sulphur or nitrogen have an unpleasant, sour smell.

The cloud-point of oil occurs at the temperature where solid paraffin waxes settle out. The pour-point is the lowest temperature at which oil is fluid. The flash-point is the temperature at which the vapours rising from oil will ignite with a flash of short duration. At a somewhat higher temperature the burning-point will be reached, and the oil will burn with a steady flame. These last two measurements are a clue to the hazards involved in storing and

handling the petroleum. The first two are important in transporting the crudes; pour-points can vary from 90°F. to —70°F. and lower.

## PETROLEUM'S ORIGIN

Petroleum deposits almost always occur in sedimentary rocks, and the abundance of petroleum in the United States is thus due in part to abundance of marine sedimentary rocks. The same applies to the regions around the Gulf of Mexico, the Persian Gulf, and the Caspian Sea. Conversely, Brazil, formed almost entirely of igneous rocks, is poor in petroleum.

Petroleum is almost always associated with sediments of marine origin and its water content is very similar to sea water. Few petroleum deposits are found in continental sediments and probably the bulk of petroleum, if not all, is marine in origin. Yet strangely enough the first theories of petroleum origin propounded an inorganic source. These theories have not yet been completely refuted, though they are no longer considered probable.

The chief support for inorganic ideas came from the laboratory, where hydrocarbons (methane, ethane, acetylene, benzene) were made from inorganic substances. There is no field evidence, however, to show that a similar process took place in nature. Moreover, the laboratory methods required substances (free alkali metal or metallic carbides) that are unknown in nature. Other evidence against the theory includes petroleum's optical rotary power, a characteristic confined almost exclusively to organic matter; its absence in association with vulcanism, volcanic material or igneous rocks except in very rare cases; its synthesis in the laboratory from organic substances.

The hydrogen content of petroleum is much higher than in most organic substances and it is, of course, possible that this hydrogen is inorganic in origin, acquired from the rocks through which the oil passes. It is also possible that the hydrogen was added by organic processed, one example being bacterial action.

There are three main reasons for favouring the theory of organic origin of petroleum:

First, there is the widespread and abundant source necessary for petroleum origin which is provided by the huge quantities of organic matter that occur in the sediments of the Earth, and the fact that hydrogen and carbon predominate in such remains. These two elements are also produced by the life processes of plants and animals.

Secondly, porphyrin pigments and nitrogen are present in most crude oils. As both are present in organic matter, a relationship is indicated.

Thirdly, there is the optical rotary power of petroleum. Except for cinnabar and quartz, inorganic substances do not possess the ability to rotate the plane of polarisation of polarised light (light vibrating in a single plane). This capacity is thought to be the result of the presence of cholesterol, which is present in animals, plants and crude oil.

Other facts indicating organic origin are: almost invariable association with sedimentary rock; the similarity of associated substances in petroleum and organic ash; the presence of microscopic organic fragments in the oil; and the presence of hydrocarbons, almost identical with petroleum hydrocarbons, in the modern sediments of the Gulf of Mexico.

In marine sediments such as limestone, sand or clay, the carbon and hydrogen necessary for the formation of hydrocarbons can be supplied by living organisms or, more precisely, most probably by their fats, waxes and resins, which are particularly rich in hydrogen and which resist decay more than the carbohydrates and proteins.

Which then are the organisms capable of producing the vast quantity of fatty substances required for the formation of the thousands of millions of tons of petroleum which have

out that as far as its chemcial composition is concerned, petroleum is not very far removed from coal, and that there are certain bituminous coals (bogheads and cannel coals) which are difficult to classify as coal rather than petroleum, and vice versa. Such a close chemical relationship can only be explained by a closely related origin, and the formation of petroleum may have resembled that of coal, which we have already examined. If petroleum contains more hydrogen than coal, this may be because it is derived from organic substances (fats, waxes, resins) richer in hydrogen than the lignin and cellulose which served as raw materials for coal.

Let us now consider how fatty matter of predominantly planktonic origin could have transformed itself into petroleum. The open sea, swept by the ocean currents, would not offer conditions favourable to its transformation. Bays, gulfs, coastal lagoons, enclosed seas like the present Caspian and Black Seas, and submarine basins with poor water circulation like those off the California coast were needed to enable petroleum deposits to form because another equally important condition was the absence of oxygen. It is clear that in contact with air organic matter will decompose and disappear altogether. Decomposition of organic matter in the presence of abundant oxygen produces as end products carbon dioxide and water. To get hydrocarbons from decay of organic matter the process must be arrested at some point short of completion. The most probable location for this is in a reducing zone in which there is also moderately rapid to rapid sedimentation. One can imagine the decaying material mixed with clay particles and constituting an organic ooze impregnated with sea water, a kind of brine in which bacteria of a special type (anaerobic bacteria) gradually reduced their oxygen content (in most petroleum only up to 4 per cent oxygen is present, whereas organic matter contains about 15 to 35 per cent), and consequently raised the proportion of carbon and hydrogen, which subsequently combined into hydrocarbons. Bacteria of this kind have been found in marine sediments and even in the salt water of petroliferous sediments. They do not differ essentially from the bacteria which, at the bottom of existing swamps, cause the formation of methane or marsh gas. A 'syrup of corpses', slowly maturing under the effects of weather, pressure, temperature, catalysts, radioactivity, and bacterial action, this became what is today petroleum.

## SOURCE ROCKS AND RESERVOIR ROCKS

Although petroleum is regarded as a normal constituent of sedimentary rocks of marine origin, it must not be assumed that every sedimentary rock, no matter where it occurs, contains workable deposits. To do so would be to ignore the lightness and volatility of this mineral oil. Petroleum has taken advantage of any permeable rock pointing towards the surface of the ground to rise as high as it could. Hydrocarbon seepages, asphalts found on the surface of the soil, bituminous shales, bituminous limestones, and fossil paraffins occurring in certain areas all furnish proof of this. Sometimes they indicate underground deposits of petroleum; often they are proof of evaporation and natural exhaustion over the ages.

The exact mechanics of oil migration are not yet fully understood. There appears to be a natural migration from fine-textured source rocks to coarser reservoir rocks with larger pore spaces. The main forces behind oil migration are: the compaction of the sediments under an overburden and during diastrophism, the relative buoyancy of oil and gas, capillary action of water which moves into rocks with very small pores and displaces any oil present into coarser adjacent rocks, and water flushing and carrying along patches of oil as it moves through the subsurface rocks.

Oil leaving a source rock tends to migrate upward in the direction of lowest pressures. This migration may be deflected and finally stopped by an impervious rock layer, called an oil trap or cap rock under which the oil accumulates. Salt, shale, gypsum, and dense limestones are often cap rocks. Oil traps are divided into three main classes: structural traps, where the rocks have been deformed by diastrophism or intrusion to form arch-like inverted containers; stratigraphic traps, which are the result of lateral and vertical changes in the permeability of the reservoir rock;

Top
An oilwell being drilled in the Waha oilfield, Lybya.

Bottom
A production well at Agha Jari, Iran.

accumulated in sedimentary rocks over the centuries? Possibly they were the myriads of microscopic animals and plants which inhabited the upper layers of the sea — plankton, and so on.

The inferior organisms are not, however, the only ones to live in such formidable quantities. It has been calculated that if the catch of fish in the North Sea were to be amassed during only one millenium, it would be sufficient to produce more than one hundred million tons of petroleum. Pursuing this theory further, we can ask what are the North Sea and a millenium compared with the immense surface area of the oceans of Earth and the immense duration of the geological eras?

If further arguments are needed in favour of organic origin, it may be pointed out that petroleum always contains nitrogen, sulphur and phosphorous in the form of organic compounds like porphyrin. Porphyrins are formed from the green colouring matter (chlorophyll), or from the red colouring matter (hemin) of blood. They occur in crude oil as complex hydrocarbons which oxidise readily. In crude oil, the vegetable porphyrins are more abundant than the animal porphyrins. It can also be pointed

combination traps, which are any combination of stratigraphic and structural traps.

Most oil pools form in combination, under certain essential conditions. These are listed here:

Where there are sedimentary rocks, predominantly marine and unmetamorphosed;

Where there is an impervious rock layer, called the cap rock, overlying the oil-bearing stratum;

Where a reservoir rock both porous and permeable (commonly a sandstone or limestone) contains the oil;

Where there is a reservoir trap, often produced by folds, faults or domes, or by variations in porosity and permeability;

Where there are adequate source rocks in which the petroleum formed, generally believed to be a shale or clay.

In practice, traps are rarely so simple. Often anticlines are combined with faults and other complexities, from which geologists must disentangle the fate of the migratory petroleum. To speak of wells, then, as 'pockets' or 'pods' of petroleum is inexact. The petroleum is not free in the underground cavities but enclosed within the permeable reservoir rocks, impregnating the minute pore spaces where it is zoned with gas and water into three layers, according to their specific gravities. The gas usually lies at the top, petroleum in the middle section, and the salt water in the lower part. If the impervious cap is pierced directly into the oil, the liberated petroleum may gush out under the pressure of the confined gases and in this way saving a costly pumping operation.

## HOW PETROLEUM IS PROSPECTED

When Edward Drake sank the first oil-well at Titusville in 1859, nothing was known of prospecting methods or drilling techniques. A modest wooden structure supporting a vertical ram, a second-hand motor and a series of iron pipes constituted the entire equipment. Today no well is sunk without a very detailed preliminary study of the bedrock. Nothing is left to chance. The drilling engineers have the most efficient modern technical equipment at their disposal, and wells can be successfully sunk to over 20,000 feet, sometimes under the most difficult conditions. Further improvements in techniques can be expected for the future.

For long the search for petroleum was based solely on the surface evidence of seepage of oil, release of gases, and the presence of bitumen-impregnated sand or limestone. Such evidence is often misleading, however. It does not necessarily indicate the existence of petroleum at depth. We have already seen that such evidence may be nothing more than traces of former deposits now exhausted, or emanations of very distant deposits. Conversely, the absence of such indications does not necessarily prove that no petroleum lies hidden beneath an impermeable stratum.

Let us suppose that petroleum is being prospected in a virgin area, an area which is known only to possess conditions favourable for such exploitation. The first step is an aerial reconnaissance of the area, when photographs may reveal a number of potential oil trap structures, such as faults and anticlines. Thus informed, the

*Left*
Drilling an oil well. 1. digging a guide-bore pit; 2. erecting a steel structure or derrick; 3. a drill pipe to drill the well; 4. drilling bit; 5. rotary table to move the rod; 6. square connection to link rod and rotary table; 7. and 8. motor-driven winches to operate rotary table; 9. Further pipes added as drilling progresses; 10. the well, completely lined internally with steel tubing; 11. water used to bring excavated material to the surface by circulating through the drill pipe, rising in the circular space between pipe and tubing and ascending into the sifting tank; 12. pump to return the current; 13. current, containing mud returned to drill pipe by way of mud pipe; 14. an injection head; 15. a flexible shaft; 16. hook to hold previous parts suspended; 17. mobile pulley to operate the hook; 18. cable from which mobile pulley is suspended.

*Right*
A rig in the Swan Hill area of northern Alberta, Canada.

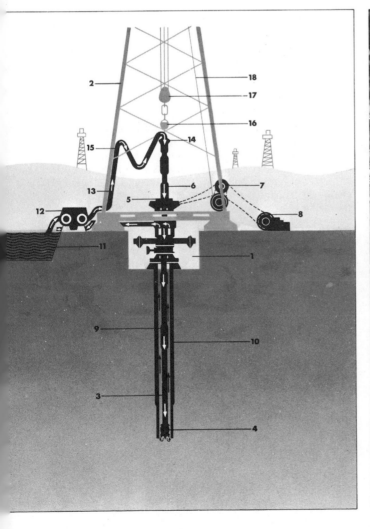

geologists delimit the area to be prospected, survey it thoroughly, and draw up a geological map showing the distribution of the rock formations. Rocks outcropping on the surface of the ground are examined, marked on the map, and frequently give the first inkling of the structure of the bedrock.

When preliminary survey suggests favourable conditions, its findings are investigated by the geophysical methods already described. Thus a more precise knowledge of the subsoil is obtained. If both the original field surveys and the geophysical methods yield promising results then examination by coring may be decided upon before final drilling.

The ground is first drilled with a core drill, and a continuous rock core obtained section by section. Age of the rocks is determined by the Foraminifera present. Precise analyses of rock specimens taken from the core determine porosity, permeability, and capillarity, and indicate the potential productivity of the future oilfield. Today, mechanical sampling is supplemented and even replaced by electrical and radioactivity records made as the drilling progresses. These provide additional accurate information about the workable strata.

Core holes can be drilled to depth of more than a thousand feet, though this type of drilling is usually confined to lesser depths. A 'slim-hole' rig (6 —7 inches in diameter) yields a continous core at a much smaller cost than the conventional coring rig. Its relative cheapness has resulted in its application in drilling for production wells in some areas.

## DRILLING

Preliminary investigation of an area may last for several years before the geologists decide on the most favourable site to drill. Even so, there is only one chance in nine (in the United States) that the drilling will strike sufficient petroleum to be commercially interesting. If it should strike a potentially profitable pool, the chances of immediately striking and releasing crude oil are one to three against In the other two cases, it will strike either the gas above the oil, or the salt water on which it rests. It is readily seen that capital must often be expended without yielding any positive result. Only the giant present-day trusts can afford to sink large sums (at least a quarter of their profits) in development programmes.

## SINKING AND OPERATING

The first step in sinking a well is to erect a steel framework or derrick, a kind of miniature Eiffel Tower with a height of 100 feet to 200 feet (average height in the United States is 176 feet or over). This supports the pulleys from which the drill pipes are suspended.

Early wells were sunk by ramming (percussion or cable-tool drilling). This vertical movement was largely replaced by rotary drilling, a method in which the bit turns at 50 to 300 revolutions a minute. Different bits (blank, insert, saw-tooth, cracker jack, diamond cast set, tri-cone) are used for different needs, some for soft rocks, some for hard, others for coring. Their function is to abrade and crumble the rock. Speed of penetration may vary from 8 inches an hour in very hard rocks to 100 feet in friable ground. Drill collars add extra weight and help the bit to withstand compression shock as the operation proceeds.

A continuous stream of mud is pumped into the hollow drill pipe and bit while the rotary table turns. One or more openings at the end of the bit allow the mud to rise between the drill pipe and the sides of the well. This fluid's function is manifold: to lubricate and cool the bit while it is operating, to transport excavated material to the surface, to act as an eroding agent in soft rocks, and to strengthen the sides of the well by plastering them with 'filter cake' which seals the hole and prevents drilling fluid from being lost in porous rocks. Unfortunately, besides sealing off water, it may also seal off gas and oil, and the pay horizon (oil formation) may be missed. When circulation of the fluid stops, it gels, and causes the bit to stick.

The great advantage of the drilling fluid is that its weight allows it to enter high-pressure gas and oil reservoirs without explosions or blow-outs dangerous to life and property.

Drilling mud must be 'tailor-made' for the conditions expected in each new well. Factors such as the density, viscosity, weight, gel strength, filtration and salinity are carefully controlled.

To make the mud gel when circulation is stopped, bentonite or other clays are added. If there are not enough clays, solids or colloids present, then fluid is lost, particularly in work on cavernous rocks or dry porous sands. Lost circulation is a serious and expensive problem; it can be overcome by adding fibrous or bulky materials such as special cements, sawdust, beet pulp, tattered sacks, cellophane, and shredded plastic foil.

Drilling through salt beds brings its own particular problem: loss of drilling fluid through solution. To counteract this tendency the fluid is completely saturated with salt before the operation is begun.

Where high-pressure zones are expected, high-density fluids are used to prevent blow-outs, and weighting materials, such as barite, are added. This roughly doubles the weight of the mud, and is satisfactory for most deep drilling.

Special equipment and technicians are required to handle the drilling fluids and costs are often high. In the United States, about one million tons of barite, half a million tons of bentonite, and a quarter of a million tons of other clays are used in drilling each year.

There are two main types of drilling fluid: water base and oil base. Water base muds are the commonest. Oil base muds are more expensive and are used mainly for low-pressure reservoir drilling where weighted fluid is not needed. Oil base fluids are also used in drilling through clays, salt and anhydrite, and in recovering rock cores in their native state. Combination oil and water base drilling fluids have also been developed. They are more easily handled than straight oil base fluids and are also non-inflammable.

In some places rotary drilling makes use of air or gas instead of drilling fluid. In this method there is no lost circulation, drilling speed is increased tremendously, effecting a 20 — 60 per cent saving in drilling costs, and identification of gas and oil horizons is easier. Disadvantages are: the very fine powder-size of the rock cuttings; reservoir pressures likely to be encountered must be low; the rocks being drilled should not show any tendency to slough-off; water-bearing strata must be few in number and volume; fire and explosion hazard is high.

On its return to the surface the drilling fluid passes into a tank for sifting and recompounding before re-use. Frequent analyses of the cuttings (samples) provide information about the nature of the excavated material and the ground's richness in hydrocarbons.

In the course of drilling the bit has to be replaced fairly frequently, depending on the hardness of the rock. Often replacement must be repeated two or three times a day. This involves drawing up the whole set of pipes, three or four sections at a time, and unscrewing each in turn until the drilling bit itself emerges. Once this has been changed, the set is reassembled and lowered again. The pump is restarted and again injects mud until it flows into the sifting tank. At this point the engines are engaged, the transmission chains tighten, and the rotary table begins to revolve again, starting the drilling bit on its excavation task.

For technical reasons cable-tool (or percussion) drilling is rarely performed beyond 5,000 feet. With rotary drilling there is no known lower limit to the depth. The main limitation on deep drilling is economic; when the cost of drilling is more than the potential return, the proposition is no longer practical. Rotary drills have other advantages over cable-tool drills: their greater speed in soft rocks (2,500 feet per day), the ready availability of cores and electric logs, larger completion diameter and lower cost per foot drilled.

Rotary drilling has replaced cable tool in most areas. In time, perhaps not too far in the future, it may be replaced by turbo-drill. Pioneered in California over thirty years ago but perfected by the Russians until it is almost standard equipment in the U.S.S.R., the turbo-drill is reported to be to twelve times faster than the rotary. The bit is rotated solely by the hydraulic power of the mud stream pumped down through a non-rotary pipe. A turbine located in a cylinder below the drill pipe converts the hydraulic power of the mud stream into rotary power. In the United States today a Russian turbo-drill with French modifications is being used for test drilling.

**Testing.** After drilling and on temporary completion of the well, a drill stem test is usually made to determine the fluid content of the rock and check the reservoir fluid pressure. This is done by lowering an empty string of pipe fitted with a valve until it stops just short of the oil formation. When the valve is opened, the reservoir fluid flows into and up the drill stem pipe. In general the amount of fluid entering the pipe is proportional to the permeability of the rock. If the flow is too slow, then shooting (exploding nitroglycerine), acidising and hydraulic fracturing may be employed to increase permeability by cracking. Sometimes these techniques convert a dry hole into a producing well.

When drilling and drill stem testing is finished and any necessary treatments have been applied to increase oil flow, then well completion takes place. This includes all the measures required to turn the test well into a commercial well. If the oil rock is competent, the procedure is simple and involves only the running of casing from ground surface to the top of the oil horizon and cementing it in place. Inside the casing and down to the bottom of the hole runs a smaller diameter tubing, open at the bottom for the oil.

Many oil rocks do not hold together well enough to use this open-hole technique. There are several methods of correcting this. A gravel pack may be placed in the bottom of the hole to prevent collapse of the walls yet still permit oil flow. Sands loosely cemented by plastic may be used in similar fashion. Preperforated pipe sections may be inserted at the bottom of the hole, or the hole may be ceased solidly to the bottom and then perforated by a device that shoots bullets or shaped charges through the casing into the rock at the level of the oil horizon. The well is then ready.

Records of drilled wells date from 221 B.C. (in China: 450 feet, for brine) but the date and the depth of the first is unknown. In the United States wells over 2,000 feet deep were drilled by A.D. 1854. A record 25,340 foot well was drilled in 1958 but was dry, and the deepest productive well is 21,443 feet, in Louisiana. New records will continue to be established because only expense and complete penetration of sedimentary rocks limit the depth of drilling.

**Directional and underwater drilling.** In certain circumstances special equipment has to be used. Fairly short drillings can be made by portable, drilling rigs, derricks transported on trucks, together with winches and all accessory apparatus. For very deep drillings, on the other hand, robot drilling rigs are used. In this the operator simply manipulates the rig by levers and controls.

**Deviation.** Controlling the direction of the drill bit presents a complex problem. Petroleum engineers have not only developed techniques to prevent 'crooked holes', that is, holes which are not vertical, but they also have a technique for diverting the direction of holes from the vertical. This is used in the exploitation of petroleum deposits situated under, say, a built-up area or a mushroom-shaped dome of salt or under the sea.

The present technique is first to prevent or control deviation by using the correct combinations of weight on the bit, revolutions per minute, and type of bit. If the deviation cannot be controlled in this way, a vertical course can be maintained by the use of a 'whipstock', a wedge-shaped device placed in the drill hole to force the bit back in a desired direction.

Directional drilling is invariably started vertically and continued so down to a depth of several hundred feet. A new drilling bit fitted with a whipstock — a long, slender, steel wedge — is then lowered into the hole. Sliding against the wedge, the bit is deflected from its original direction to open up a new hole at an angle some two to three degrees different from the first. Drilling with the normal bit is then resumed for another several hundred feet. Again a whipstock is lowered and again it deflects the bit by two or three degrees. And so it continues.

The simplest method of checking the inclination of a well consists in lowering a glass bottle half-filled with hydrofluoric acid to the desired depth. The bottle tilts and the acid corrodes the glass to a certain level; measurement of the deviation of this level from the axis of the vessel is a simple matter. Instruments called inclinometers indicate deviation away from the vertical, and direction in relation to the meridian.

Directional drilling permits the use of one well in drilling several that radiate from the bottom of the vertical portion. In this way, a single drill site can serve for many wells in expensive offshore operations, or in rugged terrain or in an exclusive residential area.

In underwater prospecting a portable seismograph records vibrations from man-made explosions as they bounce back from buried layers of hard-rock on the sea bed. This data helps to determine the nature of the hidden layers.

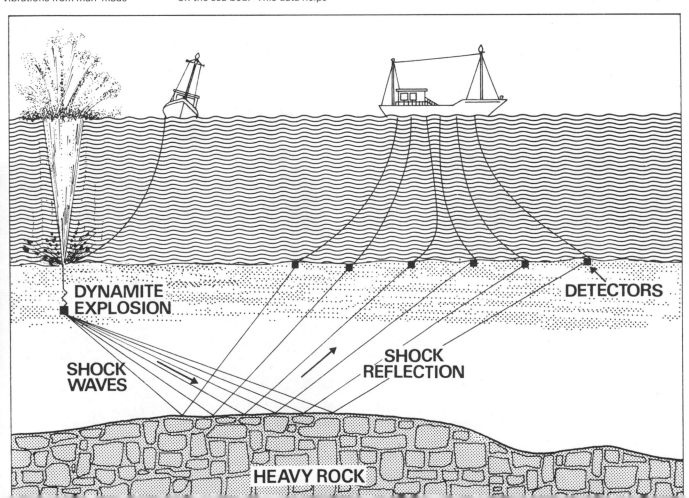

DYNAMITE EXPLOSION

DETECTORS

SHOCK WAVES

SHOCK REFLECTION

HEAVY ROCK

**Underwater operations.** Directional drilling is practised extensively in exploiting underwater coastal deposits. For deposits further out to sea a very recent and highly expensive method has to be employed, and derricks mounted on specially modified ships or erected on platforms constructed at sea must be used.

One such 'artificial oil island' is that lying some eight miles off the Mississippi delta. Its steel platform rests on foundations 50 feet high. It can withstand winds of 160 miles per hour and waves 30 feet high. It contains two stages, and the whole structure is more than 170 feet high. In addition to the derrick and the apparatus, the 'island' can comfortably accommodate some fifty men; it possesses fresh water and fuel tanks, provision stores and a generating station, and is linked by telephone with the mainland.

The problems to be solved in underwater drilling are many: winds and tides have to be countered, and fixed immersed pipes must be laid to bring the petroleum to the surface. New advances in offshore drilling equipment include drilling from a floating vessel, and completion of oil wells on the ocean floor by remote control from the surface. A new type of drilling vessel incorporates an automatic pilot (similar to that in aircraft) to keep it in position without the use of anchors while drilling goes on. Adaptable to depths as great as 1,000 feet these advanced methods will allow exploration of the continental shelf, which is a potentially rich field for exploration. It is believed that continental shelves have a joint surface area equivalent to that of Africa, and that they contain twice as much petroleum as all the continents combined.

**Oil Flow.** Accuracy in drilling and a measure of luck combine to give a producer well in which the oil will rise and flow out. However, water and gas move through rocks with greater ease than oil. When pressure at the bottom of a well is lowered abruptly by permitting the oil to flow out rapidly, the overlying gas moves down and the brine below moves up, before the oil from the margins of the pool moves in toward the well. Then, instead of pressing evenly against the oil, the water beneath may rise up in a cone just under the well and it may by-pass and seal off large segments of the pool. The well quickly produces more water or gas than oil.

In the early days of petroleum, the underground factors were not understood and wells flowed freely until pressures were reduced. Today gushers occur only rarely, because oil men are careful to prevent gushing and its resulting waste. Oil flow is controlled by a choke valve at the top of the well, and gas and water press evenly on the oil to produce a flow from the sides of the pool to the well.

Oil movement is dependent on three different types of forces, or drives. These types are:

**Dissolved gas drive.** Oil always contains some gas dissolved in it. Because it is compressed it seeks to expand and carries the oil with it in the direction of pressure release, that is, toward the well hole. This effect is similar to the effervescent overflow which results when a bottle of warm agitated soda water is uncapped. Oil recovery is low when dissolved gas is the only drive, because the oil-gas ratio cannot be controlled and pressure drops quickly. Recovery may be less than 20 per cent.

**Gas-cap drive.** In most wells dissolved gas is supplemented by a quantity of free gas which has collected above the oil because of its lower density and greater buoyancy. The compressed gas cap expands into the porous oil rock as pressure is eased and drives the oil into the wells located near the bottom of the oil layer. By producing only from beneath the gas cap, a high gas-oil ratio can be maintained until the producing life of a field reaches exhaustion. The recovery in a gas-cap field is usually 40 per cent to 50 per cent.

**Water drive.** Beneath the oil there is normally a large quantity of salt water and as pressure is released it, too, moves into the oil rock and flushes the oil out before it. This is the most efficient of the natural drives. If the oil removal rate is handled efficiently, flushing out the edges of the pool as well as the centre, recovery under certain favourable conditions may be as high as 80 per cent.

Other drives, generally of minor importance, may also be present. In an oil column several thousand feet long gravity drive may become very important. In fields utilising this, almost every

274

*Top left*
An off-shore oil well in the
Gulf of Mexico.

*Bottom left*
This charge of nitramon will be
detonated later by the shooter.
The balloon acts as a marker.

*Right*
A charge of explosives being
detonated by shooter. A
recording device aboard the ship
records the accompanying
sound waves.

drop of recoverable oil can be eventually brought to the surface.

As soon as a new field is opened the petroleum engineer sets about determining which of the drives or combinations of drives will yield the energy to move the oil. Geological and geophysical surveys provide data which show, in a general way, the kind of structure to be encountered, and give clues to the type of drive possible. More complete and more accurate data come with the sampling and testing that accompany the drilling.

**Promoting flow.** A few oils are too heavy and viscous to flow into wells and must be heated below ground to increase fluidity. Generally this is done by lowering a heating unit into the well and if this increases the flow by five or more barrels per day it is considered successful. The heating may be done electrically or by circulating hot water. Sometimes heating *in situ* is done by a 'hot foot'. In this technique the oil is chemically ignited at the bottom of one well, the fire is maintained by forcing compressed air down the hole, and the heat of combustion drives the oil through the rocks into the surrounding wells.

Sometimes because of clogging by wax, asphalt, etc., the wells may cease to flow. New channels are then opened by controlled explosions of nitroglycerine or by pumping in strong acids. Thus by application of modern methods a much larger proportion of oil is extracted from its reservoir than in earliest times. The amount left behind is only about 20 per cent; thirty years ago almost 75 per cent was considered unrecoverable.

**Secondary recovery.** In addition to regular production, many fields undergo secondary recovery. After a field has ceased to produce by natural flow and by pumping, it may be 'repressed' by restricting or stopping output while air or natural gas from which the gasoline content has been removed is injected until the desired pressure is reached and production is renewed. The field may also be artifically flooded by pumping water down a number of the old wells into the reservoir. This flushes some of the remaining oil ahead of it into the other wells.

Well shooting is the common preliminary to many of the secondary recovery system. Acidising is practised too. This consists of pumping acid (generally hydrochloric, with an additive to deter corrosion of equipment) into the reservoir. Developed early in the 1930s, this method led to a secondary flow of oil in many limestone fields which occasionally equalled or exceeded initial production. Hydraulic fracture, the newest technique in the field of secondary recovery, was applied 30,000 times in the first five years after its introduction in 1949. In this method a fluid (water, crude oil or acid, or combinations of all three) is pumped into the reservoir. The liquid, carrying with it a 'propping agent' (usually sand), is pumped into oil formations under the surface at pressures as high as 10,000 lb. per square inch. This produces new cracks and enlarges existing ones in the reservoir rock. As the fluid is removed, it leaves the sand behind to prop open these fractures.

These are the techniques used to give new life to otherwise dead fields and to increase initial production of new fields. 'Shoestring' and shallow sand pools in Kansas and Oklahoma, whose production had fallen to 4 barrels or less a day, were restored to a daily output of 50 to 100 barrels by their application. Initial but short-lived daily production figures in some old wells have been as much as 600 barrels. In the Permian Oil Basin of the United States, 13,000 fracturing operations increased the known recoverable oil reserve of the area by 120 million barrels.

**Other products.** The oil and gas reach the surface as a frothy mixture which is directed into a separation tank. The oil is pumped into storage tanks and the gas is treated in natural gasoline plants for the removal of any remaining liquids. The dry gas is used in many ways: some is burned to supply power in the oil field; some is pumped back to help maintain pressure in the reservoir; the

balance is sold or, where unsaleable, is burned off. In the last few years natural gas recovery has had an increased importance because it is being used more extensively in the manufacture of chemicals and is replacing artificial gas in the home.

In addition to natural gas, oil fields have other products. Hydrogen sulphide is the most abundant of the impurities. A small quantity in natural gas is desirable, for its disagreeable odour is a warning of leakage, but when large quantities are present the gas is hard to market. Such 'sour gas' is used to manufacture carbon black or is processed for sulphur recovery. Sometimes helium is present in a commercially useful quantity. Carbon dioxide is found in some fields and is recovered and used in the manufacture of 'dry ice'. In fields where it occurs in large quantities, the wells and surface pipes are thickly coated with frost, even in warm areas.

Salt water is produced in vast quantities, especially as an oilfield becomes exhausted. The corrosive action on pipes and and fittings is a major problem. In some pools sand comes up with the oil and this too must be disposed of.

From the storage tanks the petroleum is conveyed to the port of embarkation, or directly to the refineries. It is beginning its slow progress through the pipelines, the furnaces, distillation and reaction columns and refrigerators of the refineries.

In the search for and discovery of oil the sequence of events is, first locating a potential field, and then drilling a test well. But in this process there are a multitude of variables: geologic, economic and human. Each prospecting operation uses different methods.

The factors which determine whether a prospect is feasible are many. First and most important are the geological considerations: the trap, its type and size; and the probable producing horizon, its porosity and permeability. If the preliminary data are favourable further checks are made. The geological data are carefully examined, and geologist and geophysicist evaluate the prospect. Economic factors are next in importance to the geologic ones and may even outweigh them on occasion.

The discovery of an oil pool is not enough; the pool must be large enough to make a profit beyond all doubt. Small pools are often turned down by large companies who operate on the basis of millions of barrels per year and whose expenses therefore are proportionately high. The small prospects may be workable however by individual operators whose requirements and expenses are smaller.

The greater the expected return, the less important it is that all known geological conditions be favourable and the greater the amount of money which may be invested in preliminary surveying and testing. The economic factors that have to be considered in judging a prospect are numerous.

**Price of crude oil.** The price of the petroleum at the well and of gas or any accessory product is a major factor. Usually the entire income of a field is derived from the crude oil. All costs incurred up to the time when the oil enters the pipeline must be subtracted from the sale price in order to calculate the profit. Sale price at the well varies with demand, with the type of oil, distance to be

transported and type of transport available. These factors control the price, and each is considered when a new prospect is explored. For example: a prospect near a company's own pipelines is more desirable than a remote prospect and this may to some extent outweigh geological factors; prospects in a heavy oil area are less desirable than those in a light oil area — especially in regions with a high demand for gasoline. Exploration activity varies in direct response to the price of oil and gas; as the price rises, small prospects and prospects with doubtful data or high lease costs may begin to look more attractive.

**Geological and geophysical costs.** The outlay involved in locating a prospect may fluctuate greatly. Occasionally all the necessary data can be found in technical journals, records of government surveys, or university geology theses; in this case the costs are small. On the other hand, an area may have to be explored by all known methods and may involve years of research. Weeks, months and even years of work by seismic crews sometimes result in the location of only a few questionable prospects. Other types of survey may precede or follow; their cost is so tremendous that, comparatively, the cost of drilling is negligible.

**Lease costs.** These are usually: initial cost of the lease; royalty on the oil produced or a proportion of oil set aside for the owner; and the rental for maintaining the lease. Initial costs vary, rising in areas close to producing fields.

**Drilling costs.** These, too, vary — mainly according to depth to be drilled but other factors exert some control: the nature of the rocks—hard or soft; the number of tests to be made; the preparation of site; distance from water and other supplies; the method of transport; cost of building a road to the site. Generally, drilling a shallow prospect of, say, 2,000 feet is more attractive than drilling a deep one of 15,000 feet or so. A shallow prospect needs fewer favourable geologic factors, but a deep prospect with indications of a large potential is more attractive to large operators than several small good prospects.

**Production costs.** These must be anticipated for the life of the field and are an important consideration. Undesirable factors, such as loose sand in the oil and large amounts of hydrogen sulphide gas, both of which damage equipment and necessitate costly repairs, may cause a prospect to be abandoned. The questions must be asked — how good is the natural drive and how much is the probable recoverable reserve? The cost of the royalty is another relevant production factor. In the United States the primary royalty is about 12·5 per cent, though this may vary, expecially if brokers are handling the deal. In some foreign countries royalties and other profit-sharing expenses or taxes may consume 50 per cent of net profit and more. Early closing of a field can result from royalties overshadowing other operating expenses.

In addition to all these factors there is also the personal element, for the operator works on past experience. One man may be cautious, another may take great risks. Personal factors may take the form of deeply rooted geological prejudices — often opinions with no real evidence to support them. 'No source rocks' kept operators out of the very productive east Texas region for years. 'Red sedimentary rocks and oil don't occur together' slowed down development of petroleum in rich central and western Oklahoma.

Freedom of exploration and unrestricted production lead also to better results. United States is an example of this, where the tremendous production is due to almost uninhibited search and drilling. The end result of freely testing old and new geological ideas is more and bigger discoveries. Many sites originally tested

*Left*
The oil rig *Orion* in the North Sea. To date this rig has discovered two oil fields below the bed of the sea.

*Below*
Butane storage spheres at Fawley Esso refinery.

and shown as dry have, under a new operator with a different or more modern slant, been drilled at a slightly different spot and proved productive.

## PETROLEUM DERIVATIVES

Petroleum is too complex a mixture to serve a single purpose. In order to make it useful it is necessary to separate it into its numerous constituents, and to fractionate it into groups of hydrocarbons having certain average properties suitable for particular purposes. The aim of refining is to achieve this fractional distillation and to purify the products until their composition is exactly what is required by the consumer.

Before studying the methods of refining it may first be useful to set out some of the products sought. These petroleum derivatives are extremely numerous and frequently have most unexpected uses.

**Petroleum gases.** The lightest gases, methane and ethane, normally serve as fuels in the refineries and may be burnt as waste products if not saleable. Propane and butane, which are denser, are first scrubbed and then liquefied under pressure and stored in tanks before being placed in steel bottles for industrial and domestic use. Butylene, produced in the course of various refining processes, is the raw material used in the synthesis of rubber. Several methods have been introduced in recent years to produce gas from liquid refinery products including light distillates and naptha. Experiments are now under way to extend this to crude oil.

**Motor fuels.** Known as petrol in England and gasoline in America, these are the ideal fuels for spark-ignition engines. The different needs of aircraft, cars, motor-cycles, vans and light tractors are matched by the various qualities of the fuels produced in the refineries. Aviation fuels are those which demand the highest standard compounds. As we shall see presently their octane

index must be particularly high. A product known as lead tetraethyl is also added to the motor fuels to improve their performance in automobiles.

**Special oils.** This is the group obtained immediately after the motor fuels during distillation. These oils have many uses: lighter fuel, dry cleaning, solvents for cleaning products (wax polish, floor polish, metal polish), solvents for paint, solvents for rubber.

**Kerosene.** Formerly kerosene or lamp-oil was used only in lamps, oil-stoves, incubators, railway signals, etc. Today it is produced for use in spark-ignition and jet engines. In addition, it is a vehicle for toxic products (pyrethrum, D.D.T.) in insecticides and fungicides.

**Gas-oils.** Gas-oils have a slightly brownish colour and a moderately strong odour, and are slow to evaporate. Diesel engines have made them outstandingly important as fuels for trucks, heavy tractors, motor railway engines, locomotives, trawlers, lighters, and industry. They are also cracked to obtain gasoline, and in the enrichment of hydrocarbons in coal-gas.

**Lubricating oils.** These are excellent for all types of engines and machine tools, etc. A wide range of lubricating greases and oil for printers' ink are made from them.

**Petrolatum.** Specially refined and deodorised, these oils are used in salves, cosmetics, rust preventives, and even cable coatings.

**Fuel-oils.** Distillation residues, the fuel-oils are semi-fluid, blackish, sooty products used as substitutes for coal in industrial furnaces and central-heating boilers, in locomotives and in ships. They offer various advantages over coal; ease of transport and storage, controllability and cleanness. On combustion they leave almost no

281

waste. Lastly, they permit the liberation of the soft coals so indispensable in the iron and steel industry.

**Petroleum waxes.** These are the white products (paraffin, vaseline) extracted from distillation residues. Their uses are extremely varied: candles, wax polishes, chewing gum, pharmaceutical products, textile waterproofing, leather dressing, impregnation of packing paper, 'tropicalisation' of products for warm and humid climates, detergents, canning and etching.

**Petroleum Coke.** One of the most useful by-products of oil refining processes, petroleum coke, can be obtained in two forms, as thermal (green) coke which is sold as low ash fuel, and in the calcined form which is used in the aluminium and steel industries and also in the nuclear industry.

**Asphalts.** Extremely viscous or semi-solid residues, asphalts' adhesive and impermeable qualities make them especially suitable for paving roads, coating roofs, and impregnating wood; they are also used as electrical insulators.

**Carbon blacks.** The final residues from petroleum distillation are carbonaceous substances known as carbon blacks. They are used in the manufacture of printing inks and photograph records. Added to rubber, they prolong its life. A tyre not reinforced with carbon black has a very short life, say, 2,000 miles: the same tyre containing carbon black will last some 30,000 miles.

These, then, are the direct petroleum derivatives. The list of secondary derivatives, the countless articles manufactured from special fuels and oils in every conceivable industry, is still longer.

## THE FORBIDDEN CITY

The greatest precautions are taken at refineries to make sure that unauthorised visitors are kept out, and that even authorised ones are not admitted without making sure that they carry neither lighter nor matches.

Another characteristic feature of the petroleum industry is its discretion. At no time does one catch a glimpse of the 'black gold' which flows in a continuous stream through hundreds of miles of pipelines. The workers, too, remain practically unseen. The visitor merely catches an odd glimpse of the chemists in their laboratories, and of the checkers taking constant readings from countless dials which give remote control of the process taking place in the closed vessels.

A modern refinery extends over several acres and is divided into sectors or units, each devoted to a particular process or to the storage of a certain product. There are fractional distillation, reforming, cracking, rectifying, scrubbing, and transforming units.

**Atmospheric or 'straight-run' distillation.** When crude oil emerges from the storage tanks it is pumped into the fractional distillation unit, the refinery's tallest unit. Some of its columns are reserved for atmospheric distillation, others for vacuum distillation.

Heated to about 680°F. in the gas furnaces, the petroleum first reaches the lower third of the first atmospheric column, which is divided into compartments by superposed, perforated plates. Fractional distillation then begins. The lighter and more volatile hydrocarbons rise to the upper part, while those that are heavier and less volatile collect in the lower part. Exchanges take place between the various compartments. On rising, a volatile mass may shed its less volatile elements. These sink down again, only to be gathered up once more. Equilibrium is never attained. Strictly speaking, then, only gases and gasolines are recovered from the top of the first column. The residue, taken from the lower part, is reheated and injected into a second atmospheric column whose primary purpose is the extraction of naptha, kerosine, and a part of the gas-oil.

**Vacuum distillation.** The new residue at the bottom of the second column is a mixture of gas-oil, fuel-oil and bitumen. This must be distilled under vacuum in a large fractionating column, having been reheated to a temperature of 680°F. Under reduced pressure

the hydrocarbons dilate and separate more completely than under atmospheric pressure. Gas-oils, a series of distillates corresponding to fuel-oils or kerosene, and a residue rich in asphalts are obtained.

Emerging from the distillation unit we thus find the various groups of hydrocarbons mentioned in connection with the petroleum derivatives which have been roughly separated: gases, gasolines, naphthas, kerosene, gas-oils, fuel-oils, asphalts. All in impure forms. The gases still contain gasoline, and the gasolines contain gases. The naphthas contains gasoline and kerosene, and so on. Because of this and the small amount of gasoline recovered (less than 25 per cent of the crude) it is necessary to pass these products through special rectifying and filtering units.

**Filtration of gases.** The gases given off from the top of the first atmospheric column or recovered during subsequent operations must be processed to eliminate methane and ethane on the one hand, and gasoline on the other, and to obtain separately propane and butane. This end is achieved with a series of steam-heated rectifying columns where fractionation takes place. After filtration the propane and butane are liquefied under pressure and stored in tanks — cylindrical tanks for propane, with pressure stablised between 2 lb. and 4 lb. per square inch, and spherical tanks for butane with pressures of $\frac{3}{4}$lb. to $1\frac{3}{4}$lb. per square inch.

**Thermal cracking and reforming.** A twofold problem arises in connection with the gasolines. To meet the ever-increasing demand by simple distillation would require such large quantities of crude oil that the market would be upset with the accompanying over-production of middle and heavy products. Yet the gasolines derived from crude oil (straight-run gasoline) and those obtained by filtration of the gases are poor-quality, low-octane spirits. It is possible to use them in fairly strong spark-ignition engines, but they are quite unsuitable for motor-car engines and, particularly, for aero-engines. Under heavy pressure in the cylinders their combustion is irregular. Peroxides form and explode, causing 'knock'. The power developed by these fuels is smaller, and their action is harmful to engine parts. Knocking can be prevented by adding hydrocarbons of the aromatic series, lead tetraethyl, or octane.

Thus, the twofold problem consists in producing from a fixed quantity of crude oil an ever-greater proportion of gasoline, with an increasingly high octane index. We have said that kerosene. gas-oils and fuel-oils can be partly 'cracked' into gasolines, in other words their heavy hydrocarbon molecules can be broken down into light hydrocarbon molecules. This is done by placing them in coil furnaces, raising their temperature above 750°F., and subsequently conveying them under a pressure of 10 lb. to 17 lb. per square inch into reaction chambers, where the molecules are broken down. The gasoline yield from this thermal cracking is 55 —70 per cent for kerosene, 40 — 55 per cent for gas-oils, 30 — 40 per cent for fuel-oils. The residues are bitumens and carbonaceous products similar to coke.

Another method is thermal reforming. Whereas cracking produces gasoline from distillation products which do not originally contain any, reforming simply increases the octane index of existing poor-quality gasolines. The starting-point for this particular process is generally supplied by the naphthas, products obtained in atmospheric distillation. The temperatures and pressures required are higher than those needed for cracking.

From the straight-run, cracked and reformed gasolines the refiners compound the different mixtures sold as proprietary brands.

**Catalytic cracking.** The preceding gasolines and super motor-fuels are too unstable and too readily oxidised for use in aircraft. The refiners have accordingly evolved new processes which yield greater quantities of gasoline by cracking the heavy products, and also increase the octane producing isobutane, isopentane, isoheptane and iso-octane in branched-chain types and in the cyclic group. Both 'isos' and 'cyclos' are excellent for aircraft engines.

Cracking of this kind can take place only in the presence of catalysts, that is, substances capable of starting chemical actions

The regenerator of the catalytic cracking plant of the Pernis     Refinery in the Netherlands

without themselves undergoing appreciable change. Without discussing the various processes or the catalysts employed in any detail, let us for a moment deal with catalytic cracking proper, a process which modifies the properties of the gasolines and increases their quantity by cracking gas-oil. The process was developed in 1933 by a French engineer, Houdry, who thus made possible the tremendous production of aviation fuels during the Second World War. His patent is still widely applied in refineries throughout the world. The method consists in heating to 900°F. the substance to be treated, and circulating it in reaction chambers where the catalyst (aluminium silicate) is disposed in an even layer on the surfaces touched by the gas-oil vapours. In another process, the catalyst is in powder form and circulates with the gas-oil. Its action is thus much stronger. Carbon settles on the catalysts, and must be burned in contact with air for the product to be regenerated.

**Treatment of the white products.** The principal products discussed so far, whether derived by direct distillation, reforming or cracking, contain impurities which make them unsuitable for consumer use. These must be eliminated before storage, and a section of the refinery is given over to filtration which makes impurities soluble. At this stage sulphuretted hydrogen and the evil-smelling sulphur compounds (mercaptans) are removed.

**Treatment of the black products.** This name is generally given to the products obtained by vacuum distillation (fuel-oils and asphalts). From them the refiner must extract lubricating oils, petroleum and paraffin. The processes employed are varied. Certain methods consist in treatment with propane, furfural (a substance extracted from oat flakes) or phenol; the object of the treatment is to remove the aromatic hydrocarbons, the resins and gums. The oils are decolourised by filtration in columns of activated earth. Paraffin is removed by various solvents; one that is frequently employed is methyl-ethyl-ketone, commonly known as M.E.K. The paraffin is subsequently extracted by chilling and by centrifuging. White and crystalline, it is thrown against the wall of the revolving machine, and later removed by counterpressure in the upper part of the apparatus.

## HOW PETROLEUM IS DISTRIBUTED

**Pipelines.** How is crude oil flowing from a well conveyed to refineries and its products distributed to the consumer?

As petroleum is a liquid, it has always been possible to transport it by pipeline. As early as 1865 a first pipeline, 2 inches in diameter and 5 miles long, was constructed between Oil Creek and Kittaning for the transport of the Pennsylvanian petroleum. It carried 800 barrels a day. In 1879 the first long-distance pipeline was constructed, between Corryville and Williamsport, Pennsylvania, and later continued as far as Bayonne, New Jersey. It had a six-inch diameter and could deliver 10,000 barrels a day. Since then pipeline technique has made tremendous progress. Conveying oil today is a complex system of pipelines, tank-ships, barges, railroad tank-cars and tanker trucks, distributing crude oil to refiners and the finished products to distributors, supply points, service stations, airports and homes. The mainline of the network is the pipeline. Steel pipes up to a yard in diameter run for thousands of miles through the most densely populated areas, as in America, or across the loneliest of deserts, as in the Middle East, Powerful machines and experienced technicians make pipeline laying a straightforward matter today. Several miles of pipelines can be laid through the middle of the desert within the space of a single working day.

The initial step is to plan the pipeline's route by detailed study of maps, air photos and ground surveys. Once its course is determined hundreds and sometimes thousands of land owners must be approached to obtain the right of way. Then come the surveyors, followed by the bulldozer crew and construction men to mark, clear and grade the route. The bulldozer is followed by a second powerful machine, the ditching machine, which gauges out the deep trench that will take the pipeline. The sections of pipe are aligned end to end along the trench, where teams of workmen couple and weld them. A second team cleans the pipes and coats them first with cold and then with hot enamel. Others wrap them in tar-lined paper. Next tractor-mounted derricks lift the serpent-like structure in 20 to 40 foot sections and lower it gently into the trench. Finally trench and pipe are covered in with earth and levelled off. Most pipelines are buried, but can be found above surface level.

Oil does not flow through the pipeline of its own accord. Slight differences in level between starting- and finishing-points and the viscosity of the product make spontaneous motion difficult; a pump station must be then installed at intervals along the route. In level country the stations are usually 35 to 75 miles apart; in hilly country they are even closer.

In the field crude oil is carried through flow lines into field storage tanks. Gathering lines connect these with a main line which carries the oil to a waterside terminal or a refinery. Gathering lines are usually small, averaging two inches in diameter. The main trunk lines may be a yard across. In the United States the term 'Big-inch' applies to any pipe over one foot in diameter.

Crude oil and finished products are conveyed through different pipelines, though in each case separate shipments are fed in one after the other without much intermixing. In this way a continous stream is formed and maintained. A kerosene buffer is set between certain finished products to prevent mixing.

How do the shippers know when a particular shipment from a refinery has reached its delivery point? Speed of flow and point-to-point distance are known from pumping station data. Travel time can then be calculated. Shortly before expected arrival time, samples are drawn and tested, and in the case of crude oil the specific gravity is checked. This goes on until the expected product arrives and is directed into the waiting tanker. This too is done by timing; for example, 250,000 barrels represent a stream 330 miles long in a 16-inch pipe and take 4 days and nights of continuous pumping to deliver.

Pipelines gradually acquire a sludge coating which is periodically cleaned off by a 'go-devil', a device that operates a number of whirling arms or flanges that scrape the inside of the pipe.

'Pipeline walkers' monitor the route, checking for leaks and exposed lines. Low-flying aircraft are widely used for this, reporting on the easily spotted breaks and exposed pipes.

Today's major pipelines are found in the Middle East, the United States, and Russia. The network of the Iraq Petroleum Company comprises two double lines set out in the shape of a Y, starting at Kirkuk and terminating at Tripoli (Syria) in one direction, and at Haifa (Israel) in the other. Their respective lengths are 530 and 620 miles. A village has grown up round each of their dozen pump stations, complete with radio station and aerodrome. A pipeline of larger diameter is in course of construction between Kirkuk and Baniyas. The British Petroleum Company has a giant pipeline 800 miles long and 36 inches wide, with an annual flow of 25 million tons, conveying petroleum from Iran and Kuwait to the port of Tartus. Further to the south lies the famous Tapline (Trans-Arabian Pipe-Line) owned by the Arabian American Oil Company. This giant pipeline runs right across the Arabian desert, a distance of 1,100 miles, and transports petroleum from Saudi Arabia and Bahrein to Saida. In this way several ports along the coasts of Israel and Syria annually ship almost 50 million tons of crude oil from the immense reserves of Persia, Iraq and Saudi Arabia. The remainder of the production is shipped from various ports of the Persian Gulf and must make the lengthy journey around Arabia by way of the Suez Canal. This shows clearly the economy afforded by the pipelines, despite the high cost of initial installation and operation.

In densely populated areas where road, rail and river networks are developed to the maximum, the same problems do not arise. In fact, large-scale pipeline systems for crude oil and finished products have proved profitable only in the United States, where over 400,000 miles of pipeline now link the great petroleum areas of the south-west with the industrial regions of the north-east and north-west. The finished product lines are a relatively recent development, dating from around 1930. They still comprise only a small portion of the vast pipe network but they employ the most modern techniques — much of the sampling is done automatically and the valves are worked by electric remote control.

Where oil-tankers have tonnages that prevent them from

Oil and water pipes, all leading from Esso's refineries to the marine terminal at Southampton.

entering certain ports, they anchor at sea and are loaded and unloaded by sea-lines (pipes running out to them). Outside the United States more oil, both crude and refined, is moved by tanker than by any other means.

**Storage.** A few words, finally, about the storage tanks for petroleum and its derivatives. In the refineries and ports these tanks must have the greatest possible capacity, limited only by conditions of construction. The commonest are the cylindrical tanks with fixed roofs for products of slight volatility (petroleum, fuel-oil) or with flexible roofs for products subject to constant volume variation (gasolines). The tanks for liquefied gases, and in particular for butane, which is at a lesser pressure than propane, are generally spherical. This shape offers minimum surface area for the maximum volume.

Lastly, tanks shaped like drops of water (Caquot tanks) may be used in some areas for storing gasoline, gas-oil and fuel-oil under pressure. Whatever their shape, tanks are constructed from welded sheet-metal plates; capacity is between 10,000 and 20,000 tons.

## THE WORLD'S DEPOSITS

In its world distribution oil is quite irregular and unequal. Roughly 77 per cent of world oil has been produced by the western hemisphere and 23 per cent by the eastern. North American production has run to about 67 per cent, of which 63 per cent has been United States production.

The major petroleum deposits of the world are found in sedimentary basins and troughs, generally in the vicinity of land-locked seas occupying intercontinental depressions in the Earth's crust. The major areas of this type are: the neighbourhood of the Mediterranean, Red, Black and Caspian Seas and the Persian Gulf, which occupy the depressed area between Africa and Eurasia; the Gulf of Mexico and the Carribean Sea, lying in the land-encircled depression between North and South America; the environs of the shallow, island-dotted seas lying between Asia and Australia. Much remains to be explored in the last region. An area almost unexplored but showing much surface promise is the land-locked Arctic Sea, occupying the depressed segment between North America and Eurasia. Many smaller troughs and sedimentary basins also contain petroleum deposits.

The United States have been the greatest petroleum producers in the world since the mid-nineteenth century. Twenty-six out of the fifty States possess fairly rich deposits. Texas, Oklahoma and California alone supply 54 per cent of the total American production. After the United States come the countries of the Middle East with the extraordinary rich deposits of Persia, Iraq and Saudi Arabia lying round the Persian Gulf in unbroken line, not forgetting those of Kuwait and Bahrein. Although exploitation in Persia began only in 1908, and in 1945 in Kuwait, total production at present approaches 338 million tons and makes the Middle East the world's second biggest producer. South America comes third, thanks chiefly to the contribution of Venezuela. Russia, in its western part, occupies fourth place in world production. Despite the resources of the Urals and of Siberia, which are believed to be considerable, it is still the Caucasian deposits which provide the essential output. Next Africa, Indonesia (Borneo, Java, Sumatra) and certain Central European countries add their modest contribution to the giant production of petroleum. The figures for 1963 were as follows:

| North America | 452·6 | million tons | |
|---|---|---|---|
| (United States | 414; | Canada 38) | |
| Middle East | 338 | million tons | |
| South America | 230 | " | " |
| U.S.S.R. | 227 | " | " |
| Africa | 52·250 | " | " |
| Far East | 31·1 | " | " |
| W. Europe | 18.3 | " | " |
| Total | 1,350 | million tons | |

While current production is one matter, future production is another, and a very important one. For this reason the geologists are striving to assess the amount of the world oil reserves, careful to distinguish between proved and probable reserves. Proved reserves, which alone are certain, at present promise 20,000 million tons, enough to meet the present level of demand for another two centuries. Curiously, reserves in various parts of the world in no way corresponds to present production. North America finds itself easily displaced by the Middle East which, from all points of view, is the undisputed petroleum centre of the future. Everything contributes to it: the extensiveness and richness of its deposits, and proximity to the larger consumer countries. Moreover an oilwell in the Middle East produces an average of 1,000 tons per day, whereas corresponding production in Venezuela is only 30 tons, and in the United States $1\frac{1}{2}$ tons. Finally, the development of the system of pipelines will soon make the entire production of the

Middle East accessible in the Mediterranean itself, through a string of ports along the coasts of Israel and Syria.

As well as classification according to current and also future production, there is a third, based on each country's consumption of petroleum products. Here again the United States lead. In France only one-quarter of the total energy consumed is derived from petroleum; in the United States the proportion is more than one-half. Extensive mechanisation in America, particularly in agriculture, accounts for the high total. The largest petroleum producer, she is also a major market. Her exports are offset by imports, so that she is actually an importing nation.

Petroleum is found in rocks ranging in age from the Pre-Cambrian to Pleistocene. It is as irregular in its time distribution as it is in its geographic distribution. Some geologic periods, such as the Pennsylvanian, Cretaceous and Tertiary, contain large quantities; others, the Cambrian, Silurian and Jurassic, almost none at all. Conditions for deposition and accumulation have been more favourable during some periods than others.

## OIL IN FRANCE

France is in a paradoxical position. Its numerous sedimentary regions (the Paris basin, the Aquitaine basin, the Rhone valley, Limagne, and Alsace) theoretically ought to contain petroliferous deposits. Several of the regions do, in fact, offer much surface evidence in the form of bituminous shales, seepage of crude oil and the occurrence of natural gases. Yet effective exploitation is limited to Pechelbronn-Mulhouse in Alsace, Lacq and Saint-Marcet in Aquintaine, and Parentis-en-Born in the Landes. It is too soon to say whether there is a real lack of petroleum or whether the search for it is inadequate. Systematic prospecting is of comparatively recent date in France, and the number of drillings made is still very small. Only heavy capital outlay devoted to this purpose will produce certain results. In this respect the example of the United States is both encouraging and discouraging. If the United States do indeed lead in the production of petroleum they owe their position largely to the tenacity with which they have persevered in their drilling programmes—over three million holes since the beginning of the century (470,000 are working today).

The oldest oilfield in France is that of Pechelbronn (Alsace) which extends roughly from Pechelbronn as far as Mulhouse, in a sand dating from the mid-Tertiary period. In the absence of any gas to force it to the surface, the oil has to be pumped out. Exploitation is by means of wells and galleries giving access to the lenses of petroliferous sand. It has not been possible to replace this antiquated method of exploitation successfully by drilled wells, although these have been installed since 1882. A new deposit was discovered in 1953 at Staffelfelden, to the north of Mulhouse. The crude oil is refined at Merkwiller, and the products are subsequently sent down the Rhine as far as Strasbourg for storage.

The fissured sandstone of the Cretaceous Period at Saint Marcet in France yields what is called a wet gas, one that is mixed with easily condensable gasoline. The entire reserves are estimated at 140,000 million cubic feet and it is probable that other similar deposits will be found locally. It is probable that by that time other similar deposits will be in production. It is hoped Saint-Marcet is merely a pointer to the future in this region at the foot of the Pyrenees.

Further west, in the Lacq, reserves in the Basses-Pyrenees are estimated at three million tons. The petroleum of this field is the heaviest kind, containing only 10 per cent gasoline and larger quantities of gas-oil and fuel-oil. The more recent deep wells seem to offer only gas. Aquitaine, then, appears to possess a fairly shallow petroleum deposit (2,000 feet) and a very deep gas deposit (13,000 feet).

Another company, Standard Francaise des Pètroles (Esso-Standard), operates in the Landes. Since 1951 some 7,000 square miles have been prospected. So far only one productive anticline has been discovered, that of the Etang de Biscarrosse.

Petroleum is also being prospected in French West Africa, in Morocco (around Rharb), Algeria and Tunisia. The area of North Africa in general has in recent years brought a number of discoveries, possibly of sizeable extent. Currently France receives crude oil from Iraq (via Tripoli), Kuwait and Venezuela. Its principal oil ports are Lavéra, Sète, Bordeaux, Donges, Le Havre and Dunkirk The petroleum is then transported by pipeline to refineries situated near La Mède, Berre and Lavéra.

Le Havre is the starting-point of a 150-mile pipeline which serves the refineries along the lower course of the Seine as far as Rouen, at Gonfreville, Port-Jérôme, Notre-Dame-de-Gravenchon, La Mailleraye, Petit-Couronne and Petit-Quevilly.

## OIL IN BRITAIN

Oil and gas shows at a number of places in Britain led to extensive exploration by the D'Arcy Exploration Company beginning in 1936. The shows occur in the south, for example, as oil seepages in sandstones at Osmington Mills near Weymouth, as impregnated sands near Pevensey, Worbarrow and Lulworth Cove, and as natural gas at Heathfield in Sussex; in the Midlands as seepages near Coalport, in the peat-bogs near Formby in Lancashire, and in numerous coal mines on both flanks of the Pennines; and in Scotland as seepages in shales to the south-east and south of Edinburgh.

In the first place borings were made in known anticlinal structures which might be oil-traps, such as Portsdown near Portsmouth, Kingsclere near Newbury, and Poxwell near Weymouth. These were unsuccessful, while others at Dalkeith in Scotland and Aislaby in Eskdale yielded some natural gas. Later, attention was directed towards determining possible oil-traps in Carboniferous rocks beneath the cover of Permian and younger strata in Lincolnshire and Nottinghamshire. Geophysical exploration (mainly seismic) revealed possible anticlines, one south of Lincoln and another at Eakring in Nottinghamshire. The latter was drilled and, in 1939, oil was struck in a sandstone of the Carboniferous series at 1,912 feet. The first well on test gave a field of 12 tons a day. The outbreak of war added impetus to the search and numerous drillings were made and wells initiated. Several anticlinal structures were proved but the field is restricted in area, and from more than 250 wells drilled the production in the first few years averaged 80,000 tons a year.

The only other field to yield oil was that at Formby, in Lancashire, where oil was struck at 125 feet. The trap is associated with a fault in Triassic sandstones capped by impermeable boulder clays of Pleistocene age. More than fifty shallow wells have been drilled in this small field which, in the 1940s, was producing an average of 2·5 tons of oil a day but the supply ended in 1962.

Although the results of extensive prospecting have so far been disappointing, the search for oil in Britain continues, and with some success in North Sea areas.

## OIL IN THE UNITED STATES

Currently the world's leader in production and consumption is the United States. Its leadership in production will probably be taken over in the future by the Middle East and, in time, by the U.S.S.R.; South America and Africa, too, will move ahead.

The search for petroleum in the United States is restricted only by the rules applying in other fields of enterprise. Anyone can look for oil and drill for it if he has enough money. The main regulatory measures concerning petroleum are directed toward conservation of reserves and economic balance. There are laws which cover, pro ratio, the amount of petroleum production allowed per well, per group of wells or per acre.

There are many types of company in the American oil industry: independent producers who produce and sell only crude; independent refiners who purchase crude in the open market, refine it and market the finished product; integrated organisations covering all aspects from exploration to retail sales; and organisations making foreign markets their main interests. In addition to these major groups there are independent service companies who explore, drill, test and evaluate property under contract. And there are also oil organisations which search for favourable territory.

There are about twenty large petroleum corporations and thousands of small ones — the large companies working more efficiently, the small taking on propositions that are impractical

The 4,200-foot jetty at the Southampton marine terminal can accept ships of 65,000 tons fully loaded.

for their larger competitors. Minor companies produce about half of the nation's crude and own about 65 — 76 per cent of all the wells. Most large companies are integrated, though they tend to specialise in one branch or another of the industry: production, perhaps, or refining and selling.

The United States has produced over 60 per cent of the world's total production, despite the fact that the country contains only 10 per cent of the world's petroliferous rock formations. This is probably because the industry has free competition and is unhampered by other forms of government control. Very poor showings have been made in the past in countries where oil was government-owned and government-produced, where outside explorers and investors were barred, or where stock control of a major oil company was government-owned. Today over $1\frac{1}{2}$ million workers are employed in the American petroleum industry (an increase of about 40 per cent since World War II, with a 50 per cent increase in output per worker over the same period).

Out of about eleven petroleum provinces three or four produce the bulk of the nation's oil:

**The Appalachian province.** The oldest oil-producing area is the Appalachian geosynclinal province where production is found in slates from western New York south-westward into Tennessee. Located here is the Bradford field of Pennsylvania, in which there are six different oil-producing horizons, all of Devonian age. The oil accumulation is governed equally by lithology and structure, occuring in a sandstone which intertongues with the shale and limestone of a gently folded structure. The oil of this province is paraffin base with a large content of high-grade lubricating oil constituents. The fields of New York and Pennsylvania are all undergoing secondary recovery, aided by modern machinery.

**The Gulf coastal plain.** This region includes parts of Texas, Arkansas, Louisiana, Mississippi, and the adjacent shelf. It probably contains the largest reserve of oil in the United States. Production comes from shallow and deep wells. The oil occurs in stratigraphic and structural traps in Tertiary rocks. Most of the production is from sandstones, but limestones and salt dome caps also yield large quantities.

**The Pacific geosynclinal province.** The important fields in this province are mainly in the south, in the area around Los Angeles. The petroleum occurs mostly in complexly folded and faulted Tertiary sediments. Much of it comes from great depths and from offshore pools which are tapped by directional drilling. The oils, except for the naphthas from the Kettleman Hills area, are some of the heaviest known. Some are heavier than water and many have to be heated underground to make them flow at all.

**Mid-Continent province.** This area includes much of Oklahoma, north Arkansas, Kansas, north Louisiana, and part of Texas. The small and often quite different units are lumped together because of geographic proximity. Production in the different sections may be from folds, domes, faults and other structural traps, and from stratigraphic traps. The crude is paraffin or mixed base and occurs in rocks ranging from Ordovician to Tertiary.

Other notable United States oil provinces include the Cincinnati Arch, producing paraffin oils from anticlines and lensing sands in Devonian, Mississippian and Pennsylvanian rocks; the East Interior Basin with a production similar to that of the Cincinnati Arch; the Rocky Mountain province, producing small disconnected pools and yielding asphaltic to paraffin base oils from anticlines and 'shoestring' sands both Jurassic and Cretaceous in formation.

# Hydro-Electric Power

There is no doubt that the pioneer of dam building was the beaver with its clever constructions of all shapes and sizes, some as much as 500 yards long and 10 to 12 feet high. Seen from above, the beaver's dam is rectilinear at points where the current of the river is not too strong; upstream, in stronger currents, the dam is convex. The beaver's materials are pieces of wood (at times of considerable weight) which are floated into place. Unable to secure the wood in position with stakes, it weights it down with stones and mud. Branches are then intertwined and the spaces between them gradually filled with a mortar of clay, sticks and small stones which are impermeable and resist erosion. If necessary, the beaver even digs spillways for its pools in time of flood, and canals for the transport of its materials.

It is certain that many of man's lake-dwellings of the Neolithic age were constructed on artificial lakes. Such are the terramare of Northern Italy and the crannogs of Ireland and Scotland. Much later, causeways and dykes were erected as protection against floods. Raised embankments were constructed to keep the rivers from flooding towns and irrigation works were built.

Over the centuries, as dykes and dams were designed and built to ensure a systematic distribution of river water, the water was made to serve as the driving force for mills. Dams for the mills of olden times were built obliquely across a river, raised the water level and led to a mill-race where the falling water set in motion the water-wheel which turned the millstones.

From the fourteenth century water mills in Europe were no longer used solely for grinding; they became the source of power for varied industries. The water-wheels were set side by side across the flow of the stream, the level of the latter being itself reduced in successive falls. The wheel, together with its mechanism, developed about 30 horsepower.

The idea of driving a generator and producing electric current by water power did not follow until much later. Finally from this came the extraordinary growth of hydro-electric power at the beginning of this century. Today, as we shall see, the need for hydro-electric power is critical in many areas and has brought about the construction of giant dams and reservoirs.

Irrigation and the production of electricity are but two of the purposes of dams. Another, that of navigation, is achieved by the division of an excessively fast-flowing stream into different levels separated by locks. These locks also enable shipping to cross the watersheds between two hydrographic basins or two oceans.

Irrigation, electrification, and navigation form a trinity, each with a different relative importance depending on the area in which it is applied. Whatever the circumstances, all dams constructed nowadays use the natural power of the water-fall to generate electrical energy.

Technical skill in hydraulic engineering has advanced enormously in the few years. The actual constructional work is preceded by long years of geologic, hydrologic, biologic, and economic investigation, and the majority of projected dams are first built in the form of models on which various reactions are studied experimentally since these cannot always be evaluated theoretically.

Hydro-electric generation is not without its opponents. They are the anglers, canoeists, and nature-lovers who view the marring of the rivers and beauty-spots with misgivings. The dams, they allege, lower the water level downstream and raise it upstream — to the detriment of the flora and fauna — and interfere with the migration of the fish which are not always able, even with the aid of fish ladders, to surmount the obstacles erected in their path. The problem was debated at the Congress for the Protection of Nature at Caracas, Venezuela, in 1952. The authorities concerned had little difficulty in demonstrating the essential need for the hydro-electric harnessing of running waters, and the evident exaggeration on the part of the opponents of hydro-electric power.

## GENERATION

**Dams and Reservoirs.** The harnessing of water power involves the construction of a dam across a river. A long dam, sometimes known as a dyke, is often built of stone or earth. Its foundations must be secured by a very wide base, while a clay centre or slabs of concrete applied as facing ensure its impermeability. Such dams, reaching anything up to 300 feet high and 12 miles long, are found in the United States and Russia. It should be noted that these are irrigation dams, used to generate electrical energy only as a secondary function. Power dams proper are normally built of concrete and are either solid gravity dams or arch dams.

A solid gravity dam is so named after its considerable mass which enables it to resist the pressure of the water. Seen from above, it is rectilinear; its cross-section is triangular. Owing to the quantity of masonry necessary and the high cost of construction the solid gravity dam is now generally rejected in favour of the arch dam which has a relatively thin wall (convex upstream when seen from above and in cross-section). The saving in masonry is considerable, yet the arch dam's resistance is equal to that of the solid gravity dam. In certain circumstances, a multiple arch dam is used. Here, the whole dam is concave upstream, but divided into a series of convex arches.

The most favourable site for the construction of a dam is a flat-bottomed valley with steep sides narrowing so abruptly that only a few hundred feet or even as little as fifty feet separate the two banks. If the narrowing is due to a zone of hard rock — as it generally is — it is possible to close the gap with a dam of some 600 to 1,000 feet, forming a supply reservoir of very great capacity upstream. The highest dam planned to date is that at Super-Dixence (Switzerland) which is to be 930 feet high.

The High Dam which — has been built at Aswan to harness the waters of the River Nile is a typical example of this type of rockfill dam. Egypt is fortunate in having quantities of sand and granite close to the canal site sufficient to build the dam.

**Generating Stations.** The purpose of the supply reservoir is twofold: first, to store the water from the melted winter snows and the spring and autumn rains in order to regulate the flow and the generation

The human figure gives some idea of the size of the enormous water gate at the Chastang Dam, France

of electricity; secondly, to create a high level in the reservoir so that the ensuing drop will create a suitable head for the operation of the turbines.

The hydro-electric power station may be situated at the base of its dam or incorporated in its structure. Here the effective power is created by the difference in height between the level of the water in the reservoir and the level of the water turbines inside the reservoir. The power station may, on the other hand, stand at a lower level, sometimes a considerable distance away from the dam. In such circumstances, the water from the reservoir is first conveyed to a point above the power station by a flume cut in the form of a tunnel into the side of the gorge, and is then introduced into a pressure pipe leading to the turbines. This is one way of increasing the working head.

In both arrangements just described the water leaves the power station by a tail-race and may sometimes be passed subsequently through one or more other plants downstream. The surplus water is drained off along spillways which start at the crest of the dam and are sometimes shaped like a ski-jump.

One must distinguish between high and low heads. The former are found in the upper sections of the valleys, where gradient is greatest but flow weakest. Conversely, low heads correspond to the lower reaches, where gradient is slight but flow considerable. In theory, at least, the result is the same.

A low-head installation is nearly always 'streamlined', in the sense that its flow is too great for part of it to be stored at the beginning of the water-catchment. Navigational considerations, moreover, do not allow lock-feeds to modify the flow to any great extent. As a result, in time of flood the plant may not be able to maintain a difference in level between the upstream water and the downstream water. The plant is then said to be 'drowned'. Paradoxical though it may seem, operation must cease in case of an excess of water.

Let us consider some of the serious problems which are usually found in the operation of hydro-electric plants.

Mud, sand, gravel, floating matter, drifting ice and fish may block the flumes. There is also a danger of the reservoirs gradually silting up. Consequently from time to time the flumes must be drained for inspection and cleaning. The flumes, which start some distance from the bottom, are fitted with gratings to hold back the largest objects.

'Water hammering' may occur in the pipelines as a result of variations in the operation of the turbines. This problem is at present being solved by fitting each head-race with a surge tank, a kind of vertical pipe into which, at each closing of the turbine gates, the water is released into instead of reversing and flowing back into the race.

Variations in current consumption may occur. Less electricity is consumed in the summer than in the winter, yet the potential production of the plant is greatest in the spring, after rains and the melting of the winter snow. Then again, there are off-peak periods during the night, when there is a surplus of current. Three factors make this problem less serious today than it once was and will make it less and less important in the future: the possibilities for storing water in larger and larger reservoirs, the interconnection of plants dependent for their supplies on glaciers and those dependent on rainwater, and finally the combining of hydraulic and steam plants.

**Turbo-alternators.** In hydraulic installations the water issuing from the pipes falls upon the paddles, blades or buckets of turbines, which are basically highly efficient mill-wheels. Each turbine is combined with a generator which produces the electric current. Together, turbine and generator constitute what is known as a turbo-generator.

A recent innovation for the utilisation of very low heads is the use of hydro-electric block units, that is, turbo-generators enclosed in an impermeable casing and allowing immersion. These units can be mass-produced on a few standard patterns and installed at small cost in river dams of flat country. The block unit is placed

at the junction of a vertical (intake) shaft and an oblique (output) pipe. It can also be installed in a siphon set across the dam and operated by means of a small vacuum pump. The prototypes of these two installations are the French plants at Castet (Basses-Pyrenees) and Maignanerie (Mayenne). It is also proposed to use them in tidal power stations, a number of which are in the planning stages in Europe.

**Transformers.** The current produced by the generators is a triphase current of 50 or 60 cycles per second and 15,000 volts. Such a current could not be transmitted far without heavy losses. It is therefore necessary to step up its voltage and at the same time decrease its strength. Accordingly, voltage is usually raised by transformers. In France, for example, it is raised to 225,000 volts, the normal voltage of the high-tension networks of Electricité de France. In both Europe and America lines can be found which operate at lower voltages; still others are designed for as much as 400,000 volts. Before use, all these high-tension currents must be converted by stepdown transformers, which reduce them first to 60,000, then to 5,000, and finally to 220 or 110 volts by a gradual process.

Irrespective of voltage, an alternating current cannot economically be transmitted for more than 600 miles, or pass through underground or submarine cables. It has, in fact, an unfortunate tendency to set up in its own cable an induction current in the opposite direction, thus increasing the resistance of the conductor. In addition, contact with the Earth or with salt water produces important variations in the Earth's magnetic field and rapid wearing of the cables. In future transmission between Norway and Great Britain, Sweden and Denmark, it is planned to have recourse to direct current, but at voltages in the order of a million volts. It now appears possible to achieve such boosted direct currents for transmission over thousands of miles by means of mercury-vapour rectifiers fitted with special valves (ignitrons, thyratrons).

**Work and Power.** So far we have spoken of the strength and voltage of currents. Next we must deal with the important concepts of energy, work, and power. Energy is the capacity to act, the capacity for work in a person, an element or a machine. Energy is equivalent to force. Over-simplified as this definition may be, it has the advantage of leading directly to the idea of measurement.

In practice, energy consists in displacement. A certain mass is raised, drawn, pushed or rolled, depending upon the circumstances. The product of the mass by the distance it is displaced expresses, by definition, the work accomplished. Without entering into the mysteries of mechanics and, above all, without getting embroiled in definitions of the centimetre-gram-second and foot-pound-second systems, we shall confine ourselves to stating that the unit of work most commonly employed in England and the United States is the foot-pound.

One foot-pound, as the term implies, is the work required to raise a weight of 1 pound through a distance of 1 foot or, and this amounts to the same thing, the work recoverable by dropping 1 pound from a height of 1 foot. A crane typifies work by the first method, a waterfall the second.

The concept of work or, more precisely, mechanical work is not sufficient. It is necessary for the work to be accomplished in a given time. A tortoise covering a distance of ten yards performs the same amount of work as a hare of similar weight which covers the same distance, but the hare performs the work at a much faster rate. Accordingly, the two factors work and time have to be linked just as force and distance were previously linked. This new combination is called power.

An old unit of power is the horsepower. It is equivalent to 550 foot-pounds per second. Today it has been replaced by the kilowatt (kw), which is equivalent to 1·34 h.p. (horsepower). For example, a crane raising 100 kilograms through 1 metre in 1 second possesses power of approximately 1 kilowatt.

Yet another unit of measurement is used in industry: the kilowatt-hour (kWh). This is the work performed in 1 hour by a

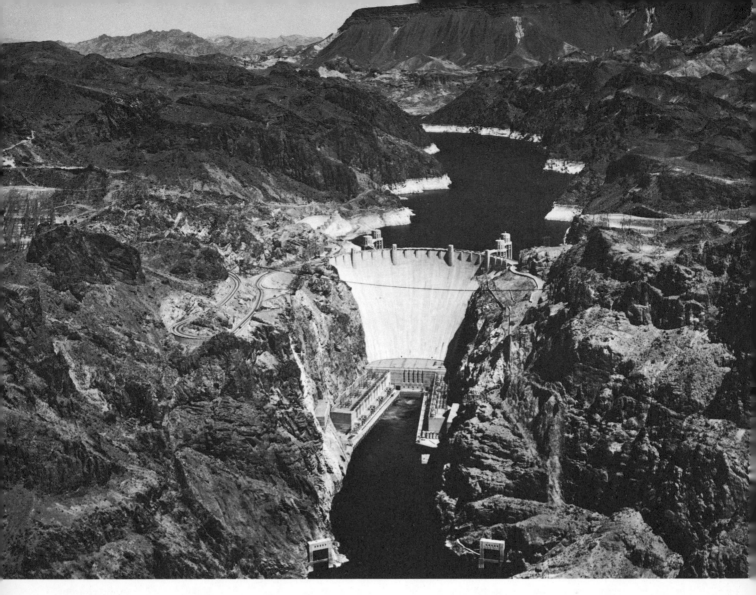

machine having a power of 1 kilowatt; for example, a crane with a 100 kilogram load operating uninterruptedly through 1 metre for 1 hour, or a waterfall precipating 100 litres of water per second for 1 hour from a height of 1 metre. The capacity of great hydro-electric dams is assessed in terms of kilowatt-hours. For instance, the Donzère-Mondragon dam of France has an annual output of 2,000 million kilowatt-hours. Thanks to the use of the common measurement represented by the kilowatt-hour, it is possible to compare even the most dissimilar sources of energy. We can say, for example, that a ton of coal represents some 1,500 kilowatt-hours, whereas a ton of crude oil represents twice that figure, and that Donzère-Mondragon saves an annual consumption of 130,000 tons of coal.

Still bearing in mind that all recently constructed dams are linked with one or several electric power stations, the main purpose of some of them is irrigation. India, China, Egypt, and Brazil are at present all engaged in irrigation programmes of the barren regions of their countries.

The High Dam at Aswan—has been built by Egypt in coopera-tion with the Sudan which will receive some of the water. Although its primary purpose is to provide irrigation for two million acres of land in Egypt and five million in the Sudan it is also expected to produce 10,000 million kWh of electricity annually by 1972.

## UNITED STATES

In the United States, large-scale dam construction has been carried out by federal, state and local governments and by commercial companies. These dams serve diverse purposes: irrigation, flood control, hydro-electric power and provision of drinking water. Whenever possible the dams are constructed to serve more than one purpose: flood control and irrigation being teamed up, or power and irrigation or water supply. Generally hydro-electric

power dams do not serve satisfactorily for flood control because to maintain a constantly even power supply, the water behind such dams is kept at a fairly high, steady level throughout the year. On the other hand, in a flood-control dam the water level is held as low as possible to ensure maximum intake capacity for water as it rises during the flood season.

The Tennessee Valley Authority was created by Congress in 1933 to take over Wilson Dam, a steam plant and a nitrate plant at Muscle Shoals, Alabama. The TVA was also given authority to develop the Tennessee River for navigation, flood control, and hydro-electric power. In addition to Wilson Dam, TVA has since acquired four other dams, built twenty, and has gained control of six more. As a result, the Tennessee is navigable from the Ohio to Knoxville, a distance of 650 miles. The dams have averted all but minor floods and still permit the river to support a freight traffic of more than two billion ton-miles per year. The hydro-electric resources are almost completely developed, and in order to meet power demands, including government atomic plants, industrial and civilian uses, the Authority has expanded output by building steam plants, which produce 64 per cent of the 10·4 million kilowatt capacity of the system. Atomic plants and other Federal agencies use more than half of the power produced.

The key structure in the development of the water resources of the Columbia River Basin in Washington is Grand Coulee Dam, built by the Bureau of Reclamation, 240 miles east of Seattle. The dam is 550 feet high, 4,173 feet long, and contains nearly 10·6 million cubic yards of concrete. The spillway is 1,650 feet wide and twice as high as Nigara Falls. Two power-houses contain eighteen generators of 108,000 kw capacity and three smaller units with a total capacity of 1·9 million kWh which hold all records for the production of electric power. A substantial amount of this power is used to bring water out of the Columbia River valley through twelve huge penstocks for irrigation.

*Left*
Looking up the Colorado River towards the Hoover Dam. The power-house is the U-shaped structure at the top of the dam.

*Right*
An aerial view of the Parker Dam on the Colorado River.

One hundred and thirty miles north-east of San Francisco, is Coyote Dam, built by the U.S. Army Corps of Engineers and completed in June 1959. This earth and rock-fill dam is 160 feet high and 3,500 feet long and will store 122,500 acre-feet of water as a flood control project. Lake Mendocino, behind the dam, is 3 miles long and 1 mile wide. Dams near Porterfield on the Tule River and Visalia on the Kaweah River are also flood control dams.

The Bureau of Reclamation also built Hoover Dam on the Colorado River below the Grand Canyon. Completed in 1936, this is 1,244 feet long and 726 feet high, the highest dam in the United States. It creates Lake Mead, which extends 115 miles up the river and stores nearly 30 million acre-feet of water, making it the largest artificial lake in the country. The eighteen generators have a total capacity of nearly $1\frac{1}{4}$ million kw, sufficient to supply a city of over seven million people. The generating equipment is operated by two power companies of southern California.

Many federal projects are under construction to increase the hydro supply, which in 1964 accounted for 172,898 million kWh out of a total of 969,509 million kWh (about 20%) of electricity generated from public supplies. The new plants under construction will further increase this capacity. Among these projects are: Glen Canyon Dam, Arizona, 580 feet high and 1,500 feet long, to generate 900,000 kw; Fleming Gorge Dam, on the Green River, 490 feet high, and 1,180 feet long, which will have a capacity of 108,000 kw; Navajo Dam, on the San Juan River in New Mexico, which is to be 406 feet high and 3,700 feet long and will be used to irrigate more than 110,000 acres, mostly in the Navajo Indian Reservation; California's Feather River project will take flood waters from the northern part of the State and carry them to the dry valleys to the south and will include 740 feet high Oroville Dam, north-west of Sacramento; and Hell Canyon project on which Brownlee Dam on the Snake River, completed in 1959, is the first of three dams.

Co-ordinated development of the Missouri River Basin to provide flood control, irrigation, electric power, and navigation improvements has been under way since 1946. The work is being done by the Corps of Engineers and the Bureau of Reclamation, which have completed or nearly completed thirty-two multipurpose dams. Four dams on the Missouri (Fort Garrison, Fort Peak, Fort Randall, and Gavas Point) are producing about three-quarters of a million kilowatts. Another dam, Oahe, in South Dakota, is completed and storing water. Its generators produce nearly 600,000 kw.

The Louisiana Power and Light Co. at Little Gipsy, Louisiana, started operation of a 325,000 kw automatically controlled station in 1960. The pressing of a button by means of an electronic computer began the first of a series of 800 steps required to put the plant in operation. Over four hours are required to complete the operation, and about the same period is needed to stop it. The controls are operated from the main offices of the company at Pine Bluff, Arkansas. The cost of the automatic controls added 2·5 per cent to the cost of the plant but this is more than compensated for by the saving in man-power.

The energy produced in the Federal power projects is distributed through three Power Administrations: the Bonneville, South-Western, and South-Eastern. These are administered by the U.S. Department of the Interior, responsible for building and maintaining the transmission systems delivering wholesale power at the lowest possible rates to public utilities, co-operatives, and private agencies.

The Bonneville Power Administration is the marketing agency for power from seventeen Federal hydro-electric projects with a combined capacity of 7·8 million kw. These are the Albein Falls, Big Cliff, Bonneville, Chandler, Chief Joseph, Cougar, Detroit, Dexter, Grand Coulee, Hills Creek, Hungry Horse, Ice Harbour, John Day, Lookout Point, McNary, Rosa, and The Dalles. Work

on these projects was continued during the 1960s. Bonneville Power Administration has nearly 8,000 miles of high voltage transmission lines and 194 sub-stations which in 1959 delivered 27·7 billion kWh to 114 customers.

The South-Western Power Administration sells the power generated at Federal projects built by the Corps of Engineers in Arkansas, Oklahoma, and Texas. These have a total capacity of over a half million kilowatts, with proposed construction to add another 90,000 kw by 1961, but projects being planned will nearly double the capacity. The present projects are Blakely Mountain, Ark.; Bull Shoals, Okla.-Texas; Ft. Gibson, Okla.; Narrows, Ark.; Norfolk, Ark.; Tenkiller, Okla.; and Whitney, Texas. The proposed new reservoirs are Zardanelle, Ark.; Eufaula, Okl.; Greers Ferry, Ark.; McGee Bend, Texas; and Table Rock, Ark.-Mo. South-Western Power operates more than a thousand miles of high voltage transmission lines and has 20 sub-stations. It sells over two billions kWh per year valued at over 13 million dollars.

The South-Eastern Power Administration distributes electrical energy from Federal Projects in Alabama, Florida, Georgia, Kentucky, North Carolina, South Carolina, Tennessee, and West Virginia. The following plants have a total capacity of 1·3 million kw: Allatoona, Ga.; Burford, Ga.; Center Hill, Tenn.; Cheatham, Tenn.; Clark Hill, Ga.-S.C.; Dale Hollow, Tenn.; Jim Woodruff, Ga.-Fla.; John H. Kerr, Va.-N.C.; Old Hickory, Tenn.; Philpott, Va.; and Wolf Creek, Ky. Three plants with a capacity of 520,000 kw are under construction as follows: Barkley, Ky.; Walter F. George, Ga.-Ala.; and Hartwell, Ga.-S.C. The South-Eastern Power markets over 4 billion kWh per year.

## RUSSIA

Almost as extensive as the hydro-electric and irrigation developments of the United States are those of Soviet Russia, which plans for the conversion of vast areas of land around the Black Sea, the Caspian Sea, the Aral Sea and Lake Balkhash into areas of rich cultivation. At the end of the Tertiary period this region was occupied by a large sea above which only the Caucasus and the Carpathian Mountains stood out. Subsequently the Sarmatic Sea, as the geologists call it, gradually disappeared. Its surviving traces are still in the process of disappearing. Today, the level of the Caspian Sea is 87 feet below mean sea level. To make this Aralo-Caspian lowland productive, intensive irrigation and electrification are required.

The first dam to be built was across the Dnieper, in 1939, and was known as Dneprostroi. Destroyed during the Second World War, rebuilt in 1947, and now flanked by giant power stations, this dam raises the lowest water level of the river 127 feet and irrigates 2,500 square miles. Another dam has been constructed across the Dnieper at Kakhovka.

Across the Don stretches the Tsimlyanskaya Dam which has an enormous reservoir, 125 miles long, with a capacity of $16\frac{1}{4}$ million acre-feet of water. This reservoir forms the heart of a network of irrigation canals designed to carry water to the 2,000 square miles of the former Rostov steppes. A 60-mile-long canals links this reservoir with the Volga, near Volgograd, thus opening a passage from the Black Sea to the Caspian Sea.

The Volga project includes the largest hydro-electric installation in Russia. Some fifteen great dams and power stations harness this river and its tributaries; among them are those at Rybinsk, Shcherbakov, Ivankovo, Gorodetz, Gorki, Kuibyshev, Balakovo, Volsk and Volgograd. These dams and power stations develop an electrical power estimated at 40,000 million kilowatt-hours.

Two rivers flow into the Aral Sea, the Amu-Darya and the Syr-Darya, which have for years been the object of local irrigation plans. At present there are large areas in the provinces of Samarkand and Bukhara already under irrigation and suitable for cotton growing.

The Amu-Darya once flowed into the Caspian Sea. Its earlier

valley (Uzboi) is still clearly visible and has been made to serve as the principal bed for the great Turkmenistan canal, 700 miles long, cut between the Caspian and the Aral Sea. Apart from its value as a route for commerce, the canal has brought about the cultivation of at least 31,000 square miles of the steppe-lands.

The Russian hydro-electric programme will also utilise three great Siberian rivers, the Irtysh, Ob, and Yenisei, which flow into the Arctic Ocean. On completion, dams and power stations at Bukhtharma, Novosibirsk and Irkutsk will provide 90,000 million kilowatt-hours annually.

The world's greatest reservoirs, in decreasing order of size (given in million acre-feet) are: Salto de Aldeadavilla, Spain (150), Kariba, Rhodesia (140), Wainganga, India (33·3), Lake Mead, Arizona-Nevada (32·47), Glen Canyon, Arizona (29·88), Oahe Lake, South Dakota (23·6), Garrison, North Dakota (23), Fort Peck, Montana (19·4), Pine Portage, Canada (11·14), Roosevelt Lake, Washington (6·4).

## FRANCE

In trying to express the relative value of hydro power in France it is important to place this source of energy in its right perspective. Great Britain for example generates almost twice as much electricity as France but hydro power supplies only two per cent of her production.

In 1960 the total amount of primary energy used in France was approximately 130 million tons of coal equivalent (this unit is convenient for comparison purposes and represents the energy released by burning one ton of average quality coal). About one quarter of this total was used to make electricity in thermal power stations and the output of hydro power stations for that year supplied the balance of almost 50 per cent.

Hydro power is, therefore, of great economic significance in France as a source of energy. In 1963 hydro generation amounted to 43,400 million kilowatt hours (kWh) against 44,900 million generated in thermal power stations.

The harnessing of this energy raised many technical and economic problems which have been solved one by one over the years, and various stations have been developed to take advantage of the differing water resources which are available. The aim is to use as much as possible of the available water under the greatest possible head in order to extract the maximum amount of energy from the potential amount represented by the fall of this water from its source to the sea, and to do this at the exact time that the energy is required. It is possible, for instance, to create high level reservoirs and release given amounts of water at the right time through turbines driving generators situated in steps along the rivers, right down to sea level. These may be called *seasonal reservoir power stations*.

A large river with a small gradient may also be used, but in such cases it is not usually possible to store the water and the uninterrupted flow is therefore used to generate electricity all the time. This is done in *run of river stations*.

Between these two extremes there are many possible variations where the amounts of water available, and the storage capacities, dictate the methods of operation to be adopted in an integrated electricity supply system. These intermediate stations may be called *short storage power stations*.

One great advantage of hydro power arises from the fact that, whilst electricity cannot be stored because it has to be generated at the instant it is required by the consumer, water when it is stored can be released in large quantities at short notice merely by the operation of valves and thus can provide a convenient means to generate power to meet large fluctuations in the demand for electricity. This can also be done in thermal power stations by keeping them running constantly at a reduced capacity which can then be boosted according to demand, but the electricity generated in the meantime is unused and the cost therefore is considerable. The precise control of hydro power, on the other hand, makes it possible to prepare without much difficulty for impending variations in the system so that the generating facilities, whether they be thermal,

Damat Cambange, Dondo-Angola.

hydro or nuclear, may be deployed to their best possible advantage.

France is particularly fortunate in her hydro resources and the following table shows how the hydro generating facilities can be subdivided into various types of stations as follows:

Electricity Generated in 1963

| Type of Station | Millions of kWh | % |
|---|---|---|
| Seasonal Storage Stations | 9,500 | 22 |
| Short Storage Stations | 11,200 | 26 |
| Run of River Stations | 22,700 | 52 |
| Total | 43,400 | 100 |

The creation of Electricité de France in 1946 as a nationalised undertaking responsible for the co-ordination and integrated operation of the whole French electricity system has made it possible, by introducing a unified control over a complicated supply system, to make the best possible use of energy resources for the purpose of generating electricity as a whole, and in particular the best possible utilisation of hydro resources in France.

Geographically speaking, France has been subdivided into three hydro generation areas:

1. The Alps, which includes all the rivers deriving their water from the Alps such as the Rhône with its many tributaries and the Rhine, only a small section of which flows along the Franco-German frontier between the Vosges and the Black Forest.
2. The Massif Central, which covers the centre, the northern and the western parts of France and includes the basins of the Seine and the Loire and the right bank tributaries of the Garonne.
3. The Pyrenees, where the water from the northern slopes is collected by the Garonne and its left bank tributaries, and also the Adour.

The respective contributions of these areas in terms of electricity generation were as follows:

Electricity Generated in an Average Year

| Area | Millions of kWh | % |
|---|---|---|
| Alps | 28,300 | 65 |
| Massif Central | 9,000 | 21 |
| Pyrenees | 6,100 | 14 |
| Total | 43,400 | 100 |

**The Alps.** The Alps of the Savoie and the Dauphiné are the highest in Western Europe and they provide an immense natural reservoir of water in the form of snow and ice. Man has been able to take advantage of this by harnessing the water released by the glaciers.

The most important developments, from the point of view of the amount of electricity generated, are the four stations of the Compagnie Nationale du Rhône: Bollène, Génissiat, Châteauneuf-du-Rhône and Logis-Neuf, which can generate in a year of average rainfall 2,110, 1,658, 1,630 and 1,170 million kilowatt hours respectively which accounts for 15% of the total hydro power of France. A fifth station in that group, Beauchastel, was commissioned in 1963.

These stations are part of an elaborate scheme to harness the Rhône for the threefold purpose of navigation, irrigation and generation of electricity. When the scheme is complete, Switzerland will be connected to the Mediterranean by a navigable waterway, several other major hydro stations will be producing electricity and the whole of the Rhône Valley will benefit from irrigation schemes.

Next in size come the stations along the part of the Rhine which forms the Franco-German border. This mighty river carries such a large amount of water that even with the relatively small head over that section, it can still be used to generate electricity in a number of large run of river stations, five of which are now completed and in operation. They are listed below, together with their respective average annual production:

Fessenheim (1,015 million kWh), Marckolsheim (1,000 million

*Top*
The pump-fed power station Hohenwarte 11 in the Gera District of Germany. This is one of the largest power stations in Europe.

*Centre*
Warragamba Dam and power station, Australia.

*Bottom*
With the level of Stocks Reservoir, Yorkshire, at its lowest ever, stones that were once the village of Dalehead were revealed.

298

kWh), Ottmarscheim (975 million kWh), Kembs (894 million kWh), and Vogelgrun (765 million kWh).

These stations together generated about 14% of the total hydro power for France in 1963. A seventh station, Rhinau, with an ultimate average annual production of 908 million kWh was operating at half capacity at the end of 1963 and at full capacity in 1964. The sixth station, Gerstheim, was already under construction in 1963 and started generating in 1966. Its average annual production should be 745 million kWh. This development has also rendered the Rhine much more navigable over that stretch.

The remainder of hydro generation in the Alps (almost 40% of the total for France) is provided by a large number of hydro stations on the main and secondary tributaries of the Rhône, the most important of which are La Bathie Roselend (1,000 million kWh), Malgovert (720 million kWh), Serre-Ponçon (700 million kWh) and Randens (797 million kWh). A large station, Salon-St.-Chamas, is under construction and will provide 885 million kWh in an average year from 1966.

**The Massif Central.** The Massif Central Area, leaving out the northern and western parts of France which are mostly in low lying country and therefore unsuitable for hydro generation, includes the catchment areas of the Dordogne and the right bank tributaries of the Garonne, the Seine and the Loire. Here are to be found the earlier large hydro stations with seasonal reservoirs, and their function is, to this day, to regulate the frequency on the French electricity supply system. The largest of these stations are:

Brommat (635 million kWh), L'Aigle (460 million kWh) and Chastang (417 million kWh).

The total amount of electricity generated in this area for 1963 was 9,000 million kWh or 21% of the total for France.

**The Pyrenees.** The last area, the Pyrenees, contributes 6,100 million kWh or 14% of the total for France by means of a large number of medium and small hydro stations in some very interesting developments.

In order to co-ordinate the use of the available electrical energy mentioned above with the other sources of energy provided in various areas of France such as the North Industrial Area, the Paris Area and the recently discovered natural gas resources of the South West, Electricité de France operates one of the most modern and extensive high voltage transmission networks in the world. In the winter, when large quantities of energy are required in the industrial centres and the populated centres of France, large quantities of electrical energy from thermal sources are transmitted over this network towards the south, whilst in summer when large quantities of water are available electrical energy from hydro power stations is transmitted to the north. The bulk transmission of electricity is carried out mostly at the extra high voltage of 380,000 volts.

## OTHER EUROPEAN COUNTRIES

The Netherlands and Denmark have practically no hydro-electric power resources. In Britain, hydro-electric power is developed on only a small scale except in the Highlands of Scotland. At the end of 1963 the North of Scotland Hydro-Electric Board operated 50 large stations with a total capacity of 958 megawatts. The major schemes are situated at Loch Sloy (Dumbartonshire), Pitlochry, Breadalbane and Lawers (Perthshire), Glen Affric, Garry and Moriston (Invernesshire), Glascarnoch, Luichart and Torr Achilty (Rosshire), Shira (Argyllshire) and Shin (Sutherland). The complexity of these schemes is well illustrated by that at Breadalbane, which was inaugurated in 1953. The project, with a capacity of 88 megawatts (a megawatt is a unit of one million watts), comprises in all:

A dam, 96 feet high, across the River Lyon near Lubreoch, to increase the size of Loch Lyon;
An aqueduct from Glen Lochy, carrying water into the enlarged Loch Lyon;
A power station on the River Lyon dam;
A dam, 92 feet high, on the Allt Conait, to enlarge and unite Lochs Daimh and Glorra;

Dams to utilize Lochan Breadlaich and Monachyle Burn;
The creation of a completely new loch by damming River Lednoch.

The Glen Affric scheme, — commenced at Fansakyle in 1952 comprises;
A dam 120 feet high, at Loch Mullardoch;
A tunnel taking water from this dam through an underground power house into Loch Benevean;
A dam, 86 feet high at Loch Benevean, and a power station.

In contrast, Sweden, Norway, Switzerland and Italy are richly endowed. In Norway hydro-electric power is especially important and production practically doubled in the five years ending 1962 to 37,624 megawatts, providing 99·7 per cent of the country's electrical supply. Sweden has a much lower capacity than Norway, but began to develop her resources earlier and in 1963 nearly 95 per cent of her total production of 40,000 million kWh came from hydro production. The trunk transmission line between Harspranget, in the far north, and central Sweden was the first in the world to operate at a voltage of 380,000.

Switzerland's potentialities are estimated at 26,000 megawatts, of which 21,154 had been installed at the end of 1962. In addition to present installations, which include the Mauvoisin project (having the highest arch dam in the world: 777 feet high, 1,700 feet long and an output of 352·5 Mw), giant schemes are planned, like that of the Super Dixence, which will impound 90,000 million gallons of water behind a dam 930 feet high. At its crest it will be 2,300 feet long and 74 feet wide; its base will be 670 feet thick. In other words, the dam will be a gravity type. It will impound meltwaters from the glaciers in the valleys of Zermatt and Arolla.

Italy is one of the European countries most advanced in harnessing its water power potential; 81 per cent of the total electric energy generated comes from water power and three-quarters of this percentage is derived from the Alps. Output in 1961 stood at 42·6 billion kilowatts, mainly due to a number of new power stations. Among the principal hydro-electric stations which came into service in 1961 are Varzo Crevola (21,500 kw), Torbola (110,000 kw), Guadalami (53,000 kw) and Taloro in Sardinia which has a productive capacity of 80 million kilowatts.

In Spain the production has doubled every seven years and in 1962 was equal to that of Switzerland. The Los Pearos project produces 187·2 megawatts, and five power stations are to be built on the river Navia in Northern Spain, including the Granadas de Salime scheme (120 megawatts) and the Silvon scheme (60 megawatts). In Finland the hydro-electric output of 11,094 megawatts in 1962 amounted to 95 per cent of the total electricity production, and Austria in 1962 was producing more than 17,000 megawatts of hydro-electric power.

In contrast to this West Germany draws only 18 per cent of her electric supply from hydro-electric stations although large dams are being constructed to increase output. In East Germany the great dams on the upper Saale contribute to the intense industrial electrification of the middle Elbe basin. In Czechoslovakia hydro-electric power accounts for 15 per cent of the current used. Eleven dams are being built in the Vah valley, and plans have been drawn up for the electrification of the Hron and Hornad valleys which should greatly increase the supply. In Hungary dams on the Tiszat and the Danube have increased the production of hydro-electric power but as there are no high mountains the supply will always be inadequate.

## THE BRITISH COMMONWEALTH

Hydro-electric power accounted for some 90 per cent of the total electricity generated in Canada in 1963. Output amounted to 8,592,827 kWh out of a total 10,360,264 kWh generated. There are large stations on the Colombia river and in New Brunswick, and electricity from these is sold to the U.S.A.

Australia, on the other hand, is poorly supplied with water and hydro plant represents only 23 per cent of the total generating capacity. The only region high enough to receive reliable rainfall is the mountain chain stretching from New South Wales to Victoria. The hydro-electric potential of this area is considerable and plans have been formulated to develop more than 3,000 megawatts within

the next 25 years. The two major construction projects in this area are the Snowy Mountains and Kiewa schemes. In Tasmania, however, hydro-electric resources have been estimated at about 50 per cent of the total Australian hydro-electric potential, although this does not relate to that exploited.

In New Zealand 81 per cent of the total electricity production of 1,860,380 kilowatts is generated by hydro power, and a further 6 per cent from the geothermal steam at Wairakei. The largest station is at Benmore in the South Island, with a generating capacity of 540,000 kilowatts. New Zealand's first underground station is to be built soon at Lake Manapouri in the South Island and will have a capacity of 700,000 kilowatts.

In the North Island the backbone of the system is the Waikato River with eight generating stations. New Zealand engineers are now turning their attention to ways of connecting the northern and southern networks, for although the South Island has much the greater hydro-electric potential, 70 per cent of the consumers live

in the North. Work began in 1961 on laying three cables, one a spare, between the islands and should be completed in 1965. Five new schemes, with a total generating capacity of 582,000 kilowatts will go into service between April 1968 and October 1972. The most interesting will be the one using gas-turbines drawing on the natural gas resources of Kapuni.

Three Commonwealth projects of more than usual interest deserve special mention. The Owen Falls hydro-electric scheme in Uganda, East Africa, was formally inaugurated in April 1954. A gravity type dam, 100 feet high and 2,500 feet long, controls the flow of the Nile at a point about two miles from its outlet from Lake Victoria and provides electric power for domestic and light industrial use throughout eastern and central Uganda. A high-voltage transmission line takes power from the dam into Kenya. At full output, the capacity is 150 megawatts. The height of the dam, as originally proposed, would increase the level of Lake Victoria by 1·3 metres above the highest recorded level, so converting

the lake, the second largest in the world, into a reservoir 26,000 square miles in area. It has been estimated that it would take eighteen years for the lake to reach its new level. In the meantime, proposals to increase the height of the lake by an additional metre have been made, the object being to ensure additional water storage capacity sufficient to meet all foreseeable needs for irrigation in the Sudan and Egypt.

The River Zambesi, forming the boundary between Zambia and Rhodesia, was dammed at the point where it flows through the Kariba gorge. The dam, an arch type, is 400 feet high and will hold up a lake 175 miles long by 40 miles wide at its widest part, covering about 2,000 square miles. The hydro-electric scheme will have an output of 1,500 megawatts intended mainly for industrial development in the two countries. It was officially opened in May 1960, having been completed according to schedule in spite of severe flooding during 1959, when the coffer dam was overtopped. Possibly the feature which has appealed most to public interest is the

rescue operation which was organised as Lake Kariba began to fill up. 'Operation Noah', as it was called, was made necessary by the reluctance of game animals and birds to leave familiar haunts, being marooned as the flood waters rose.

In Ghana a dam and hydro-electric station are being built on the River Volta at Akosombo. The dam will be 370 feet high and 2,100 feet long, impounding a lake 3,275 miles square containing 120 million acre-feet of water, the largest artificial lake in the world. It is estimated that with the completion of the first stage of construction (four out of six generators) 512,000 kilowatts of electricity will be generated. It is intended that 300,000 kilowatts of this will be used by the aluminium smelter at Tema, which is developing in co-ordination with the power-station, and the rest will be distributed over the national grid to the major part of Southern Ghana. By the end of 1963 40 per cent of the work had been completed, and arrangements were being made for the re-settlement of the 80,000 people who would be affected by the flooding of the Volta Lake.

# Future Sources of Energy

The immense industrial advances made in the western world during the last century which led in turn to higher standards of living were all based on man's ability to harness power. Watt's steam engine and Arkwright's 'Spinning Jenny' have their places in the history of man's progress. At first, power was entirely represented by water power or by the utilisation of steam. Later, steam and electricity became the principal sources of power, both being generated by burning coal. The twentieth century saw the beginnings of the utilisation of oil as a fuel. Today the countries of the western world are dependent on vast quantities of coal and oil in order to maintain their industrial progress; the industrialised world operates largely on a two-fuel economy.

Man, having through history exploited first the richest and most accessible sources of raw materials, is now faced with exploiting the less rich and less accessible sources, and this, coupled with industry's ever-increasing demand for raw materials, has meant that more and more power has to be utilised to extract and process these raw materials. In other words, man is having to dig deeper, work poorer seams, and carry the raw material obtained further than ever before.

This will mean that in time our existing sources of power will be hard pressed to meet the increasing demands on all sides. This has already happened in some countries, and other forms of power-production have been utilised: natural gas in the United States, hydro-power in the United States, Great Britain, Sweden, etc. The trend has been particularly marked in countries that are not well endowed with indigenous sources of coal or oil but which are able to exploit other forms of power-production. Coal and oil are assisted then by hydro-power, natural gas, wood, peat, geothermic steam, etc., but although these sources may make significant contributions to particular fuel economies, their overall contribution to the world's energy economy is small, and often insignificant. For example, in 1961 the United Nations Statistical Year-Book itemised total world energy production (in terms of millions of metric tons of coal equivalent) as follows:

| | |
|---|---|
| Coal | 2,164 |
| Petroleum | 1,487 |
| Natural Gas | 677 |
| Hydro-electricity | 90 |
| Total | 4,418 |

Coal and oil contribute, therefore, some 85 per cent of total production. In North America, where both hydro-power and natural gas deposits are well developed, the contribution of coal and oil is 65 per cent of all production. In Western Europe coal itself represents nearly 80 per cent of the total. The contribution of hydro-power and natural gas is a fairly recent development.

Will these fuels be able to expand sufficiently? There is no doubt that production could be increased, especially where there is incentive to do so in the form of high prices. Equally certain is the fact that there is a limit beyond which it is neither physically possible nor economically feasible to expand production. There is however, no immediate prospect of a shortage of sources of power. It was fashionable some years ago to prophesy 'gaps' between future demand and future supply. However, the normal economic forces, given time to operate, will effect a balance between the two, and there can certainly be no suggestion that the world faces a fuel famine in the near future. A recent estimate made in the United States suggests that the recoverable reserves of coal, oil and natural gas throughout the world amount to 14 million million tons of coal equivalent. At present rates of consumption, and allowing for projected increases (statistically, consumption doubles every twenty-five years), this should be only one-sixth exhausted in one hundred years.

Nevertheless considerable attention is being paid to the discovery and development of new sources of power. There is a desire to find an alternative to the 'big two', especially in countries which have no indigenous sources of either; and an attempt to reduce power costs in face of the increasing price of conventional fuel. If, therefore, one or more of the possible future sources of power proves a cheaper method of supplying energy than coal or oil, their contribution could be very significant; they could, indeed, revolutionise the whole pattern of energy production. There is also a desire, for economic and political reasons, to introduce some flexibility into the almost universal two-fuel economy.

A study of the available sources of power makes it clear that man has scarcely begun to tap the vast potential power resources of this planet.

## WIND POWER

Man's earliest use of natural resources of energy for mechanical power was undoubtedly his employment of animals to supplement his own strength. Next he turned to the wind and to flowing water. The wind has been used to drive sailing ships for centuries; until the steam engine appeared in a practical form towards the end of the eighteenth century watermills and windmills were virtually the only source of stationary power.

It is known that windmills and watermills have been in use since 250 B.C. or earlier. But while watermills have developed to the point where hydro-electric stations of over a million horse-power capacity are being built, wind power has lagged behind. As electric power became more readily available, windmills fell into disuse, though they were commonly used for grinding corn up to the end of the last century. Up to the Second World War small wind-driven machines with rotors in the form of propellers or multi-blade wheels were probably the only practical use of wind-power remaining. They continue to be popular for providing electric power or for pumping water in isolated areas, particularly in Australia.

The war drew attention to the advantage of wind-driven plants for generating electricity. Operation on a relatively small scale, they would be less liable to major disruption than would normal power stations. Research has continued because of the obvious economic attraction of a form of power which, like solar power, is free for the taking. Modern utilisation of wind power is mainly

The parabolic mirror of the solar furnace at Mont-Louis.

*Top*
A windmill at Mykonos, Greece.
Until late in the eighteenth
century windmills were one of
the few sources of power.

*Bottom*
Sailing boats have made use
of the wind's driving power for
many centuries.

aimed at electricity production, and plant of this kind is in operation in many countries, including Australia, Canada, Denmark, France, Germany, South Africa, Russia, the U.K. and the U.S.A.

The basic elements determining the energy obtainable are the wind speed, the size of the area swept by the rotor, and the conversion efficiency of the plant. Whatever the chosen diameter of the rotor, electric windmills are designed to give their maximum power capacity at a chosen or 'rated' wind speed. Energy in a wind velocity above the rated speed is wasted, and energy input and conversion efficiency of the plant fall rapidly when the wind drops below the rated speed. In theory windmills should extract close on 60 per cent of the power in the wind. In practice losses in the rotor, in the gearing and in the electrical equipment reduce efficiency to about 40 per cent.

The energy in the wind increases in cubic proportion to its velocity. Below 18 miles an hour there is very little energy. The North of Scotland Hydro-Electricity Board is testing a 100 kw wind-powered generator in the Orkneys where average wind velocity is about 30 miles an hour. A survey has disclosed several hundred possible sites in different parts of Britain with a total potential capacity of several million kilowatts, capable of saving two million tons of coal a year.

Whether wind power is likely to be an economically attractive means of generating electricity in Britain must await evaluation over the experimental period in the Orkneys. There seems little doubt that it is a method of definite value in areas where fuel costs are high. One example is the Negev, in Israel, where there is interest in the use of wind power for electricity production and for pumping irrigation water.

## TIDAL ENERGY

The energy involved in tidal movements is immense and inexhaustible. It is estimated, for example, that tidal energy enters the English Channel from the Atlantic at the rate of 180,000 megawatts and that the greater part of this is wasted by friction on the bottom of the Channel. It is natural that man should try to discover means of harnessing and exploiting this energy. A number of studies have been made in several countries, including Canada, Great Britain, United States, Brazil and New Zealand.

The energy of the tide can be harnessed and converted to electricity by a process similar to that used in the production of hydro-electricity. Let us first examine the nature of tides themselves. The size of the tides is the difference between 'high' and 'low' water and is controlled by the relative position of the Moon and the Sun; the maximum tide occurs in the spring. The size of tide varies with the season and with geographical location. There are, for example, only negligible (and, therefore, unexploitable) tides in the Mediterranean Sea, whereas the western coast of Europe, subjected to the Atlantic, enjoys high tides, particularly the coasts of England and France facing on to the English Channel. However, the Atlantic coasts with their two tides each day, each of equal size, are by no means typical; in other parts of the world tides of very different kinds occur at different intervals. The fact is that there are relatively few sites throughout the world that are suitable for large-scale tidal power plants, but there would appear to be scope for small-sized plants at numerous points along coastlines where the tide reaches a magnitude of round about 30 feet. Tidal energy may be harnessed as follows:

**One-way cycle.** This is the simplest arrangement and entails the closing of an estuary or bay by a barrage to create a large basin; the potential energy of the water is utilised as the basin empties. The cycle takes place in three stages: the filling phase, in which as the tide rises the sluices are opened and the water flows into the basin; the waiting phase, in which the tide has reached its maximum and the sluices closed to await the final generating phase. When the most suitable moment has arrived (based, for example, on maximum production or demand) the water is fed through the opened turbines which produce power under the load of water offered by the difference in water level between the basin and the sea.

**Two-way cycle.** Naturally it would be much better if the turbines could generate electricity during both filling and emptying stages. With a single barrage, the water falls during the generating phase first from the sea to the basin and then from the basin to the sea. In this case, special arrangements have to be made to maintain the water flow in the same direction through the turbines. A two-cycle scheme could also be achieved by building two lines of sluices to seal off the basin, with the station parallel to the banks between the two. Both these schemes involve much greater expenditure than the simple one-way cycle, but the very simple layouts used in modern turbines are compensating factors.

**Other Schemes.** Use could also be made of two adjacent basins each using the simple one-way cycle with one (lower) basin operating during the filling stage, the other (higher) operating the emptying phase. If both basins are served by plants electrically linked the electricity produced would be more or less continuous. Consideration has been given to many other cycles employing two, three or even more basins, so arranged and linked that their output is continuous and allows of flexibility. Such schemes involve complicated and very expensive arrangements, chiefly because of the scale of civil engineering work involved or the size of sluices required. An interesting method of improving upon the efficiency of these cycles would be the use of the turbines as pumps during the period of slack water at high and low tide. The net gain by employing pumping is fairly considerable, but it is not yet fully established that the pumping could, in fact, be satisfactorily carried out. Such a system would make for maximum flexibility independent of seasonal tides and would be opened as a pump during periods of low demand for electricity and as a generator during periods of peak demand. The pump-storage scheme at Ffestiniog in Wales, based on two reservoirs, produces 300,000 kWh during peak hours.

The type of hydro-electric plant used in tidal power schemes was devised only after considerable effort. Much research has been carried out in France, and a full-scale experimental scheme at the Rance estuary in Brittany was commissioned for 1966—67. It is anticipated that it will provide for an average annual production of 554 million kWh. Apart from this scheme studies are also being made in Spain, Wales, Germany, Brazil, New Zealand, the Pacific and Arctic coasts of the U.S.S.R., the Netherlands and India. In Great Britain a great deal of attention has been paid to a scheme on the Severn Estuary from as early as 1918. In 1933 a further study envisaged a one-way cycle with pumping, the barrage or dam also serving as a road and rail viaduct. The scheme was abandoned because of the lack of economic incentive but was revived ten years later when coal was scarce and its price rising. A commission headed by Lord Brabazon reconsidered the scheme, and ultimately recommended construction of a single basin operating on discharge only. Several problems, recognised by the Commission, prevented the work being put in hand. One is the problem of large-scale silting up, aggravated by the proximity of several ports: Bristol, Avonmouth, Cardiff and Newport. Moreover on present-day coal prices it would not be economic. A Canadian scheme for the Petitcodiac and Memramcook estuaries was also abandoned on economic grounds. A scheme in the U.S.A. at Passamaquoddy operating on the one-cycle principle, has proceeded by fits and starts since 1935, when it was started mainly to alleviate local unemployment. It received official backing from the U.S. Government who regarded it as a means of studying tidal energy. The scheme was abandoned the following year and later was resuscitated in 1952 and 1955, when studies were made by the U.S. Corps of Engineers.

Thus though many sites have been studied and schemes suggested, no work has proceeded beyond this point except at Rance, where construction was authorised in 1956. Other schemes have been abandoned on economic grounds, for a tidal energy project would cost more than the value of coal it would save. Schemes of this kind involve high capital costs to seal off the large areas required, and even then they provide only an intermittent source of supply of electricity determined by tides and seasons and quite unrelated to peak demands. While there are ways in which these problems could be overcome, still further expense would be involved. Tidal energy remains a great potential source of electrical power.

## HYDRO-ELECTRICAL POWER

Falling or flowing water was the earliest of the natural sources of energy to be harnessed for providing power, but some two-thirds of the water power at present in use has been developed in the last fifty years. The reasons are partly economic. On the technical side, improvements in hydraulic turbines have made it possible to use almost any head of water. At the same time, better transmission cables and other equipment have enabled power to be transmitted over long distances — up to 30 miles or more. The importance of this development is significant when it is realised that hydro-electric sites are usually remote from areas of commercial or industrial activity.

From the economic point of view, modern industry has created a demand for large blocks of cheap electricity which can sometimes be satisfied more readily by water power than by any other means. This applies only to countries with the requisite physical characteristics (the most obvious is adequate rainfall). Equally, a region with pronounced topographical features has many advantages in providing high-level sites where water can be collected and stored. These sites can be chosen for their proximity to lower points to which the water can be led. These geographical factors are of great economic importance, since the horse-power which can be developed by each cubic foot of water depends directly on the working head, that is, on the distance the water has to fall before entering the turbine. At the same time, the higher the head, the less water has to be stored to produce a given quantity of electricity. Countries with heavy rainfall, mountainous areas, and highly industrialised regions are very well placed for hydro-electric development. In Europe hydro-electric schemes are being advanced most vigorously in Norway, France, Switzerland, Italy and Scotland. At Lac Fully in Switzerland there is a scheme which uses the highest head on record — 5,412 feet, developing a pressure in the pipeline of 2,260 lb. per sq. inch.

At the same time, low and medium head sites can be, and in many places are successfully utilised for hydro-electric generation.

The production of hydro-electricity continues to rise (it is now about 500,000 million kilowatt-hours annually), and so great is the increase in the world's total consumption of electricity that the proportion of hydro-electricity to other means of generation is steadily declining and is now only about ten per cent of total consumption. There seems little doubt that this will be the pattern of the future. More hydro-electric sites will be brought into operation but there simply are not enough of them for hydro-electricity ever to contribute more than a relatively small proportion of total needs.

A new development is compressed-air storage. During off peak periods electrically driven compressors are used to compress the air into hard rock caverns and at times of peak demand it is supplied to the combustion chambers of gas turbines coupled to generators. Techniques for sealing fissures in the caverns, to reduce leaks, are being studied.

## THERMAL ENERGY OF WATER

The feasibility of exploiting this type of energy dates back about forty years when it was demonstrated in France that it was possible to produce energy from a small temperature difference between two vast bodies of water. The principal features of thermal energy from the seas are the enormous size of the bodies of water involved and the fact that the conditions allowing the extraction of energy (e.g. temperature differences at different levels) are constantly restored. Temperature differences are found where there is a warm surface layer of water heated by the Sun and, deep below, a mass of very cold water of greater density which streams from the polar regions, or by an interplay of several factors, such as coastline configuration in conjunction with current movements and wind. Here is a constant, reliable and entirely inexhaustible source of power.

The simple method of utilising this energy might consist of deflecting a flow of warm water into a neighbouring expanse of cold water so as to melt ice and facilitate navigation or to alter the climatic conditions of the area. This operation could be carried out with two adjacent bodies of water or by pumping warm water from the deeps up to frozen surface water. More sophisticated methods could be devised to convert the latent energy into mechanical or electrical energy, and in this way fresh water can be produced as a by-product. This could be of considerable significance in arid areas. The plant required to produce electricity from this source consist of an evaporator, a generator and a condenser. Warm salt water from the surface is led into a low-pressure container in which a vacuum is created; in this vacuum some of the warm water turns to vapour. This vapour or steam is then sucked into the condenser which is cooled by cold water pumped up from the deep cold layer. The steam, on its way through the condenser, is fed through a low-pressure turbine-generator which produces electrical current. Depending on the type of condenser employed, the steam may be distilled either as salt water or fresh water.

Stations for the production of thermal energy power could be land based or floating. The practicability of the land based station would depend on the close co-existence of bodies of cold and warm water lying off a steeply shelving coastline, otherwise the cost of pipe-laying to pump up cold water artificially — would tend to be prohibitive.

No station utilising the thermal energy of the sea on an industrial scale has yet been built, although a small-scale plant exists at Abidjan on the Ivory Coast of Africa, and in California there is a small plant producing fresh water. Nevertheless, over the forty years or so since such schemes were first demonstrated, experiment and small-scale trials have shown that this source could be utilised and could be used to provide or, perhaps more important, to produce power and fresh water in the arid regions of the world. It has also been suggested that such schemes could be assisted and improved by utilising solar energy to increase surface water temperature and by adding surface coatings of oil or plastic sheeting to inhibit evaporation and cooling at night. It has also been suggested that the hot coolant water from nuclear power stations could act as the warm source in a station of this type.

However, the principal drawback remains one of cost. Efficiency of a station utilising the thermal energy of the sea depends upon and varies with the temperature difference between the two bodies of water. If the purpose of the station is electrical power, the temperature difference must be about 20 degrees Centigrade. This might necessitate pumping from great depth; furthermore, to pump these enormous quantities from great depths would itself require power. It is difficult to estimate costs for such schemes and difficult to establish its economic feasibility, but under certain conditions and especially when water is produced as a by-product, the scheme is certainly not without its attractions. In the vast oceans there lies yet another source of immense inexhaustible power.

Increasing attention is being paid in New Zealand to the production of geothermal electricity. The active volcanic region in the North Island produces hot steam which is tapped at Wairakei, near Lake Taupo. This station began operating in 1958 and accounted for 6 per cent of the total power produced. It is now being expanded to add 90,000 kilowatts to its capacity, bringing the total output to 250,000 kilowatts in 1971. It is expected that the revolutionary process of flashing steam from hot water instead of allowing this to go to waste as it turns to steam at the well heads will be used to provide much of the extra capacity. It is believed that geothermal stations will ultimately supply all the base-load power for the North Island, and that hydro-electricity will then be used only for peak loads.

## SOLAR ENERGY

On Mount Shinobu in Japan there is a completely unattended ultra high frequency repeater station, part of the national telephone system. The power to drive it is derived from 4,320 silicon cells having a total active area of 2·7 sq. metres. These extract energy from the Sun and convert it to electricity in much the same way as a photo-electric cell moves a needle on a dial to indicate the strength of the light for photography.

Silicon is prepared as a semi-conductor, that is, a material with the property of conducting electricity and insulating it. If the material is prepared in such a way that the current is carried by

negatively charged particles (electrons) it is said to be 'n-type'. Conductors can also be made to act as if they were positively charged, when the material is called a 'p-type' semiconductor.

In practice, a material with both properties is made, and when light falls on the cell a voltage is developed at the junction between n-type region and the p-type region. The cells are popularly known as silicon solar batteries and were discovered by Bell Telephone Laboratories in 1954, in the course of a transistor research programme. They are attractive because of their relatively high efficiency compared with other methods of converting sunshine into useful power. Even the earliest silicon cells had an efficiency of six per cent; those in use today achieve nearly 15 per cent, and it is believed that a maximum efficiency of 20 per cent is possible.

At Mont-Louis in France the Sun's rays are caught in a steerable mirror, which reflects them into a parabolic mirror. This concentrates the rays into a furnace where the effects of high temperatures on various material is studied.

Older methods of converting solar energy involved the use of thermocouples — two unlike metals joined in a hoop. When one joint is placed in the sun and the other kept cool, an electric current is produced. Despite recent improvements in thermo-electric materials, however, maximum efficiency is still only about 0·6 per cent. Hence it seems certain that the silicon cell and a possible improvement using gallium arsenide will for some time lead in the conversion of sunlight into electric power.

An advantage of silicon solar batteries is that they respond to light of wavelengths covering the whole spectrum and part of the infra-red. Where weight is vital, as in space exploration, for which these batteries have been successfully employed, cells can be constructed to develop over 100 watts per pound. So far they are expensive but there is hope that with new manufacturing techniques and applications, the cost will decrease. If this happens they could well be used by individual households to generate the electricity required for domestic use. Fortunately they are particularly suitable for use in conjunction with storage batteries, since the voltage output remains fairly constant in spite of fluctuations in the brightness of the light. Clearly, storage is essential, since no electricity could be generated during darkness or when there was rain or heavy cloud.

Further progress is necessary before this can be achieved in the British climate, but it seems likely that this approach of small individual solar 'power stations' may have considerable value as a domestic supplement to electricity drawn from the national grid. Large central solar power stations would seem to be ruled out, certainly in smaller countries, because the area they would occupy would be about four times that of a conventional or a nuclear power station. If this difficulty could be overcome economically prospects would be good, for over England alone the Sun provides enough energy every day to meet the world's need for power. But there seems little likelihood of this energy being tapped on a large scale in the foreseeable future.

Of course, in places with a warm climate, houses can be designed to be heated by direct use of sunlight, some of which is absorbed into heat storage devices for use when the Sun is not shining. This can be improved by incorporating a heat pump into the system to extract heat from the surrounding atmosphere. Other direct but limited applications of solar energy are in use in tropical and subtropical areas for cooking, water heating, distilling sea water for drinking, and for pumping, especially for irrigation. Heat engines in which solar heat is used to generate steam have been tried but have never proved economically successful.

## NUCLEAR ENERGY

Perhaps the most immediately available new source of energy is nuclear energy, or atomic energy as it is known to the layman. This energy is derived from the central core or nucleus of the atom, and to understand how it is made available for power production

*Top*
Calder Hall, the world's first commercial-scale nuclear power station.

*Bottom*
A general view of the thermo-electric plant at Larderello.

*Left*
The installation of a reactor vessel head during the construction of the nuclear power plant at Rowe, Massachusetts.

*Right*
The Dounreay fast reactor, showing the top of the stainless steel pressure vessel in position in the reactor vault.

some elementary knowledge of the structure of the atom is necessary.

The atom consists of negatively charged electrons orbiting round the nucleus, which normally has sufficient positive electric charge to neutralise the negative charges associated with the electrons. The nucleus itself is composed of protons which have one positive charge and neutrons which have no electric charge. The proton and neutron masses are about equal and are 1,840 times more massive than the electrons, so that most of the atomic mass rests in the nucleus. The number of protons in the nucleus decides the chemical element to which the atom belongs. The number of neutrons may vary to give the various isotopes of the chemical element; these isotopes are chemically identical but because of the different number of neutrons in the nucleus they are of different atomic mass. For example, the uranium atom, the heaviest to occur in nature, has 92 protons in the nucleus. But in nature there are two isotopes of this element; in one there are 143 neutrons, in the other 146, as well as the 92 protons in the nucleus. These have respectively a total of 235 and 238 particles in the nucleus and are referred to in the physicist's shorthand as $U^{235}$ and $U^{238}$. $U^{235}$ constitutes about 0·7 per cent and $U^{238}$ about 99·3 per cent of the naturally occurring uranium.

Reactions involving the nuclei of atoms which may release useful energy are of two types: fission and fusion. In the fission process a large nucleus is induced to split into two or more smaller nuclei; in the fusion process the particles of two light nuclei are rearranged to form a larger nucleus. Nuclear power stations have been constructed using the fission process but so far the only method of releasing usable quantities of energy from the fusion reaction is in the hydrogen bomb, where the energy release is uncontrolled.

At present all nuclear power stations in operation or under construction use the fission chain reaction in $U^{235}$, the only natural isotope which will sustain a chain reaction. When a neutron strikes a $U^{235}$ nucleus, the nucleus may split into parts, releasing energy and more neutrons. If one of the neutrons released from every fission causes another $U^{235}$ nucleus to split, a chain reaction is set up and energy is continuously released, becoming available in the form of heat.

When the neutrons are released by fission they have high speed and are more likely to be captured by a $U^{238}$ nucleus than by one of $U^{235}$. If the neutrons are slowed down they will be more likely to cause $U^{235}$ fission. So, to produce a chain reaction the neutrons must lose much of their original energy, or some of the $U^{238}$ must

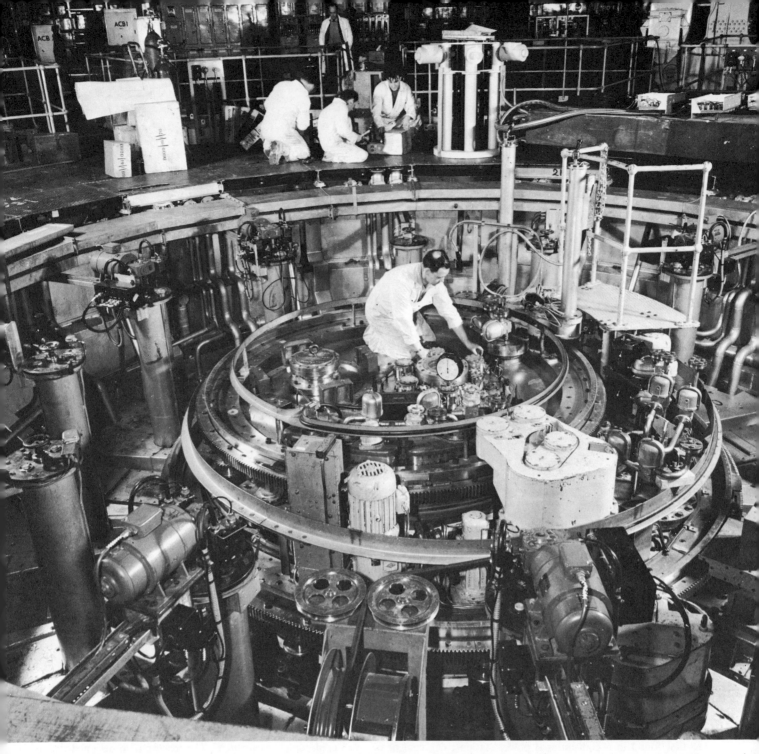

be removed. If, however, a $U^{238}$ nucleus captures a neutron, it converts by radioactive decay (i.e. it emits small particles) to plutonium, an artificial element which, like $U^{235}$, can sustain a chain reaction. Materials capable of sustaining fission chain reactions are said to be fissile.

In uranium reactors both methods of ensuring a chain reaction have been used. Thermal or slow neutron reactors have in their cores a material known as a moderator, introduced to slow down the neutron speeds; the most common moderating materials are water, heavy water and graphite. Fast neutron reactors have no moderator so that the neutrons have velocities close to those at which they are emitted from the fission process. Fast reactors must have a high concentration of fissile material and, where uranium is used, most of the $U^{238}$ must first be separated from the fissile $U^{235}$; thermal reactors may use uranium with a fairly low concentration of $U^{235}$. The fuel is placed in the reactor in many forms, either as a metal or as some chemical form of uranium. The fuel element varies according to the reactor and includes plates, single rods and clusters of rods.

Control of a nuclear reactor is exercised by use of rods made of neutron-absorbing materials such as boron or cadmium; these are inserted or withdrawn from the reactors as necessary. When the control rods are inserted into the reactor core they absorb neutrons which would normally cause fission and then chain reaction is inhibited. Withdrawal of the rods leaves more neutrons available to cause fission, so that the chain reaction increases in intensity and so power production is increased.

The heat liberated in the fission process is removed by a coolant in the form of a gas, a liquid or even a liquid metal. The coolant's function is to transfer the heat from the reactor to the boilers or heat exchangers where the steam that drives the turbo-generator is produced.

A consideration of the existing possibilities for the component parts of a reactor will soon show that there are many versions of reactor design. The selection of any given reactor will depend on the many problems that must be considered before construction begins. First of all the intended function of the reactor can have a great influence since certain reactor types, while useful experimental tools, are not suitable for power production. The behaviour of the many materials that constitute the reactor must be taken into account in the design; the canning, moderator and coolant must be chemically compatible with each other. The canning which protects

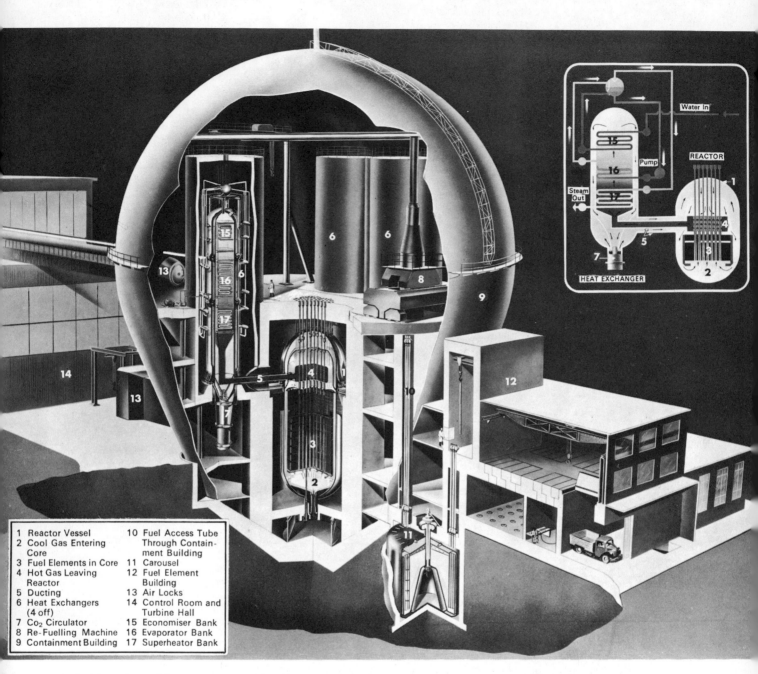

| | |
|---|---|
| 1 Reactor Vessel | 10 Fuel Access Tube Through Containment Building |
| 2 Cool Gas Entering Core | |
| 3 Fuel Elements in Core | 11 Carousel |
| 4 Hot Gas Leaving Reactor | 12 Fuel Element Building |
| 5 Ducting | 13 Air Locks |
| 6 Heat Exchangers (4 off) | 14 Control Room and Turbine Hall |
| 7 $CO_2$ Circulator | 15 Economiser Bank |
| 8 Re-Fuelling Machine | 16 Evaporator Bank |
| 9 Containment Building | 17 Superheater Bank |

the fuel from the coolant must obviously be compatible with the fuel.

Much of the plant used in a nuclear power station is similar to that found in a conventional power station using coal or oil. The major difference lies in the method of heat production.

The world's first commercial-scale nuclear power station at Calder Hall, Great Britain, came into operation in October 1956.

The reactor designed for Britain's nuclear power stations uses as fuel uranium metal rods $1\frac{1}{4}''$ in diameter enclosed in magnesium alloy cans. Some thousands of these fuel elements are set in channels in a cylindrical block of pure graphite 30 to 50 feet across. This assembly of stacked uranium and graphite is enclosed in a steel pressure vessel.

When the chain reaction starts, the fuel gets hot and is cooled by carbon dioxide under pressure which passes through the channels in the reactor, out to a boiler and back through the reactor again. Steam from the boiler is used to turn a turbo-alternator which generates electricity, using the same methods as any conventional power station.

The Shippingport nuclear power station in United States is based on the pressured water reactor (P.W.R.), so called because the functions of moderator and coolant are effected by water kept under pressure to prevent it boiling. The fuel is enriched uranium clad in a zirconium alloy. It is worth while considering the special advantage of breeding that is offered by a fast reactor. Round the

fissile core of a fast reactor there is placed a breeding blanket of $U^{238}$ which captures those neutrons that escape from the core; thus plutonium is formed. The reactor is so designed that more plutonium is formed in the blanket than fissile fuel consumed in the core. This is the phenomenon referred to as 'breeding'. The plutonium may be used later as fuel for other reactors. In this way the fast reactor offers a better exploitation of uranium resources, since it provides a method of eventually fissioning all the atoms of uranium, instead of only the $U^{235}$ atoms.

An experimental reactor of this type went critical in Dounreay in Scotland in 1959. It is used to provide the information and operating needed to design a full-scale power-producing fast-reactor and the components can easily be changed to test different designs.

Britain announced the world's first nuclear power programme in 1955, and stations opened at Berkley and Bradwell in 1962. The high temperature reactor operates at even higher temperatures than the Windscale prototype, and research into this has been conducted at the nuclear experimental station at Winfrith for many years.

The pressurised water reactor is incorporated in the nuclear power stations at Shippingport and Dresden, U.S.A., and is used in the propulsion units of the U.S. nuclear submarines, the U.K. submarine *Dreadnought*, the U.S. ship *Savannah*, and the U.S.S.R.

*Top right*
A typical control room of the experimental boiling water reactor nuclear plant at Lemont, Illinois. The man at the console controls the reactor.

*Bottom right*
An aerial view of Calder Hall Nuclear Power Station. Since 1959 the station has been supplying over one thousand million units of electricity to the national grid each year.

ice-breaker *Lenin*. The small size and great reliability of this reactor type is its major attraction for ship propulsion.

Economic consideration have influenced the various countries' plans for tackling the nuclear power challenge and this is reflected in the choice of reactor and the form of each country's nuclear power programme. In countries with very large resources of cheap fossil fuel, such as coal or oil, there is not the same pressing need to develop nuclear power fully that exists in countries where such fuels scarce and must be imported. For example, in the U.K. where it is foreseen that the exploitation of coal resources is approaching its possible maximum, effort has been concentrated to obtain economic nuclear power as quickly as possible. In the U.S.A. with its large resources of cheap coal and hydro-electric power, technological effort has been spread over a wide front so that the full potentials of many reactor systems can be appreciated for future possible use.

The division of power generation costs into the three components — capital charges, fuel replacement charges and operating costs — will largely decide the most suitable reactor type. Operating costs are unlikely to vary and the other two components of cost are so related by the nature of reactors that, in general terms, the more highly enriched the fissile fuel the smaller the reactor. Thus a reactor using highly enriched fuel will have lower capital cost but greater fuel replacement charges than a reactor using low enrichment

fuel, since fuel enrichment tends to be an expensive process. A compromise must be effected between these conflicting factors according to each reactor's particular situation and function. For example, where a nuclear power station is to be used for continuous, base-load operation, high capital costs may be tolerated, while similar costs would be prohibitive in a station destined for intermittent use. Here higher fuel costs can be tolerated if the advantage of low capital charge is gained.

At present there is no nuclear power station in operation or under construction that is competitive with the lower costs of conventional power but it is ensivaged that nuclear generating costs will be lower than coal-fired generating costs in the U.K. by the 1970s. Other parts of the world give varying estimates for this stage, and in some places it is unlikely that such economic parity will ever be reached.

The future for fusion power is not so clear-cut since at the moment this possible power source is only in the fundamental experiment stage, and it will be several years before it will be possible to say whether the method is capable of exploitation and whether it will be economically competitive with other sources of power. But as the raw material for this method of nuclear power production is heavy hydrogen, or deuterium, which forms one part in five thousand of all the virtually inexhaustible hydrogen in oceanic waters, the prospect is one of unlimited power supply.

# The Past

## Our Place in the Universe

Fable makes fun of the astrologer who, looking up into the sky, fell into a well. Yet the stars have much to teach those who can understand or interpret their 'obscure clarity'.

The mind is bewildered by the vast number of stars and by their distance from the Earth. In the northern hemisphere alone three thousand are visible to the naked eye; 100 million can be seen with the aid of the huge Mount Wilson telescope in California; photographic plates reveal millions more — some so distant that they appear as clouds of pale grey light. To the luminous and visible stars must be added the almost incalculable number of those which are extinct and invisible, revealing their existence only by the attractive force they exert. The development of radio astronomy has revealed powerful, yet invisible transmitters of radio radiation in the sky, some at the greatest distances to which optical telescopes have penetrated. These are being investigated by new radio telescopes under construction at Cambridge and at the University of Sydney, where the Mills Cross Telescope will be the largest in the world, with a range of 1,000 million light years. These telescopes will offer the opportunity not only of seeing in great detail the parts of the galaxy which are hidden from optical view by galactic dust, but also of providing new information on the galaxy itself.

Vast though their numbers seem, stars form but a very minute part of the universe. To be more exact, they constitute the accumulation of stellar material in which we live. In the language of astronomy, they constitute our nebula, our galaxy, our island-universe. Let us consider the galaxy of which the Sun and its planets are a part. Its roughly lenticular or saucer-shaped form explains why when we look at it edgewise more stars are seen than when we look at its flat side. The impression is of a luminous train, a girdle of stars of which we might be the centre. Indeed, astronomers once thought that the Earth and the solar system were the exact centre of our galaxy, called the Milky Way. It is now known that the solar system is approximately two-thirds of the way towards the edge. The stars in this galaxy are concentrated near the centre.

No ordinary unit of measurement is adequate in evaluating interstellar distances if a clumsy string of zeros is to be avoided. Much less cumbersome is the astronomer's unit based on the distance travelled by light in one second (186,000 miles). At 92,900,000 miles from the Earth the Sun is little more than 8 light-minutes away ($8 \times 60 \times 186,000 = 89,289,000$). This is infinitesimal compared with the 52 light-months which separate us from the nearest star, the 140 million light-years from the most distant star, and the 3,600 million light-years from the farthest known galaxy.

To say that a star is, for example, one million light-years from the terrestrial globe means that its light takes one million years to reach Earth. The light received today left the star when man's ancestors still went about naked and animals now extinct roamed the Earth. It is possible that the star has now disappeared, that it is now itself extinct or broken up. We continue to see it even though it no longer exists; its past is our present.

The Sun is but one star among others in a galaxy of 2,000 million similar visible members, a star which has nothing special about it.

Fossils of the Earth's past

Around the Sun nine known planets revolve perpetually. In order of increasing distance from the Sun, which provides them with heat and light, they are Mercury, Venus, Earth, Mars, Jupiter, Saturn, Uranus, Neptune, Pluto.

The dimensions of the solar system are such that the mind finds it difficult to grasp the significance of actual comparisons. Reducing size to more comprehensible proportions, let us suppose the Sun to be represented by a pumpkin. Mercury would then be represented by a piece of lead shot 16 yards away, Venus and Earth by small peas 38 yards and 54 yards away respectively, while Pluto, the outermost of the planets yet discovered, would be represented by a cherry at $9\frac{1}{2}$ miles.

What, then, is Earth within this group of planets? Nothing more than one planet among the others, average in its dimensions and in its position. Perhaps it is because of its average state that Earth still has sufficient water and oxygen to support life as we know it. Man is an inhabitant of a minute planet lost in the immensity of space; he is infinitesimal in stature and length of life; physically, he is scarcely distinguishable from the animals which surround him, his only special circumstance is his power of thought which has allowed him to have some conception of his universe.

Now that we know something of the present condition of the Earth, let us attempt to reach back into its history. Like the archivist-palaeographer, we shall have to study its sediments and its fossils. We shall consider the work of the geophysicist, geochemist, geologist, petrographer, stratigrapher, palaeontologist, and biologist.

Two problems immediately occupy our attention; that of the origin and evolution of the terrestrial globe, and that of the origin and evolution of life. The very beginnings of our study, they are the least well-established and the most doubtful of all the chapters of Earth's history.

At the present time astronomy is in a paradoxical position. While astronomers are able to predict with the utmost precision the date and duration of an eclipse or the passage of a comet across the sky; while they can construct complex apparatus of marvellous precision, they are still uncertain about the origin of the solar system.

### THE HISTORY OF THE EARTH

Direct evidence of the earliest history of the planet on which we live was lost in the profound changes that took place before the crust of the Earth was formed. Instead of evidence we have only questions. Was the protoplanet incandescent like the Sun, or was it cool, much as it is today? What was its size? Its shape? Was it a gas, a liquid, or a solid?

As Earth is one of nine planets in the solar system, its origin must be bound up with the origin of all the other elements in this system. Therefore it is the astronomers and the astrophysicists that we turn for clues about its birth.

A German philosopher, Kant, in 1755, and a French mathematician, Laplace (1796), published similar ideas about the origin of the solar system. They pictured as a starting point a gaseous mass at a very high temperature, a sort of giant Sun stretching beyond the orbit of the most distant planet, and possessing a uniform, rotational movement. Its form was that of a discus or saucer, the central part a nucleus of condensed material contrasting with the diffused material near the rim. They supposed successive equatorial rings to have become detached from the gaseous mass by centrifugal force and to have condensed finally into the planets. The theory explains certain characteristics of the solar system: the planets' increase in mass and decrease in density with distance from the Sun, and their rotation in almost the same plane and generally in the same direction as the Sun itself. Finally, the satellites would be formed by the same process, at the expense of the still hot and liquid planets.

To illustrate this theory, the physicist Plateau devised a simple experiment — too simple perhaps — in which a drop of oil was put into suspension in a mixture of water and alcohol of equal density. The drop was then made to rotate about itself more and more rapidly. It was observed to flatten at the poles and to release in its equatorial plane a ring which resolved into small droplets. The experiment was, however, inconclusive, for the droplets never joined to form a single sphere comparable with a planet. Further, the experimental conditions were in no way comparable to the conditions required by the theory of planetary rings. A drop of liquid of uniform density, subject to capillary forces, to surface tension, and to hydrostatic pressure of the surrounding medium must necessarily be a rough and inexact picture of a primitive nebula.

It is possible today to oppose Kant's and Laplace's theory with arguments that destroy it completely. In the first place it is difficult to explain why the direction of rotation of Uranus and certain satellites of Jupiter, Saturn and Neptune should be the reverse of those of other members of the solar system. It is equally difficult to explain the preservation of Saturn's ring as a unit. Finally, and most important, physicists have established that the rings were formed of a gas which was millions of millions times more rarefied than the Earth's atmosphere. This gas would have diffused into space instead of condensing into planets, and this is exactly what astronomers are observing in the rings emitted by certain stars in the constellation of the Pleiades.

**Intruding Star and Twin Planets.** Two American geologists T.C. Chamberlin and astronomer F.R. Moulton, and two Englishmen, astronomer James Jeans and geophysicist Harold Jeffreys, suggested that all the planets, their satellites, and the planetesimals (solid bodies much smaller than planets) were part of the Sun round which they now resolve. Their theory postulated that the solar system was formed by a star which, passing close to the Sun, dragged from it a sufficient amount of matter to form the planets. It is possible to imagine the intruding star approaching close enough to the Sun to induce a gaseous bulge on its surface. This bulge would then follow the movement of the intruding star, at the same time becoming elongated and incurved. Just as an oceanic tide has a corresponding antipodal tide on the globe, so the 'tidal filament' induced on the side nearest to the intruding star had its corresponding antipodal tide on the other side. The two antipodal bodies were incurved in opposite directions and assumed the shape of an 'S', with the Sun

*Left*
A star cluster enveloped by a nebula (NGC 6611).

*Top right*
The size of the stars and planets in the solar system.

*Bottom right*
Diagram illustrating the theory of the twin planets.

at the centre, Being electrically charged, they coiled in a spiral (again with the Sun in the centre) and quickly broke into planets situated in pairs at equal distances from the centre and rotating in the same orbit. Thus there were two planet Mercuries, two Venuses, two Earths, and so on. One of the two 'twin planets' — that which formed in the 'tidal filament'—turned towards the intruding star, which was bigger than the planets, and ended up by being captured. In this way the number of planets was reduced by half; the two Earths, for example, became one. This theory has tended to be discredited in recent years.

If we assume that these tidal filaments, or gaseous bulges, detached from the Sun by the intruding star, were thinned out towards their extremities and swollen in the centre, it can be understood how the planets at the present time increase fairly regularly in mass from Mercury to Jupiter and decrease from Jupiter to Neptune. (Pluto is not included as there is at present some uncertainty regarding its diameter.)

The difference between these two ideas is that Chamberlin and Moulton supposed that the materials pulled from the Sun by the passing star were planetesimals. The infall of the planetesimals into the nuclei of the planets would eventually sweep most of the smaller bodies from the space within their orbits. Jeans and Jeffreys suggested that the passing star pulled out long incandescent streamers of gaseous matter.

If, further, it is assumed that the density of the constituent materials of the tidal filaments dropped in proportion to their distance from the Sun — that at the base, for example, they were constituted of heavy metallic vapours, and at the extremities of the lightest gases from the surface of the Sun — we have an explanation of the planets' decrease in density the farther their orbits are from the Sun. There are acceptable reasons for the few exceptions which have been established (too low a mass for Mars, too low a density for Saturn).

The different characteristics of the planets today was determined to a large extent by the physical state of the twin planets at the time of their collision. The outer planets (those farthest away from the Sun) collided in a gaseous state. This resulted in a number of satellites for each planet: twelve for Jupiter, ten for Saturn, five for Uranus. The inner planets — those nearest to the Sun — collided in a liquid state and so the number of pieces dragged off was either restricted or non-existent, and these planets had few satellites — none for Mercury and Venus, one for the Earth, two for Mars. They were also denser.

Not all satellites would have the same origin. Some would be formed from material dragged from the mother planet. Others would be captured from outside. Asteroids are minor planets, over one thousand in number, which revolve around the Sun in orbits between those of Mars and Jupiter, and of various eccentricities. The gravitational fields of the two planets have a strong effect on asteroid movement, and it has been suggested that eight of Jupiter's twelve satellites and the two satellites of Mars are merely captured asteroids.

As for the Moon, opinions are divided. Is it a small planet captured by the Earth, or was it born of the same substance? Is it in fact the sister or the daughter of Earth? If we favour the second hypothesis we can put forward the theory that from the twin Earths which collided tangentially in the liquid state a filament of superficial materials was dragged off to form the Moon. This filament would have taken the form of a bulge and then a pear before finally becoming detached from the Earth, and as it was dragged off would have raised the temperature of the Earth by 1,800°F. (982°C.) Without being able to date this event precisely it can at least be said that it must have happened more than 4,500 million years ago, long before the Earth's crust was solid. This hypothesis is more in keeping with modern trends of thought than that suggesting that the Pacific Ocean basin represents the scar left when the Moon was snatched from the Earth.

**The Contracting Protostar.** Another approach to the origin of the solar system is a comparison of solar and planetary compositions which must have been almost identical originally. It is true that the Sun, the stars nearest to it, and interstellar matter have almost the same composition. Kuiper (1957) of Yerkes Observatory has summarised the current theories about possible events in a contracting protostar, originally part of the Milky Way.

At some stage the contracting protostar left behind a small amount of its gaseous mass which later formed the planets. These gaseous protoplanets were developed from a rotating nebula surrounding the Sun. As the Sun became luminous only near the end of its period of contracting, the temperature of the nebula was low. Heat was lost by radiation into interstellar space until the absolute temperature dropped to about 50° Kelvin (—223°C. or—433 F.). Because of the low temperature, the nebula assumed the shape of a thin disc with an increased density. Later, gravitational instability broke it up into separate clouds. Thus the protoearth was formed as a gaseous disc in which some of the gases (ammonia and water-vapour) were frozen out under low temperature. Other gases present were hydrogen, helium, neon, and methane. The Moon was formed at the same time as the Earth, not as a true satellite but as a double planet, similar to a binary star. The size and mass of the protoearth at this time were hundreds of times greater than today.

The Sun continued to contract for the next eighty million years or so, finally becoming a bright star and bringing changes in the entire solar system, including the protoearth. Interplanetary space was swept clean of gases by streams of ionised particles much as the ionised tails of comets are swept away from the Sun today. In the absence of interstellar gas the young Earth was bombarded by a powerful ultra-violet radiation, and temperatures became

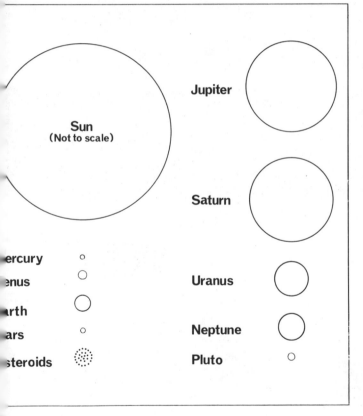

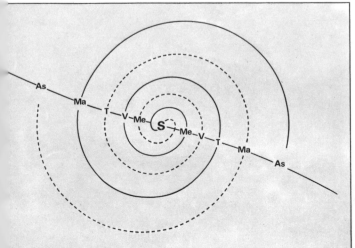

comparable with those of today. Most of the heavy, inert gases were lost while the Earth's exosphere was still several hundred times higher above the centre than it is now.

The material at the centre gradually heated up following compaction and radioactivity. Finally, the formation of a molten core of iron became possible. Further geo-chemical separations took place during the melting period, a phase lasting about 1,000 million years. Then the mantle began to solidify, but only the outer 500 miles cooled to the freezing-point. Radioactive elements were separated from the silicates and concentrated in the lower part of the crust where they acted as sources of vulcanism and diastrophism.

Kuiper states, 'Some idea of the evolution of the new atmosphere can be derived from the basis of the processes operating in the present atmosphere. Ammonia, water-vapour, and methane (in this order) are decomposed photochemically in the higher layers of the atmosphere to which solar ultra-violet radiations can penetrate. These processes are in the part irreversible because hydrogen will be diffused upward and lost at exosphere level. Nitrogen and oxygen cannot escape and must accumulate or be used in chemical combination. Carbon dioxide and carbon monoxide are are prominent constituents of volcanic gases, and the vast quantities of carbonate rock produced during geologic time suggests that this always been so. If the present rate of oxygen production from water-vapour by photolysis is compared with the average rate of volcanic carbon monoxide production, it is found that the latter is somewhat too great to permit oxidation of all carbon monoxide to carbon dioxide. On that basis, no free oxygen would have been present in the atmosphere until photosynthesis by plants appreciably increased the oxygen production rate.'

It has been pointed out that an atmosphere oxygen content of only 0·001 percent (compared to the 21 per cent in our present atmosphere) would produce an ozone layer half as thick as the present layer to provide an effective shield against the ultra-violet radiations dangerous to living cells. In this way, the stage was set for life on Earth, which began some time after the formation of the crust (about 4,600 million years ago) and before the time when the oldest fossils are estimated to have lived about 2,600 million years ago.

## OUR ULTIMATE DESTINY

Just as the origin of our planet was an inseparable part of the origin of the solar system, so Earth's future must be linked with that of the solar system. In seeking to determine what lies in store for our planet we must look for causes outside the Earth.

If the Sun retains its present brightness and relative position with the planets, life on Earth will go on much as it does today. The loss of gases will be about the same, and will be limited to hydrogen and helium, which are formed about as fast as they are lost. In spite of the decomposition of water-vapour and the loss of hydrogen, the oceans will not disappear. Although helium is being formed by radioactive decay, the decay will slow down the melting of rocks in the Earth's crust. Vulcanism and mountain building will occur with decreasing frequency. Ultimately the mountains will be eroded and their sediments will build the edge of the continental shelves seaward to fill the oceans, possibly resulting in a universal sea. On the other hand, the thin crust will be lifted isostatically until the equilibrium of low lands and vast seas tends. resemble the Low Countries on world-wide scale. The duration of this flattening process will involve a period roughly double that taken to produce Earth's present condition.

The Moon's distance from the Earth will change during this great period of time, affecting both the tides and the rotation of the Earth. We are still left to decide whether the Moon will ultimately complete its recession from the Earth only to be destroyed on its return to the Earth's proximity.

Because of the vast time-scale involved in such possible changes on the Earth, let us consider events that could occur in the solar system or galaxy. If the Sun should encounter another star, the solar system would be at once involved in a holocaust that would probably vaporise or disintegrate all solid and liquid elements now in existence. However, the chances of such a collision are too remote to be seriously considered.

Let us consider, too, the Sun's history. Astronomers have studied the evolution of the many island-universes (other star systems scattered throughout the cosmos beyond the confines of our own galaxy), and from them have discovered much that is relevant to the study of the Sun's destiny. Knowledge of the size, brightness, radio-activity, and other physical and chemical characteristics of other island-universes can be applied to our own system. After contracting from a protostar the Sun again began to expand to its present dimensions, and as it grew larger so it became hotter. This pattern is expected to continue until the Sun becomes a giant star, though not as large or as bright as the Mo giant, Antares, which is 3,500 times brighter than the Sun and 480 times greater in diameter. The mean absolute temperature of the Earth, now slightly above 279° K. (6°C.), will increase threefold until in about 2, 000 million years' time the oceans will boil. All water eventually vaporise and enter the atmosphere, and although the increase in the Sun's temperature will cause a one-hundredfold increase in radiation, steam will form an effective blanket around the Earth.

Ultimately the Sun will pass through a white-dwarf stage, the characteristics of which are small size, low luminosity and extremely high density. The atmosphere of steam will cool and condense to form the oceans once more. As cooling continues the oceans will freeze, leaving Earth a sphere completely covered with ice and snow, with much the appearance of Greenland today.

Will the dwarfed Sun and its frozen planets be able to enter another nebula, or is it possible that the expanding universe, after reaching its ultimate position, will repeat in reverse order the series of events through which it has previously passed? Whether Earth and other members of the solar system are destined to an icy or a fiery end is of little practical concern to us, for tremendous dimensions of time and space would be involved. Possibly the universe is not proceeding from an unknown beginning to an inexorable end. It may be that conditions will swing forward and back like some giant pendulum. It may be that the Earth will preserve its habitable environment for an incalculable period of time. Known facts can give us little lead here. This much we can be reasonably sure of — the conditions under which life now flourishes will continue for a very long time.

**The Moon and the Earth.** The primitive crust of the Earth, which was and still is very thin compared with Earth's radius, must have been reshaped, reworked and remelted many times before it became thick enough to resist the pull of the internal tides. And it must have looked vastly different during the first periods of its existence.

The Moon will give us some idea of the young world. A telescope or even the naked eye shows the lunar surface to be dotted with dark, more or less circular areas called 'seas' covering almost a third of the visible hemisphere. Some are of colossal dimensions; the Sea of Rains, or Mare Imbrium, has a diameter of 750 miles and an area of about 400,000 sq. miles. After the seas, in order of importance, come the cirques, craters and craterlets. The smallest craterlet exposed at the present time is no more than 300 yards in diameter, and the total number of these features visible is over 30,000.

There was a time when lunar craters were thought to be craters and shell holes caused by the impact of falling meteorites. It must be admitted that the resemblance between the smallest moon crater and the Meteor Creator of Arizona or Chubb Crater in Canada is remarkably close; they display the same flat bottoms and steep ramparts. However, it is not possible to use this hypothesis to explain the existence of cirques and 'seas' several hundred miles in diameter. Meteorites or asteroids big enough to produce such craters would have brought about the complete destruction to the Moon. Moreover, the hypothesis of a lunar bombardment does not explain the alignment of the craters along cracks and fractures.

A plausible explanation of the Moon's relief is that it is the result of a long series of volcanic eruptions. Beneath the first lunar crust, the molten materials were subject to the attraction of the

*Top*
Tycho with its central column.
*Bottom*
Close-up photograph of the crater

Copernicus, one of the most prominent features on the face of the Moon.

Earth and gave rise to uplifts related to tidal rhythms. Enormous swellings, commonly called blisters or bumps, were formed. While some had only a fleeting existence, others burst and then foundered after the lava inside them had escaped. This violent action produced the 'seas', craters, and other features characteristic of the Moon.

It should be noted that this type of vulcanism is in no way comparable with the vulcanism known on the Earth today, which is based essentially on the escape of water-vapour under pressure. In primitive vulcanism, probably the only process was the breaking out of internal magna under the attractive forces of neighbouring bodies.

Another theory has been put forward to explain lunar relief and the polygonal distribution of cracks. It is suggested that convection currents similar to those seen in a pan of boiling jam may have been produced in the viscous material of our satellite. Ascending warm currents and descending cold currents would then have delimited polygonal areas of 'cells' whose outlines were preserved on the surface of the Moon after cooling. In the chapters on vulcanism and the movements of the Earth's surface we saw that similar convection currents are today believed to be responsible for these phenomena.

The development of the very characteristic relief of the Moon gives us an indication of what must have taken place on the Earth about 3,000 or 4,000 million years ago. The primitive crust of the globe must have been the theatre of innumerable eruptions and subject to internal convection currents which produced great fractures. The difference is that the absence of atmosphere and water on the Moon has preserved the craters and fractures intact, while on the Earth the same formations disappeared under the combined effects of vulcanism, isostasy, and the great influence of erosion and sedimentation.

Is it possible that traces of Earth's 'primitive lunar relief' still survive? Certain geophysicists believe that the vast oceanic troughs of the Pacific, Atlantic, Indian and Arctic Oceans are ancient 'lunar seas' dating back to the 'lunar period' of the Earth. Such a theory of the permanence of the essential of terrestrial relief is, of course, opposed to the theories of continental founderings and continental drift.

**The Original Crust.** No vestige of the original crust of the Earth has yet been found. How, then, is it possible to describe, analyse, weigh, or measure materials to provide some basis for comparison with the Moon? The oldest known rocks in Australia and Africa once thought to have been the 6 fundamental gneiss' are now known to be intrusive rocks — rocks intruded as magna into even older rocks, eroded eons ago, which may in their turn have been intrusive into still older rocks. The supposed fundamental gneiss occurs in continental 'Shields' thought to be the roots of ancient mountain systems. They are therefore pieces of historical evidence pointing to events which we know took place repeatedly in the last 500 million years. It was in the course of these events that the astronomical history of the Earth gradually evolved into the geologic history.

By radioactive methods periods of instrusion have been dated in the Canadian Shield alone, and doubtless others will be found in time. How many periods of mountain building will be discovered in the known ancient Shields—Australian, Baltic, Brazilian, Canadian, Guianan, Indian, and Siberian—is still to be demonstrated.

## THE AGE OF THE EARTH

**Geologic Time.** Before the introduction of radioactive methods to the measurement of geologic time, the age of the eras, periods and epochs was expressed only in relative terms, based on the rate of weathering, biologic evolution, marine invasion, loss of heat by the Sun or Earth, sedimentation in geosynclines, accumulation of salt in the seas, crustal folding, the rise of mountains, and so on. None of these methods can be applied singly over wide areas, nor can they be correlated to build up a complete picture of Earth's history, and not until it was discovered that radioactive minerals disintegrate at fixed rates to form new minerals and one or more decay products, was there a means of ascertaining the absolute age of the different divisions of geologic time. Uranium ($U^{23}$), for example, distintegrates into lead ($Pb^{206}$) and 8 atoms of helium ($He^4$) in a little more than 4,500 million years. As particles of radioactive minerals enclosed in the rocks would start to disintegrate at the instant of crystallisation, the age of the mineral must bear a fixed relation to the ratio of helium to uranium and to the ratio of uranium-lead to uranium. From the known rate of disintegration and the amount of lead produced since the rock crystallised, it is possible to calculate the date of its formation. If, therefore, we know that 1,000,000 gr. of uranium will produce 1/7,600 gr. of lead every year, then age equals pb/U × 7,600 million years. In the same way the actinium and thorium families can be used to determine the age of enclosing rocks.

## A TERRESTRIAL CALENDAR

**Relative Age.** If we know that an event fell in the reign of King George V, we immediately conclude that it took place later then events happening in the reign of George IV, and earlier than those under George VI. On a much larger scale we talk of eras and ages: the Christian era, the atomic era, or the Stone Age, the Iron Age, and so on. In all these, objects and events are given a relative age, not an absolute age expressed in millennia, centuries, years, and days. The same principles obtain in geologic events, for determination of absolute age is of recent application.

Rocks and their fossils for long enjoyed only a relative chronology, quoted in terms of a scale of eras, periods, epochs and ages similar to the eras and ages in human history. The position of sedimentary rocks in this scale was achieved by patient study of their relationship to older or younger rocks, and in the case of fossils it was necessary to study their degree of evolution.

It can be seen that, on this scale, Earth's history is divided into six major eras. The Azoic era, the oldest, represents the unknown period during which the crust was formed, before any known life forms existed. This was followed by the Archaezoic, Proterozoic, Palaeozoic, Mesozoic, and Cenozoic eras. The first three from which the record of life is lacking or very scanty, are commonly grouped together, making up the Pre-Cambrian. Generally. eras are separated by major geologic events such as the development of mountain chains, the emergence or submergence of continents, the appearance or disappearance of an important group of fossils. In turn, each era is divided into periods, sometimes into subperiods, and finally into epochs. Thus the Palaezoic era comprises the Cambrian, Ordovician, Silurian, Devonian, Carboniferous, and Permian periods. In the United States the Carboniferous is split into the Mississippian and the Pennsylvanian.

Sub-division of the era, etc. can be illustrated diagrammatically as follows:

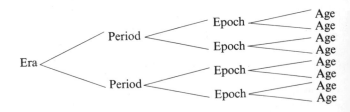

In a similar way, the rock formations can be divided into groups, systems and series.

Two sets of terms are employed for each geological division because, for clarity, it is necessary to discriminate between the strata themselves and the time intervals they represent. Thus, an era is the duration of a group; a period of a system, and so on. Rock formations are usually given a local name: for example, the Penrith sandstone, the Borrowdale volcanics and the Skiddaw slates.

**Absolute age.** If we take a Cambrian trilobite, a Cretaeous ammonite, and an Eocene cerithium, we see that on the relative scale the trilobite is the oldest since it belongs to the earliest period of the Palaeozoic. The ammonite is not so old as it belongs to the Upper

# GEOLOGIC TIME SCALE

| Era | Period or System | | Epoch or Series | Physical Events & Fauna | Million Years Ago |
|---|---|---|---|---|---|
| **CENOZOIC** | **QUATERNARY** | **NEOGENE** | **RECENT** | Glaciers melted; milder climates. Many mammals disappeared. | |
| | | | **PLEISTOCENE** | Glaciation; fluctuating cold to mild climates. Most invertebrates living species. Dominance of large mammals. Man. | **1** |
| | **TERTIARY** | | **PLIOCENE** | Continued uplift and mountain building. Climate cooler. Mammals reach peak in size and abundance. | **10** |
| | | | **MIOCENE** | Uplift of Sierras and Rockies. Moderate climates. Rise of grazing mammals. | **25** |
| | | **PALAEOGENE** | **OLIGOCENE** | Lands generally low. Rise of Alps and Himalayas began. Volcanoes in Rockies area. First sabre-tooth cats. | **40** |
| | | | **EOCENE** | Mountains eroded, many lakes in western North America. Climates mild to very tropical. All modern mammals present (first horses). | **60** |
| | | | **PALAEOCENE** | Mountains high, climates mild to cool. Primitive mammals, modern birds and new invertebrates. | **70** |
| **MESOZOIC** | **CRETACEOUS** | | | Lands low and extensive, mild climates. Last widespread oceans. Flowering plants expand rapidly. Giant reptiles become extinct; modern insects. Ammonites die out. Foraminifera. Period closed with Laramide Revolution (Sierra uplifted). | **135** |
| | **JURASSIC** | | | Continents low; large areas of Europe covered by seas. Climates mild. Mountains from Alaska to Mexico rise; eruptions and intrusions in the north-west. Dinosaurs, marine reptiles, ammonites and belemnites abundant. Ginkgos, conifers and cycads. | **180** |
| | **TRIASSIC** | | | Continents mountains, large areas arid. Eruptions in eastern North America and New Zealand. First dinosaurs and marine reptiles. First hexacorals, last conodonts. | **220** |
| **PALAEOZOIC** | **PERMIAN** | | | First mammal-like reptiles, other reptiles diversified. Many marine invertebrates became extinct, last trilobites and hexacorals. Period ended with Appalachian Revolution. | **270** |
| | **CARBON- IFEROUS** | **PENNSYLVANIAN** | | Lands low, covered by shallow seas or extensive coal swamps, climates warm. Amphibians and reptiles reach large size; large insects, scorpions, cockroaches. Fusuline foraminifera abundant. | |
| | | **MISSISSIPPIAN** | | Widespread seas retreated as result of mountain building in eastern United States, Texas, Colorado. Climates warm. Crinoids dominate; amphibians, sharks and bony fishes spread. Insects developed wings. | **350** |
| | **DEVONIAN** | | | North America low and flat, but mountains and volcanoes present in eastern United States and Canada. Europe mountainous with arid basins. Fishes dominant, first amphibians, many brachiopods, corals, bryozoans and blastoids. First ammonites. | **400** |
| | **SILURIAN** | | | Continents relatively flat, mountain building in Europe; climates mild. Much salt deposited. Eurypterids and corals abundant. First air breathers; lycopod plants. | **440** |
| | **ORDOVICIAN** | | | Continents low with shallow seas, mountains rose at close in Europe and North America. Abundant graptolites, trilobites, nautiloids, cystoids; first ostracods and conodonts, seaweeds and algae. Climates appeared uniformly mild. | **500** |
| | **CAMBRIAN** | | | Extensive seas in major synclines on all continents. Climates mild. Marine invertebrates and algae abundant. Trilobites dominant. All animal phyla probably existed. | **600** |
| **PRE-CAMBRIAN** | **PROTEROZOIC (ALGONKIAN)** | | | Shallow seas in geosynclines. Climates warm and moist to dry and cold. Glaciation in eastern Canada. Bacteria, marine algae, worm burrows sponge spicules. Probably mosts phyla lived but left no record. Few fossils. Iron ores of Lake Superior formed. | **1000** |
| | **ARCHAEOZOIC (PRIMITIVE LIFE)** | | | Extensive mountain building with intrusions and eruptions. Iron deposits formed. Earliest known life, blue-green algae, fungi. Graphite and carbonaceous shales in Australia and Canada. Carbon in Rhodesian rock. | **2000** **2600** **3000** |
| | **AZOIC (WITHOUT LIFE)** | | | Formation of the Earth's crust. No rocks have been found, therefore cannot be dated by any known method. | **3500** **4000** **4500** **5000** **6000** |

The series of the systems in the Mesozoic and Palaeozoic are usually designated Lower, Middle, and Upper, but are given provincial names in many areas; thus the Chester series (Upper Mississippian) of the Mississippi Valley is the equivalent of the Visean (Lower Carboniferous) of Europe. The corresponding time terms that are applicable are Early, Middle, and Late. The accuracy of dating decreases from the younger to the older periods, especially in the Pre-Cambrian. Beyond the age of the oldest known fossil (about 2,600 million years) life had its origin. The dates preceding this even can only be conjectural. Time estimates after Arthur Holmes, 1960 a revised geological time scale. Edinburgh Geol. Soc. Trans., v17, pt. 3, pp. 183–216.

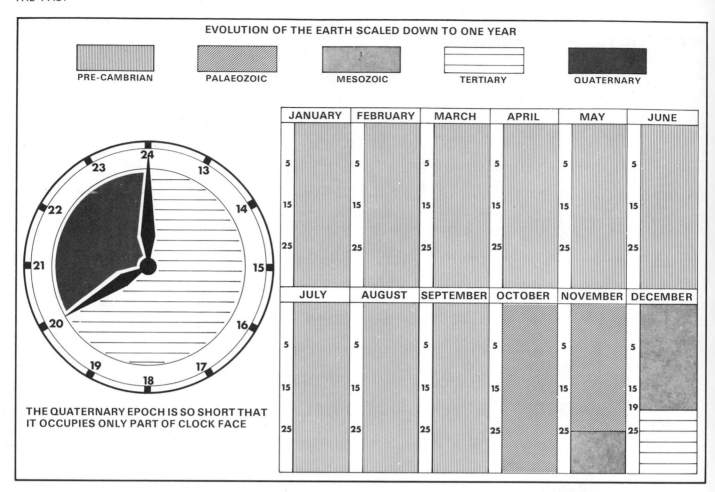

EVOLUTION OF THE EARTH SCALED DOWN TO ONE YEAR

PRE-CAMBRIAN    PALAEOZOIC    MESOZOIC    TERTIARY    QUATERNARY

THE QUATERNARY EPOCH IS SO SHORT THAT IT OCCUPIES ONLY PART OF CLOCK FACE

Mesozoic period and the certithium is the youngest, dating from the Tertiary.

Thanks to radioactive methods, much more accurate dating is now possible and it can be established that such and such an epoch lasted 20 million years, and that a fossil is 350 million years old.

It is known that elements of all matter are made up of atoms, extremely small particles invisible under the microscope and formed, like the solar system, of a central nucleus around which rotate negatively charged electrons. The nucleus in its turn is an aggregate of particles of which some are neutral (neutrons) and others positively charged (protons). In a stable element there are always as many electrons as protons, so that their charges cancel out. In a radioactivity element, on the other hand, protons and neutrons continually escape, and this process modifies the structure of the atom and leads to the transmutation of matter. The disintegration of uranium into lead illustrates this process.

Primary uranium ($U^{238}$), the heaviest and most complex element known, loses an atom of helium ($He^4$) every 4,560 million years to form $U^{234}$. The process continues, and a little over a quarter of a billion years two more helium atoms are lost. The loss does not stop here, but continues until, after passing through a number of stages (some of very short duration), lead ($Pb^{206}$) and a total of eight atoms of helium are released ($U^{238} - 8\,He^4 = Pb^{206}$).

Uranium-235, the form associated with atomic energy, disintegrates into lead-207. Another radioactive element, thorium-232, disintegrates into lead-208. These various forms of lead are known as isotopes, a term which can be described as follows: the chemical properties of an atom, or rather of a number of identical atoms great enough to permit the determination of these properties, depend primarily upon its atomic number. Atoms, or collections of atoms, which have the same atomic number but which differ in other respects, in mass or stability perhaps, are called *isotopes*. For example, the investigation of natural radioactivity revealed that radium C, radium E, thorium C and actinium C were all identical in their chemical properties with bismuth (atomic number 83), but differed from it and from one another in radioactive properties and origins. To say, therefore, that the various forms of lead derived by the disintegration of uranium-238, uranium-235 and

thorium-232 are isotopes means that they are identical chemically and cannot be separated by chemical methods of analysis; they can be distinguished from one another only by a special apparatus called a mass spectrograph.

Returning to the determination of the absolute age of the strata of the Earth we note that the disintegration of uranium or thorium is a phenomenon whose mechanism has now been clearly determined. It is known, for example, that one gram of uranium-238 produces in one year $0.014 \times 10^8$ grams of lead-206 and $1.2 \times 10^{-8}$ cubic millimetres of helium ($10^{-8} = \dfrac{1}{100,000,000}$). So, if a specimen of rock contains 100 grams of uranium$^{238}$ and 7 grams of lead-206 the proportion of these two elements can be calculated and used to date the rock. As 100 grams of uranium produce $1.4 \times 10^{-8}$ grams of lead per year, it has therefore taken 500 million years to produce the 7 grams of lead in that specimen. Actually the calculation is not quite so simple but to go into greater detail is unnecessary here.

Sometimes the 'helium method' is preferred to the 'lead method', with the relative proportions of uranium and helium used to give a dating. This method, however, is less accurate because some helium (a light gas) may have escaped from the rock.

The 'strontium method' makes use of the transformation of rubidium into strontium, during which $Rb^{87}$ changes into $Sr^{87}$ by the emission of a $\beta$-particle. The rate of decay is very slow, for the half-life of $Rb^{87}$ is about 60,000 million years.

The 'lead method' is the surest and can be applied over the longest period of time. It acts as a natural clock that has functioned without irregularity since the solidification of the Earth's crust more than 4,500 million years ago. From it we learn that the oldest known fossil, the imprint of an algae in a graphite limestone, may date back 2,600 million years. The figure of 4,500 million years for crustal solidification may have to be increased if one gives credence to the discovery, in South Africa, of a mica more than 3,500 million years old. The lead method shows, too, that the beginnings of the Pre-Cambrian, Palaeozoic, Mesozoic, Tertiary and Quaternary eras are respectively 3,000 million, 600 million,

**Right**
Each stage of the disintegration of uranium is represented by a circle. The half-life of each stage is given by the arrow connecting the circles. The small circles indicate the loss of helium. One million grams of uranium will produce 1/7600 grams of lead in one year.

**Far right**
Rhythms of Earth history showing the 'pulsations' of sea level, mountain building, and climate. After Umbgrove (1947).

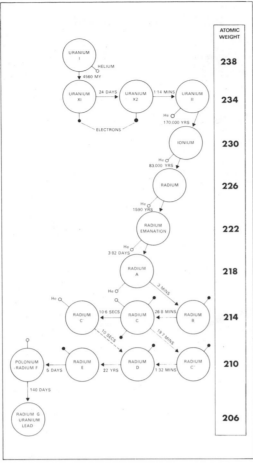

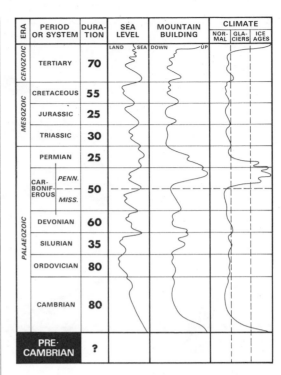

| ERA | PERIOD OR SYSTEM | | DURA-TION | SEA LEVEL | MOUNTAIN BUILDING | CLIMATE | | |
|---|---|---|---|---|---|---|---|---|
| | | | | LAND ⎸ SEA | DOWN ⎸ UP | NOR-MAL | GLA-CIERS | ICE AGES |
| CENOZOIC | TERTIARY | | 70 | | | | | |
| MESOZOIC | CRETACEOUS | | 55 | | | | | |
| | JURASSIC | | 25 | | | | | |
| | TRIASSIC | | 30 | | | | | |
| PALAEOZOIC | PERMIAN | | 25 | | | | | |
| | CAR-BONIF-EROUS | PENN. | 50 | | | | | |
| | | MISS. | | | | | | |
| | DEVONIAN | | 60 | | | | | |
| | SILURIAN | | 35 | | | | | |
| | ORDOVICIAN | | 80 | | | | | |
| | CAMBRIAN | | 80 | | | | | |
| | PRE-CAMBRIAN | | ? | | | | | |

270 million, 70 million, and 1 million years old, approximately.

For the Quaternary and more particularly for the 15,000 years since the retreat of the ice-cap, a new radio-active 'clock', based on the disintegration of carbon, is employed. It is known that two carbons exist in nature: ordinary carbon (carbon-12), which is stable, and radioactive carbon (carbon-14) which is progressively destroyed with the emission of radiations. In the course of their existence all living things absorb both forms, and contain at the same time carbon-12 and carbon-14; for every atom of the latter there are 1,000,000 million atoms of carbon-12.

On the death of an animal, carbon ceases to be renewed in the body, and the carbon-14 begins to diminish by disintegration, following a known pattern. By the end of 5,600 years half has disappeared, three-quarters after 11,200 years, seven-eighths after 16,800 years, and so on. The progress of this disintegration can be measured by diminution in the intensity of emitted radiation, using a very sensitive Geiger-Muller counter similar to that used for other radioactive elements. Specimens of bone, wood, or old tissue to be examined are carbonised and reduced to pure carbon. It is the carbon that is put into the counter to enable the age of the specimens to be determined.

The carbon-14 method, applied in the United States since 1948, has already produced surprising discoveries. Not only has it determined the age of certain sarcophagi, mummies, and ancient monuments, it has also dated charcoal found in prehistoric caves, fossil pollen grains, and other items. The 'carbon clock' has only one disadvantage: it can go back only about 26,000 years.

It is all very well to write of eras lasting millions of years, but the mind cannot conceive of time on such a scale. Other methods, however, can suggest the proportions of the years intervening between our own times and the formation of the Earth's crust. Suppose, for example, that today you are celebrating your fortieth birthday and that these 40 years represent the 4,000 million years of geological history. On this basis, life began when you were 14, the Pre-Cambrian era when you were 20, the Palaeozoic when you were 35, the Mesozoic at 38, and the Tertiary at 39 years 4 months. As for the Quaternary, you came into it only four days before your present birthday, and the Ice Age ended about two hours ago.

## PRE-CAMBRIAN ERAS

These began when the Earth's solidified crust buckled up in places and hollowed out in others, bringing together the oceanic waters in the low regions and the fresh waters in the higher regions. Then erosion started, and with it came sedimentation. Finally, life in its simplest form appeared.

The Pre-Cambrian era is often divided into three periods: Azoic (without life), Archaeozoic (oldest life), and Proterozoic (when most of the known phyla developed).

A particular terrain can be described as Pre-Cambrian only if it is normally found below rocks that are dated by their fossils as Cambrian. In Brittany, for example, the Saint-Lô phyllites are overlain by Cambrian conglomerates and shales. Similarly, in Shropshire on the border between England and Wales, and in the north-west Highlands of Scotland, gneisses, schists, conglomerates, grits and slates are referred to the Pre-Cambrian because of their relation to superincumbent Cambrian strata. In the Central Massif there are immense stretches of gneiss and mica schist that are probably Pre-Cambrian, but definite proof is lacking.

Determining the distribution of land and sea during the Pre-Cambrian is obviously a difficult task. One has only to think of the geographical changes which must have taken place in the course of 2,000 million years or more, and of the rarity of unaltered and unmoved sedimentary rocks which alone would enable us to say, 'There was the sea, there the land, and there fresh water'. We can, however, be almost certain of the existence of emerged continents. At the end of the Pre-Cambrian era there must have been six — the Canadian, Russo-Scandinavian, and Sino-Siberian continents in the northern hemisphere, and the Brazilian, African, and Australian in the southern hemisphere. The best-known 'shield areas' are those in the north, and it will be observed that they form the stable platforms of North America, Europe and Asia.

**The Canadian Shield.** One of the main present-day outcrops of Pre-Cambrian rocks is found north of the Great Lakes in Canada, with Hudson Bay as the centre of the region. On the edge of this vast area, which also comprises Greenland, Spitzbergen and the

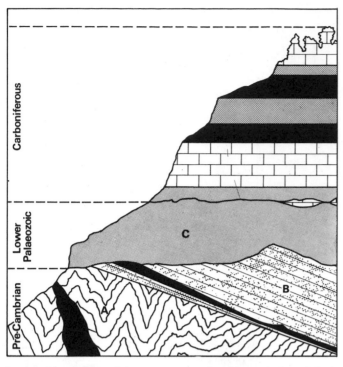

Section through rocks of the Grand Canyon. Note the unconformities at A, B, and C separating rocks of different ages

extreme north of Scotland, the Pre-Cambrian terrains disappear beneath the more recent rocks, beneath the Cambrian in particular. In the United States, Pre-Cambrian rock has been found in borings at a depth of about 6,000 feet and they have been brought to light through the agency of that huge natural cut, the Grand Canyon.

**Russo-Scandinavian and Sino-Siberian continents.** The Baltic Sea occupies the centre of the former Russo-Scandinavian continent which comprises Sweden, Finland, and Lapland, where there are Pre-Cambrian outcrops. In the west the rocks plunge beneath north Germany, and to the east beneath the horizontal strata of the Russian platform. Large lakes, like Onega and Ladoga in the Soviet Union, form a girdle comparable to the Canadian Great Lakes.

The former Sino-Siberian continent, whose limits are as yet imperfectly known, outcrops to the north of lake Baikal in the Irkutsk region. It disappears in the south beneath the Siberian plain of which it forms the base.

**Pre-Cambrian rocks.** Pre-Cambrian rocks are essentially highly folded gneisses and mica shists with much of their 30,000 to 50,000-foot-thick mass containing igneous intrusions. They are undoubtedly old sedimentary rocks metamorphosed in the roots of geosynclines. This metamorphism decreases in the passage to the youngest beds which are made up of thousands of feet of limestones, dolomites, sandstones, conglomerates and shales preserving to greater or lesser extent their original sedimentary characters.

Pre-Cambrian terrains also contain important ores which form the basis of some of the mineral wealth of the United States, Sweden and Finland. Around the Great Lakes outcrops contain iron, copper, nickel, cobalt, lead, zinc, and silver. Mica is worked in the vicinity of Ottawa, talc near Toronto, and native copper near Lake Superior. Veritable mountains of iron oxide are found in the United States, Canada, and northern Sweden.

The Pre-Cambrian had desert climates which have left their traces in red and brown sandstones, and glacial climates which are indicated by polished and striated pebbles. Ancient mudcracks prove the existence of shallow seas, while algae deposits (probably formed by lime-secreting seaweeds) are found in limestones of Glacier National Park, Montana; ancient lava flows are evidence of volcanic eruptions.

**Pre-Cambrian mountains.** Numerous mountain chains must have been uplifted and worn down during the Pre-Cambrian eras. In the Grand Canyon, 3,000 to 6,000 feet of Palaeozoic sediments are cut by the river, and in the deepest parts Pre-Cambrian gneisses and mica schists can be seen. There are several series separated by unconformities as shown in the diagram opposite. Often a series like B begins with a conglomerate whose pebbles are derived from the preceding series.

The sequence of events can be reconstructed as follows: deposition of the beds of A on the bottom of the geosyncline — formation of the first mountain chain — erosion of these mountains and production of a peneplain — return of the sea over the land thus planed down — deposition of beds B — formation of another chain of mountains — erosion of these mountains, and so on.

Each series, A, B, C, corresponds to a cycle of geologic phenomena. We are touching here on one of the most grandiose conceptions in the history of the Earth, that of its immense duration. The fact that in Finland and in America there are six superimposed Pre-Cambrian series means that in each, six huge ranges of mountains as high as the Alps and the Himalayas were developed in succession and then destroyed. The uranium 'clock' indicates that these events took place between 2,000 million and 700 million years ago; 700 million years is the approximate age of the Huronian chains (the last formed), which must have extended from the Lake Huron region of America right up to the Hebrides off the extreme northern coast of Scotland. It is possible that offshoots of the Huronian chain reached Brittany. Here the phyllites of Saint-Lô and the shales of Granville and Douarnenez could have belonged to the Huronian Ranges. Pre-Cambrian vulcanism is shown by the flows of porphyry in the region of Tréguier and Paimpol.

**The first fossils.** Pre-Cambrian rocks contain such fossils as algae, radiolaria, sponges, worm-like animals, and biothiopods (lampshells). The Lamballe shales of France, for instance, contain siliceous debris of radiolaria and sponges. The Pre-Cambrian fauna must have had a certain richness which metamorphism or the lack of hard parts prevents us from appreciating to the full. Their evidence of the series of the Pre-Cambrian is so deteriorated and so obscured that we can no longer decipher it. We do not know and probably we shall never know what the first inhabitants of the Earth were like. The oldest fossils are 2,600 million years old. Out of the 4,000 million years that have passed since the formation of the Earth's crust, the oceans and the continents — and probably also the earliest life—more than 1,400 million years, although the exact number is not known, remain indecipherable.

# The Palaeozoic Era

We have seen that rocks can be described as Pre-Cambrian only if they are overlain by rocks which are dated by their fossils as Cambrian (early Palaeozoic). Thus, one of the criteria of a Palaeozoic rock is that remains of once-living organisms should be present in sufficient numbers and in a sufficiently high state of preservation to be recognisable without doubt. Another, but not quite so reliable, criterion is the grade of metamorphism (or degree of change from the original form of the deposit). Increasing grades of metamorphism can be illustrated by the sequence mud — clay — shale — slate — phyllite — mica schist; chemically these rocks are identical but the amount of compaction and cementation, and the development of constituent minerals in recognisable crystal form increases with the degree of change from mud to mica schist. Rocks of this composition are called argillaceous; generally speaking, they are represented by shales and slates in the Palaeozoic, as opposed to mica schists in the Pre-Cambrian. The corresponding sandy rocks are described as arenaceous; present in the Palaeozoic as compact grits and sandstones (more or less coarse-grained, respectively), in the Pre-Cambrian they have usually been recrystallised into quartzite. Calcareous rocks which occur as limestone or dolomite in lower grades of metamorphism change to marble as the degree of metamorphism increases.

A century and a quarter ago two Englishmen began a study of the Early Palaeozoic rocks in Wales. Sir Roderick Murchison, recognising the great age of the rocks he was studying compared with other strata that had been described in Britain, named them Silurian. Adam Sedgwick, working much farther north, found an even older stratum which he called Cambrian. However, on tracing the two new rock systems into the same area it was found that Murchison's Upper Cambrian formations were the same as Sedgwick's Lower Silurian farther south. In 1879, these formations were renamed Ordovician, after an early Celtic tribe known as the Ordovices.

Thus was established a principle of geologic nomenclature that is now almost universally followed. Periods, systems, epochs, and so on are usually named after a place or region where the formations were first described or where they are well exposed. The commonest exception is the naming of beds, zones, or formations after a fossil found in rocks of a particular area whose presence in them is unique.

The Palaeozoic era is the period during which Earth's physical and chemical make-up gradually became hospitable to animals and plants, recognisable today from their fossils. From the beginning of the Cambrian to the end of the Permian, the world's geography, its fauna and its flora differed vastly from those with which we are familiar. The Palaeozoic began 600 million years ago, at a time when the Huronian mountain chains were almost completely worn down. It ended 225 million years ago at a period when the Hercynian mountains were in their turn, in the process of being demolished.

The Palaeozoic era is divided into seven periods: the Cambrian, Ordovician, Silurian, Devonian, Mississippian and Pennsylvanian (Carboniferous), and Permian in order of progressing age.

## THE PALAEOZOIC IN NORTH AMERICA

The Palaeozoic history of North America began with the oceans inundating two narrow areas (geosynclines) running parallel to the east and west coasts, from which they spread periodically to cover a large part of the continent. The eastern geosyncline is called the Appalachian, after the mountains which rose there at the close of the era. The western geosyncline, the Cordilleran, included the region of the present Rocky Mountains. The Late Cambrian seas overflowed the boundaries of these great troughs to spread widely over the continent, with the greatest submergence occurring during the Ordovician, when more than 60 per cent of the present land surface was covered by great continental seas. There was some aridity in conjunction with the partial withdrawal of the seas in the Silurian period. This is shown by the great salt deposits in the rocks of this age in Michigan and New York.

The early Palaeozoic rocks differ from most of those on which they were deposited. The platform on which the Cambrian formations were laid down was made up of the eroded roots of the ancient mountains of the Pre-Cambrian, the Killarney of the Lake Superior region and the Karelian of Scandinavia. The rocks involved in these ancient mountain uplifts were strongly folded and contorted, and so intensely metamorphosed that most of them are now crystalline. Deep-seated igneous rocks are common. The Palaeozoic and younger rock formations are composed of sedimentary strata in which bedding is the dominant feature and, except where disturbed (by folding and faulting) during later mountain-building movements, the beds are horizontal, or nearly so.

Thus we have the basis for the separation of geologic time into eras. The mountain-building movements which closed the eras elevated the great masses of sediments in the geosynclines, often folding, faulting and intruding them with igneous rocks until a great mountain system was formed. As they rose above sea level these lands were immediately attacked by the agents of erosion, which broke up and transported the rock debris into new troughs of deposition. The pressure and heat of folding and intrusion changed much of the rock material into new minerals some of which are economically important. Minerals that were once sparsely scattered through the rocks became concentrated into the rich ore deposits mined today (gold and platinum placers, iron ores).

During and immediately following igneous intrusions, hot waters and gases invaded the older sedimentary beds and by solution and chemical action formed new minerals. Ultimately much of the land was reduced to an almost level surface (a peneplain) on which the layers of sediments of the next era were deposited in a nearly horizontal position. At the plane of contact between older and younger rocks we find that the respective beds meet at an angle, called an angular unconformity. In this way the eras are separated by unconformities that are, with minor exceptions, world-wide in application. The older rocks are more severely metamorphosed and more structurally complex (being faulted and folded) than the younger rocks lying above the unconformity.

Thus the ultimate basis for separating geologic time into logical units is the result of great earth movements which reflect changes in the relative elevation and position of continents and ocean basins, which in turn changed the environments in which all forms of life must live — or die.

Most of the elements of change occurring throughout the geologic history of the Earth were thus intimately related. Life flourished in the oceans at a period when they were widespread and the climates mild. The elevation of lands and the withdrawal of the seas were accompanied by mountain building. As land areas developed, profound climatic changes occurred which followed each other in a somewhat rhythmic or cyclic fashion. As we go down the scale of magnitude in these cycles we find the intervals become shorter, and the areas influenced smaller or more local. Individual beds of formations are no longer world-wide in extent. Deposited in dissimilar environments, beds of the same age may not resemble each other at all. Their correlation must be made by the use of fossils, radioactive minerals, and other methods of dating.

A classic example of dissimilarity is found in the Pennsylvanian rocks of the United States. In the east the beds are largely conglomerates, sandstone, shales, coals, and a few freshwater limestones; in the Mississippi Valley, beds deposited in fresh water alternate with marine limestones and shales; farther west, in Kansas and Oklahoma, the formations are largely marine with few coarse-grained or freshwater beds.

The physical events at the beginning of the Cambrian brought to a close the long periods of emergence during which Pre-Cambrian land areas were subject to deep erosion. Early Cambrian deposits, the Waucobian, cover only about one-fifth of the North American continent, but they are about a mile thick in the Cordilleran geosyncline at Waucoba Springs, Nevada. These sediments were mostly clastic (composed of broken pieces of older rock) and have now become quartzites. The intervening area covered by the central states contains a few Lower and Middle Cambrian deposits, but the great inundation of the Ordovician was heralded in the Upper Cambrian when much of the continent received deposits. On the whole, the period in North America closed quietly except for a small uplift, the Vermont disturbance, which is shown by a minor unconformity in the basic structure of the rocks.

Ordovician deposition was continuous in the areas of the two major Cambrian geosynclines where black graptolitic shales are found. Great thicknesses of dolomitic limestones were deposited in the extensive central areas. However, St Peter sandstone was formed in the interior of the United States in the advancing Middle Ordovician sea when the residual sands of the large central land area between the geosyncline were reworked. At this time the northern end of the eastern land mass, Appalachia, started to rise, as did the Cincinnati arch and, later, the Ozark dome. Extensive ash falls in eastern United States and Canada indicate that vulcanism accompanied the crustal disturbances at this time. The ash beds (bentonites) furnish accurate marker-beds that are useful for correlation throughout the area of ash fall. They occur from Quebec to Alabama, and as far west as Iowa. Finally, the Taconic disturbance in New England and eastern New York resulted in the unconformity which separates the Ordovician from the Silurian.

In the east and north-east, the Silurian sediments are mostly sandstones and shales. To the west, the formations become thinner and are made up of limestones and dolomites. Two distinctive and striking features characterise the formations. One is the occurrence of abundantly fossiliferous thick beds of dolomite in the thin-bedded Middle Silurian (Niagaran) formations, especially in Illinois, Indiana, and Wisconsin. These masses, some of which are five or six miles in diameter, are coral reefs which were built up as much as 1,000 feet from the floor of the ancient ocean. Their unstratified beds contain corals and other invertebrate fossils common to reef environments. Some of the reefs are economically important as they contain large deposits of petroleum.

The second distinctive feature, found in the Upper Silurian of Michigan and New York (Cayugan), is the thick salt beds deposited in shallow bays which must have existed when the nearby lands were deserts. In this area rock salt covers over 100,000 square miles, and in the Michigan basin it reaches a thickness of 1,800 feet. Much of this is being economically exploited.

*Left*
Limestone pavement at the foot of Ingleborough, Yorkshire.

*Below*
Silicified trees and burrs exposed at the cliff edge east of Lulworth Cove, in Dorset.

*Top right*
Root of a Pennsylvanian (Coal

Measure) tree lying in a bed of shale that once was soil at Joggins, Nova Scotia.

*Bottom right*
A slab of Ordovician slate in Pennsylvania. The diagonal bands mark the original beds which were cut by cleavage planes as the rock (originally shaley) was changed into slate.

In North America the Silurian closed quietly, except near the northern border of the continent. However, as in the preceding period, there was some volcanic activity in eastern Canada where lava beds about 4,000 feet thick are found in Quebec and New Brunswick, and there was folding as far west as Alberta.

Elsewhere, the Silurian period was marked by one of the greatest mountain-building disturbances in the world, the rise of the Caledonian mountains in western Europe, Alaska, Greenland, and much of Asia. Instead of the quiet transition from the Silurian into the Devonian, as in America, we find strong crustal movements which moved the older and Silurian beds in Norway and part of Sweden eastward for many miles along thrust-faults. The Caledonian movement affected a great arc about 4,000 miles long, from Wales across the Arctic to Greenland.

Because of the lack of strong mountain-building movements over most of North America at the close of the Silurian, the early Devonian formations were deposited conformably upon the Silurian deposits. In New York State there is an almost complete Devonian section. Intensively studied, it has became known as the type or reference section in North America. In that area the threefold division (Lower, Middle and Upper) commonly used in other Palaeozoic periods is divided further into many other formations. Most Early and Middle Devonian formations are richly fossiliferous, especially in western New York. However, in the east, Appalachia again began to rise in the Middle Devonian, and by Late Devonian the uplift became the Acadian disturbance which produced a mountain range running through the Maritime Provinces of Canada and southward to North Carolina. Great thickness of lava and ash accumulated, and deltas covered much of New York and Pennsylvania. These higher lands produced many clastic, non-marine formations, because of the more vigorous erosion along the borders of the new uplift. The intervening area is an excellent laboratory in which to trace changes from a non-marine environment to a typical marine environment in sediments which cover a comparatively short distance.

Another distinct type of formation was deposited in the Upper Mississippi and Ohio valleys during the Upper Devonian period. These are the thin-bedded Ohio and New Albany black shales deposited in great thicknesses in Ohio, Indiana, and Kentucky, and their equivalents in adjacent areas. The lack of common index fossils found in normal marine limestones and shales has made it difficult to correlate these formations. However, small tooth-like fossils of doubtful zoological affinity — the conodonts — can be found in fairly large numbers, and correlations based on these have been made. The close of the Devonian was marked by a general emergence of the continent, with some volcanic activity in south-eastern Canada and New England. There was some crustal disturbance near the Ozarks in Missouri.

Together, the Mississippian and Pennsylvanian systems of North America cover the Carboniferous period of Europe, though the boundary between them does not exactly match that of the Upper and Lower Carboniferous of Europe. It is now known that the top of the Chester (youngest Mississippian) lies above the base of the Upper Carboniferous (the Tournaisian).

Throughout most of Mississippian time seas continued to flood the continent, and in the type areas of the Mississippi River valley a thick succession of limestones is found, with some shales in the upper and lower parts and a few sandstones alternating with the younger formations. The faint beginnings of the end of the Palaeozoic are found in the red Mauch Chunk sandstones and shales in Pennsylvania and West Virginia, and in the thick clastics that were produced by the erosion of Llanoria, a small continent south of the Ouachita Mountains in Oklahoma, Arkansas, and Texas.

The rapid oscillation of strand lines which began in the Late Mississippian time produced the cyclic sedimentation through the Pennsylvanian system favourable to the existence of coal swamps, especially in the Central Interior and mid-continent regions. Thick clastic sediments were deposited in the geosyncline, particularly in the Appalachian and the Ouachite. The world's most extensive deposits of bituminous coal (about a quarter of a million square miles) are found in these areas. Coals of this age are also found in the Saar, Ruhr and Donetz coal basins and in the coalfields of Britain, Silesia, and Belgium.

Throughout the Pennsylvanian, local crustal movements occurred over much of the Earth, first in one place and then in

another. Indications of these movements are particularly marked in the Early, Middle and Late Pennsylvanian in Kansas, Oklahoma, and Texas, and reach a climax in the Permian with the Appalachian revolution that closed the Palaeozoic Era. This unrest resulted in low, widespread lands which were warm and moist, and therefore favourable for the rapid development of land plants. Except for some salt and gypsum deposits indicating local aridity in the West, climates were mild and equable throughout much of the Pennsylvanian period.

The final period of the Palaeozoic, the Permian, although named after the province of Perm in north-eastern Russia, is best represented in the abundantly fossiliferous strata in the southern Mid-continent area (Texas and New Mexico) where several thousand feet of marine strata are well adapted to a study of marine sedimentary environments. In western Texas, the basin, shelf, and intermediate deposits can be studied under almost ideal conditions, for the formations were uplifted with little folding or faulting. There, in their natural relationship, can be found the even-bedded limestones of the deep basins, the steep slopes from the shelf into the basins on which the coral reefs were built, and the shallow-water shelf deposits — each with its varied fauna so characteristic of environment.

Marine formations with fossils similar to those in Texas are found throughout the Rocky Mountains area, from California and Nevada to Alaska. The Late Permian fossils found in Oregon and Washington show a strong resemblance to those found in Asia. In eastern North America only small remnants of rocks thought to be Permian are found. Those in the upper Ohio valley and in Nova Scotia are all that remain of once extensive deposits that were continuous with the marine deposits farther west.

Mountain building in America caused folding, faulting, and uplift of the sediments in the Appalachian geosyncline, reducing the area covered by the once horizontal beds by as much as 200 miles. It is believed that the young Appalachian Mountains thus formed rose to heights comparable with the Rocky Mountains.

Other parts of the world, as well as the 2,000-mile chain of the Appalachians, experienced intensive mountain building. Today widely scattered folded areas are found throughout Europe — in Wales, France (the Vosges), Germany (the Black Forest, Harz, and Thuringian ranges), Czechoslovakia (the Erz, Riesen, and Sudeten), and Russia (Urals). These ranges may have been connected with the Palaeozoic mountains in Asia north of the Himalayas; westward they may have extended through northern France into England, Ireland, and the Atlantic.

## THE PALAEOZOIC IN BRITAIN

The planed-down Pre-Cambrian continent was invaded by the sea in early Cambrian times and remained covered throughout the Lower Palaeozoic. During the Cambrian, Ordovician, and Silurian much of the British Isles was covered by a geosyncline running north-east and with shallow 'shelf' seas on its margins, particularly over the Welsh Borderlands and the Midlands of England. The seas' many withdrawals and renewed invasions are shown by a number of unconformities, and the ancient shoreline remains between Builth Wells and Llandrindod Wells.

The main Cambrian outcrops are found in the Harlech dome and Caernarvonshire areas, with others in the Welsh Borderland, Pembrokeshire, Nuneaton, and extreme north-western Scotland. The sediments were either sandy or muddy, the latter being converted into high quality slates by the earth pressures during the Caledonian period of mountain building.

In the Ordovician and Silurian, sediments can be divided into two main groups: the shallow-water or 'shelly' facies which consist of limestones and shales with abundant brachiopods and trilobites, and the deeper water or 'graptolitic' facies of a mainly muddy and sandy composition. The muds are not thick, and are found, for example, on Wenlock Edge near Church Stretton, Shropshire; the sands are often very thick and form the Denbigh Moors and a great belt of country in central and south-west Wales.

North Wales and the Lake District were the centres of much volcanic activity in the Ordovician era. More than 10,000 feet of rhyolitic and andesitic lavas and ashes were poured out, often as submarine flows. These rocks gave rise to the magnificent craggy scenery of Borrowdale and Scafell, of Snowdon, Cader Idris and Conway Mountain.

The Silurian closed with a period of rapid sedimentation when thousands of feet of sediments were deposited in Denbighshire, the Lake District, and Moffat in — Southern Scotland.

Mountain-building movements had been proceeding throughout the Lower Palaeozoic with an unusually intense phase at the end of the Ordovician, but the main Caledonian uplift occurred at the end of the Silurian, lasting in Scotland into the Old Red Sandstone (Devonian) period. The great mass of sediments, up to 45,000 feet thick, were folded, faulted, and thrust into a north-to-east range of mountains which stretched from Ireland to Norway. This was the Caledonian range, which takes its name from Caledonia, the old name for Scotland. Erosion has been at work for 330 million years since these mountains were formed, and there is little left today. The maximum altitude of the Scandinavian Alps is only 8,140 feet in the Jotunheim, but because of their latitude they have several important glaciers. The Highlands of Scotland and the Grampian Mountains reach only 4,406 feet at the highest point, the summit of Ben Nevis; Snowdon, in Wales, is 3,560 feet; in Ireland, Carrantuohill is 3,414 feet high.

Large-scale maps reveal another feature of the Caledonian lands; they are highly fractured in all directions and divided like an old mosaic. Deep rectilinear troughs are found side by side with high peaks. The most remarkable of the troughs, the Caledonian Canal, crosses the Highlands of Scotland from one side to the other and contains Loch Ness, famous for its legendary monster. It is precisely this juxtaposition of peaks and deep valleys which gives the Caledonian countries a mountainous aspect not dependent on altitude alone. For example, the Snowdon massif in Wales — at its summit only 3,560 feet high — gives such an illusion of high, mountainous country that the Llanberis district has been styled the 'Welsh Chamonix'.

On the coasts, the indentations are even more striking, for whether we are dealing with the Norwegian fjords or the Scottish firths, we see in them the combined effects of submergence of the coast and of erosion. The coastal archipelagos (the Hebrides, Orkneys, Shetlands, Lofoten) are lands which have been broken and cut up into small pieces. On the Isle of Skye, for example, it is impossible to travel in a straight line for more than three or four miles without coming upon an arm of the sea. To sum up, old lands, eroded, fractured, foundered—are the Caledonian lands.

The folding in the central part of the geosyncline in North Wales, the Lake District and the Southern Uplands was intense, often isoclinal (the term used to describe very tight, compressed, concertina-like folding in which the limbs of the folds are parallel). Much cleavage was produced, but in the 'shelf' areas (the Welsh borderland, for example) folding was more gentle. It must be stressed that the cessation of downwarping of the geosyncline and of sedimentation occured at different times in different places. Mountain building is not a momentary paroxysm, but a long-continued process. In the Southern Uplands continental conditions typical of the succeeding Devonian period prevailed during Late Silurian times. On the Welsh borderland transition was gradual, and such conditions were not established until well into Devonian. The period of transition is marked by a change from grey or greenish sediments to red.

The palaeogeography of Britain was thus completely changed. Areas once covered by the sea became ranges of high mountains from which were derived vast quantities of pebbles and sand. These were carried down into the intermountain basins, or 'cuvettes', in South Wales, the Midland valley of Scotland, and the Moray Firth to give red conglomerates, red feldspathic sandstones, and red shales. Occasionally limestones (cornstones) were precipitated in shallow lakes. In lower Old Red Sandstone time great thicknesses of lavas, mainly andesites (light brown or light grey, fine-grained rocks composed of feldspar and hornblende), were poured out in the Midland valley of Scotland and in the Glen Coe/Ben Nevis region.

In the Devonian period graptolites became extinct, fishes flourished, and a land flora became established. While Britain north of the Bristol Channel lay above the sea, a geosyncline

Volcanic activity during the Ordovician produced craggy scenery in the Borrowdale area of the English Lakes.

stretched from southern Ireland across Devon and Cornwall into the Ardennes in France. Here shales, limestones, and grits containing marine fossils were deposited. These sediments are quite distinct from the Old Red Sandstone of the north, though in the marine Devonian there are some thin bands of Old Red Sandstone type. Long-continued erosion wore down the Caledonian mountains, and the vast northern continent became low lying at the beginning of the Carboniferous. A marine invasion flooded most of Ireland, Scotland south of a line from Stonehaven to Oban, and much of England and Wales, except for two big islands in Wales and East Anglia.

Carboniferous history is known in a wealth of detail, mainly because its rocks are of considerable economic importance and therefore closely studied. The story differs from place to place, but generally — certainly over northern England and the Midlands — greyish limestones were deposited in the early shallow seas, limestones such as those found in the Ingleborough district of Yorkshire, in Derbyshire, and in North Wales. Toward the end of the Lower Carboniferous conditions changed rather rapidly. Shaley and sandy beds gave way to coarse, massive grit bands, the Millstone grit which now forms the moorlands in the Ilkley, Harrogate and Lancaster areas. The grit is believed to be a vast composite delta built by sediments brought down from the northern continent by a Carboniferous 'river'. The low-lying ground thus formed at or near sea level was covered by vast swamps whose deposits now constitute the Coal Measures. Only 2 to 3 per cent of the total thickness of strata is coal; the rest is shale, mudstone, and sandstone. Occasionally the sea flooded the coastal swamps, and thin marine bands were deposited above the coals. Some of these marine bands can be traced over the whole of Britain and into northern Europe.

The Hercynian earth movement at the end of the Carboniferous produced intense folding on an east-west axis in southern Ireland, Devon, and Cornwall, and there was intense thrusting in the western end of the big syncline of the South Wales coalfield. In the north, the north-south Pennine anticline and the east-west Howgill-Cleveland anticline broke up the once continous coal measures into the four principal coalfields of Cumberland, Northumberland and Durham, Lancashire and Yorkshire, Nottinghamshire and Derbyshire.

## THE PALAEOZOIC IN EUROPE

Rocks of this era are found in scattered outcrops throughout Europe, from the Iberian Peninsula in the west, through France and Germany, and into Russia in the east.

In Spain the great massif known as the Meseta is composed of Archaean crystalline rocks, Palaeozoic sediments and eruptive rocks, partly concealed by later formations but characterised by the absence of marine Mesozoic except at the margins. The Meseta is a fragment of the Hercynian mountain range, and fossil evidence indicates the presence of Cambrian, Ordovician and Silurian strata which extend through Andalusia, Estremadura, Castile, Salamanca, León and Asturia. East of Almadén the rocks are pierced by Quaternary basalts, and the complex lead-silver zone of Linares and La Carolina is emplaced in these rocks.

Similarly, in France, remnants of the Hercynian range occur in the Ardennes, the Armorican Massif, and the Massif Central; both Archaean and Palaeozoic rocks are exposed. In the Armorican Massif, the Palaeozoic formations were derived from the destruction of the Caledonian ranges and are mainly shales with thin bands of sandstone and conglomerate. The Archaean is represented by crystalline schists, mica schists, gneiss, quartzite, granite and granulite. The Breton landscapes reflect the presence of two types of bedrock: crystalline rocks and sandstone and shales. The harder rocks give rise to the uncultivated hilltops that are often covered with forests and heath. The shales have been worn into valleys which are much cultivated. The sandstones correspond to periods of shallowing and partial emergence of the Hercynian geosyncline, the shales to periods of deepening. The fact that limestones, so well developed in the Belgian Ardennes, are entirely lacking in Brittany is explained by the proximity of the Caledonian ranges which were being eroded. Deposits of mud and sand do not allow reef-building organisms to develop.

The Massif Central has been even more deeply eroded than the Armorican massif. Metamorphism and intense erosion have brought about the disappearance of most Palaeozoic rocks.

The Palaeozoic of Germany occurs as a 'floor' under considerable areas of the territory and is largely concealed by later formations, particularly in the north. The rocks have been folded along WSW-ENE axes and are exposed near the Rhine and in the Bohemian Massif, the Frankenwald, the Thuringenwald and the Harz. Cambrian, Ordovician and Silurian rocks occur on the east side of the Frankenwald. Carboniferous and Devonian rocks are found south-east of the Bohemian Massif. As in England, Permian rocks were not involved in the folding affecting the older Palaeozoic formations on which they rest unconformably.

Russia-in-Europe is characterised by vast plains which reflect the occurrence of Archaean and Palaeozoic rocks in what was probably a continuous platform. The Palaeozoic strata are still horizontal over enormous areas, lying in much the same positions as when they were first laid down. In the Urals, however, the rocks are folded. Russia thus shows far less mountain building than western and southern Europe and central Asia. The Upper Permian rocks are noteworthy in that they contain fossil plants and animals like those of the corresponding formation in South Africa — known as the 'Gondwana' assemblage. In the Urals the Palaeozoic rocks are folded on a longitudinal axis which exhibits crystalline rocks at its core and is faulted on a large scale. Devonian rocks here are of marine origin. The folds die away westward in low parallel chains known as the Parma. There is heavy copper mineralisation in the Permian rocks west of the Urals but the main metalliferous veins of the area are in the faulted zone of crystalline rocks to the east.

**Hercynian ranges.** The Hercynian ranges can be followed as an immense belt that stretches over the whole surface of the Earth. In America, they can be found in the Appalachians and a part of the Rockies. In Europe, we have Cornwall, the Ardennes, the Rhenish massif, the Massif Central, the Vosges separated from the Black Forest by the Rhine valley, the Harz (which gave its name to the Hercynian mountains), Bohemia, the Urals, the Spanish Meseta, the Maures, Esterel, parts of the Pyrenees, and so on. In Asia, there are the Altai Mountains spreading fanwise around the ancient islands of Gobi and Tibet. In Africa, there are the Moroccan Meseta and the mountainous massifs of the Sahara, in the southern hemisphere, the Australian Alps.

**Hercynian coalfields.** The practical interest in these worn-down mountains or peneplains is that they contain an important proportion of the coalfields being worked today — either in isolated basins of limited extent and lacustrine origin, like those of the Massif Central, or in basins formed by arms of the sea. The latter extend over such wide distances that they constitute 'belts', like that which can be followed from England through Germany.

Although Wales belongs largely to the Caledonian zone, its southern part is of Hercynian origin and possesses the important coalfield whose beds later plunge beneath the London basin and the Straits of Dover. Preliminary borings for an English Channel tunnel project have found the field on both sides of the Channel. This long belt comes to the surface again in eastern Flanders and extends from there without a break to Aachen, passing en route a number of famous coal-producing localities: Bethune, Lens, Mons, Charleroi, Namur, and Liège. After a break of 65 miles the coal appears again on bank of the Rhine, in the Ruhr basin.

The basins just mentioned make up a line 500 miles long. A second belt exists further to the north, extending from the middle of England (Manchester, Sheffield) to the northern part of Silesia with a projection under the North Sea and the Low Countries. Finally, the Saar coalfield is the beginning of a third belt south of the other two, which must have its continuation beneath the Paris basin. Coal has been found in borings at Nancy but is much too deep to be workable. It is possible that one day at depth the link will be found between this coal belt, or others further south, and the coalfields of Silesia (the richest in Europe) in Poland.

In general, the coalfields formed within the Hercynian lands, or intra-Hercynian basin, are highly folded and difficult to work. In the Franco-Belgian coalfield, for example, there are at least three parallel folds separated by faults and bounded on the south by a fault called the South Fault (Faille du Midi). There is a large number of seams (156 at Mons), but because the basins in which the coal formed were spasmodically deepened none is very thick. There was a series of cycles, each comprising an initial phase of swampy vegetation, then a phase of peat-making from the plant debris, a phase of downward sinking of the ground and, finally, invasion by the sea and attendant sedimentation. Each coal seam represents one complete cycle.

In Belgium the coal is near the surface; in France it is covered by thick overburden which demands mines of an average depth of 1,450 feet. Here again we see the consequences of a deeper sinking of the basin in the west than in the east and of sedimentation more pronounced in France than in Belgium.

For the most part the coalfields of the Massif Central are small, having originated in lakes. They are elongated in two directions, along the Armorican folds (Commentry, Ahun, Brive, Figeac coalfields) and the Vosgesian folds (Autun, Le Creusot, Bert, Roanne, Saint-Etienne). This offers the strongest evidence that the Massif Central is the zone of 'turn-round' of the ancient Hercynian folds which are first aligned north-west to south-east from Wales and Brittany, and then turned south-west to north-east toward the Vosges and Black Forest. A coal trough can be followed for 125 miles from Decize to Carmaux and then through Saint Eloy, Champagnac, and Decazeville. The coalfields of the Massif Central were developed within the folds of a range of mountains.

In contrast with the folded intra-Hercynian coalfields are those that formed outside the Hercynian ranges, and sufficiently far away from them not to have been involved in the folding. Their seams are horizontal and easy to work. These fields include those in Scotland (Glasgow), the north of England (Newcastle), and the United States.

## THE PALAEOZOIC IN AUSTRALIA

Through much of its geologic history Australia was probably part of a continent which extended far enough northward and eastward to include New Caledonia and New Zealand. Over the Great Western Plateau early and Middle Pre-Cambrian rocks are widely exposed, highly folded, and metamorphosed. After uplift and erosion, Upper Pre-Cambrian deposits were laid down in the Flinders and Amadeus troughs that stretch from the Adelaide area through central Australia to the north-west coast.

Unlike most of the rest of the world, Cambrian rocks here follow uniformly above Upper Pre-Cambrian. In the Flinders trough a ten-mile thickness of sediments was deposited by the Middle Cambrian, when extensive folding took place. The

Tasman geosyncline came into being in the Cambrian, and during the Ordovician received deposits over the eastern third of the continent. The Amadeus trough also received sediments. South of the MacDonnell Ranges there is a 6,000-feet-thick Ordovician section of sandstone, shale, and limestone.

Fossiliferous Silurian rocks were all laid down in the Tasman geosyncline. Typical genera included *Favosites* (honeycomb coral), *Spirifer* (a brachiopod), and *Dalmanites* (a trilobite). At Heathcote, north of Melbourne, sediments are 20,000 feet thick. Vulcanism (in Mount Wellington) and aridity occurred at the end of the period, particularly in Victoria and New South Wales.

No folding separated the Devonian from the Carboniferous. Rocks of the latter period were laid down principally in the Tasman geosyncline and in the Westralian geosyncline, near the present coasts of Western Australia; some continental deposits are also found. The Lower Carboniferous closed with much folding, uplift, and igneous intrusions from Cape York to southern Tasmania. Many of the intrusions are economically important as they contain gold, molybdenum, and other metals.

The Upper Carboniferous was marked by an extensive south-to-north glaciation that finally covered most of the continent northward to 17°30'. This continued into the Permian. Subsequently Australia's most important coal deposits were laid in Permian formations. These have been exploited, particularly in New South Wales, where one seam is 32 feet thick.

## THE PALAEOZOIC IN NEW ZEALAND

The only Palaeozoic rocks in the North Island of New Zealand are some Permian limestones at Whangaroa Harbour and Bay of Islands. On the west coast of the South Island some greywacke (hard grey silty sandstone) and argillite are thought to be Pre-Cambrian. Cambrian rocks (limestones) containing trilobites are known only from Cobb Valley, near Takaka. This region has more extensive Ordovician outcrops, many of which have fossil graptolites. New Zealand has no Silurian horizons, and the only known Devonian outcrops are near Reefton and southwest of Motueka, along the sides of the Mount Arthur Range. Carboniferous of Permian rocks are exposed across the South Island as the northern limb of the Southland Syncline. Severe and extensive volcanic action is shown in the miles-thick Te Anau system of breccias and tuffs.

## RED BEDS AND SALINE DEPOSITS

We may remind ourselves that the Palaeozoic witnessed the formation and subsequent destruction of two mighty mountain systems: the Caledonian and Hercynian. The former was built during the Cambrian and Silurian, to be destroyed during the Devonian; the latter was formed in the Carboniferous and the Permian, only to be eroded in the Triassic. Destruction began soon after formation; in fact, it could be said that the two processes were almost contemporaneous. The smallest degree of relief is sufficient for the agents of erosion to begin to work, and as the relief increases so the agents increase in number and intensity. To the normal erosive work of surface water, there is added the transporting power of young streams and glaciers and the destructive power of frost and thaw. The agents of chemical weathering (oxygen, carbon dioxide, and water) attack the minerals of the rocks, dissolving parts of them and breaking the remaining pieces into easily transported particles. In this way hard rocks such as fresh granite eventually succumb to chemical weathering to form sand and clay. Other constituents dissolved in the water are carried to the sea where calcium, with carbon dioxide, forms the calcite of the limestones and the shells of the marine invertebrates. The sodium, magnesium and other chemicals combine to form the minerals which gave and still give the seas their saltiness. It is the continued action of the agents of erosion that, in the end, breaks up and planes down even the highest mountains.

Although the destruction of the Caledonian was separated from the destruction of the Hercynian chains by an interval of about 100 million years, both were characterised by the same sedimentary and climatic phenomena. In the Permian as in the Devonian, pebbles, sands, and clays were predominant over limestones. Glacial, torrential, and fluvial erosion carried so much detrital material into the sea that reef-building organisms were confined to limited areas of marine waters. Further, the presence of iron oxide generally coloured sediments of land origin red. There are two principal oxides of iron: the anhydrous form (hematite) and the hydrated form (limonite). The former gave the sands, sandstones, and clays of the Devonian and Permian their characteristic red colour. These deposits were formed under a dry climate that also produced extensive saline deposits.

Silurian beds of gypsum and rock salt, called saline, cover an area of more than 100,000 square miles in the Great Lakes area of the United States, in Michigan, Ohio, Pennsylvania, and New York. In the Devonian there is the Old Red Sandstone of North America, Greenland, Spitzbergen, and the British Isles. This sandstone reaches a thickness of 23,000 feet in Scotland. Such depths could only be formed when the sedimentary basins receiving clastic sediments sink about as fast as the detritus is laid down by rivers. It was in the Old Red Sandstones that armoured fishes and eurypterids (which belong to the same family — Arthropoda — as the trilobites) were first discovered. The Permian includes the New Red Sandstone of England, the Rotliegende of Saxony and Thuringia, the red shales of the Dalius and Cians gorges (Alps Maritimes), the red sandstones of the high plateaux of the central Sahara, and so on. Permian salt deposits in North America (Kansas to New Mexico and western Texas) cover about 125,000 square miles. One of the world's largest mines of rock salt and potash salts is at Stassfurt, Germany.

## GLACIAL DEPOSITS

There is a tendency to suppose that the great extension of ice sheets was unique to the Pleistocene, and contemporary with the first men. The Quaternary glaciations are certainly the best known and those which most clearly affect us, but it should not be forgotten that many other glaciations took place in terrestrial history. We have already indicated some during the Pre-Cambrian era. In the late Pre-Cambrian, probably the Algonkian (Proterozoic), glacial deposits in the form of tillites (hard, consolidated boulder beds) are found in many parts of the world — North America, Scandinavia, India, Africa, China, and Australia. These deposits contain glacially polished, striated pebble, and ground rock, probably formed under conditions similar to those found in the ice-cap at present covering Greenland. Other glacial deposits are found in the Palaeozoic.

Clays with striated boulders are intercalated in the Cambrian sediments of Norway, China, and Australia. Glaciation during the Carboniferous occurred in peninsular India, Australia, South Africa and Brazil, all of which possess extensive areas of striated, glacially polished rocks. In a number of places there are moraines and erratic boulders. Striated rocks, transported and abandoned by icebergs, are often scattered amid fossiliferous marine beds which determine their age and confirm that the glaciers reached down to the sea-shore just as they do today in the United States, Greenland, and Spitzbergen. Evidence of continental glaciation in the Permian has been found in Australia, Brazil, India, and South Africa.

In this account of Palaeozoic geology we have seen the application of the principle 'the present is the key to the past'. Mountain-building movements, erosion, volcanic eruptions, desert climates, and glacial episodes are revealed by the rocks. Geologic processes that can be seen in action at the present day were at work in the Palaeozoic and their effects are recognisable. A Palaeozoic limestone does not differ much in appearance from one of Tertiary age—except in fossil content.

In many parts of the world the close of the Palaeozoic and the opening phases of the Mesozoic were not dissimilar. In Britain, for example, the desert conditions of the Permian were followed without a climatic break by the desert conditions of the Triassic. Elsewhere marine conditions prevailed from the Permian through into Triassic times. Thus the Palaeozoic merged into the succeeding Mesozoic as part of a gradual process without the sharp break implied at first sight by the use of two different names.

# The Mesozoic Era

The Mesozoic Era, or Age of Reptiles, which lasted for approximately 155 million years, followed the Palaeozoic. During this era the physical make-up of the Earth began to look more as it does today. Chief mountains formed were the Sierra Nevada, the Andes, and the Pyrenees. Generally, however, it was a relatively quiet period following the formation of the Hercynian mountains in Europe and the Appalachian Mountains in North America.

The era witnessed enormous progress in the world of living creatures. Among the marine invertebrates, molluscs were dominant. On land, the reptiles, notably the dinosaurs, held sway. In the plant kingdom the ancient plant types which had dominated the Carboniferous forests were gradually replaced by conifers and their allies, and by flowering plants.

The Mesozoic is divided into three periods: the Triassic (45 million years), Jurassic (45 million years), and Cretaceous (65 million years). Each of these was concluded by a period of crustal disturbance. The Palisades disturbance ended the Triassic; the Nevadan terminated the Jurassic; and the Laramide Revolution closed the Cretaceous.

## THE MESOZOIC IN NORTH AMERICA

The Appalachian Revolution which closed the Permian Period and the Palaeozoic Era uplifted, folded, and faulted the seven-mile-thick Palaeozoic deposits in the former Appalachian geosyncline and formed the original Appalachian Mountains, which may have been higher than the present Rockies. The Mesozoic Era in North America was characterised in the east by the erosion of the young Appalachian Mountains, and in the west by broad shallow seas.

**Triassic in the East in America.** The Triassic opened in eastern North America with erosion starting to wear down the Appalachians, and its rivers deposited mud and sand in a narrow trough which ran along the eastern side of the mountains from the Carolinas to Nova Scotia. A thick blanket of red beds, called the Newark series, consisting of reddish-brown sandstone, shales, and some conglomerates was deposited in the trough, which subsided and eventually collapsed under the weight, forming a series of grabens (downfaulted valleys) into which sediments continued to pour.

One graben, the Connecticut trough, 25 miles wide and 100 miles long, extended through western Massachusetts and central Connecticut. Triassic deposits in the Connecticut Valley are from 10,000 to 13,000 feet in thickness; in New Jersey they reach 20,000 feet; farther south, in Virginia and North Carolina, they thin out to 2,000 or 3,000 feet.

Dinosaurs left thousands of footprints in the ancient sands and mud flats that now form the Newark series in the Connecticut Valley and New Jersey. Fossil ripple marks and mud cracks

Dolomitic formations in the Austrian Tyrol.

are commonly found, and sometimes even the impressions of Triassic raindrops. Farther south, small coal beds (once mined at Richmond, Virginia) record the existence of cycads, ferns, amphibians, and small mammal-like reptiles. The record, though scanty and poorly preserved in continental deposits, heralds the great development of the dinosaurs in later periods.

There was considerable volcanic activity in the eastern part of the United States, when dark igneous rocks called 'trap' were extruded as lava flows and intruded into the sediments as sills. Lava flowed over the Connecticut trough three times, leaving a sheet 500 feet thick, which is now exposed at the Hanging Hills of Meriden. Extrusive flows are recognised by the breaking (metamorphism) of the lower rocks and the scoria (sponge-like lava) on top of the flow. The heat of intrusive sills altered both upper and lower enclosing beds. The prominent Palisades of the Hudson is a huge sill, 170 million years old, from which the overlying strata have been eroded. The Watchung Mountains of New Jersey are also sills. In addition, at various places intrusive magmas (molten rock) cut across beds to form dykes and veins, such as those bearing chalcocite at Griggstown and Arlington, New Jersey.

**Triassic in the West.** In the West the Cordilleran geosyncline, which had been repeatedly invaded by Palaeozoic seas, subsided again in the Triassic. Marine sediments were deposited in Oregon, Washington, Idaho, Nevada and California. In some areas the limestones and shales reached a thickness of 25,000 feet. Throughout the West there was considerable volcanic activity, especially in British Columbia where the much-disturbed Triassic formations contain an abundance of volcanic rocks.

Early Triassic seas invaded Utah and Idaho by way of an arm of the Pacific which advanced across California and Nevada, and may have spread eastward as far as the Black Hills (Dakota). In South Dakota gypsum deposits possibly represent shallow bays which formed evaporation pans at the edge of these seas. In the middle of the period, however, the seas almost completely withdrew, except in California and Nevada. Throughout the Triassic the country bordering the seas in the east (the present Plains and Rocky Mountain states) was a vast, featureless lowland on which sand and mud were deposited by sluggish streams flowing from the central part of the continent.

These sediments are in some ways the most interesting of the Triassic formations. Oxidisation of the different mineral constituents has made them some of the most vividly coloured rocks in the world. Spread widely throughout the Colorado Plateau and the Rocky Mountain region their colours are the most intense in the Painted Desert, below the steep cliffs in the south-western part of Zion National Park, and in the vicinity of Monument Valley on the Arizona-Utah border.

The Moenkopi, the lowest Triassic formation on the Colorado Plateau, averages 500 feet thick in Arizona, increasing in Nevada to 2,000 feet. It is a continental deposit of red and brown sandy shales, siltstones and sandstones, together with layers of gypsum and salt. The remarkably even beds were deposited by shallow, inland bodies of water. Good examples can by seen near the

Hopi village of Moenkopi, Arizona, and in Capitol Reef National Monument, Utah.

Uplift and erosion followed; gravels were spread by torrential streams to an average thickness of 50 feet over some 75,000 square miles on the Colorado Plateau. These gravels have been consolidated into the thin Shinarump conglomerate. Petrified logs are often found in this formation which caps various mesas (flat-topped ridges) in Arizona, New Mexico, Nevada, and Utah.

The last Triassic formation of the Colorado Plateau is the Chinle, a series of clay-shales, sandstones, impure limestones, and conglomerates deposited in stream channels and on outwash plains. These beds are soft and under erosion often give rise to a badland topography. The famous Petrified Forest in Arizona is found in this formation. Here, huge logs of conifers washed down from upstream were buried, became petrified, and have now been exposed by erosion. A few petrified stumps have also been discovered in their original locations, and a little house built out of pieces of petrified wood by ancient Indians has been partly restored by the National Park Service. The climate during the late Triassic was tropical with occasional intense dry seasons. Dense forests grew along the rivers, and some trees with swollen bases apparently grew in swamps. In this formation are found many ferns and plants such as cycads (thick-leaved, hollow-stemmed) and *Calamites* (giant horse-tail). Reptiles include the 25-foot, crocodile-like phytosaur, known as *Machaeroprosopus.*

**Palisades disturbance.** The Triassic in America ended with the withdrawal of the seas and the Palisades disturbance in the East, where there was considerable block-faulting and tilting of formations deposited on the eroded surfaces of the complexly folded older rocks. The same disturbance was responsible for the present outcrops of the Watchung Mountains in New Jersey, the Palisades along the Hudson River, and the Holyoke Range in Massachusetts.

**Jurassic in the western interior.** In the Jurassic, as in the Triassic, the eastern part of the United States was not covered by seas. The block mountains and the Appalachian ranges were worn down by erosion; large areas were reduced to a peneplain. There are no known exposures of Jurassic rocks east of the Great Plains.

The Rocky Mountain area was separated from the Pacific borderland by a land mass undergoing erosion which furnished sediments for streams flowing both to the east and to the west. The land mass induced precipitation from the clouds moving eastward, resulting in extensive aridity and great deposits of wind-blown sand in the Great Plains area. First the reddish Wingate and, above it, the brick-red and white Wingate and, above it, the brick-red and white Navajo sandstones of Arizona, New Mexico, Utah, and Idaho were laid down. The Navajo sandstones reach a thickness of 3,500 feet in southern Nevada, and from them is carved the Rainbow Natural Bridge; so are the Great White Throne and Angel's Landing in Zion National Park. The

Vermilion Cliffs of Arizona and Utah are also of Wingate formation.

In the late Jurassic a subsidence of this land area again brought a shallow sea from the Arctic into the Rocky Mountain region, in which the Sundance formation (mostly sandstones and shales, with a few beds of limestone) was deposited. Covering a wide area of the Great Plains, it is generally a few hundred feet thick. In southern Idaho, however, its sediments are 6,000 feet thick.

After the withdrawal of the Sundance Sea much of the southern Cordilleran area became a well-watered lowland on which the Morrison formation was laid down. This was a continental deposit of shales, siltstones, sandstones, and some conglomerates with a few thin beds of limestone. The Morrison is a colourful formation of white, red, brown, greenish grey, purple, and black. At the time of deposit the climate was tropical, vegetation luxuriant, and dinosaurs abundant (69 species). Genera include *Apatosaurus, Brachiosaurus,* and *Allosaurus.* The formation averages less than 400 feet in thickness but covers some 100,000 square miles from Montana to northern New Mexico. It is generally regarded as a Late Jurassic formation although its plants are much like those of early Cretaceous times.

**Jurassic along the Pacific coast.** The Jurassic rocks of California and south-western Oregon consist of great thicknesses of volcanic material interbedded with marine sediments. The Horseshoe Bend Mountains in Mariposa County, California, show some 12,000 feet of lava flows, tuffs, and breccias. At one point along the Merced River pillow basalts (formed by underwater eruptions) are piled up 1,400 feet deep. In eastern California the Mariposa slates are interbedded with volcanic material to a depth of 10,000 feet.

In the Coast Ranges of California the Franciscan series of organic, stream-borne, volcanic sediments are well developed. These beds were deposited in a sinking geosyncline along the line of the present Coast Ranges. Most of the sediments came from a land mass or chain of high islands to the west, called Cascadia. Above these, the Knoxville series of shales (with some sandstone and limestone) was laid down. Together, the Franciscan and the Knoxville series are 25,000 feet thick. Much volcanic activity is indicated by the prevalence of lava flows and tuffs interbedded with marine rocks; in fact, much of the material of the marine sediments was of volcanic origin.

**Nevadan disturbance.** A new series of crustal movements began as the Jurassic drew to a close. The newly deposited sediments in the western geosynclines were uplifted, crumpled, folded and faulted, producing large over-turned folds and thrust-faults. These movements, known as the Nevadan disturbance, involved areas of British Columbia and the Cascades of Washington, Oregon, Idaho, Nevada, and eastern Calfornia, and to a lesser extent the Coast Range area. Igneous batholiths, great instrusions of molten rock (magma) having the chemical composition of granite, were thrust up in the deformed sedimentary rocks. They gave off ore-forming solutions, carrying gold, silver and many other metals, which invaded the host rocks along joints and fissures to form an extensive mineralised belt along the Sierras. The famous Mother Lode of California, a narrow band of gold-bearing veins nearly 130 miles long, and several mining areas in western Nevada were formed by these late Jurassic intrusions.

**Cretaceous in the East.** The basal Cretaceous formations rest upon metamorphosed Pre-Cambrian rocks, on the bevelled edges of strongly folded Palaeozoic formations, or on eroded Triassic rocks. Cretaceous deposits in North America can be closely correlated with their counterparts in Europe.

By the beginning of the Cretaceous, the Appalachian Mountains were eroded almost to a level plain. Slight warpings or gentle up-lifts rejuvenated the eastward-flowing streams which deposited their sediments over the coastal plains to form the Potomac series, now exposed along the Atlantic coast in a narrow belt from New Jersey to Georgia.

In the eastern United States the continental border was depressed. The shore-line of the Cretaceous sea corresponds roughly to the Fall Line from New York to Georgia. The Potomac formations include both continental and marine sediments, indicating that there was some oscillation of the strand line during the period. The basal beds are river deposits on which was laid down a mixture of clays, sands and marls, some containing marine fossils. Their total thickness is about 1,000 feet.

The first formation of this Potomac group is the Patuxent, deposited by streams that meandered over the low coastal plain and consisting largely of sands and gravels with some beds of white, red, brown, or yellow clay. Some stumps and lignite (soft brown coal) are the probable signs of former swamps.

The next formation is the Arundel clay which includes some lignite and iron ore. Today the clays are used commercially in pottery and brick manufacture. The Patapsco formation overlying the Arundel contains red and brown clays and sands. These three Lower Cretaceous formations are best developed in Maryland.

Upper Cretaceous coastal plain formations are the Raritan and Magothy, the Matawan and Monmouth. The first two are river or estuary deposits and consist principally of unconsolidated sands, gravels, and clays.

The Matawan and Monmouth formations are derived from the first deposits made in an eastward-lying sea that covered the site of the former continent of Appalachia. Both formations are sandy and contain a great deal of glauconite, or green sand, which is used as a commercial fertiliser.

From Georgia the Cretaceous formations cover a broad belt across Alabama, where they swing north to southern Illinois. West of the Mississippi River the formations pass south-west in an ever-broadening band across Arkansas, Oklahoma, and Texas into Mexico. The basal deposits are coarse sandstones and conglomerates, covered by 1,000 to 3,000 feet of marine sandstones, shales, and chalky limestones.

**Cretaceous in the western interior.** In Early Cretaceous times a widespread sea advanced northward through Mexico and by the mid-Cretaceous had engulfed most of the southern United States. A long trough, the Rocky Mountain geosyncline, located where the Rockies now stand, continued to subside, and through it the seas finally reached the Arctic. Subsidence kept pace with deposition until sediments lay 30,000 feet thick near the centre of the trough, in north-eastern Utah. In the lower Mississippi Valley area, however, there was a temporary withdrawal of the seas, dividing the period into a Lower and Upper Cretaceous. Withdrawal was followed by a rapid readvance which reached as far north as St Louis.

In the Rocky Mountain region, from Mexico to north-western Canada and Alaska, Cretaceous formations now crop out in a broad belt 700 miles wide. Both marine and non-marine formations are found in abundance. The latter include lake, stream, and swamp deposits — conglomerates, sands, shales, and thick coal seams. The marine beds contain many oyster shells and other invertebrates similar to those of the Gulf Coast.

The earliest Cretaceous rocks in America are the Comanche series whose fossiliferous limestones reach a thickness of 4,000 feet in Mexico and 15,000 feet in southern Texas. Near the shore-line of the Comanchean sea in Iowa, limestone gives way to shales and sandstones. In the Bisbee area of Arizona the sea received sediments from the Mesocordilleran highlands to the west. Erosion was unusually active and almost 5,000 feet of limestones, shale, sandstone, and conglomerate were laid down. In Canda, sediments of this age were deposited in the Rocky Mountain geosyncline by a sea that came down from the Arctic. In Alberta, the fresh-water Blairmore and Kootenai deposits contain important coal beds.

From time to time during the Upper Cretaceous the coastline shifted, and along the edge of the Rocky Mountain geosyncline marine, lagoonal and continental deposits are found. Four main groups of beds are to be distinguished in the Upper Cretaceous. The Dakota group of dark brown, resistant sandstone of marine and continental origin is the earliest. The next group, the Colorado, consists of the dark grey, marine Benton shale, the thin-bedded Greenhorn limestone, and the Niobrara chalk. This last bed is famous for its fossils of the flightless, loon-like bird, *Hesperornis,* of the giant flying reptile, *Pteranodon,* and of the large marine lizard, *Musasaurus.*

Following these comes the fine, dark marine Pierre shale. Usually about 2,000 feet thick, in some places it reaches a maximum of 10,000 feet. On top of this lies the Fox Hills sandstone, also of marine origin. Together, these two layers constitute the Montana group. The final division of the Upper Cretaceous, the Laramine, consists of non-marine deposits several thousand feet thick, laid down as the Cretaceous sea retreated. These deposits, which contain many dinosaur fossils, are thickest in the Laramie and Big Horn Basins of Wyoming, the Denver Basin in Colorado, and are also found in Alberta.

**Cretaceous along the Pacific.** Several embayments spread from the Pacific across parts of the west coast area, and from Alaska to southern California. They covered the down-faulted blocks of the Nevadan disturbance with coarse sediments up to 50,000 feet thick, derived largely from sandstone and shales eroded from the uplands formed during the disturbance.

The chief units can be recognised. The Shastan series of the Lower Cretaceous is represented by deposits 20,000 feet thick in the northern Coast Ranges. Farther north there are Upper Cretaceous rocks 50,000 feet thick on Vancouver Island, and 11,000 feet deep on Queen Charlotte Islands.

**The Laramide Revolution.** The Cretaceous sea was the last great expansion of an inland sea in North America. It stretched from the Arctic Ocean to the Gulf of Mexico and was as much as 1,000 miles wide. No younger marine deposits have been made in the central part of the continent, and the end of the period also marked the end of the Mesozoic Era. As in the closing of other eras, there were great crustal movements involving uplift, folding, mountain building, and vulcanism. This, the Laramide Revolution, gave rise to the Rocky Mountains and caused vast changes. The Rocky Mountain geosyncline, which had been periodically receiving sediments for more than four hundred million years, was folded, faulted, and finally arched upward or overthrust to the east. Some overthrust faults have been traced for more than 200 miles along a line from Utah into Canada. The Lewis overthrust, for example, on the eastern border of Glacier National Park, thrust the Pre-Cambrain thirty-five miles over Cretaceous rocks. Although there was extrusive volcanic activity in most of the western states, many granite batholiths were intruded into overlying strata. The Idaho batholith, now exposed over a large area in central Idaho, is one, and the Boulder batholith underlying Butte, Montana, another.

**Cretaceous deposits of economic value.** Cretaceous formations are host rocks to many valuable and useful materials, most important of which are petroleum and natural gas. The East Texas field is the largest producing oilfield in the Americas. Other Cretaceous oilfields are found in Louisiana and Mississippi. In Mexico, in the prolific Tampico area, was found the largest producing well ever drilled. Much Venezuelan oil comes from Cretaceous rocks.

Coal is found in Cretaceous rocks in the Rocky Mountain region, from Arizona and New Mexico into Canada. Beds are commonly more than ten feet thick, but the coal is not of such a high grade as the Pennsylvania coals. In central Colorado, however, an igneous intrusion changed the coal to anthracite ( a very hard variety), and in Alberta strong folding produced a higher grade coal. Most of the best Cretaceous coal beds are found in the formations of the Montana group.

In addition to fuels, many metals are found in Cretaceous rocks, bauxite, antimony, vanadium, gold, silver, copper, and zinc among them. Important non-metal products are clays, cement rocks, glauconite, chalk, building stone, and glass sand. Ground water supplies from the Dakota sandstone of the northern Great Plains and the Trinity sand of Texas have industrial and domestic use.

## THE MESOZOIC IN BRITAIN

Mesozoic rocks in Britain lie south-east of a line drawn from the mouth of the Tees to Torquay, with the addition of a big tongue of Triassic deposits north-west of this line, in Cheshire and Lancashire. A good geologic map will show the outcrops of the various sub-divisions swinging across England, with the youngest system, the Cretaceous, coming in at the south-east.

These Cretaceous formations have a particular claim to fame, for it was from them that William Smith, the pioneer of English stratigraphic geology, collected the evidence on which he based his principles for correlating the beds. A surveyor of the 1790s who laid out routes for canals in the Bath district, he noted that particular types of rocks there displayed characteristic fossils. When he travelled farther afield into the Midlands, he found similar phenomena and was able to formulate his two principles: that of super-position, and that of the identification of formations by index fossils. The names Corallian, Kimmeridgian, and Liassic (all Jurassic) are based on the terms Smith used for sub-divisions of these rocks.

The Hercynian uplands of Wales, south-west England and the Pennines had been planed down by the beginning of the Mesozoic. The Triassic sediments deposited were fine-grained red sandstones and marls with some salt beds, as in Cheshire. Near the top of the Triassic, the marls became greenish in colour, heralding the approach of a big marine invasion in the Lower Jurassic. In detail, the distribution of land and sea was variable, but in general, Wales, Scotland, and the south-west peninsula were land, while London and East Anglia, together with the Ardennes area on the Continent, constituted a big island (finally inundated during the great marine invasions of the Middle and Late Cretaceous).

In the Jurassic the Midlands of England were covered with a fairly shallow sea, with even shallower belts near the Mendips in Somerset, Moreton-in-the-Marsh in Gloucestershire, and Market Weighton in Yorkshire. The earliest marine rocks are the Rhaetic series, in which there is a thin bone bed made up of fragments of fish, amphibians, and reptiles. Above this comes the Lias, where blue-grey limestones alternate with dark shales. Both are familiar to visitors to the South Wales coast near Barry, to Lyme Regis in Dorset, or to Whitby in Yorkshire. The Lias also contains the iron ores of Frodingham, Banbury, and Cleveland.

The succeeding Inferior and Great Oolites are variable clays, sands, ironstones, and shelly limestones. They form the Cotswold escarpment which dominates the Severn Valley. Some, like Bath stone, are good building stones, others, like the Northampton ironstone, are workable iron ores —, some for brick-making.

The next group, the Oxford Clay, is the basis of a considerable brick industry that stretches northward from Bedford to Peter-borough. The clay is thick and contains a fair proportion of organic material which burns when the kilns are heated and makes the clay to a certain extent self-firing. The pits abound in fossils, which include some fish and a wealth of reptiles, dinosaurs, croco-diles, ichthyosaurs, and plesiosaurs. The clay underlines the broad valley, formerly densely forested and now good pasture land, which includes Chippenham, Oxford, the Fens, and the Vale of Lincoln.

The Corallian series is distinguished by several fringing coral reefs that occur at places as far apart as Steeple Ashton in Wiltshire, Upware in Cambridgeshire, and in Yorkshire. Many of the corals

**Top**
A stream deposit of pebbles and clays in Washington, D.C. Although this material was piled up in early Cretaceous times and is over 100 million years old, it has not hardened into stone.

**Centre**
The Seven Sisters cliffs on the Channel coast of England, laid down during the Cretaceous period.

**Bottom**
Hogback ridges of Cretaceous formations outcropping in parallel bands around Sheep Mountain uplift, Wyoming.

are found in their original position and are accompanied by fossil sea-urchins and clams, while around the reefs wave-cut debris is found. The succeeding thick, blue Kimmeridge clay is known for its bands of bituminous shale, used locally on the Dorset coast as a fuel even though it burns with a strong smell and leaves much ash.

After the deposition of the Kimmeridge clay, the Jurassic sea retreated. The Portlandian and Purbeckian formations which followed were confined to the south of England, in particular to the Isle of Purbeck and the Island of Portland. The Portlandian contains the famous Portland stone, an oolitic limestone and easy to work. Quarrymen divide the Portlandian from bottom to top into 'Best Bed', 'White Bed', and the 'Roach'. The Best Bed has few fossils, the White a few more; the Roach contains the internal casts of two typical fossils, the 'Portland screw' (a gastropod) and the 'horse's head' (a clam). The rock itself is a rough limestone useful in coarse constructions such as harbour defence works. The beds are all marine, but the Purbeck marks a change in conditions. Its beds include lagoonal, freshwater, and even terrestrial deposits. Gypsum, worked at Battle, in Sussex, was precipitated in the lagoons, and shales were deposited in the freshwater lakes. Continental conditions are indicated in the 'Dirt Beds' (thin seams of dark, shaley rock) and in the famous 'fossil forest' of Lulworth, where fossil tree stumps and fallen trunks can be seen at very low tide. These interesting formations can be studied along the Dorset coast from Swanage to Durdle Dor. Their fossil content ranges from butterflies, dragonflies, and ants (in the insect beds) to bones of turtles and crocodiles.

There is only a slight break between the Jurassic and the lowest beds of the Cretaceous, the Wealden. Deposited in the Wealden Lake and in much the same area as the Purbeckian formations,

they are freshwater deposits and contain dinosaur (*Iguanodon*) remains and footprints, lignite bands, and freshwater clams and snails. The lower part, the Ashdown sandstone, underlies the central part of the Weald of Sussex, Surrey, and Kent.

While southern England was covered by the Wealden Lake, a sea encroached from the north into Yorkshire and Lincolnshire. As this movement continued, the lake itself was flooded by the sea from the south. Gradually the London-Ardennes island was submerged. The earlier Lower Greensand sea did not quite drown it, but the later Gault sea did. Today the Gault Clay is continuous from the edge of the Chiltern escarpment to the foot of the South Downs. This great Cretaceous marine invasion reached its maximum when the Chalk was deposited. The Chalk, or variations of it, now extends from County Antrim, in Northern Ireland, and western Scotland to the Haldon Hills of Devonshire. It may once have covered Wales too.

The Chalk is a very pure limestone but with impure bands such as the Chalk rock, the Belemnite marl, and the harder Melbourne rock. The Chalk rock occurs close to the northern outcrop of the Cretaceous from north Dorset to Cambridgeshire and Norfolk. It is a band of hard, cream-coloured limestone containing green-coated phosphatic nodules and grain of green glauconite (an iron-potassium silicate mineral). The Belemnite marl is widely distributed and occurs at the base of the middle section of the Upper Cretaceous; it is called after the fossil of the same name. The Melbourne rock is a hard nodular limestone, thickest in Kent and Sussex, where it is also known as the Grit Bed. In its upper parts the Chalk contains layers of flint nodules which are black or dark brown when broken, and at its margin often has rounded grains of embedded quartz. The main constituent is a very fine-grained

Flamborough Head, a Jurassic formation on the Yorkshire coast.

calcium carbonate, derived probably in the main from shells of tiny one-celled Foraminifera. As very little sedimentary material was washed into this shallow sea, it is likely that the Chalk was formed in a sea surrounded by deserts. In places, however, boulders up to a foot across are found embedded in the fine-grained Chalk. How they were carried there remains a mystery to the present day.

## THE MESOZOIC IN FRANCE

**The Paris basin.** At the beginning of the Mesozoic the site of the Paris basin was not differentiated from the surrounding regions. With the Ardennes, the Armorican massif, the Massif Central, and the Vosges, it formed part of the immense peneplain that resulted from the erosion of the Hercynian chains. In the Triassic, however, the eastern section of the basin sank, and part of the Vosges Mountains and Lorraine were inundated. In the Jurassic, the marine invasion grew; one arm of the sea stretched from the Vosges along the edge of the Ardennes; another skirted the northern edge of the Massif Central; a third came from England. The three arms eventually met in the Paris basins, to which the Aquitaine basin and the Rhône basin were later joined as they in turn were inundated. Once again France became an archipelago where coral reefs flourished. The Mesozoic Era ended with a general retreat of the sea toward the north.

The Paris basin is a good example of a basin of sedimentation. Since its emergence from the sea in the Cenozoic era, erosion has removed an enormous quantity of rock deposited during the Mesozoic. The hard rocks, which have suffered less from erosion than the softer ones, stand out in relief. The result is that the basin today looks like a series of broad, shallow bowls laid one inside the other. The limestone rim of each layer dominates the clayey part of the basin on which it rests, producing a series of cliffs leading up to Paris.

**Eastern Paris basin.** In the east, the Paris basin begins with the Vosges sandstone derived from the Ardennes and Rhenish mountains. Characteristically, the coarse sandstones are overlain by fine red sandstones. Toward the west, the sandstones merge into shelly limestones, and then marls containing some salt give rise to the rock salt mines at Vic and Dieuze.

The first cliff is the *côte de Moselle*. Iron ore (Lorraine oolitic ore) outcrops at the foot. The *côte ce Moselle* is made up of oolitic and crinoidal (sea-lily) limestones of the Lower Jurassic, above which fertile clays and marls are found. This land, called the Woëvre, is narrow and comes to an abrupt end at the foot of a new series of heights, the *côte de Meuse*. This is a barrier-reef where warm-water corals, and oolitic and crinoidal limestones are the common rocks. The top of the *côte de Meuse* is covered by marls and clays which lie beneath the marly, fine-grained, cavern-riddled lithographic limestone of the *côte des Bars*.

At the end of the Jurassic much of the sea withdrew.

The Cretaceous period opened with another advance of the sea that initially deposited the marls, clays, green sands, and marly chalk of *Champagne humide* (moist Champagne). The artesian ground water found 2,000 feet beneath Paris originates here.

To the north, *Champagne humide* passes into a porous sandstone known as *gaize*. On the west, it is bounded by the dry, porous and fertile cliffs of the chalky Champagne which mark the beginning of the Upper Cretaceous white chalk (Senonian stage). Water in this area soaks quickly into the ground and later gushes forth in springs in the level valleys of the Seine, Marne and Aisne.

**Western Paris basin.** In Normandy we can observe, though not too clearly, the complete succession of Mesozoic rocks. At May-sur-Orne, Triassic and Lower Jurassic rocks rest directly on the eroded folds of the Armorican shales and sandstones. Lower Jurassic (Liassic) marls are found in the Vire valley. The *campagne de Caen* is made up of oolitic limestones contemporary to those of the *côte de Moselle*. Bayeux gave its name to the Bajocian stage of the Jurassic. In the *pays d'Auge*, clays of the same age as the Woëvre form the cliffs of the Vaches-Noires.

Beyond the *pays d'Auge* the clays dip beneath the Chalk which forms the plateaux of upper Normandy on both sides of the Seine.

At Rouen the *côte Sainte-Catherine* is composed of marly chalk rich in the fossils of such marine invertebrates as ammonites, belemnites, and echinoids. In the *pays de Caux*, one enormous block of chalk, the white chalk is thicker and outcrops in cliffs in the Seine valley as well as on the Channel coast, presenting an uninterrupted face more than 300 feet high and 80 miles long. Picardy and Artois, northerly prolongations of Caux, are also formed by a block of chalk with overlying fertile sediments. On the borders of the *pays de Caux* and Picardy, however, there is an 'inlier' (an outcrop of older rock) in the form of an eroded anticline. This is the *pays de Bray*, made up of the same sands, clays, and marls as *Champagne humide*.

Another eroded anticline can be seen in the Boulonnais. It is, however, cut through by the English Channel and reappears as the Weald in the south of England. Cap Gris-Nez is made of Jurassic limestone. At Marquise the folding was intense, and Palaeozoic rocks (the same schists and marbles as in the Ardennes) appear at the surface.

The Aquitaine basin of Mesozoic rocks (limestones and chalk with thin marls) lies south of the Paris basin and extends as far as the Pyrenees. A branch of the Hercynian chain formerly extended over the site of the Pyrenees but was worn down during the Jurassic. A new geosyncline developed which underwent successive deepenings and uplifts. A most important uplift took place in the Cretaceous and gave birth to the present Pyrenees. It left a sea on the French and Spanish flanks into which sediments from the new mountains were deposited. These sediments, poorly consolidated sandstones, marls, shales and clays, are called flysch.

**The Rhône Basin and the Western Alps.** Much of the area of the present Alps was a geosynclinal sea in the Jurassic. The geosyncline consisted of the Dauphiné geosyncline on the French side, and the Piedmont geosyncline on the Italian side of the present Alps; the two were separated by the Briançon geanticline (uplifted arch) which followed much the same line as the present Franco-Italian frontier. In the Jurassic, the Dauphiné geosyncline extended as far as the Massif Central. In the area now bounded by the Rhône, Isère, and Durance rivers, it constituted the basin of sedimentation known as the Vocontian trough.

Black shales, followed by limestones, accumulated in the Dauphiné geosyncline. Now uplifted, they represent the entire Middle Jurassic and attain a thickness of 5,000 feet in the Basses-Alpes, where they form the impermeable black lands of the Durance basin, famed as examples of excessive erosion. The limestones date from the Upper Jurassic and Lower Cretaceous.

In the Piedmont geosyncline, which was more constant and deeper than the Dauphiné, subsequent metamorphism turned the deposits into schists. They are penetrated by peaks and steep massifs (Mount Viso) of greenish volcanic rocks.

The Briancon geanticline stretching between the two geosynclines was formed by a series of shoals, sometimes raised above the sea or covered by a few score feet of water at the most. Today it is uplifted and the rocks of the anticline are found from the Briançon region to Dauphiné (Vercors) and Provence (Ventoux, Coudon). They are coral limestones followed, from the Cretaceous onward, by Urgonian limestones.

**Eastern Alps.** The Eastern Alps (that part of the range oriented in an east-west direction and lying east of the Upper Rhine Valley) are composed of limestones. The dominant elements are Triassic, consisting of reef, crinoidal, and oolitic dolomites (limestones rich in magnesium carbonates) alternating with marls. Where they have not been protected by overlying limestone beds, erosion has carved the marls into gentle slopes, valleys, and plains. Above them stand the steep walls of dolomitic limestone.

**Franco-Swiss Jura.** In France, Switzerland, and southern Germany, there is a series of plateaux and folded zones composed for the most part of Jurassic rocks and lying in front of the western Alps. This Alpine apron is the transitional region between the structurally complex Alps and the basins of sedimentation west and north.

Limestones similar to those of the *côte de Moselle* form the arid, fissured plateaux of the Jura tableland. They display deep valleys, caverns, and springs. In the west, the Jura ends above the Saône valley in an abrupt cliff (Revermont). At the foot of the cliffs, saline Triassic rocks (Lons-le-Saunier) come to the surface in an anticlinal 'botton-hole', or inlier.

The Jura offers many attractions to the amateur geologist; its structure is simple, its deep valleys contain fine geologic sections, its rocks are rich in fossils, and its coral reefs are arranged in a clearly demarcated series, running from north to south.

Beyond the Rhine, the Franco-Swiss Jura is matched by the Swabian-Franconian Jura. The latter, however, displays uniform beds and lacks folds. It, too, is a geologist's paradise, where all the Jurassic formations are found resting one upon another in perfect order. Certain beds of the area are particularly rich in fossils; such is the Boll shale, which is impregnated with bitumen and has preserved admirably ammonites complete with their mother-of-pearl, belemnites with their ink sacs, and ichthyosaurs with their muscles and skin. Another fossiliferous bed is the lithographic limestone of Solenhofen in which fossil jellyfish, crustaceans, insects, fishes, and flying reptiles, as well as *Archaeopteryx,* the oldest and most primitive bird, have been found.

## THE MESOZOIC IN AUSTRALIA

**Triassic.** Queensland has twenty-two workable Triassic coal seams. At Ipswich and Mount Crosby fine volcanic dust has preserved thousands of fossil insect-wings in perfect order. Triassic deposits are also found in New South Wales and in South Australia. At Copley there is an open-cast, sub-bituminous coal seam forty-eight feet thick. Triassic coal accounts for most of the Tasmanian output.

**Jurassic.** In the Jurassic nearly one-half of Australia was covered by the huge, freshwater Lake Walloon which left many lacustrine deposits. In Queensland it covered the present Great Artesian Basin. Australian Jurasic (like Triassic) strata are generally horizontal or only gently warped, but in south-east Queensland they are strongly folded. Imposing features of the Tasmanian landscape are the Jurassic dolerite dykes and sills, some 1,000 feet thick.

**Cretaceous.** Portions of Lake Walloon still existed in the early Cretaceous. But in the Great Artesian Basin most Lower Cretaceous rocks are marine shales. During part of the period a great continental sea covered much of Australia. Subsequent uplift replaced this in part with the freshwater Lake Winton, which covered the middle of the Great Artesian Basin.

Fossiliferous marine beds have yielded the giant 31-inch-wide ammonite *Tropaeum imperator,* and marine reptiles 50 feet long. In the Late Cretaceous there was local glaciation in the south-west part of the Great Artesian Basin. At the end of the period there was folding, faulting, and overthrusting in the Tasman geosyncline, together with some igneous intrusion, followed by uplift and erosion.

## THE MESOZOIC IN NEW ZEALAND

Mesozoic greywacke, a hard, dark-coloured, coarse-grained sandstone or grit, most of it devoid of fossils, is common in New Zealand. Found mainly in the mountain chains from Wellington to north of Lake Waikaremoana, these deposits accumulated in the New Zealand geosyncline, a trough that extended north-west to New Caledonia. Many fossiliferous Triassic and Jurrasic strata are exposed in Kawhia Harbour. Cretaceous rocks in North Island are found chiefly at North Auckland and from East Cape to Cape Palliser on the east coast. Hard Tatai (Lower Cretaceous) sandstone and volcanic agglomerate make up the bold peaks of Mounts Hikurangi, Aoranig, and Tatai.

In the South Island undifferentiated Mesozoic greywacke covers a large area from Marlborough to northern Otago. Mountain building in the Late Cretaceous squeezed much of this into folds which now make up the major part of the Southern Alps. Mesozoic volcanic rocks are exposed at Red Rocks Point, near Wellington, and in the Tararua and Ruahine Ranges.

Most South Island rocks that are classified as Triassic and Jurrasic are found in Southland and southern Otago. In the Hokonui Hills there are 30,000 feet of well-exposed, sharply dipping Triassic and Jurassic sediments. The greatest expanse of Jurassic rocks in New Zealand lies between Gore and the south-east corner of the South Island, and a petrified Jurassic forest can be seen near Waikawa. The limited West Coast Cretaceous beds are terrestrial; some contain coal. The Clarence Valley contains the thickest Cretaceous marine beds.

## THE MESOZOIC IN OTHER PARTS OF THE WORLD

At the close of the Cretaceous all continents experienced similar events, for not only were the Rocky Mountains folded and uplifted, but the Andes as well, thus forming a continuous chain of mountains from Alaska to Cape Horn. Land areas on both sides of the Pacific were affected, for the Great Antilles, Antarctica, and north-eastern Asia were also being folded. Great lava flows poured out over hundreds of thousands of square miles in the Deccan region of India, and may even have spilled across Arabia and the Arabian Sea into Africa.

A brief glance at a geologic map of the world will show the vast extent of Mesozoic rocks. In Europe the Triassic forms the largest part of Bavaria, Württemberg, Saxony, the Dolomitic Alps (Tyrol), and the northern part of Russia; the Apennines, Carpathians, and Dinaric Alps are Cretaceous.

In India and South Africa, clays and red sands known as the Gondwana and Karroo beds were deposited in the Late Permian and Triassic. In them are found fossils of a reptile family known as *Theriodontia* which shows anatomical characteristics more mammalian than reptilian: it is considered to be the 'missing link' between reptiles and mammals. The limestones which form the central Saharan plateaux were deposited in the Cretaceous and much of the desert's sand comes from worn-down Cretaceous sandstones.

China and Indochina have coal basins that date from the early Jurassic. Their fossil flora contains the more ancient types (ferns, calamites, trees such as *Cordaites* and the giant club-moss *Lepidodendron,* and modern types (cycads, conifers). Anthracite is worked in the Kebao, Hongay, and Dongtrieu basins. Mesozoic coal, too, is found in Siberia, Japan and some parts of Europe.

In many parts of the world the close of the Mesozoic was marked by the end of the stable conditions which had prevailed during the Cretaceous. In America, the Laramide mountain-building movements gave rise to the Rocky Mountains: in the Alpine region (the Pyrenees, southern Europe, Asia Minor, Iran, the Himalayas, Burma, Indo-China and the East Indies) movements began which culminated in the Alpine orogenesis of the Tertiary period; in Burma, which escaped the more severe mountain-building movements of the Alpine area, there was withdrawal of the sea. In contrast therefore with its opening phases following on from the Palaeozoic without a break, the end of the Mesozoic was marked by a distinct change of conditions leading into the Cenozoic. In America and Europe dominantly marine deposition of sediments gave way to estuarine and freshwater beds with relatively minor marine incursions. Consequently the fauna preserved — are very different from these of Mesozoic times.

The Meije Massif seen from La Grave. It lies amid slopes of black calcareous shales (Liassic).

# The Cenozoic Era: Tertiary Period

The most recent era of the geologic time scale, the Cenozoic, is composed of the Tertiary and Quaternary periods, the first lasting 69 million years, the latter no more than one million years. Thus the Cenozoic is a little less than half the duration of the Mesozoic (155 million years) and less than a fifth that of the Palaeozoic (375 million years).

At the beginning of the Cenozoic the continents had almost reached their present form, except that North America was still joined to Europe by way of Greenland, and North and South America were still separated from each other. Shallow continental seas that were widely developed during the Mesozoic had largely disappeared. The climate was generally mild, but was becoming cooler at the Poles. The migration of coral reefs towards the Equator indicated that present climatic zones were beginning to be established. It was a period of mountain building and vulcanism.

The Tertiary witnessed the rise of the more advanced species of the animal and vegetable kingdoms. Side by side with a large number of marine invertebrates lived an ever-increasing quantity of fishes similar to those of the present day. The great reptiles were replaced by birds and by mammals whose constant blood temperatures made them less susceptible to climatic variations. Flowering plants became the dominant species.

## THE TERTIARY IN NORTH AMERICA

Marine Tertiary sedimentation in North America was restricted to the fringes of three coastal areas: the Atlantic, the Gulf, and the Pacific. The Western Interior region was the site of continental or non-marine deposition. In western North America intensive and extensive vulcanism covered hundreds of thousands of square miles with lava flows and ash falls intermingled with various water- and wind-deposited sediments. The Laramide Revolution at the close of the Cretaceous had given birth to the Rockies and other mountains. This stimulated erosion which, during the Tertiary, laid down great fossil-rich alluvial deposits over wide areas.

Lava flows were numerous in the Columbia and Snake River valleys, where dozens of individual flows alternating with soils and their plant remains reach a total thickness of 5,000 to 6,000 feet. Widespread vulcanism was only one phase of the world-wide mountain-building activities of the Cenozoic. Mountain building of unusual magnitude further raised the Cascade Ranges in the Miocene times. In addition to folding and faulting them, Cenozoic mountain building also elevated the affected areas. As a result, the Rocky and Appalachian Mountains were pushed higher above sea level and were subjected to renewed erosion which produced deep canyons and reshaped the formations. Well-known examples of relatively recent uplift and consequent erosion are found in the National Parks of the Grand Canyon, Bryce Canyon, Rocky Mountains, and Great Smoky Mountains.

Tertiary seas covered only about five per cent of the present continental area. On the Atlantic and Gulf coasts the formations of this period dip gently toward the sea, and average half a mile thick. From New Jersey southward into Mexico, the Tertiary seas deposited a belt of poorly consolidated shales, sandstones, and limestones. These rocks have produced as much petroleum as those of all other ages put together, much of it from the great salt domes of Texas, Louisiana, and Mexico. The domes also produce salt and sulphur. The Eocene Green River shales, the oil shales of Colorado, Utah, and Wyoming, are potential sources of oil of staggering proportions. They cover an area of more than 16,000 square miles in these three states, and in some places are 3,000 feet thick. Tertiary formations probably contain about sixty per cent of the world's oil.

**Early Tertiary deposits.** The divisions of the Tertiary period are called epochs. The earliest is the Palaeocene (10 million years in duration). Succeeding that, in order, are the Eocene (20 million years), Oligocene (15 million years), Miocene (15 million years) and Pliocene (9 million years). The English geologist, Sir Charles Lyell (1797—1875), noticed that some fossils of the older Tertiary beds resembled living creatures. He also noted that the younger beds showed more recent living representatives than the older beds and so on. At the top of the section he found that as much as fifty per cent of the fauna from the youngest beds in the Tertiary resemble species living at the present time. So he applied to these epochs terms that reflect the increasing abundance of modern species in the younger beds. Thus, Palaeocene (from the Greek) means 'Ancient Recent'; Eocene, 'Dawn of the Recent'; Oligocene, 'a Few Recent'; Miocene, 'Less Recent'; and Pliocene, 'More Recent'. The Pleistocene, in the succeeding Quaternary Period, means 'Most Recent'.

Except where they are overlapped by younger deposits, deposits of early Tertiary age are found along the Atlantic coastal plain from New York to Florida. From Long Island to Virginia the deposits, mostly Eocene marine lime-bearing and green glauconite-bearing sands, are thin and outcrop only in a narrow belt. The Palaeocene is not exposed but has been found in some wells near the coast. To the south the formations become thicker, (up to 1,000 feet) and cover a wider outcrop area. There are also formations of Palaeocene, Eocene, and Oligocene ages composed of sands, shales, and marly limestones. Near the North and South Carolina border is an area of uplift almost at right angles to the trend of the shorelines. If this feature, known as the Great Carolian Ridge, was covered by sediments in the early Tertiary, they must have been subsequently removed by erosion before late Tertiary seas spread over the region. The strand lines of these seas are not definitely known because their edges have been eroded.

Deep drilling along the Gulf Coast has revealed scattered uplifts on which erosion had removed all or part of the Cretaceous rocks. At these points the Tertiary beds are separated from older rocks by an angular unconformity. In other places the contract between Mesozoic and Cenozoic rocks is conformable. One such uplift, the Jackson in central Mississippi, is an important gas-producing structure. There the basal Cenozoic formation, the Midway (Palaeocene), and younger Cenozoic beds were deposited on the eroded edges of the older Cretaceous and Jurassic formations. Gulf Coast Tertiary formations dip toward the sea in a gradient of about 35 feet per mile, also becoming thicker in

The sandstone cliffs of Capitol
Reef, Utah.

the same direction. These are about 25,000 feet thick, 10,000 feet
of which belong to the Early Tertiary. The oscillatory nature of
the seas is shown by the shore-lines in the Mississippi embayment.
There the Palaeocene sea reached southern Illinois, and at the next
advance, in the Eocene, came almost as far. In the Oligocene,
however, the sea barely reached central Louisiana. An east-west
trough, the Barton geosyncline, runs along the northern edge of
the Gulf of Mexico. Downwarping of the Gulf began, it is thought,
in the Cretaceous, and sinking continued during the Cenozoic to
produce the present considerable depth. Florida has 5,000 feet of
Tertiary sediments of all epochs. Some of these are 'Coquina
limestone', made up of shells and shell fragments.

The Palaeocene is distinguished by the Fort Union formation
in the northern Plains states and Alberta. Its continental sediments,
largely of sandstone and shale, vary from 2,000 to 5,000 feet thick,
and contain coal. These beds display a characteristic mammalian
fauna in the Big Horn Basin of Wyoming and in the San Juan
Basin of New Mexico.

Among the widespread continental deposits there is an isolated
remnant of marine strata from the Palaeozoic age in western North
Dakota (the Cannonball beds). There, 200 to 300 feet of dark
shale with oyster shells, Foraminifera, and other marine animals
dovetail with plant-bearing, non-marine deposits, all lying on top
of Late Cretaceous rocks. This, it is believed, represents the last
remnant of the great inland sea which once stretched from the
Arctic to the Gulf.

Eocene formations of continental origin occur in Wyoming,
Utah, Colorado, and New Mexico. The Lower Eocene Wasatch
beds reach a thickness of 2,000 feet in the Big Horn Basin. *Eohippus
(Hyracotherium)*, forerunner of the horse, is one of many mammal

fossils found. Spectacular Bryce Canyon in southern Utah is the
result of the erosion of the beds.

In the middle Eocene the Green River Basin in south-western
Wyoming and north-western Colorado received into its freshwater
lakes some 2,000 feet of very finely laminated shales, deposited
over a period of 6,500,000 years. Each very thin layer is supposed
to represent the deposit of one year. The overlying Bridger beds
of south-western Wyoming and north-eastern Utah have been
eroded into extensive badlands. The Uinta beds are similar to the
Bridger, and complete the Eocene section in the western interior.

In the Oligocene, continental deposits were widely distributed
in the northern Plains states. The White River beds, 200 to 600
feet thick, are composed of clay, silt, and volcanic ash, and contain
many mammalian fossils. Rain erosion has carved these beds
into the Badlands of South Dakota.

Along the Pacific coast the erosion of the Sierra Nevada and
Coast Ranges furnished thick sediments for the downwarped
areas, particularly the Great Valley in California. Some places
were depressed between faults, some even below sea level, as in
Death Valley. Both marine and continental deposits are found
along the present-day Pacific coast in these fault troughs. A marine
invasion of the Pacific coast in the Miocene resulted in deposits
up to 12,000 feet thick in southern California. One of the groups
of beds, the Monterey, consists largely of shales containing diatoms
(fossil single-celled organisms having a silica shell); it is an important
oil-producing horizon.

**Late Tertiary deposits.** On the East Coast, the highly fossiliferous
marine beds of the Miocene extend from Massachusetts to Florida,
while those of the Pliocene are confined to a strip south of the

Carolinas. Together, these formations consists of 600 to 800 feet of sands, clays, and marls. Because of the fluctuating nature of the shorelines, many raised beaches and wave-cut terraces mark the location of ancient strand lines.

Unusual and distinctive features of Cenozoic stratigraphy are the salt domes of the Gulf Coast area. Numerous plugs of salt, often more than a mile in diameter, penetrate Palaeocene and younger rocks over a wide area. The strata along the flanks of these domes are arched upward. Large quantities of oil and gas have been found in the porous rocks of the arched strata and in the cap rock over the salt plugs. The salt is probably older than the Cenozoic strata, originating in the extensive salt beds of Jurassic age. Its density was less than that of the enclosing rocks and its fluidity under the pressure of its heavy overburden permitted it to be squeezed out of its original beds and moved upward along faults or fissures. In addition to salt and petroleum, much of the world's sulphur is mined from the cap rock over the domes.

In the Late Tertiary almost all the gently sloping plains of the Rocky Mountains received a thin covering of sediments from the streams that flowed from the mountains.

The Miocene contains the Arikaree group of outwash sands and gravels. Near Agate, Nebraska, they contain vast quantities of bones of the extinct rhinoceros, *Dicerat; herium*.

The loosely consolidated Ogallala sands, gravels, silts, and clays, with a thickness of up to 500 feet, are significant deposits in the Pliocene strata.

**The Appalachian Mountains.** Almost every great mountain system in existence today was formed by or was involved in the crustal deformations of the Tertiary.

The original Appalachian Mountains had been reduced by the end of the Cretaceous to what has been called the Schooley peneplain. Exceptions in the north were the White Mountains of New Hampshire, the Adirondacks, and Mounts Katahdin and Cadillac in Maine, whose tops remained elevated above the near-level surface. In the south the erosion remnant of the Great Smokies still rose 2,000 to 3,000 feet above the surrounding lowlands. The Susquehanna, Delaware, and Potomac rivers meandered across the Schooley peneplain in their passage to the sea.

During the Cenozoic, a general uplift raised this peneplain, and started a new period of erosion. The rejuvenated rivers entrenched their meanders, and the major rivers, such as the Delaware, continued to wear down even the most resistant rocks (producing, for example, the Delaware Water Gap). Erosion wore down the softer Palaeozoic rocks more rapidly than the resistant formations, and tributary streams developed in these valleys. The general arrangement of the Palaeozoic formations with parallel hard and soft beds resulted in the canoe-shaped valleys and whaleback ridges typical of the Appalachians today. Where the uplift was too rapid for smaller streams crossing hard ridges to continue cutting their channels, the water gaps became wind gaps as the streams were forced to follow new courses. At one period uplift stopped long enough to produce the Harrisburg surface in the lowlands, but resumed again at intervals until the original Schooley peneplain had been raised to 4,000 feet above sea level along the central axis of the uplift arch. The Schooley peneplain is identified today by the level summits of the various Appalachian ridges.

**The Rockies and other western mountains.** At the end of the Cretaceous, the Laramide Revolution upwarped the original Rocky Mountains and downwarped various basins between the principal ranges. By the Miocene, extensive erosion had filled them, buried some of the lower ranges, and worn down most of the original Rockies to the Overland Mountain erosion surface, a rough peneplain 2,000 to 3,000 feet above sea level. Across this graded surface rivers flowed irrespective of the directional pattern of the ranges buried beneath.

The Pliocene uplift continued into the Pleistocene and gave the Rockies their present height. It rejuvenated the rivers, many of which have now cut deep canyons in the underlying rocks, often at right angles to the general direction of the mountains. Well-known examples in Colorado are the Arkansas River flowing through 1,600-feet-deep Royal Gorge, and the Gunnison River flowing through the 2,400-feet-deep gorge of the Black Canyon.

The Colorado Plateau is a relatively level region in western Colorado, New Mexico, eastern Utah, and Arizona that has been uplifted to heights of 5,000 to 11,000 feet. In the western part of the region several plateaux rise, step-like, to the east, in the spectacular escarpments of Vermilion Cliffs, Pink Cliffs, and White Cliffs in southern Utah.

The most recent (Pleistocene) uplift to rejuvenate the rivers in this area is responsible for the development of Zion Canyon and the world-famous Grand Canyon of the Colorado, which measures up to 10 miles wide, 1 mile deep, and 217 miles long.

The Basin and Range Province lies between the Colorado Plateau and the Sierra Nevada. After the Laramide Revolution it was mountainous, but erosion before the Miocene had greatly reduced the elevation of its mountains. During the Miocene a great deal of block-faulting occurred which uplifted the present Basin Ranges. At the same time erosion quickened, and alluvial fans up to 3,000 feet thick were developed in several places in southern Nevada. Much of the sediments of clays and silts with interbedded layers of borax and gypsum were deposited under arid conditions similar to those of the Great Salt Lake area today.

The Sierra Nevada Mountains, originally uplifted during the Nevadan Revolution in the late Jurassic, had been largely peneplained by the Miocene. Faulting then occurred, raising the eastern front of the Sierra earth-block to heights of 13,000 feet or more. From this great, tilted block, 100 miles wide and 400 miles long, the present Sierra Nevada Mountains have been eroded. The highest is Mount Whitney (14,555 feet) in the southern Sierras.

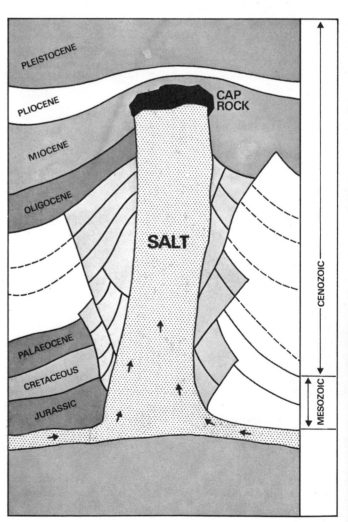

Diagram of a hypothetical salt dome in the Gulf Coast; length of section about 30,000 feet. Note how the salt was squeezed out of the Jurassic salt bed and forced upward, both doming and faulting the strata through which it flowed.

Uplift of the Coast Ranges which began in the Miocene reached its maximum in the Middle Pleistocene. The ranges were folded and thrust-faulted, and developed a highly complicated structure. The Cascades were pushed up in the early Pliocene, following the great basalt flows of the Columbia plateau.

During this crustal unrest one of the world's great faults, the San Andreas, received most of its lateral displacement. The fault can be traced for more than 600 miles, from Point Reyes, north of San Francisco, south-eastward into Mexico. It is known to have had a horizontal displacement of more than 175 miles during the last 25 million years, and possibly twice that distance since the beginning of the Cretaceous. As explained on page 181 earthquakes are produced by movements along faults in the rocks of the Earth's crust. In 1906, portions of the San Andreas fault moved as much as 21 feet, producing the earthquake that destroyed San Francisco.

Elsewhere in North America the Tertiary saw the uplift and development of many of the present erosional features of the Cascades in the north-west; the rise of the Taconics, Green and White Mountains in New England; the Ozarks, Ouchita and Wichita Mountains in Missouri, Arkansas, and Oklahoma; and, in the far north-east, the coastal mountains of Labrador and the mountains of Baffin Island.

**Vulcanism.** The Tertiary was a time of great volcanic activity in western North America. Much volcanic material was piled on top of the arched lavas that formed the bulk of the Cascades. Many now-extinct or dormant volcanoes, such as Mounts Shasta, Hood, St Helens, and Rainier were then active, as was Mount Lassen, the only currently active volcano in the United States. The San Francisco Mountains in Arizona were raised, together with many smaller craters and cinder cones in Nevada, Arizona and New Mexico. Eruptions continued from the Pliocene well into the Pleistocene; small cones continued to be raised into the Recent.

In the Cordillera we find a rock section primarily made up of volcanic materials, chiefly lava flows. In the Rockies lava flows and ejected rocks reach a thickness of several thousand feet. The base of a Miocene volcano contains the rich gold-bearing veins of Cripple Creek, Colorado. The lava flows covering the Columbia Plateau, an area of 200,000 square miles, are some of the most extensive in the world and are separated by beds of re-worked volcanic ash, sands and clays. They originated largely from fissures in and before the Miocene. In places this section measures 10,000 to 12,000 feet thick.

Evidence of vulcanism is abundant in Yellowstone National Park. Yellowstone Canyon, for example, is cut to a depth of 1,200 feet in volcanic debris. Seventeen forests on the slopes of Amethyst Mountain were successively buried beneath 2,000 feet of lava and ash. The famous petrified forests of Yellowstone National Park stand in beds of volcanic agglomerates, a mixture of coarse fragments, ash, and cinders that reached maximum thickness in the Absaroka Mountains.

The Obligocene John Day beds of Oregon are largely volcanic. Most of the deposits, whose thickness varies from 2,000 to 3,000 feet, were subsequently covered by the Columbia River lava flows. The middle beds contain a rich mammalian fauna. Near Florissant, Colorado, volcanic ash of Oligocene age buried many

Large columns of Tertiary lava near Vantage, Washington.

thousands of leaves and insects and preserved their impressions with extraordinary fidelity and fine detail on fine-grained, thin-bedded shales.

## THE TERTIARY IN BRITAIN

The basin of Tertiary sedimentations around Paris extended into Belgium and southern England to form the Anglo-Gallic or London-Paris basin. In Britain the out-crops of Tertiary rocks now occupy two synclines that were formerly joined, the London and the Hampshire basins. There are no Tertiary outcrops north of a line from the Wash to Torquay.

In Eocene times south-east England was a shallow sea fringed on the west by the delta flats of a low-lying land mass that included present-day Cornwall, Wales and northern England. The early Eocene sea did not advance far into the London basin. Overlying the Thanet sandstone come the Woolwich and Reading beds. These are probably of delta origin in the west around Reading, lagoonal near Woolwich, and completely marine in east Kent. The Reading beds of unfossiliferous sandstone contributed to the stones of Stonehenge. The pebble beds make up the Hertfordshire puddingstone, which is composed of rounded black flint fragments in a pale matrix.

A further marine invasion, reaching as far west as Hungerford in Berkshire, led to the deposition in the east of the sandy and pebbly Blackheath beds, but west of Croydon these disappear. The Blackheath beds were followed by the stiff blue London Clay of marine origin. In Essex it reaches a thickness of 600 feet, but westward thins to only a few near Hungerford. It has a variable fauna of crabs, fishes, crocodiles, and birds, as well as fragments of vegetation. The presence of almonds, magnolias, and laurels, together with the palm *Nipadites* (now found only in Malaya) indicates a tropical climate. The London and Paris basins are of the same age.

The clay gradually passes upward into the Bagshot beds, which are continental deposits consisting of pale sands with occasional beds of pipe clay and pebbles; there are only a few fossil plant remains. The beds, in part, give rise to the fine heathlands around Bagshot, west of London; they also cap Highgate Hill and Hampstead Heath.

In the Hampshire basin, as in the London basin, the various members of the Early Tertiary thin out toward the west. The cliff sections of Alum Bay and Headon Hill in the west of the Isle of Wight exhibit most of Britain's Lower Tertiary strata. The Reading beds are the lowest, and rest directly on the Cretaceous chalk. Next comes a rather loamy equivalent of the London Clay which, in turn, passes into a thick series of variegated sandstone referred in part to the Bagshot beds.

Elsewhere in the Hampshire basin, the Bagshot sands have seams of pipe clay which have given rise to local pottery industries. The Bracklesham and the Barton beds, showing marine origin in the east and continental in the west, constitute the upper part of the Bagshot sandstones. They contain various species of nummulites (single-celled organisms with a limey shell) which can be used for correlation with the rest of the Anglo-Gallic basin. In addition, the Barton beds include more than 500 species of snails and clams.

At the end of the Eocene, continental conditions prevailed in Britain, and the top of the Eocene and the bottom of the Oligocene form a single lithological unit: the Headon beds, named after Headon Hill. In the Oligocene Bembridge limestones are found freshwater snails such as *Limnaea* and *Viviparus* as well as the bones of swamp-dwelling mammals. Thin beds containing marine fossils indicate temporary marine invasions, showing that the land must have been very low. Finally, in the Oligocene Hampstead strata there are beds of special character such as the 'Crocodile Bed', the 'Mammalian Bed' (with bones of *Palaeotherium, Anthracotherium* and *Hyaenodon*), the 'Insect Bed', and the 'Waterlily Bed'.

While sedimentation was quietly taking place in the Eocene in the South, an area from northern Ireland and western Scotland to the Faeroes, Iceland, Jan Mayen, and Greenland was the scene of

great volcanic activity. This was the British-Icelandic, or Thulean, volcanic province. Some of the lavas, largely basaltic, were poured out of fissures, but most came from large, Massif Central type volcanoes whose complex roots can still be seen in Ardnamurchan, Slieve Gullion, Mull and Skye. In certain places at least 6,000 feet of lavas were piled up, and the flows originally spread over a region of some 500,000 square miles. In the Giant's Causeway in County Antrim, Northern Ireland, and Fingal's Cave in Scotland, the columnar jointing of the basalts is spectacular. The volcanic episode closed with the emplacement of dykes. On Arran, Scotland, a very small area contains several hundred dykes.

This volcanic activity can be dated as Eocene by plant remains found in a few localities in the thin lenses of sandstones, shales, and fine-grained lacustrine limestones lying between lava beds. The most famous plant bed, at Ardtun on the Isle of Mull, contains fragments of sequoia, ginkgo, magnolia, oak and lotus. One tree trunk, carbonised by the heat of the lava, was found standing in the position it occupied when it was engulfed by the molten stream some sixty million years previously.

Rocks that can with certainty be referred to the Miocene do not occur in Britain. At this period, however, Alpine folding reached its maximum and was felt in southern England, where a series of east-west folds was produced. The best developed is that running through the Isle of Wight and the Dorset coast, where horizontal beds have been pushed into vertical position. From the harder rocks has been fashioned some fine coastal scenery, such as the Needles in the Chalk and Durdle Dor in the Portland limestone.

The synclines of the Hampshire and London basins and the anticline of the Weald show only gentle folding. Erosion of the Weald has produced the well-known dip and scarp topography of the Downs and the Surrey Hills. London owes part of its water supply to this folding. The synclinal Chalk, Upper Greensand, and Thanet sands are sandwiched between the Gault Clay, below and the London Clay above. Thus water which falls on the North Downs and the Chilterns collects in the basin underneath London.

Pliocene sedimentary rocks are found only in East Anglia. These marine gravels and shelly sands are collectively spoken of as the 'Crag', the basal bed of which contains sharks' teeth, primitive elephants, whales and seals. The fauna in the 'Crag' becomes, with decreasing age, representative of colder climates, thus heralding the approach of the low temperatures of the Pleistocene. With a narrow sinuous outcrop containing minerals typical of the Ardennes, the Chillesford Crag is believed to represent one of the mouths of the ancestral Rhine.

## THE TERTIARY IN FRANCE

**The Paris and Aquitaine basins.** During the Eocene and Oligocene the Paris basin was invaded alternately by the sea or by freshwater lakes and lagoons. The sea deposited sand, lime-muds, or marls; the lagoons, clay, marls or gypsum, depending upon the salinity; and the lakes deposited limestones.

The earliest Eocene beds are marine and are followed by lignites and clays. These in turn are overlain by the Cuise sands laid down during a new marine invasion. The Cuisian sea retreated and returned during the Middle Eocene, giving rise to wide-spread limestones, the *calcaire grossier*. The main outcrops of this formation are in the northern part of the basin (Soissonnaire and Valois) but isolated occurrences show that it was deposited over an area that reached as far as Brussels.

The *calcaire grossier* is followed by an alternation of marine sands and lacustrine limestones known as the Middle Sands. During Upper Eocene times (the Ludian stage) the Paris basin was the site of shallow lagoons in which evaporation of salt water laid down large deposits of gypsum or 'plaster stone'. One of the finest quarries is at Cormeilles-en-Parisis (Seine-et-Oise), while at Lagny the gypsum has been transformed into alabaster.

The gypsum beds are overlain by marls and limestones laid down in Early Oligocene times (the Sannoisian stage). The most important formation is the Brie millstone. The last marine invasion took place in the Middle Oligocene (the Stampian stage) when oyster-bearing marls were laid down. After the deposition of the marls, the sea retreated; and the region it had occupied was covered by huge sand dunes, up to 20 miles long and from 600 to 1,300 feet high. These are known as the Fontainebleau sands or Upper Sands of the Paris basin. During the succeeding stage of the Upper Oligocene (the Aquitainian) the lakes increased in extent and the Beauce limestone was deposited. At one time this limestone covered all of Brie and extended northward to Paris and southward into Sologne and Touraine.

In the Touraine and Aquitaine basin, rocks of the Eocene and Oligocene are dominantly calcareous: the Blaye limestone corresponds to the *calcaire grossier,* the Saint Estèphe limestone to the gypsum, and the 'starfish' limestone to the Fontainebleau sands. These are the limestones of the Bordeaux vineyards.

The Early Miocene of the Paris basin was marked by deep erosion of the Upper Oligocene limestone and the deposition of the Orléanais sands (the Burdigalian stage). Younger Miocene deposits may occur but have not been identified with certainty. Finally, the Pliocene is represented by river gravels which are found at high elevations above the modern Seine and its tributaries: some of the gravels contain material derived from the Fontainebleau region. In Aquitaine the sea withdrew in Miocene times and lagoons were formed in which a shell-marl overlain by soft sandstones and limestones was deposited.

## MOUNTAINS

The Pryenees were built in three stages: initial folding in the Cretaceous, major folding in the Eocene, and further folding in the Oligocene. The rocks involved in these movements were the Cretaceous flysch and thick nummulitic limestones of Eocene age.

In the Miocene and Pliocene, great quatities of stream detritus were washed down to the foot of the Pyrenees and now constitute the plateaux of Lannemezan and Ger, which are areas of pebbles and clay cut into the valleys of rivers such as the Save and Garonne. These pateaux rise gently to heights of 1,800 to 2,200 feet and cover an area of 700 square miles.

**The Western Alps and the Rhône basin.** Uplifting in the Alpine region became active in the Tertiary, and extended into the Dauphine and Piedmont geosynclines. In the Eocene, an Alpine mountain chain was formed in what is now Dauphiné and Briançonnais. The uplift forced the sea back towards Italy into the Piedmont geosyncline. By the end of the Eocene, the chain had been worn down until only the massifs of Mont Blanc, Pelvoux, Mercantour, and Esterel remained above the sea. In this sea, limestones and marls were deposited, forming a flysch.

In the Oligocene the flysch became sandy, with bands of volcanic ash. In places it developed conglomerate horizons (as in the needle-shaped peaks of Arve), the material being derived from the high relief regions to the east where the Piedmont geosyncline had been uplifted. The Eocene chain was succeeded by an Oligocene chain of which Monte Viso, Gran Paradiso and Monte Rosa were the principal summits.

In the Miocene a trough of sedimentation became established along the border of the Alps. This sub-Alpine trough is the last relic of the Dauphine geosyncline. Submergence began with the Rhône basin and the Vienna basin. Later these two arms of the sea joined across Switzerland and Bavaria. The sediments varied from place to place and consisted of marine or freshwater 'molasses', similar to the soft sandstones and limestones of Aquitaine. The sub-Alpine trough 'molasses' consist of soft, light greyish sandstones and conglomerates which harden on contact with the air. Clayey molasses give rise to typical Swiss pasture lands. Calcareous molasses occur in white cliffs, sculpted by erosion into *alpilles.*

At the end of the Miocene and the beginning of the Pliocene, molasses beds of the same age were folded, faulted and overthrust. These movements elevated the sub-Alpine chains or the pre-Alps (Vercors, Grande-Chartreuse). Thus the Alps were built from the Eocene to the Pliocene, while the Pyrenees were formed from the Cretaceous to the Oligocene. Only 10 million years have elapsed since the completion of the Alps, but as many as 40 million since the final uplift of the Pyrenees.

After the formation of the sub-Alpine chains, the Rhône basin was completely raised above sea level before being invaded again in the Pliocene by a long and narrow arm of the sea, called the Rhodanian gulf. This reached to within thirteen miles of Lyons and had a series of small off-shoots corresponding to the present-day deep valleys of the Durance, Drôme and Isère. The deposits were mainly shales. In the north, the Rhodanian gulf acted as an overflow for a lake at Bresse, where marly deposits reached a thickness of 330 feet. Near Lyons the deposits were cut by a valley whose flanks had a series of waterfalls that gave rise to the travertine of Meximieux and a characteristic Pliocene flora. In the south, the Alpine streams poured in masses of sediments. The Durance, in particular, spread its debris over the whole region between Orange and Arles (Petite Crau) in deposits reminiscent of those of the plateau of Lannemezan.

**The Alps.** The Alps include crystalline massifs, like that of Mont Blanc, which are remnants of the old Hercynian chain. Over and around the massifs, which were temporarily submerged, were deposited Mesozoic sediments (black shales, Tithonian limestones, Urgonian limestones) and Tertiary sediments (flysch). In time, these sediments were folded, faulted and in some cases overthrust.

**Pre-Alps.** If we explore the Alps, starting at Lyons and travelling in the direction of Grenoble, we come first to the pre-Alps, seen in the cliffs of Vercors and Grande-Chartreuse beyond Rives. The cliffs are separated from the surrounding plain by a transverse valley, the *cluse* of Voreppe, through which flows the Isère. Other pre-Alps include the Bauges, between Chambéry and Annecy; the Bornes or Genevois between Annecy and Geneva; the Dévoluy and Diois, south of Vercors; Baronnies, still further south; and finally the pre-Alps of Vaucluse, Digne, Grasse, and Nice. All are cut by transverse valleys or *cluses.*

While the southern pre-Alps are relatively wide and form low chains extending in all directions, the northern pre-Alps are narrow and bounded by high limestone ramparts. Their folds have a north-south orientation which prevails also in the western Alps and the Jura.

The most southerly chains of the pre-Alps run into the Pyreneo-Provençal chains whose east-west orientation contrasts with the north-south direction of the sub-Alpine chains. The complex contorted structure characteristic of the Mesozoic and Tertiary formations of Provence is the result of the interaction of folds in two directions.

**Alpine furrow.** Whichever valley he follows, the traveller will end up in a longitudinal furrow, some 2 miles wide and about 125 miles long, which follows the inner margin of the pre-Alps. Grenoble is situated at the elbow formed by the Isère where it flows out of this trough into the cluse of Voreppe. Towards Chambéry, the trough is dominated on the west by the limestone wall of Grande-Chartreuse and to the east by a marly basement on which rise the jagged granitic peaks of the Belledonne range.

**Hercynian massifs.** Within the Alps there is a series of Hercynian massifs beginning with the Franco-Italian Mercantour massif and its snow-capped peak of Argentera (11,050 feet). Separated from this by the broad gash of Ubaye and Embrunais comes the Pelvoux massif culminating in the Meije (12,935 feet) and the Ecrins (13,350 feet). Two northerly offshoots of this massif constitute the Grandes-Rousses and Belledonne massifs. Finally there is the massif of Mont Blanc (15,625 feet). The series is continued in Switzerland by the Aar Massif and Glarus Alps. All these mountain massifs are made of crystalline rocks (granites, gneisses, and mica schists).

**Flysch nappes.** While broad passes cross the pre-Alps and give access to the Alpine furrow, the entrances to the high Alps are narrow, impressive gorges, like those of the Drac, Romanche, Arc and Upper Isère, with sides rising between 3,000 and 6,000 feet. Each valley is a separate district with little communication with its neighbours. Two, the valleys of the Arc and the Isère, are cut into

*Top*
The Lac Noir in the Haute Alpes
is a typical glacial lake.

*Bottom*
Lava formations, Dimmuborgir,
Myvatn, N. Iceland.

the resistant Carboniferous shales comprising the Vanoise massif.

A nappe is defined as a large body of rock that has moved far from its original position, either by overthrusting or recumbent folding. Beginning in the south of the area there is first the Ubaye nappe, which is the longest and most important. Its 38-mile front is characterised by the coral limestone of the Grande and Petite-Séolane and by Mont Morgan, the head of the nappe, situated where the Durance and Ubaye join. In the interior of the nappe, flysch overlain by a black sandstone (the Annet sandstone) can be seen on the shores of the lake of Allos.

At Embrun in the Durance valley and Barcelonette in the Ubaye valley there are deep cuts, or 'windows', in the nappe of flysch through which one can see the thrust resting on its basement of black shales found in the Dauphine geosyncline.

To the north, in the Romanche valley, the flysch nappes reappear to form the massif of the Aigullies d'Arves (11,300 feet). A fine-grained flysch is found to the south-east of Briançon, particularly in the Izoard Pass and the Casse-Déserte which it overlooks. Everywhere in this desolate area there are bare slopes of yellow or white sands and large, strange-shaped sandstone 'needles'.

Farther north, in the Chablais massif, the flysch gives way in part to breccias and limestones. Here little lies in its original position; such peaks as the Dent d'Oche (7,230 feet) have been thrust about sixty-five miles north-ward. Indeed, they overrode the Mont Blanc massif itself, though at a time when the latter had not reached its maximum height.

**Glossy schist nappes.** The construction of the Simplon tunnel through the Alps revealed half a dozen superimposed nappes. These are made up of gneisses and glossy schists (composed of metamorphosed flysch) which came from deep in the Piedmont geosyncline. All the majestic peaks which dominate Piedmont, from the superb pyramid of green rocks of Monte Viso (12,480 feet) to Monte Rosa (15,060 feet), Mont Cenis (6,770 feet), Gran Paradiso (13,195 feet), and the Matterhorn (14,713 feet), are part of this zone. At their feet stretch the beautiful Lakes Maggiore, Lugano, Como and Garda whose forms are the result of faulting.

At Guillestre, in the valley of the Guil, the gorge below Monte Viso is cut into grey lustrous limestone shales alternating with greenish sheets of eruptive rocks—the region of the Piedmont nappes.

**Alpine chains on a world-wide basis.** In terms of geologic time the Alpine chains are the most recent, following the Hercynian, Caledonian, and Huronian chains. They arose in narrow geosynclines crowded with the remnants of preceding mountains, and in Europe describe a series of arcs and loops which diverge from the Alps proper:

To the west is a succession of Pyreneo-Provencal chains, i.e., the Pyrenees and Cantabrian Mountains as far as Cape Finisterre.

To the south-west a range runs from Nice across the Gulf of Lions to the Balearic Islands, thence to Spain to form the Baetic Cordillera. It ends in the Moroccan Rif, with a total course of nearly 1,000 miles.

A south-easterly line runs through the Apennines, the backbone of peninsular Italy, then curves through Sicily, is continued in North Africa by the Atlas Mountains, sinks below the sea off Morocco, and appears again for a short distance in Madeira and the Canary Islands. The overall length of this branch measures at least 2,000 miles.

To the south-east are the Dinaric Alps which end in the Pindus and Peloponnesian Mountains of Greece.

Finally there is the long series of the Carpathians, Transylvanian Alps, Balkan Mountains, and the ranges of Asia Minor and the Caucasus.

These arcs can be described as borders around sunken massifs: the Hungarian massif became the Hungarian plain; the Aegean, the Aegean Sea; the Adriatic, the Adriatic Sea and the Piedmont plain; the Tyrrhenian; the Corse-Sardinian, that part of the Mediterranean between Corsica, Sardinia, Spain and North Africa; finally the Spanish Meseta, which has not changed. Thus the Mediterranean appears as a group of foundered regions framed by recently elevated mountains. Earthquakes and volcanoes (there were once many volcanoes in Sardinia, north-west Algeria, Morocco and Spain) are evidence of the area's instability.

Outside Europe, alpine chains are of great extent and again describe arcs moulded round resistant ancient massifs. In Asia, the Caucasus and the mountains of Asia Minor continue into Afghanistan, the Himalayas, Burma, Indo-China, and the East Indies. Coastal chains form island girdles or arcs on the west side of the Pacific: the arcs of the Aleutians, Kuriles, Japan, Marianas, Philippines, New Guinea, New Caledonia, and New Zealand.

The uplift of alpine chains was spread over the Mesozoic and Tertiary. The major episodes of folding were: Sierra Nevada and the Andean cordillera, Upper Jurassic; Rocky Mountains and Pyrenees, Eocene; Alps and Himalayas, Oligocene; Apennines, Miocene. The island arcs of the Pacific are mountains in the process of formation, is shown by current volcanic eruptions and earthquakes.

**Influence of the Alpine chains.** The uplift of alpine chains resulted in the foundering of the massifs, or 'Mediterranean ovals' (Aegean, Adriatic, Tyrrhenian), which acted as moulds to the chains.

The Masiff Central, in contrast, was forced by the growing chain, and this old Hercynian massif today rises with a gentle slope from west to east and ends in an abrupt edge above the Rhône and Saône valleys. Extensive faulting accompanied the tilting movement. Certain of the faults were also accompanied by foundering which gave rise to the Limagne and the Forez plain. The 3,000 feet of lacustrine sediments in the Limagne, deposited as the basin was sinking, represent the vertical movement.

Before the formation of the Alps, the Vosges and the Black Forest formed a single peneplained massif stretching from France into Germany. Stresses associated with the rise of the Alps led to an uplift in the south of the crystalline Vosges and a down-sinking in the middle to form the Alsace plain. This latter is comparable with the Limagne and the Forez plain, even to the 4,600 feet of sediments (this time marine) deposited in the Rhenish trough. The sediments contain asphalt (Lebsann), petroleum (Pechebronn), and potash salts (Mulhouse).

Farther away from the Alps, in the Ardennes and the Rhenish slate massif, the uplift was more accentuated in the east than west. With this movement, the Meuse, an old river, became rejuvenated and incised its meanders to a depth of 600 to 900 feet.

In the Paris basin the Hercynian folds again became active under the influence of the Alpine pressures. The Paris basin was fractured and faulted. The Artois ridge is continued in the boundary fault of the Franco-Belgian coal basin; the *pays de Bray* is bounded by a fault in the north. Forges-les-Eaux (Seine Maritime) and Pougues (Nievre) are spas whose waters come up along faults.

## VULCANISM

**Volcanoes of the Massif Central.** The elevation of the Alpine chains was accompanied by volcanic eruptions, the oldest of which in Europe dates from the Oligocene. Today the process continues in such active volcanoes as Etna and Vesuvius.

Between the Oligocene and the present there were Miocene eruptions on the fringes of the Mediterranean. In this epoch volcanoes in the open sea of the maritime Alps contributed their ashes to the littoral deposits. There were also Miocene eruptions in the plain of Alsace and of the Vosges. The small volcanic massif of Kaiserstuhl, across the Rhine from Colmar, is one manifestation of this activity.

Because of their interest to geologists the Miocene and Pliocene eruptions of the Auvergne will be described in some detail.

**Volcanoes of the Limagne.** The Limagne was an eruptive centre from the end of the Oligocene to the Pliocene. On the plateau of Georgovie, for example, flows and ashes are intercalated with or rest upon lacustrine marls and limestones of Oligocene age. At the top there is a basalt flow. The relief of the plateau is a secondary feature resulting from erosion. 'Fairy chimneys', for example, are formed because of the protection afforded by the basaltic capping. Such inversions of relief are common in the neighbourhood of Clermont-Ferrand where there is another basaltic plateau.

The town of Clermont, built on an ancient volcano, is divided by a fault separating the upper part of the town from the lower. The bedrock of the town is volcanic tuff. Carbon dioxide still sometimes escapes into the cellars of the houses. On the outskirts of town and for more than thirty miles into the Bourbonnais there are many carbonic fumaroles and hot mineral springs.

**Volcanoes of Velay.** The region of the upper Loire and the Allier constitutes one of the large eruptive centres of the Massif Central. Basalt flows and phonolitic plugs from several volcanoes are found side by side.

Between the Allier and the Loire, in Velay proper, an enormous accumulation of basalt covers more than 385 square miles, and reaches a height of 4,625 feet at Mont Deves. From more than 150 vents lava flowed toward the Allier valley and the Loire. Today the beds of scoria form red, stony, barren soils covered with heath and forests of pine and fir.

The volcanic basin of Puy is in the upper valley of the Allier. Basaltic rocks occur within the town of Puy itself. St Michel rock, for example, is the exposed neck of an ancient volcano, and the rocks of Corneille and Saint Joseph d'Espaly are fragments of an old cone. Nearby is the extraordinary basaltic plateau on which the castle of Polignac stands. Here too are the famous Organs of Espaly, 485 feet long and 98 feet high.

Further east, between the Loire and the Lignon, stretches the Mégal massif. This is continued to the south-east by the Mézenc massif and then to the Rhône valley by the basaltic plateau of Coirons. In the Mézenc and Mégal massifs basalt and phonolite occur together. Gerbier-de-Jonc is the type example of a phonolitic stump of a volcanic plug, characterised by the resonant flags on the path to the summit. There are many other phonolitic peaks: Signon, Bachot, Tourte, and Mézenc (5,700 feet).

**Volcanoes of the Monts Dore.** The region of the Monts Dore (Puy-de-Dôme) is volcanic, with a deep valley cut by the Dordogne. The cones, much reduced by erosion, are variously made up of basaltic sheets, trachytic flows, accumulations of ash from explosive eruptions, or phonolitic peaks like those of Velay. The Monts Dore region presents other aspects of vulcanism: deep gashes (as in the valley of Chaudefour where lava can be seen resting on the granitic basement of the Massif Central), accumulations of scoria, crater lakes, lakes formed by the damming of valleys by lava flows, and mineral springs.

**Volcanoes of the Cantal.** During the Pliocene, small Miocene volcanoes scattered throughout the Cantal area were buried by a single immense volcanic cone that covered the whole of the present department of Cantal and reached a height of 13,000 feet. It is now much reduced by erosion and the deep valleys of Cére Jordanne, and Alagnon that have been cut into it reveal its structure. consists of an andesitic breccia composed of bombs, boulders, and volcanic ash. In the Mougude ravine, near Vic-sur-Cere, there are imprints of laurel and bamboo leaves, indicating that the climate ten million years ago was warm. Griou is a phonolitic dome around whose margins are found the andesitic domes of Puy Mary and Puy Chavaroche, and the basaltic peaks of Plomb du Cantal and Puy Violent. From the Puy du Cantal (6,040 feet), the highest point of this volcanic massif, a basaltic sheet, the Planèze de Saint Flour, spreads out in the form of a triangle.

## THE TERTIARY IN AUSTRALIA

In Australia Tertiary deposits are quite varied. Around the coast and in two ancient embayments, the Murchison Gulf and the Adelaide Gulf, sediments are largely of marine origin. The continent in general was already beginning to assume its present shape, though in the Miocene a sea covered the Nullarbor Plain to a point 200 miles inland. In the interior, lacustrine and other continental deposits prevail, many of them fossiliferous. Parts of Tasmania were under the sea during the Tertiary, and highly fossiliferous Miocene deposits are found here, especially at Wynyard. At Table Cape they have yielded bones of *Wynyardia*, which is easily Australia's oldest known marsupial.

During this period several great lakes were formed. One was Eyrian Lake, which succeeded the Cretaceous Lake Winton. In Queensland there were several, one of which left deposits 1,200 feet thick. Estuarine lakes in Victoria laid down lignites (Latrobe, Valley, Port Albert) with individual seams up to 500 feet thick.

Some Tertiary river deposits contain 'deep leads' of gold-bearing gravels. Cenozoic basalt flows, mostly Tertiary (Oligocene), cover 50,000 square miles in eastern Australia and reach a thickness of 3,000 feet.

## THE TERTIARY IN NEW ZEALAND

These rocks follow on from those of the Cretaceous without any marked break. During the Eocene, the economically valuable coal measures of the south-west part of the South Island were formed. The marine transgression of the Oligocene and Miocene submerged most of both islands except in the extreme south of South Island (Otago) where fluviatile and lacustrine deposits, frequently gold-bearing, are found. There was a general withdrawal of the sea from the South Island during the Pliocene but the North Island remained below sea level. In the Wanganui area, clays 3,500 feet thick were deposited, apparently without a break.

Vulcanicity developed in the mid-Tertiary and produced the pillow lavas and tuffs of Oamaru district and the acid lavas of the Coromandel peninsula which are associated with gold-bearing beds. A little later, alkaline lavas were extruded in the Dunedin area. In the Upper Tertiary there was much volcanic activity in the North Island: the main centres lie on a line of crustal weakness from Ruapehu to Mount Edgecumbe.

## THE TERTIARY IN OTHER PARTS OF THE WORLD

In Eocene times an enormous outflow of basaltic lava took place in the Deccan of peninsular India. This volcanic episode probably began in the Cretaceous and continued without a break into the Tertiary. The Deccan trap, as it is called, is several thousand feet thick and extends over an area of 250,000 square miles in the north-west quadrant of the country. Along the west coast it reaches the sea without any diminution of thickness from about 21°N. to 16°N. latitude, and since the lavas are certainly sub-aerial not sub-marine, it is clear that at the time of their eruption the land must have extended farther west than it does now. Flat-topped hills and deep-cut valleys characteristic of the Deccan trap are well shown on the route to Bombay from Calcutta or Madras. In spite of the great extent of the lavas there is little indication of volcanoes of 'central' type, and it is believed that the eruptions took place from fissures with little explosive action. In the absence of fossil-bearing beds between individual flows of lava, the duration of the eruptions cannot be determined with certainty.

Tertiary sediments in peninsular India are insignificant. Near the west coast there are a few small outcrops in Travancore, Gujerat and Kathiawar, and a more complete sequence in Kutch. On the east coast the Cuddalore series, ranging in age from Eocene to Pliocene, occurs in patches from Orissa to the extreme southerly tip of the peninsula.

In the north, however, Tertiary sediments are better represented. As in Europe, they show an alternation between marine and freshwater deposits. Eocene beds are divided into the Ranikot, Laki and Kirthar groups and are followed by the Nari Series of Oligocene times and the Gaj Series of the Miocene, which has yielded many mammalian remains. The first of the three stages of Himalayan uplift took place in the Middle Eocene but even from Early Eocene times there existed round the fringes of Asia a series of basins in which sediments were laid down in much the same way as in the Anglo-French basin. Two good examples are to be seen in the Gulfs of Assam and Burma. The Tertiary sediments in the Burma Gulf are over 10,000 feet thick: at the landward end they are sands with freshwater and estuarine clays; at the seaward end they are dominantly marine.

The end of the Tertiary throughout the world was not marked by mountain-building movements such as those which terminated the Mesozoic era. The Quaternary followed with an increasingly adverse climate in the northern hemisphere.

# The Cenozoic Era: Quaternary Period

The Quaternary, because it is still in progress, is the shortest of all geologic periods; it is composed of two epochs, the Pleistocene and the Recent. The Pleistocene is usually said to have lasted for one million years; in North America and Europe it was a time of extensive and repeated glaciation. The Recent is the geologists' name for the few thousand years since the Pleistocene ended.

No extensive mountain building marked the end of the Tertiary as it had marked the end of some other periods. But due to climatic changes, the Pleistocene witnessed four advances of great ice sheets across the northern hemisphere and, after the retreat of each, three interglacial stages in some of which the climate became warmer than it is today. We are now living in what may prove to be the fourth inter-glacial stage. Some scientists believe that, if conditions follow their previous pattern, the ice will return in about 50,000 years. If it does, the Recent Period could be regarded as an extension of the Pleistocene.

During the Pleistocene glaciations, pluvial periods occurred over large areas of the unglaciated zones. Pole-ward from latitudes 50° north and 50° south the climate was cold, humid, and snowy. Between these latitudes there was much rainfall. Conversely, during the inter-glacial periods of northern Europe and North America, zones near the Equator experienced dry periods. Thus the Sahara passed alternately through phases of humidity and dryness, fertility and aridity, before becoming stabilised.

Advances and retreats of the ice were approximately contemporaneous in both the Old World and the New. These glacial and interglacial periods are known in Europe and America by different names. The following table correlates the two sets of terms:

*Glacial (1—4) and interglacial (a—d) stages*

| America | Europe | Years to beginning of |
|---|---|---|
| d. Recent | d. Recent | |
| *Quaternary* | | |
| 4. Wisconsin (Mankato (Cary (Tazewell (Iowan | 4. Würm | 100,000 |
| c. Sangamon | c. Riss-Würm | |
| 3. Illinoian | 3. Riss | 300,000 |
| b. Yarmouth | b. Mindel-Riss | |
| 2. Kansan | 2. Mindel | 700,000 |
| a. Afton | a. Günz-Mindel | |
| 1. Nebraskan | 1. Günz | 1,000,000 |
| Blancan | Villafranchian | |
| *Tertiary* | | |

Though weathering acts very slowly on fresh rock material, the depth of weathering gives a relative measure of the time that elapsed between advances of the ice sheets.

Chimney Rock, Capitol Reef.

**Types of Glaciers.** There are three principal types of glaciers: continental, piedmont, and alpine.

Continental glaciers are the largest; during the height of their Pleistocene advance they covered about one-half of the area of North America and of Europe. Today they are confined to the comparatively restricted areas of Greenland and Antarctica.

Piedmont glaciers develop at the foot of mountains, sometimes where several alpine glaciers converge and amalgamate. The Malaspina Glacier at the foot of Mount St Elias, in Alaska, is a good example. It is formed by the union of eight valley glaciers and has an area of 1,500 square miles. In cold climates many glaciers flow right down to the sea.

An ice sheet is one of the most powerful forces in nature, over-riding everything lying before it — rocks, soil, and trees. The Pleistocene ice sheets therefore had a profound effect on the topography of the northern hemisphere.

## QUATERNARY GLACIATION IN NORTH AMERICA

In North America there were three centres of ice accumulation during the Pleistocene: the Keewatin, west of the bulge in Hudson Bay; the Labrador, east of Hudson Bay; and the Cordilleran sheet, which occupied the northern part of the Cordilleran belt of mountains in western Canada and north-western United States. In these areas, ice accumulated to great thickness and, in response to its own weight, moved by plastic flow to cover most of North America as far south as the Raritan, Ohio, and Missouri rivers. Till deposits formed by the melting of Pleistocene ice sheets cover nearly 5 million square miles in North America, and they cover a total of more than 12 million square miles of the entire surface of the Earth.

In Wisconsin, Iowa, and Illinois, there is an unglaciated area of 10,000 square miles, known as the Driftless Area, between the Keewatin and Labrador ice sheets. The flow of ice was deflected from this area by the deep basins of Lakes Michigan and Superior and by highlands to the north.

The Nebraskan, the first of the four North American glacial stages, may have had its centre of origin some 300 miles north-west of Hudson Bay, but it apparently received most of its moisture from the Gulf of Mexico. The ice-front almost reached the latitude of St Louis. In the succeeding interglacial (the Afton) the early forests in Iowa were conifers. Then came grasslands and oaks, and then conifers again as the cold returned.

The most extensive of the four ice sheets was the Kansan; its ice-front reached Kansas City and St Louis. In pre-Kansan days it is believed that the headwaters of the Missouri River drained into Hudson Bay. When ice blocked the bay all the Missouri waters were forced into the Mississippi. The long Yarmouth interglacial period was at first mild and moist, but later became dry, and may have included some cool periods. Iowa has Yarmouth peat beds 15 feet thick; loess was widely deposited over the Kansan till.

The Illinoian ice sheet radiated out from Labrador, reaching the Ohio River and crossing the Mississippi into Iowa and Missouri.

The succeeding Sangamon interglacial period was temperate; loess was deposited over the Illinoian drift.

One of the most remarkable incidents in the Wisconsin (the latest glaciation) was the movement of boulders by the ice from the Hudson Bay region to the foothills of the Rockies. Neither of the last two glaciations covered as large an area as the first two.

## QUATERNARY GLACIATION IN EUROPE

The Ice Age in Europe was contemporaneous with that in North America, and its manifestations were similar.

A characteristic example of glacier retreat can be seen in the valley of Argentières (Haute Savoie). Just beyond the village houses, can be seen the amphitheatre of moraines that has only recently been vacated by the ice. Beyond, on the sides of the valley, a smooth vertical area rises several hundred feet above the present level of the glacier. This was caused at an earlier stage when the glacier was larger, extended much farther down the Arve valley, and joined the Mer de Glace and the Bossons glacier.

If the Argentières glacier were to melt, the site it occupies would become a typical U-shaped valley with a flat bottom and almost vertical sides. Such valleys excavated by ancient glaciers are found in the Alps, the Pyrenees, the Vosges, the Massif Central, North Wales, and Scotland. The magnificent valley of Monts Dore (Puy-de-Dôme) is a classic example.

Europe has many other examples of the work of glaciers: The Norwegian fjords are glacial valleys which were invaded by the sea when the ice melted. A glacial moraine crosses the city of Stockholm and stretches as far as Uppsala. In their retreat Pleistocene glaciers left behind a clay mixed with sand polished and striated boulders. Such a boulder clay forms the marshy soil of Dombes, in France, and the soil of East Prussia and Finland. Terminal moraines determined the east-west direction taken by the rivers of north Germany — the Elbe, Havel, Spree, Warthe, Netze, and Vistula, each of which follows an old ice-front.

In mountainous regions the retreat of the Pleistocene glaciers is shown on a more restricted scale. Each valley was barred by a crescent moraine which retained the meltwater, forming lakes similar to those constructed by man for his hydro-electric shemes. Examples are Lakes Garda, Como, and Constance. When such a lake ultimately disappears, the morainic crescent or amphitheatre remains as an important feature of the landscape. The famous moraine at Ivrea (Italy) stretches for 13 miles across the Lombardy plain, through the middle of which passes the Dora Baltea. From its 2,200-foot crest it is possible to look down on the polished *roches mountonnées* of the old glacier. In the Vosges the glacial retreat took place in stages, a fact which explains the 'stepped' lakes of Gerardmer (2,100 feet), Longemer (2,270 feet), and Retournemer (2,400 feet); for each moraine that forms a barrier across the valley there is a corresponding lake.

## SUBSIDENCE AND RECOVERY

The weight of the ice sheets caused the land to sink as much as 1,000 feet in places. After the ice melted, this land depression formed huge lakes in North America, larger than any known today. Lake Algonquin included the present Lakes Superior, Huron, and Michigan, plus much of the surrounding territory. The Champlain Sea joined Lakes Ontario, Champlain, and the St Lawrence River Valley with the Atlantic. In the course of several thousand years the land gradually readjusted its level, the lakes were reduced to their present dimensions and their beaches raised above the surrounding land, sometimes as much as 900 feet.

From time to time the river pattern changed. While the St Lawrence was blocked by ice to the north, it drained into the Mississippi. Later it flowed through the Mohawk and the Hudson to the Atlantic. Not until all the ice had gone did it flow to the Atlantic through its own gulf. When melting did begin, the Illinois, Minnesota and many other rivers became huge rushing torrents with beds far wider than those they use today.

Because deep valleys held more ice, lobes of it remained in the basins of the Great Lakes long after the thinner ice on the edges of their basins had melted. As the glaciers retreated northward the precursors of the present Great Lakes began to form from meltwater in the farther reaches of these deep trenches. Lake Duluth at its west end was the beginning of Lake Superior; Lake Chicago at its southern end (and draining into the Mississippi) was the forerunner of Lake Michigan; Lake Maumee at its west end (and draining into the Wabash) was the ancestor of Lake Erie.

Lake Algonquin first flowed through the Chicago outlet into the Mississippi, then by Port Huron and down the Niagara into the Atlantic. Later an outlet opened near Kirkfield, Ontario, and the water flowed to the Atlantic through the Trent River and Ontario Basin. It is believed that for several thousand years

Lake Algonquin drained simultaneously through the Kirkfield, Chicago, and Port Huron outlets — or at least through two of them. Further retreat of the ice opened up a new outlet through the Ottawa River to the Champlain Sea. As the land rose, Lake Algonquin gave way to the Nippissing Great Lakes, ancestors of the present Lakes. About 3,000 years ago the present drainage through the St Clair River to the St Lawrence was established.

Niagara Falls have been produced by the Niagara River flowing over a cliff of hard Silurian dolomite, eating into the underlying soft sediments and finally causing collapse in the overlying hard rock. The cutting of the gorge began when the ice had retreated far enough to open up the outlet from Lake Erie. The whirlpool section had been partially cut by an earlier interglacial Niagara falls, then excavated by the present falls.

In North Dakota, Minnesota, Manitoba and Saskatchewan a vast lake, Lake Agassiz, covered 110,000 square miles and deposited the silt that now forms the rich farmlands of the area. Lakes Winnipeg and Winnipegosis are what was left of Lake Agassiz after its drainage into Hudson Bay was established.

Lake Bonneville, of which Great Salt Lake in Utah is a remnant, once covered about 20,000 square miles in the Great Basin. However, with the climatic changes that occurred after the glacial advance (reduction in rainfall and an increase in evaporation), the lake abandoned its outlet to the sea through Snake River and eventually became a salt lake. The old shore-lines, with their wave-cut terraces, may be seen today.

In Europe the oscillations of the sea level in Scandinavia are characteristic: first the land subsided and was transformed into an island; then uplift connected it on the north and south with Finland; renewed subsidence created the Baltic Sea; then uplift continued until the present time at the rate of three feet a century.

## DATING GLACIAL DEPOSITS

The age of relatively recent glacial and interglacial stages can often be arrived at by counting the varves (the annual layers of sediments deposited around a glacial lake). The principle was first worked out in Scandinavia where varve deposits from one-tenth of an inch to several inches thick are found. Each year is represented by a thick layer of coarse sediments laid down in summer, and a thin layer of fine-grained sediments deposited in winter. A count of these layers indicates that the ice took 13,200 years to retreat

gradually from southern Sweden to reach its present position.

The age of deposits up to about 70,000 years old can also be ascertained by a study of their carbon-14 content. Computations using this method show that the great Wisconsin ice sheet retreated from its farthest point (Metuchen, New Jersey) 12,000 years ago.

## MARINE SEDIMENTS AND RIVER DEPOSITS

At each glaciation a huge quantity of water, which originally came from the sea, accumulated and was immobilised on the land in the form of snow and ice. A fall in sea level resulted. As the sea level was lowered, the rivers had to cut deeper valleys to reach the new level at their mouths. Glaciation, lowering of sea levels, and excavation of valleys occurred simultaneously.

The three processes were reversed during the interglacial periods. The melting of the ice caused the sea level to rise and to invade the coastline and drown the mouth of rivers. This reduced the slope and the speed of rivers, causing them to deposit more sediments in their valleys. With each marine invasion the sea-shore was lined with new beaches which were abandoned at each withdrawal of the sea. In this way, raised beaches or marine terraces were formed, corresponding to the inter-glacial periods and containing shells that tell us the temperature of the period in which they lived.

Similarly, at each interglacial stage the valley bottom was filled with alluvial deposits in which the river once again cut its bed during the next glacial stage. These alternating processes produced river terraces on the valley sides; the highest are the oldest. Thus Terrace *A,* the uppermost and 150 feet above the present river level, corresponds to the Afton (Gunz-Mindel) interglacial, and represents the high marine terrace.

Terrace *B,* deposited during the Yarmouth (Mindel-Riss) interglacial and partially eroded during the Illinoian (Riss) glaciation, is contemporary with the middle marine terrace and is about 90 feet above river level.

Finally, Terrace *C,* the low terrace, is 30 feet above the river level and was laid down during the Sangamon (Riss-Würm) interglacial. Its characteristic species is *Elefhas antiquus.* Examples of each terrace in western Europe also contain worked flints, which indicate the presence of man. The number of terraces is not everywhere the same, so that a valley may posses one, two, three, or more. The height of the same terrace, too, may differ from region to region.

*Left*
Glacial till and loess in a railroad cut south-west of Rhodes, Iowa.
A. Fresh Wisconsin till.
B. Windblown Peoria loess.
C. Weathered Kansas gumbo till.
D. Unweathered Kansas till.

*Right*
A broad U-shaped valley formed by glaciation.

*Left*
A glaciated pavement showing
Hinchingham grit overlain by
boulder clay.

*Below*
Volcanic activity in New Zealand,
begun in the Upper Tertiary,
continued in the Quaternary.

## LOESS AND CAVE DEPOSITS

Eolian loess is wind-borne and comes from finely ground glacier dust, carried on to the outwash plain by glacial streams, then dried and transported by the wind. It is produced in dry seasons or in dry areas, and before vegetation reclothes the outwash plain.

In North America vast deposits of loess derived from drift left by the Wisconsin ice sheet are found in the Mississippi Valley. They cover extensive areas, particularly in the Corn Belt of Iowa and Illinois, and are often 100 feet thick. The angular chararcter of the tiny, unconsolidated grains of silt that compose the loess maintains vertical banks and cliffs for some time after exposure, as at Council Bluffs, Iowa. Typical Mississippi Valley loess is unstratified yellowish silt, half-way between sand and clay in texture. Important

loess deposits are also found in the Palouse Hills of Washington.

In France loess is found only north of the Loire and in the Rhône valley, and generally the only fossils found there are the prints of grass roots and the shells of land molluscs. Near Paris, however, the loess also contains microscopic wind-borne fossils from the chalk of Champagne, sixty-five miles away. The French loess was probably deposited in grasslands similar to the present steppes of the Soviet Union, and probably developed on the edge of the glaciers.

In Europe a distinction is made between old loess and recent loess. The former corresponds to the Mindel and Riss glaciations, thus explaining why it covers the plateaux as well as the high and middle fluviatile terraces, but not the low ones. The recent loess of the Würm glaciation covers the old loess and also the low terraces containing worked flints whose age can be determined from the character of their workmanship. Percolating waters moving through the loess have generally formed a concentration of iron oxide at the surface (hence its red colour) and a concentration of calcium carbonate lower down. The carbonate is sometimes found in strange shapes called *poupees* (dolls) in French and similarly, *loess Kindchen* in German.

In Europe and Asia, loess covers considerable areas. In France, it forms the cereal-growing soil of Beauce and the Caux district, the beet soil of Valois, Soissonnais and Picardy, and the loam of the Rhine valley. It is found also in Belgium, Holland, Germany, and China.

In limestone districts such as the Dordogne, underground caves of Pleistocene age served as refuges for animals and as dwellings for man. Today excavation reveals that their floors have been built up by the successive layers of river flood sediments, surface water debris, stalagmite deposits, and loess. Prehistoric hearths made of cinders and kitchen debris are sometimes found, together with occasional flint implements and, though rarely, human bones.

At the end of the Quaternary, in the bottom of cold swampy valleys, peat-bogs were established, like those of northern Europe and North America today. Present-day peat-bogs, in fact, are a continuation of these post-glacial bogs, and sometimes yield historic and prehistoric artefacts as much as 10,000 years old. Peat also contains pollen grains from trees that grew in the post-glacial bogs.

Microscopic examination of the grains, layer by layer, tells the story of the evolution of the forests and of variations in post-glacial climates. It tells how forests of pine and birch gave way to those of oak, elm and beech as the ice retreated.

## QUATERNARY VOLCANOES

In North America numerous cinder cones and lava flows indicate a continuation of vulcanism from the Pliocene into the Pleistocene. In the western Cordillera these youthful flows are found from Mexico to Canada. Their geologic infancy is indicated by their fresh, unweathered appearance, by the fact that they cover recent alluvium in stream valleys, and by lava-dammed streams. The lack of weathering on the surface of the flows makes them look like the recent cooled Hawaiian lavas. The cones are so fresh that they look as if they might begin erupting again at any time.

Mount Lassen, California, the only currently active volcano in the Pleistocene; glacial valleys on its slopes were filled with lava from later eruptions. The mountain once reached a height of at least 10,000 feet, but after many eruptions its core finally subsided and most of the cone above it collapsed. The result was a roughly circular pit, or caldera, 5 to 6 miles wide and 2,900 to 4,200 feet deep. This caldera is now occupied by Crater Lake.

Mount Lassen, California, the only currently active volcano in the United States was also active during the Pleistocene. The volcanic deposits of craters of the Moon National Monument, in Idaho, and the Lava Beds National Monument in California, as well as the cones of Capulin Mountain, New Mexico, and Sunset Crater which erupted in A.D. 1069, are all of Pleistocene or Recent age.

In France eruptive activity declined in the Quaternary. The Monts Dore area was once the scene of the Tartaret lava flow, 13 miles long but only 50 yards wide. The flow now dams the Couze valley, giving rise to Lake Chambon.

A new volcanic centre on the edge of the Limagne produced the chain of the Puys. It consisted of 80 volcanoes along the 20-mile-long and 3-mile-wide plateau separating the Limagne from the valley of the Sioule. The alignment of the volcanoes coincides with fault lines. Today some remain in the form of domes, others as craters.

The domes issued from the earth in a way similar to the spine of Mount Pelée in 1902. Viewed from the summit of the Puy de Dôme, the craters present a lunar landscape. There are simple craters like the Nid de la Poule, breached craters like the Puys de la Vache, Lassolas and Louchardière, and interlocking craters like those of the Puy de Dôme. These volcanoes today are only 900 feet above the plateau, but their flows are fairly extensive; the largest are La Vache (10 miles) and Côme (6 miles). A flow from Lassolas dams the Veyre and gives rise to the Lake of Aydat.

Outside France, the principal Quaternary volcanoes are found in Iceland, Italy, and Greece. In Iceland, several geysers are evidence of eruptive activity. The Alban Hills near Rome, with the crater lakes of Albano and Nemi, are the remains of a volcano comparable to Vesuvius. Near Naples the region of the Phlegrean fields includes the crater-lake of Averno and the solfatara (sulphurous gas vent) of Pozzuoli. In Greece, the Santorin archipelago is a Quaternary volcano destroyed by an explosion. Impressions of palm, olive, and pistachio leaves in the Santorin tuffs indicate a climate wetter than it is today.

Most modern European volcanoes date back to the Early Quaternary. When Etna and Vesuvius belched forth torrents of lava the chain of the Puys was lighting up the sky of the Auvergne.

## THE QUATERNARY IN AUSTRALIA

Pleistocene alluvial deposits are found along a number of Australian rivers. The Ice Age affected Australia much less than it did the northern hemisphere. An ice sheet on the Kosciusko plateau covered only 350 square miles and later retreated to a valley glacier and a cirque-cutting stage. In Tasmania, however, about half the island was affected by ice.

The world-wide rise in sea level when the ice melted, and the consequent lowering of river valleys, resulted in many drowned valleys. Some preglacial and interglacial terraces from 40 to 100 feet above sea level can be found around the coasts. The Great Barrier Reef was formed during the Quaternary, rising with the sea level to the melting ice of the last glaciation.

One noteworthy Pleistocene earth movement, an upwarping east and west from the Gawler Ranges, formed Lake Dieri, with an area of 40,000 square miles. Eyre is a descendant this early lake. In early Recent time, however, aridity set in and converted the interior into sand-ridge deserts, which cover some 500,000 square miles.

## THE QUATERNARY IN NEW ZEALAND

Volcanic activity begun in the Upper Tertiary continued; extensive eruptions of andesite and rhyolite lava took place, particularly in the North Island. Mount Egmont, for example, is built up mainly of andesite.

The present-day system of glaciers had a great extension in the Quaternary and undoubtedly reached sea level in the south-west of the South Island. On the eastern side of the Island the glaciers appear to have been confined almost entirely to mountain valleys.

## THE QUATERNARY IN OTHER PARTS OF THE WORLD

One of the most interesting of recent deposits is the laterite which caps many hills and plateau of the Deccan trap area of India, and is found extensively at a lower level on both the western and eastern borders of the peninsula. It is a surface formation caused by weathering in tropical conditions of high temperature and humidity, and owes its prominence to the fact that it hardens on exposure to the air and after hardening often resists erosion more strongly that the rocks from which it has been derived. It is composed of the hydroxides of aluminium and iron. Where these hydroxides are equal in proportion, the rock is known as laterite; where the hydroxide of aluminium exceeds that of iron, it is bauxite, and where the hydroxide of iron is predominant, the rock is an ironstone. It also occurs in Central and West Africa, Malaya, East Indies, the Northern Territories and West Australia, South America and Cuba.

# Palaeontology

## THE HISTORY

Fossils have been known to man since ancient and even prehistoric times. It is possible that the existence of sea-shells in places far from the sea first supported the idea of a universal flood, and that the discovery of strange and grotesque fossils gave rise to legends of fabulous monsters, cyclops, centaurs and dragons.

Nearer our own times, the early philosophers arrived at a conception of the nature of fossils and of the movements of the sea which allow us to observe them. Pythagoras believed in the instability of land and sea. 'I perceive,' he wrote, 'that nothing retains the same appearance for long. What was formerly solid ground has become sea; lands have arisen from below the sea; sea-shells are found far from the sea.'

The seed of the theory of evolution can be found in the writings of Pythagoras, Anaximander, Empedocles, Aristotle, and in even earlier writings. Lucretius continued it in his *De natura rerum,* and Ovid half a century later in his *Metamorphoses.*

During the long period of intellectual darkness of the Middle Ages, fossils were considered simply freaks of nature, that is, stones with an accidental or imaginary resemblance to plants or animals. The influence of stars and thunderbolts was invoked to explain their origin.

Later, at the time of the Renaissance, two great artists and thinkers of the fifteenth and sixteenth centuries, Leonardo da Vinci in Italy and Bernard Palissy in France, reverted to the opinion of the ancient philosophers. Both insisted and demonstrated that fossils are the remains of animals and plants which once lived in the places where they are found. By this token, the presence of marine fossils in land areas indicated that land and sea had not always maintained exactly the same position.

A potter and maker of china figurines, Bernard Palissy was also a most perceptive observer and a man of science. Uneducated by the standards of his day, knowing neither Latin nor Greek, he raised himself by his own efforts to a high degree of scientific attainment, and made a number of discoveries which he described in his *Récepte véritable* (1563) and *Discours admirables* (1580). Some of his opinions on fossils are expressed in the following passage, where it is clear that he refers to ammonites and sea-urchins: 'When that I had regarded well the shapes of these stones, I found that none could assume the form of a shell nor of any other creature, except the animal itself had shaped it...I maintain that the petrified fishes found in certain quarries have been engendered in that very spot, while the rocks were yet but water and mud, and which have since become petrified with the aforesaid fishes...

'I walked one day among the rocks of my own native town of Saintes, and in contemplating nature I saw in the rock certain stones made in the shape of a sheep's horn, not so long nor so curved, but similarly bent in the bow-shape, about half a foot in length. Some years had yet to pass before I understood what could be the cause of stones thus formed. But it happened that an alderman of the town of Saintes found on his farm one of these said stones which was half open and showed certain crenulations joining admirably one into the other... He made me a present of the said stone, which pleased me greatly, for it seemed to me to be formerly a shellfish of a kind no longer seen. It is to be supposed that this creature once inhabited the sea of Saintonge...

'On another occasion in the island of Oléron I saw in great numbers in the sea an armoured fish in the form of a chestnut husk. The outer part of the shell was covered with hard, sharp hairs like the prickles of a hedgehog. I caught a dozen and took them home, but was greatly surprised, for when the outside of the shell decayed, the root of each hair decayed also in a few days, and the hair fell out. When all the hair had fallen the shell was left smooth, and at the base of each root was found a boss, quite small, the which bosses were arranged in so neat a pattern as to make the shell pleasant to look on. Some time after, a lawyer showed me two stones exactly similar to the form of the shell of the sea-hedgehog. The said lawyer believed these stones to have been shaped by a mason and was much surprised when I gave him to understand that they were natural, for I had already judged them to be the shells of sea-hedgehogs which in the course of time had been turned to stone.'

Until the end of the eighteenth century, such ideas made little progress. Fossils were, indeed, collected, and described in magnificently illustrated works. Every nobleman had his cabinet of curiosities where fossils were found side by side with a mixture of objects. They were labelled with the strangest of names: screw-stones, horns of Ammon, star-stones, thunderbolts and glossopetrae. Their true significance was never realised. Voltaire, taxing his inventive powers to the full, suggested that certain fossils found in the Alps were oysters thrown aside by the Romans because they were not fresh, or cockle-shells left by pilgrims returning from Rome. In 1613, a surgeon of Beaucaire showed the bones of a mastodon, a type of extinct elephant, as those of the legendary giant Teutobochus, King of the Cimbrians, slain by Marius. A century later, in 1726, a Swiss doctor described as the skeleton of a man who had perished in the Flood the skeleton of a giant salamander found at Oeningen in Switzerland.

It was at this point that the French naturalist Buffon reasserted the observations already made by the great thinkers of Antiquity and the Renaissance:

'I have often examined from top to bottom,' he wrote, 'quarries in which the beds of rock were full of shells; I have seen entire hills made up of them. After considering this innumerable multitude of shells and other marine organisms, one can no longer doubt that parts of our Earth have been at the bottom of some ancient sea as full of marine life as our oceans are today. We are convinced that the shells of which our hills are formed belong partly to unknown forms, that is, to forms which cannot be found living in any sea... The horns of Ammon (Ammonites), in particular, are really the remains of former animals which have perished and exist no longer.'

The early or fundamental groundwork had, therefore, been prepared by the beginning of the nineteenth century.

For the publication of his great work, Georges Dagobert (Baron Cuvier), devoted himself to a thorough study of zoology and the anatomy of the living vertebrates, and recognised the fundamental principles of comparative anatomy. He was then able to start the study of fossil bones with great success, having available the collections of the natural history museum of Paris and fresh material brought in daily by workmen from the gypsum quarries. The beds of gypsum of the hill of Montmartre were actively worked during the last century, and contain a rich mammalian fauna. Little by little, Cuvier pieced together a whole series of vanished animals, of which over thirty species belonged to the order Perissodactyla, nowadays comprising horses, rhinoceroses and tapirs. One of these, *Palaeotherium*, was like a giant tapir in build, but its teeth resembled those of a rhinoceros. There were also primitive carnivores and a large lemur, *Adapis*. In 1812 Cuvier published his *Recherches sur les Ossements Fossiles*. The third edition in 1825 was accompanied by an essay on *Revolutions of the Globe*, in which Cuvier affirmed that the plant and animal kingdoms had several times been wiped out by cataclysms and recreated afresh, that no continuity existed among them, and that species were fixed and incapable of giving rise to others.

While Cuvier upheld with vehemence his theory of the fixation of species, he was opposed by one of his colleagues, Lamarck, who had also devoted himself to the study of fossils but was led to the opposite opinion. Having observed intermediate forms between extinct and living species, he believed in their transformation and linkage from one age to another. His great work, *Philosophie zoologique,* published in 1809, can be considered the basis of the theory of evolution.

After Cuvier and Lamarck had laid the foundations, palaeontological research developed all over the world. The most amazing discoveries were made in America, where palaeontologists Marsh, Cope, Osborn, and others accomplished an enormous amount of investigation. With financial help from museums, universities, larger corporations and other generous private donors the vast territories of the West and Rocky Mountains were examined. A resurrection of extraordinary animals took place, especially among the dinosaurs, the giant extinct reptiles such as *Diplodocus, Brontosaurus, Triceratops* and *Stegosaurus*. The ancestors of the horse were discovered in the West, forming links in a chain which left no doubt as to its evolution. Pioneer work was done throughout the world, and the main outlines of many fossil faunas were determined, though many wait to be accurately studied.

Today palaeontological research is going on in every country of the world. There are institutes, laboratories and academic chairs of palaeontology, just as there are chairs of zoology and botany. The study of fossils divides into two paths: the pure study of the extinct and ancient forms of life and its development through geological time, and its applied use in dating strata of the rocks. Following the second path, for example, are the research teams of oil companies working on the development of techniques for quick and easy determination of fossils.

## WHAT IS A FOSSIL?

Travelling through the countryside we may notice nothing that is unfamiliar or unusual. But if we pause a moment and look more closely we may often find strange and unfamiliar objects, seemingly alien to their surroundings. Reef corals and oyster shells may be embedded in limestones of cliffs and quarries. Fern and leaf impressions may be found in the coal seams of a mine. Lily-shaped creatures similar to the modern starfish and sea anemones may occur in the limey shales of a road-cut. These are fossils, dead for many ages and strangely out of place along our modern roads and pastures.

Fossils are either the actual remains of animals or plants that once lived on the surface of the Earth, or evidence of their existence. They provide some clue to the character of part or the whole of the organism. The branch of geology and biology devoted to their study is called palaeontology (Greek *palaios* — old, *on, ontos* — a being, and *logos* — study).

The fossils were formed by the process known as fossilisation. Preservation is of various kinds:

**Actual remains.** SOFT ANATOMY. In this form the entire body or some part of the soft anatomy is preserved. The ice of northern Siberia has revealed mammoth and rhinoceros carcasses dating from the glacial period, making it possible to study their anatomical structures, examine their blood-vessels, and remove clots of blood from them for detailed study. When thawed, the frozen flesh has often been fresh enough to feed to dogs, and has even been served at banquets. Soft anatomy is also occasionally preserved in peat-bogs, highly saline waters, desert sands and oil deposits. Fossils of soft anatomy, no matter what the type of preservation, are exceedingly rare.

HARD PARTS. Here, the soft anatomy decomposes while the hard parts remain. The result is a fossil of part of the organism, perhaps of a bone, a tooth, a shell, a carapace, or a woody tissue. This is by far the most common type of fossil, and in most cases the form of the flesh and soft organs of the animal are unknown, except for the palaeontologist's deductions based generally on the tissue impressions and on the study of comparative anatomy.

The skeleton of *Ichthyosaurus,* a
Mesozoic marine reptile.

Some pre-cambrian fossils of proterozoic, age.
A. Filled worm burrows in sandstone, Belt Series, Glacier National Park, Montana.
B. *Eozoon canadense.* Grenville Marble, Quebec.
C. *Newlandia sarcinula,* a small stromatolite, Belt Series, Glacier National Park, Montana.
D. *Lingulella,* a brachiopod, Belt Series, west of White Sulphur Springs, Montana.

**IN AMBER.** Another example of fossilisation is offered in preservation in amber. During previous geologic epochs when forests of conifers flourished, insects, spiders, and millipedes sometimes became stuck in the resin oozing out of the bark. The resin later hardened into amber, retaining the insects with almost every detail of hard anatomy in perfect preservation. The colour plate on page 393 shows four creatures trapped in amber: spiders, millipedes, ants, and flies.

**Replacement.** In this type of fossil the organism is replaced by mineral matter, sometimes by slow molecular replacement under unusual circumstances, but generally by volume for volume replacement, and most of the details of form and anatomy are lost in the process. The fossils of this type are labelled calcified (replaced by calcite), silicified (replaced by silica), pyritised (replaced by pyrite), according to the mineral substance involved. Silicified trees are found in many parts of the world — the Petrified Forest of Arizona, and the Fossil Grove in Victoria Park, Glasgow, are two such examples.

**Carbonisation.** In this process the volatile materials of the organism are removed, leaving behind a thin layer or residue of carbon. There is no new material added, for the carbon film is the carbon that was once present in the living organism. Plant remains, graptolites and fishes are the forms most commonly preserved in this manner. A remarkable group of fossils preserved by carbonisation, many showing minute details of soft anatomy, occurs in the Cambrian Burgess shale of British Columbia, Canada, deposited originally as a fine-textured mud in quiet marine water. In the shales are found jellyfish, crustaceans, sponges, an unusual array of worms, and many other animals, all preserved as films of carbon. The superb preservation of these organisms gives us a unique glimpse into the life of part of the Cambrian.

**Moulds.** In this form the organism disappears entirely, leaving nothing in the ground but its impression or mould in the material that once surrounded it. Percolating water often dissolves a fossil in this way without replacing it with any substance. The result is an external mould, similar to that produced in plaster or wax by a model-maker. If the fossil is hollow — a shell, for example — the interior may have been filled with mud which hardened before the original organism disappeared, and the final result here is an internal mould inside the external mould.

Moulds can preserve an outline of the original object with great faithfulness. The travertine deposits of Sézanne, near Rheims, are pitted with small, irregular holes. If fine plaster or other casting material is poured into the holes, the casts obtained show flowers with petals and stamens preserved, and insects with legs in perfect detail.

**Casts.** The external mould, once it is formed, may become filled with a mineral deposit or a clastic sediment (composed of broken fragments of older rocks). What the professional caster does artificially is often carried out by nature. After dissolving the fossil, water fills the mould with sediment or a deposit of mineral matter. This produces a cast or almost perfect replica of the original specimen. The in-filling material may be lime, silica, clay, sand or other substances.

**Other forms.** There are very many more fossil objects, mainly indirect forms of evidence, which give information on the organic character or life habits of plants and animals. They include tracks, trails, burrows, and coprolites (fossil excrement).

**Conditions necessary for preservation.** Only a minute portion of the life of the past is represented by fossils. Today, only a few of the millions of living organisms are being deposited under conditions favourable for their preservation as fossils. For fossilisation to take place, the following prerequisites must generally be met. First, the organism must possess hard parts. Only rarely and under unusual conditions is a soft structure preserved. Then the organism must be buried quickly and in relatively fine-grained sediment. This prevents destruction by scavengers, decay, erosion, etc.

**Regions where fossils are generally preserved.** The largest region in which fossils are preserved in abundance is that of the shallow sea-floor. Life abounds in such a zone. Here sediments are being deposited in quantity and are often fine in texture. Less frequently, fossils are found in the sediment deposits of lake and stream, though in streams erosion by running water often destroys most of the remains. Water-deposited volcanic ash is an ideal medium for fossilisation, and remarkable insect fossils showing the most minute structures have been found in such beds.

In caves and caverns, organisms or their hard parts may be preserved by coatings or fillings of dripstone. Unfortunately for the record, these organic remains are usually masked or distorted by the limey deposits. Bogs and asphalt pits are also excellent places for preservation. Wind-blown sands and loess deposits would provide ideal media for fossil preservation, but because they are characteristic of arid or semi-arid climates where life is uncommon, the fossils are relatively few. The ice of cold areas or glacial epochs is a perfect natural deep-freeze but, again, fossils preserved under such conditions are rare because life is less abundant in frozen regions, and also because the ice lasts for only a relatively short time.

## USES OF FOSSILS

The geologist and palaeontologist use fossils in many ways: as indicators of past climates and environments; in determining ancient geography — palaeogeography — certain fossils show by their presence the location of ancient seas and continents; in the study of evolution, where fossils provide definite lines of ancestry for many of the living plants and animals; lastly when their chronology has been determined, as markers for geologic time units.

Early in the nineteenth century William Smith discovered that he could identify rock formations by the fossils contained in them (*see* p. 335). He noted that each formation possessed characteristic fossils different from the fossils collected in the rock units above or below.

Smith reasoned that if the fossil assemblages in a series of rocks in one area were identical with the assemblages in a series of rocks in another area miles away then the rocks in the two areas were deposited at the same time. On the basis of fossil assemblages he divided the rock series in Wales into formations, matched them with similar ones in adjacent areas, and thus correlated and mapped the various rock units.

Other geologists used Smith's method successfully in other regions and discovered that they were able to divide the rock layers into even smaller units by more detailed examination of the fossils. Some fossils are used more often than others in correlation (the process of chronological matching of rocks). These are the guide or index fossils, which must fulfil the following requirements:

1. They must be representative of a group that was very widespread geographically but existed for only a relatively short span of time or evolved very rapidly. Index fossils therefore occur only through a short vertical range of rock layers, but in many different areas. This enables the geologist to subdivide geologic time into small units and to match rocks in many and widely separated areas.

2. For practical purposes the organisms must have been abundant when they existed so that their fossils will be common.

3. Index fossils must have a unique or characteristic appearance so that they will not easily be confused with others and will be identified readily not by a few specialists in palaeontology, but by most working geologists.

4. Lastly, for most economic geologic work index fossils should be small, preferably microscopic in size, because the specimens of rock obtained in drilling or coring, for example, are so small that they reveal only the microscopic fossils. These methods usually miss the larger fossils or include only unidentifiable portions. The protozoan order Formation fills the bill admirably and is widely used. Index fossils are indispensable guides in locating and tracing petroleum and coal-bearing rocks. They are equally useful in the discovery of other sedimentary deposits.

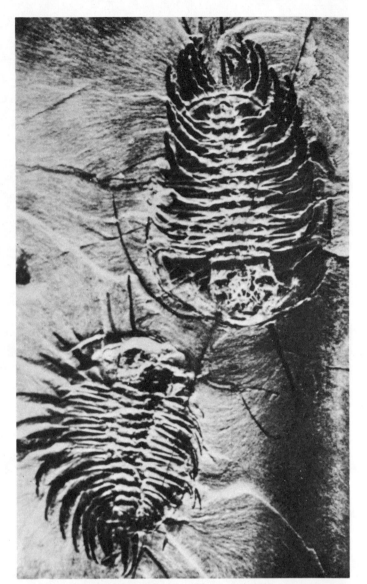

*Top*
*Olenoides, serratus,* trilobites with antennae and legs preserved, as well as their shells. Middle Cambrian, near Field, British Columbia.

*Bottom*
Coquina is a coarse limestone made up of shells. It was deposited along the coast of Florida during the Pleistocene epoch.

359

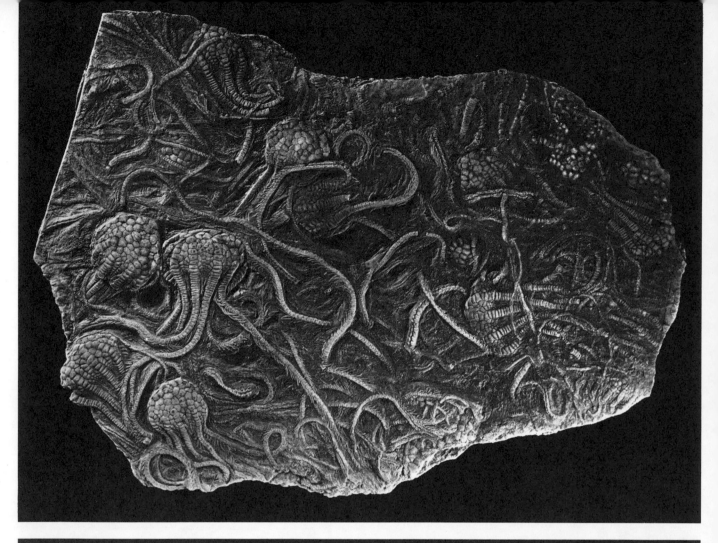

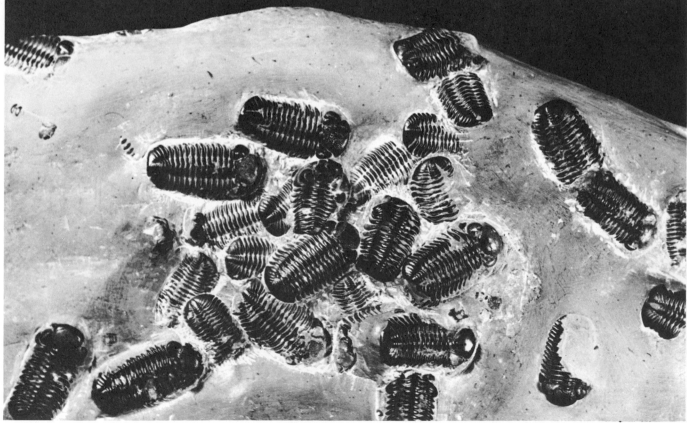

*Top*
*Uintacrinus socialis,* a stemless
crinoid that floated in a late
Cretaceous sea that covered the
western part of Kansas.

*Bottom*
*Phacops rana,* small trilobites on
a calcareous nodule found
in Middle Devonian shale in
western New York.

# HOW FOSSILS ARE STUDIED

What does the palaeontologist do with a fossil? How does he manage to extract it from the rock, identify it, and reconstruct the original form of the animal?

**Extraction of fossils.** When a fossil is enclosed in rock, the first step is to extract it from its matrix. This is comparatively easy in a loose deposit such as a friable sandstone, or a soft rock like chalk, but the operation is more delicate where the rock is hard, or where the fossil is very small. Large specimens in hard rocks are usually subjected to sudden great stress and the fossil is jerked free. If the fossil is more fragile, and firmly cemented to the matrix, it is possible to loosen it by heating the rock and then plunging it into cold water. In the case of microscopic fossils the rock is broken up and passed through a series of successively finer sieves which separate the fossils from the surrounding rock.

Remains of the matrix which cling to the fossil may be removed with a soft brush and water. Fossils which break up in water can be picked clean with a shoemaker's awl or a gramophone needle. Increasing use is being made of the dentist's drill because it enables a rapidly rotating brush to be applied to any part with concentrated and rapid effect. Occasionally a fossil can be extracted by coating the exposed part with varnish for protection and pouring weak acid over it, but this method demands great care. To obtain casts from fossil moulds they are filled with melted wax or similar material (polyvinyl chloride is increasingly used) to make a solid cast.

Many fossils are physically or chemically unstable as soon as they are extracted and must be treated quickly to avoid disintegration, sometimes, even, before excavation is complete. If they are merely fragile, a binding agent such as gelatine or a solution of collodion in amyl acetate can be brushed over the shell. Chemical disintegration occurs most frequently in fossils preserved in marcasite and these have to be soaked in ammonia, washed, and varnished.

**Reconstruction of fossils.** Invertebrate fossils are generally found whole or nearly whole. On the other hand, it is rare for a vertebrate skeleton to be found without some part missing. If the carcass of a dog, for example, floats down a river, it is carried along by the current and washed continuously against the banks. As the body decomposes, the lower jaw, the limbs and tail become detached and fall to the river bottom. Finally the trunk disintegrates. By this time the skeleton has been broken up and the fragments scattered along several miles of river-bed.

The need to reconstruct an entire animal from fragments is therefore common, and to be successful in this the palaeontologist needs a wide knowledge of the comparative anatomy of animals, both living and fossil.

In the early days the removal of the specimens from the rock and their subsequent mounting was done by the palaeontologist himself. This still holds good in small institutions today, but in the large museums the operation is carried out by highly skilled technicians called preparators. The designing or lay-out for the mounting and the final display of the specimens in life positions is a co-operative effort between palaeontologist, preparator, artist and other skilled museum workers, many of whom also take part in the expeditions which collect the fossil specimens.

A mounted skeleton is clearly more interesting and more informative than odd bones laid out in a glass case. Scientists therefore try to reconstruct not only the skeleton but also the exterior shape of the animal, and the environment in which it lived. Murals, drawings, and coloured plaster models of extinct animals ornament the galleries of museums. These restorations are interesting but must not be taken as absolutely correct and final, for each exhibit is founded, at least in part, on scientific guesswork, and will be modified in the light of new knowledge as it is acquired.

# RECONSTRUCTION OF ANCIENT LIFE

Palaeontology reconstructs vanished animals. It also attempts to understand their mode of life, habits, source of food, method of locomotion, and the surroundings in which they lived. It divides into palaeobiology, palaeoecology, and palaeogeography.

**Tracks and footprints.** The passage of an animal across wet sand or mud leaves a track which, under the right conditions, may be preserved in a fossilised state; fossilised tracks of worms, crustaceans, and jellyfish have already been identified. Certain sandstones and shales carry hollow impressions or moulds of impressions, which are the footprints of vertebrates. These give scientists a useful guide to the stature of the animal which made them.

**Food and fossil excrement.** Sometimes the remains of food — branches of trees, seeds, bones of fish, and shells of molluscs—are found in the stomachs or between the teeth of fossil animals.

Fossil excrement, too, may be found; these coprolites (Greek *kopros*—excrement, and *lithos*—stone) vary in size. One type frequently found in the chalk of Western Europe has the appearance of a whitish pine-cone an inch or two long—a shape that has often misled geologists as to their origin. In reality they are the fossil excrement of small coelacanth fishes, whose intestinal spiral valve gave the excrement its cone-shape. Under the microscope the remains of the food and the bacteria can still be seen.

**Reproduction of extinct animals.** Fossil eggs and young have also been found. Eggs of the dinosaurs have been collected from Cretaceous deposits of the Gobi Desert. Some of the largest eggs ever recorded were those of an enormous fossil Madagascan bird, the *Aepyornis*. They have a capacity of twelve pints.

Ovoviviparous reproduction is suggested by a fossil ichthyosaur, a swimming reptile dating from Jurassic times. Some specimens have been found with several tiny skeletons lying near and within the ribs. It must be admitted that it cannot be proved conclusively that the smaller skeletons are unborn young; it is possible that they might have been swallowed by the fish-eating ichthyosaur. Each fossil must be examined on its own merits to decide problems of this kind.

**Palaeoneurology.** A very recent branch of palaeontology consists in the study of the endocranium (cerebrum, cerebellum and cranial nerves) of the fossil animal, basing research on natural or artificial endocranial moulds. This reveals the degree of intelligence and the various receptive and motor faculties of the specimen.

Many mammals evolved very quickly and developed characteristic habits and abilities. Among the earliest members, the brain is small and poorly differentiated, and the olfactory lobes are dominant. In more recent forms, the cerebrum becomes more complex and dominant. This can be seen very clearly by comparing the endocranial cast of a primitive carnivore such as a Creodont with, say, a dog's skull. The differences are even more apparent on comparing primitive and modern monkey skulls.

**Climates and seasons.** In an earlier chapter we saw that drops of rain can leave impressions on mud and sand. From the shape of these impressions, round or oval, it can be determined whether the rain fell vertically or obliquely. Lightning striking dry sand can leave its path marked out by a fulgurite. These natural phenomena are pointers to the sort of climate enjoyed in a given area. Fossils, too, act as climate pointers.

The degree of atmospheric humidity is indicated by the types of animals and plants. The forests of the Carboniferous period must have been humid and swampy, since giant horse-tails and tree ferns abounded. It is hardly surprising, then, that in rocks of this era we find some of the richest coal seams. An abundance of saline deposits (gypsum, rock-salt) and beds of red rock, on the other hand, may indicate a dry climate producing an intense evaporation.

There are endless indications of climate to be found in the fossil world. In the Quaternary, for example, the presence of woolly mammoths, woolly rhinoceroses, and cave-dwelling animals (including man) characterises the glacial periods; the interglacial periods are marked by the presence of the monkey, elephant, hippopotamus and, especially, by plants of warm countries.

Information on the seasons of earlier ages can be gained from biologic processes dependent on them: the annual rings on trees, the fall of leaves in the autumn, and the alternation of flowers and fruit.

# The Life Record

In the preceding chapters we have studied the history of the Earth as revealed by its rocks, from its origin to the present day. We have followed the development of geosynclines, the deposition in them of enormous thicknesses of sediments, the rise of mountain chains and their subsequent erosion, the climatic changes which produced desert conditions on one part of the world while glacial conditions were in force elsewhere, and volcanic activity which gave rise to outpourings of lava over vast areas.

One aspect of the Earth's history has so far been mentioned only in passing. This is the story of life itself, the story of the origin and evolution of plants and animals which lived in past eras and of their fossilised remains. As with the rocks themselves, it is convenient to consider life on the Earth from its origin and to follow its development, step by step, through the Palaeozoic, Mesozoic and Cenozoic eras up to the present.

## THE ORIGIN OF LIFE

All living matter, plant or animal, is composed of complex molecules made up mainly of carbon, hydrogen, and oxygen. This matter, called protoplasm, has the ability to grow, reproduce, and respond to conditions under which it lives — its environment. We know that all living forms are derived from previously living organisms, but we do not know what protoplasm's beginning were, or when they first occurred. The origin of life, its cause, and its chemical and physical environments are still mysteries which stretch back through time for perhaps two billion years. That conditions during this remote era must have been vastly different from those of today is certain, for no one has yet observed spontaneous generation — not even of a single, simple living cell.

Biology is the science of living things — their origin, growth, reproduction, and from. Such matters had long been studied but it was not until about 300 years ago that they became the object of scientific observation. Early science believed that each distinct species of plant and animal had always had the same characteristics, having been supernaturally produced. It also believed that new life, such as mites and maggots, was constantly and spontaneously formed by natural processes. Extinct species were unknown and uncertain distinctions between many of the living species unrecognised. During the nineteenth century, however, it was noticed that certain species responded to breeding and changing environments by producing variable forms. Charles Darwin, in his *Origin of Species,* published in 1859, demonstrated that the known forms of life, far from being changeless, were subject to evolution, and that the struggle for existence led to adaption to environment and to the survival of the fittest.

The past few decades have witnessed the efforts of diverse fields of science working on the problem of the origin of life. All living matter contains three types of very large molecules; proteins, nucleic acids, and polysaccharides. Many complex organic compounds were believed to have formed while the Earth's original reducing atmosphere (a de-oxidising atmosphere in which most of the chemical reactions involved combinations with hydrogen) was changing into an oxidising atmosphere due to heavy losses of hydrogen. It has been estimated that this transitional stage, when many of the organic compounds which make up life as we know it were formed in the atmosphere and the sea, could have lasted from 2,500 to 100 million years. Just how far back in the Pre-Cambrian the Earth had a reducing atmosphere has not been determined; it might have been 1,500 million years before the beginning of the Palaeozoic. When the tremendous life development from the Cambrian to the present is considered — and we have a fairly complete record of it — it is reasonable to assume that a very long time was required to bring life from its first primitive state to its complexity at the beginning of the Palaeozoic.

Some people believe that the appearance of life was a sudden, 'improbable' event or several such events, occurring perhaps less than 1,000 million years ago, because before that time there was insufficient oxygen in the air to support life. However, highly oxidised iron in certain Archaean rocks indicates that an oxidising environment prevailed at a very remote time in the Pre-Cambrian.

Most theories on the origin of life assume that the physical setting was the saline environment of the sea. It has been concluded from geochemical, palaeontologic, and biogeochemical evidence that, as far as carbon dioxide content is concerned, the seas throughout such geologic time were comparable to those today, but how far present conditions can be carried back into the Pre-Cambrian is largely speculative. Spontaneous generation might have taken place if large quantities of organic compounds (compounds with carbon) were present in primitive oceans, and the first organisms might have obtained their basic constituents by synthesising them from carbon dioxide and water.

Another theory puts forward the coming together of the necessary organic molecules in a sea that had gradually become a 'dilute broth, sterile and oxygen-free'. In a classical experiment, a series of proteins was produced from a purely 'inorganic' system when, for the first time under controlled laboratory conditions, the synthesis of numerous complex organic molecules from relatively simple compounds of quite widespread occurrence was achieved. It has also been said that nitrogenous compounds, including amino acids, were formed by the action of high-energy ultra-violet radiation, from which Earth is shielded today by the ozone layer in the upper atmosphere.

Thus, in the early phases of the molecular stage of evolution, only simple molecules were formed. At some point in the development of more advanced stages, a molecule appeared with two new properties: self-duplication and mutation. These are characteristic of all living systems today and provide an objective definition of the living state.

Through mutation and aggregation these first 'living' molecules evolved to take on multimolecular form, with the capacity to catalyse their building blocks from simpler molecules. Primitive genes, the cells which control heredity, conferred selective advantage and complete autonomy, such as is found in the one-celled green algae. The evolution of unicellular forms into multicellular organisms may have been a quicker process than the development

Beach patterns.

of unimolecular into unicellular systems. This last may well have taken as long as a billion years.

Eventually, from the evolution of simple colonies of like cells came a division of labour among cells — cellular differentiation — to form the diversified multicellular invertebrates.

## CHARACTERISTICS OF LIFE

The biosphere, the zone of living creatures located at the boundary between the lithosphere, hydrosphere, and the atmosphere, has some of the characteristics of the three physical states of matter which typify them, drawing heavily on these states for chemical and energy resources. The most typical of living organisms are the green plants. The chlorophyll (green colouring matter) contained in their organs enables them to trap light rays from which to build up energy or synthesise a number of substances. Water and mineral salts from the hydrosphere, and carbon dioxide from the atmosphere are the starting points in the production of a series of organic materials which serve as food for the plants which eventually provide energy for herbivorous and carnivorous animals.

Organic substances are for the most part compounds of carbon, an element as plentiful in living matter and its derivatives as it is scarce in the Earth's crust — except in coal and oil which are themselves of organic origin. That is why the chemistry of carbon is generally called organic chemistry. The complexity of organic compounds is shown in a molecule of albumen which is made up of 403 hydrogen atoms, 250 carbon, 81 oxygen, 67 nitrogen, and 3 sulphur.

One of the fundamental characteristics of all living things is that of assimilation, or the transformation of foodstuffs into their own substance. Living beings can also be distinguished from the non-living by their colloidal state. Colloidal suspensions, known as gels, are neither solid, liquid nor gaseous, but are formed by suspension in water of very fine particles kept a certain distance apart by their electrical charges. The colloidal state rarely exists naturally, except in living things and in siliceous and aluminous gels.

We have seen that the role of living creatures in the evolution of the terrestrial globe is often unsuspected yet considerable. At once agents of erosion and of sedimentation, they have contributed to the destruction as well as to the building of island and continental masses. Coral and shelly limestones are the most striking examples of their building power.

## EVOLUTION OF LIFE

We shall now try to describe the series of events which may have led from primitive photosynthesis, the combination of chemical compounds in the presence of light, to present-day life forms. At some stage in the physicochemical evolution of the globe, it is postulated that intense ultra-violet irradiation brought into being, by the process of photosynthesis, huge organic molecules. These molecules could be crystallised; they were in a position to feed themselves; and they were able to multiply by a process of division. But because of a lack of oxygen in the atmosphere, they were devoid of any respiratory function. They may have had some resemblance to the ultraviruses — particularly the bacteriophages discovered in the intestines of cholera victims — and to the crystals of nucleoprotein responsible for tobacco mosaic. These subvital units are on the threshold of life without actually being alive.

The second stage brought the grouping of organic molecules into a colloidal complex and the formation round it of a membrane similar to that of living cells. In possessing a central mass and external envelope, these organites, as they are known, bore an astonishing resemblance to bacteria, except that they still had no respiratory function — one of the essential characteristics of life.

At the third stage chlorophyll appeared. This was a fundamental event which entirely revolutionised the evolution of the biosphere and the atmosphere. Because of their chlorophyll pigments, blue-green algae (Cyanophyceae), similar to modern forms, could trap luminous rays of a greater wavelength than the ultra-violet rays — mainly the blue-violet and orange-red radiations — and use them for organic synthesis. The carbon dioxide in the

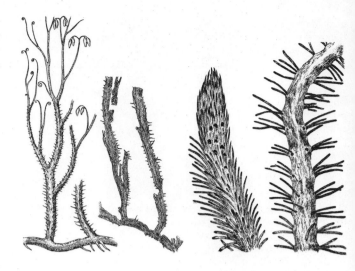

*Top*
*(Left) Psilophyton, a prickly* shrub from the early Devonian of Europe and North America.
*(Right) Baragwanathia, an* early Silurian lycopod from Australia.

*Centre*
*Megistocrinus nobilis, in a* slab of Mississippian crinoid beds at Le Grand, Iowa.

*Bottom*
*Protolindenia wittei, a true* dragonfly from Jurassic limestones of Solenhofen, in Germany.

atmosphere was split up by chlorophyll into its two elements; the carbon was retained by the plant and took part in new chemical combinations, while the oxygen was thrown back into the air as unusable waste, later to be taken up again by the nascent respiration of what can be called true living beings.

The primitive atmosphere did not contain free oxygen; it was liberated by the organisms with chlorophyll. In other words, the atmosphere in its present form evolved as a by-product of photosynthesis. If living things were to disappear from the surface of the Earth, free oxygen would soon be fixed afresh by the rocks, and the atmosphere would resume its former primitive character.

The formation of a layer of ozone at an altitude of about 20 miles produced a blanket which stopped most of the ultra-violet radiations that emanated from the Sun, and the creation of living matter by photosynthesis became impossible. On the other hand, the living beings already formed were protected against the injurious action of further ultra-violet rays and continued to thrive, as they do today.

Among the chlorophyllic beings, particular importance must be attached to the green flagellata, for they gave birth to the vegetable and animal kingdoms. Their cell, more complex than that of the blue algae, comprised a colloidal mass (which from now on we shall call cytoplasm), an enveloping membrane, and a nucleus. The nucleus could be considered an organite which controlled the cell's nutritive and reproductive functions. In the latter function the nucleus contained the chromosomes and genes responsible for hereditary character.

The question is, then, how did the evolution of animals and plants take place? The processes themselves are known even though their interactions are often difficult to distinguish. First of all the cells were grouped together, just as the large organic molecules had been grouped. The groups were then differentiated by division of functions; there were digestive cells, muscular cells, nervous and reproductive cells, and so on. Later, cells were distributed in such a way that tissues were formed; these, in turn, gave rise to organs. A sea-urchin and a worm are examples, of this stage of development and can be taken as typical of animals with radial symmetry and bilateral symmetry respectively. A creature shows radial symmetry when its organs are arranged symmetrically around a central axis; bilateral symmetry is symmetry on two sides of a plane. To take an example outside the field of living creatures, the open pages of a book display bilateral symmetry. Radial symmetry became dominant in the vegetable kingdom, while bilateral symmetry prevails in most of the animal kingdom, up to and including man.

At each stage of evolution mutations originating from sudden and chance changes in the genes and chromosomes occurred. As some of these mutations were unfavourable or even fatal, the individuals possessing them were eliminated in the struggle for life. Others were favourable or neutral; their possessors were preserved by the process of selection.

## THE MISSING FOSSIL RECORD

One of the great unsolved mysteries of science is the missing fossil record of life which evolved between 600 million and 1,500 million years ago, during the late Pre-Cambrian. The occurrence in Lower Cambrian rocks of a diversified multicellular marine invertebrate fauna, with many complex and fully developed groups, presupposes a long and complex history of evolution of which we have had only a very faint glimpse. In many places thick marine sedimentary strata (more than 5,000 feet) are known to lie beneath formations containing Early Cambrian fossils.

Diligent search in these rocks, many of which resemble the fossiliferous beds above them, has not succeeded in revealing the missing record.

Many theories have been advanced to explain the gap, but none completely clarifies the mystery. Some early opinions expressed are: that Pre-Cambrian fossils were destroyed by metamorphism; that fossil skeletons or shells could not form in a supposedly acid sea; that Pre-Cambrian fossils had no hard parts that could be preserved; that these animals lived in fresh water lacking the calcium essential for building hard parts; that the early animals lacked hard parts because they lived in surface waters where protection was unnecessary, and skeletons were not needed; that animals originated in fresh water and later migrated to the sea; and that hard parts were suddenly developed when sea-bottom habitats were adopted.

Later views suggest acquisition of hard parts under the stress of environmental change, explosive evolution in the Late Pre-Cambrian, and a shore-line and shallow-water zone which would provide ideal conditions for explosive development but would not be preserved.

Theoretically, once in existence, life must have left some record. Yet with all our understanding of the major events in organic evolution and the general framework of absolute chronology involved, the search for the remains of primeval life has met with meagre success. Because of the scant record, few geologists have concerned themselves with the origin of life. The theories concerning the earliest life records have been left to palaeobiochemistry and astrophysics, and geologic observation of the older sedimentary rocks.

The oldest red beds, which must have formed under a strongly oxidising environment, are evidence of life, even through none of the original organic matter remains. The palaeochemistry of oxygen and carbon dioxide can thus be traced back into the Pre-Cambrian red beds, limestones, graphites, and 'coals' which are difficult to explain except that they were formed by a well-organised photosynthetic life. Pre-Cambrian 'coals' containing beds of almost pure carbon are found in Michigan and Finland. Simple organisms have been found in the cherts of the Gunflint iron formation in Ontario. These fossils in uncrystallised chert reveal colonies of filamentous aggregates and abundant spore-like bodies resembling blue-green algae or aquatic fungi. The beds, thought to be 1,600 to 1,800 million years old, contain the oldest known fossils displaying definite biologic structures.

*Eozoon canadense*, the 'dawn animal' of Canada, has been the source of argument for a century. Some have called it a fossil, others have insisted that is is a banded mixture of minerals formed by heat and pressure. Similar forms collected in Ontario from unaltered rock were called stromatolites, and are thought to have been formed by algae. Other stromatolites have been found in Wyoming, New Brunswick, the Northwest Territories, and the Lake Superior region. Great numbers of fossil algae are found in Glacier and Waterton Lakes National Parks in Montana and Alberta where the stromatolites form massive beds and rocks, and are found scattered in beds of dolomite.

Animals of the Late Pre-Cambrian era are not as plentiful as algae. The Lamballe shales from the Côtes du Nord of France contain a siliceous debris of radiolaria (unicellular organisms) and sponge spines. Sponges from limestone of Ontario, worm burrows from Glacier National Park, a jellyfish from Grand Canyon, and a brachiopod resembling *Lingula* from the Belt series, Montana, almost completes the list of known Pre-Cambrian animal fossils. However, protozoans, annelids (segmented worms), and arthropods (tiny crab-like creatures) have been reported from Australian beds.

# Life in the Palaeozoic

During the 330 million years of the Palaeozoic Era, parts of the continents were inundated time after time. Animals and plants lived in these recurring seas, and the hard parts of some were entombed in the sands and muds of the ocean floor. As the sediments hardened into rock they were raised above sea level, becoming a storehouse of information for the palaeontologist who can reconstruct the environments under which these ancient animals lived by applying the principle of uniformitarianism, and observe the laws of evolution by tracing a fossil form through rocks of successive ages. For example, corals today live in warm, shallow waters of the tropical seas, so it is reasoned that fossil corals lived under similar conditions.

Marine rocks are the only Early Palaeozoic rocks in which fossils are preserved. Terrestrial deposits in those remote eras were formed just as they are today, but successive uplift and erosion has largely destroyed them. It is not until late in the Palaeozoic that extensive records of land plants are found, partly because few lived in the earlier periods.

## AN UNDERWATER EXCURSION

Let us try to picture the shores of a Palaeozoic sea. Let us try to picture too the unknown waters of a sea 300 million years ago. The first impression is of something quite ordinary. In the shallow water algae similar to modern kelps clothe the rocks. Among the algae or in the sand below water live 'shelfish' and crustaceans in profusion, or at least what we take to be such. Jellyfish swim about and worms crawl along the bottom. The rôle of the octopus is played by a mollusc with numerous arms and a shell into which it retires from time to time. The warmer part of the sea offers a profusion of coral reefs with their associated fauna of shellfish, worms, and sea lilies. Around us swim fishes and small sharks.

Palaeozoic rocks contain the first evidence of abundant marine life. In fact, the Cambrian record probably contains fossils representing all phyla of the plant and animal kingdoms. Knowing this, the great importance of the lost record of the Late Pre-Cambrian can be appreciated.

Tracing each class of plant and animal through the periods of the era from the Cambrian to the Permian, their relations with each other and their changes through time can best be understood by a systematic presentation, describing first the plants, then the unicellular animals (the protozoa), then the more complex invertebrates, and finally the vertebrates.

## FLORA

Plant life has been abundant throughout the fossil record, for the earliest records contain algae. Since animals depend upon plants for food, plants and animals must have co-existed throughout much of the span of life on Earth. Highly organised land plant forms are known from Devonian and later rocks. Algae are the only known marine plants, some of which are very well preserved as black carbonaceous films in the Middle Cambrian Burgess shale of British Columbia. Later, lime-secreting algae appeared in the Upper Cambrian and Lower Ordovician formations.

The oldest reported land plant (*Aldanophyton anti-quissimum*), supposed to be a lycopod (club-moss) from the Middle Cambrian of Siberia, has a stem containing vascular bundles and is covered with narrow, delicate leaves. This plant and the many kinds of spores found in Cambrian rocks indicate the existence of a well-developed land flora, and suggest that the first may have developed from a marine flora in the Late Pre-Cambrian.

The next oldest known land plant (*Baragwanathia longifolia*) was found in Silurian rocks in Australia. However, the first well-known fossils of land plants are found in Devonian rocks in Wyoming, eastern Canada, Maine, Cornwall, Scotland and Eire. *Psilophyton*, a prickly, leafless shrub, is found in the Lower Devonian rocks of Wyoming.

On such a desolate land, life could only become established in water, and plants in particular were semi-aquatic. At Rhynie, Aberdeenshire, Scotland, silica-bearing water from hot springs appears to have seeped into a bog and to have fossilised the plants growing there by replacing their vegetable matter with silica on an intimate scale. The plants so preserved had no roots, leaves or flowers, but they were not mosses since they had a central, well-defined sap-tube, or stomata, characteristic of true land plants. The body of the plant consisted of an underground stem, or rhizome, from which upright shoots with few branches grew. Three species are represented, *Rhynia*, *Hornea* and *Asteroxylon*, all of which are related to *Psilophyton*, in which the tips of the branches are coiled rather like the fronds of a fern.

Along Schoharie Creek at Gilboa, New York, the remains of a dense forest was discovered in Devonian rocks. The commonest tree was a giant seed-fern (*Eospermatopteris*) which grew to a height of about 40 feet, with a trunk that was sometimes more than 3 feet in diameter at its bulb-like base. The leaves grew out of the tops of the huge trunks like the fronds of a palm. These fern-like fronds were 6 to 8 feet long, with 'seed'-containing spores growing on the tip. Two more spore-bearing plants (lycopods) grew in the Gilboa swamp, one with narrow leaves, the other with wider, overlapping leaves.

Although not found in the Gilboa forest, a seed-bearing tree (*Callixylon newberryi*) grew over much of the world in the Late Devonian. Its well-preserved wood is commonly found in the Upper Devonian New Albany black shales in the Mississippi and Ohio valleys. The trunk of this ancestral conifer sometimes reached a diameter of five feet. Its slender branches bore thick, pointed leaves, but a species found in New York (*C. zalisskii*) had narrow, ray-like leaves. Many smaller plants and vines formed the undergrowth in these primitive forests.

An important event in the evolution of plants occurred at this time: the development of seed-bearing plants. Early plants like the fungi, mosses and ferns were propagated by spores — minute bodies, usually spherical. In other plants, true seeds are formed

Silicified fossil logs in the Petrified Forest, Arizona.

of two different cells and contain, in addition to a protective covering, a supply of food to support initial growth — unlike a spore, which must fend for itself after germinating.

The ginkgo, or maidenhair (the sacred tree of the Buddhists) made its appearance in the Late Devonian with the ancestral *Psygmophyllum*, but little is known about it until the Permian. The modern species reaches a height of 80 feet and a diameter of about 3 feet. Its broad subtriangular-to-lobate leaves are marked by straight veins radiating from a pointed base; its fruits look like persimmons. Of particular interest because of its specialised method of fertilisation, it is also distinctive because it is probably the oldest living tree. It is not possible to differentiate leaves of the living *Ginkgo biloba* from the fossil leaves which are found in Jurassic deposits in Alaska, Greenland, and northern Europe. Fossils of the ginkgo are also found by the hundreds in the Oligocene beds of Oregon.

## THE CARBONIFEROUS FORESTS

In the early part of the Carboniferous (Mississippian in America), land areas were restricted by widespread seas. Large trees continued to flourish, however, and we find new species of club-mosses, ferns, seed-ferns, and trees related to *Callixylon*. The conifer-like *Cordaites* appeared for the first time; it is a relatively small tree showing features that link it with the gymnosperms, particularly with the Monkey Puzzle tree *(Araucaria)* and with the Kauri Pine *(Agathis)*.

No period of plant development was more impressive than the Upper Carboniferous (Pennsylvanian) with its great swamps. The abundance of plant life gave rise to many coal beds, especially in western Pennsylvania, where there are thirteen workable coal beds, and in the Nottinghamshire coalfields in Britain, where coal seams up to 12 feet thick are worked. Pollen-like microspores were produced in such vast numbers that at times they must have turned the humid air above the swamps into a yellow fog. In places they accumulated to such thicknesses that beds of cannel coal (a soft bituminous type, which ignites easily) were formed.

The sudden rise of coal-forming plants was one of the most important events in botanical evolution. For the first time the land was dominated by plants of large size and with true roots, stems, and leaves formed of genuine tissue. Sap was drawn up from the ground through woody ducts and distributed to all their organs through canals in the inner bark. They were sufficiently numerous and well developed to give rise to considerable forests in humid regions.

The sky above the Carboniferous forests was heavy, low and the atmosphere warm and humid. Abundant rains formed shallow, stagnant bodies of water. Lakes, ponds, and swamps were numerous. The trees, growing out of the water, were not like their modern forms; branches and leaves would have a tendency to grow vertically, and the leaves, except those of the ferns, would be small. There was very little shade from the Sun whenever it pierced the clouds. Noises in the forests were in no way comparable with those of the present day. The woodpecker would not tap his beak against the bark of the trees; no birds would sing; no squirrels, rabbits, and deer would be heard. The world of air-breathers was confined to a small number of insects and one or two amphibia and reptiles.

A nineteenth-century botanist, using the imprints of leaves as a basis, distinguished a great number of genera and species of the tree-ferns to which he gave the names *Neuropteris, Pecopteris, Ondontopteris, Sphenopteris,* etc. These names of uniform termination (Greek *pteros* — fern) give a false impression of precision or rigid classification; certain plants so designated have since been recognised as much superior to the ferns in that they possess flowers and seeds. Most of the leaves attributed to the ferns are huge palm-like fronds. The growing stems were similar to those of the modern arborescent ferns and palms and their surface is covered with scars corresponding to the insertions of ancient leaves. The latest ferns have a magnificent plume at the tip of their stem, while in the middle, young leaves break out from a terminal bud. Other ferns have climbing stems or even underground stems from which the huge leaves shoot direct.

The identification of fossil plants is particularly difficult because their parts are often separated. A find of leaves still associated with fragments of stems and reproductive organs that enable us to put the plant into its exact place in the botanical classification is a

*Left*
Some plants from the coal
forests of Pennsylvanian (Late
Carboniferous) age.
A. *Alethopteris,* a seed-fern.
B. *Calamites suckowi,* internal
mould of a scouring rush stem.
C. *Lepidostrobus,* the cone
of *Lepidodendron.*
D. *Neuropteris,* probably a
seed-fern.
E. *Lepidodendron,* bark of a
tree 30 to 60 feet high.
F. *Pecopteris,* probably a
true fern.

*Right*
Three members of the *Glossopteris* flora, which spread through
the Old World during the
Permian period.

*Far right*
*Aulopora,* a tabulate coral
from the middle Devonian
of northern Michigan.

lucky find indeed. It is possible to confirm the true nature of ferns and seed-ferns only if fertile leaves with groups of spore cases are found.

*Calamites* does not exist today, yet it reminds us strongly of the horse-tails of our modern swamps. It has the same stems, formed of joints or nodes, and bears collars of branches and leaves; it has the same structure in cross-section and, in the centre, a large cavity surrounded by a row of smaller cavities. This structure is typical of plants that grow in stagnant water.

If we can imagine horse-tails enlarged from their present 3 feet to trees 60 to 100 feet high, we reproduce the *Calamites* of the Carboniferous forests. In their youth these trees had exactly the same structure as the horsetails; only as they grew older did they acquire wood and the secondary inner bark that supported them and led the sap to all parts of the plant.

The giant club-mosses *Sigillaria* and *Lepidodendron,* the main inhabitants of the Carboniferous forest, raised their tufts of leaves 60 to 100 feet above the ground. Their modern descendants, reduced to heights of a few inches, form turf in warm, moist localities, under the names of club-moss and *Selaginella.* Apart from size, the modern club-mosses preserve all the main structural features of their gigantic ancestors: stem and roots ramifying in a series of successive forks, narrow leaves attached to the stem, two sorts of spores, etc.

The lepidodendrons (Greek *lepis* — scale, and *dendron* — tree) are so called because of their lozenge-shaped leaf scars arranged in oblique series on the trunk. The trunk was divided into many branches, the last of which bore tufts of long, narrow leaves with a single vein similar to pine needles.

In the sigillarias (Latin *sigillum* — impression ) the leaf scars are hexagonal and arranged in vertical series on the trunk. There is slight branching or none at all.

To support trunks of six-foot base diameter and 60- to 100-foot height tissues must increased in thickness from year to year. There was, as we have already said, secondary bark and wood, similar to that of modern trees but lacking the spring and winter rings which correspond to seasonal alternation of moisture and dryness. This is a further proof that the Carboniferous climate was fairly uniform throughout, the conditions being extremely hot and humid.

The gymnosperms possessed flowers and naked seeds (not enclosed in an ovary like the more advanced angio-sperms). Such Carboniferous gymnosperms as *Neuropteris* and *Spehenopteris* herald the group of plants which, in the form of pines, firs, cedars and sequoias, became the dominant plant life throughout the Mesozoic era.

It is difficult to differentiate between Carboniferous seed-ferns (pteridosperms) and true ferns. Fortunately other gymnosperms have more well-defined features: Cordaites, recognisable by their large, simple leaves, and walchias, which resemble araucarias (Monkey Puzzle trees). In these plants the reproductive organs are no longer sparsely scattered on the lower surface of the leaves, but as in later species have become grouped together in recognisable flowers.

After the Carboniferous, the continents were uplifted. The climates of the Permian became more varied, and formerly common plants were able to survive only in sheltered places where climates were mild. *Cordaites* gave way to true conifers such as *Walchia* and *Lebachia.* In the northern hemisphere the Permian floras were much like those of the Upper Carboniferous (Pennsylvanian). In the southern hemisphere, however, the great forests vanished and were replaced by smaller, hardier plants that lived between glaciations. This hardy flora included true ferns, *Calamites,* and conifers. Its best-known member, *Glassopteris,* was a seed-fern with long, thick, blade-like leaves. The *Glossopteris* flora spread as far north as Siberia, but did not reach western Europe or North America.

The Hermit shale exposed in the Grand Canyon of Arizona contains a flora that grew in an inhospitable, semi-arid region. Its alternating wet and dry seasons would not support the humid swamp-loving plants; instead, blankets of algae grew on the mud flats; seed-ferns, a thick-stemmed *Sphenophyllum,* a small-leafed *Walchia* and other conifers flourished along the shores of ponds and streams.

The site and remains of a Coal Measures forest can be seen in Scotland at the 'Fossil Grove', Victoria Park, Glasgow. Stony casts of the stumps of *Lepidodendron* trees are particularly well exposed, showing bifurcating root-bearing branches on the ground.

## CLASSIFICATION AND NOMENCLATURE

Perhaps it would be useful at this point to explain how a fossil and living forms, basically alike in possessing a cell structure, are named or labelled. The study of natural history (including fossils) is impossible without a systematic arrangement for description and comparison. Proposals for classification in the field of natural history were put forward towards the end of the seventeenth century by the English naturalist, John Ray. He contended that a sound classification could only be established on a basis of bodily structure: the individuals of any one type of animal (a horse, for example) are almost identical in structure, suggesting that they all belong to one group or *species*. A characteristic of a species is that the offspring resemble one another and the parents. Furthermore, there is closer resemblance between some species than between others. For example, the horse and the donkey resemble each other more closely than they resemble the cow or sheep: the horse and the donkey should, therefore, be placed with other closely similar species in a larger group, or *genus*.

This concept of genus and species was elaborated by the Swedish doctor and scientist, Carolus Linnaeus (1707–78) in his celebrated work, *Systema Naturae*. In an edition published in 1753, he introduced the idea of giving names to both genus and species, which, following the scientific manner of the time, were in Latin. The use of two words, one to designate the genus and the other to designate the species, is called the binomial system of nomenclature: for example, *Crassostrea virginica*, the Virginia oyster. Linnaeus found that similarities could be demonstrated between groups of genera and set up broader categories called *orders* and *classes*.

Chevalier de Lamarck (1744–1829) applied the principles of Linnaeus to the classification of fossils. Working in the Paris area, he developed a system of nomenclature which, in broad outline, is still in use today.

We all know what is meant by the expression the 'animal kingdom'. As far as the fossils which make up the life record are concerned, the animal kingdom is divided for purposes of classification into groups called phyla. Each phylum is divided into classes; each class into orders, each order into families; each family into genera; each genus into species. The eminent German palaeontologist, Professor Karl A. von Zittel divided the animal kingdom into eight phyla (seven invertebrate and one vertebrate) in his book *Handbuch der Palaeontologie*. His *Grundziige der Palaeontologie* (Textbook of Palaeontology) was published in German in 1895, and the first English edition appeared in 1900, edited by Charles R. Eastman of Harvard University. This has become a standard work of reference for students of palaeontology throughout the world, and for this reason it would be both useful and appropriate if we were to follow Zittel's classification in the succeeding account of the fossil record of the Palaeozoic, Mesozoic and Cenozoic. The phyla are numbered as follows:

| | | |
|---|---|---|
| Phylum | I | — Protozoa |
| Phylum | II | — Coelenterata |
| Phylum | III | — Vermes |
| Phylum | IV | — Echinodermata |
| Phylum | V | — Molluscoidea |
| Phylum | VI | — Mollusca |
| Phylum | VII | — Arthropoda |
| Phylum | VIII | — Vertebrata |

It should be pointed out that not all naturalists and palaeontologists follow the Zittel order of classification. Wells and Huxley, for example, in their book *The Science of Life* refer to the vertebrates as the first great phylum and to the arthropods as the second: they make the observation that 'there is no sort of precedence or subordination between the phyla'.

## FAUNA

**Protozoa.** Consisting of one simple cell, the protozoans, as they are called, are the most simply constructed of all animals. They have left only a scant record in the Early Palaeozoic, probably because in those times they produced no structures capable of preservation. However, several species of foraminifers with sandy shells have been found by dissolving limestones in dilute acid and recovering the small delicate shells from the residues. From the Devonian there is an increasing record of these one-celled animals, the Radiolaria and Foraminifera. The former secreted beautiful, minute shells of glassy silica in an almost infinite variety of forms; the latter secreted calcareous shells. The Foraminifera became abundant in Carboniferous seas, and those which were restricted largely to the Pennsylvanian and Permian have been studied extensively as they are excellent index fossils. Although only one-celled organisms, the foraminifers built — and still build — shells of extreme diversity of size, shape, and the arrangement of their chambers. Their shell materials include chitin (an organic, horny substance), calcium carbonate, silica, or mere particles of cemented sand, mud, or broken shells found on the sea bottom. Among the cemented or agglutinated forms each species usually selects only one kind of material with which to build its shell.

**Coelenterata.** The Coelenterata, or Zoophytes, are free-swimming or attached aquatic animals. They differ from the Protozoa in having multicellular bodies with distinct organs and from organisms in higher-numbered phyla by the absence of a definite body chamber. The phylum is divided into two sub-phyla: the Porifera (sponges) and the Cnidaria, which in turn is split up into two classes — Anthozoa (coral polyps), and Hydrozoa (jellyfish). The Hydrozoa also embrace the important sub-classes Stromatoporoidea and Graptolitoidea (graptolites).

PORIFERA. Sponges are animals with cells arranged in three layers: an inner, an outer layer, and an intermediate jelly-like substance through which cells move in a primitive circulatory system. Some of the cells distribute food and waste materials, others digest it, and some secrete a skeleton of horny fibres or rods of calcite or silica, called spicules.

Modern sponges with their soft bodies would not leave much of a fossil record, but many fossil species built durable structures which are preserved in Palaeozoic rocks. They are usually grouped into three classes: the siliceous, the calcareous, and the glassy sponges. From the Cambrian Burgess shale comes a glassy sponge, *Vauxia*, which resembles a very small version of the modern Venus's Flower Basket. The Ordovician produced *Branchiospongia* from Kentucky and *Ischadites iowensis* from the Trenton formation of the central United States. The Silurian sponges *Astylospongia* and *Astraeospongia* occur in profusion in Tennessee and are found in other regions. A distinctive glass sponge is found in the Devonian of New York (*Hydnoceras tuberosum*). Carboniferous and Permian strata contain an abundance of both calcareous and siliceous sponges. A problematical sponge or sponge-like fossil, *Receptaculites oweni*, the 'sun-flower coral', is found in Middle to Late Ordovician formations in the central United States. Other species of fossil sponges lived in later Palaeozoic periods.

CNIDARIA. Coral seas today are found only in the tropics, for coral reefs now flourish only between latitudes 30° North and 30° South. In the Palaeozoic this was not the case; the climate was sufficiently warm for western Europe, parts of North America, and other regions of the world to become veritable coral archipelagos. The species that built the reefs were not closely related to modern forms, for their digestive cavity had four divisions instead of six or eight; in consequence they belong to the group of Tetracoralla. The reefs were so abundant that today, in the Ardennes and elsewhere, they form beds of limestone several hundred feet thick which are worked as marble quarries.

After a modest beginning in the Middle Ordovician, the tetracorals flourished throughout the remainder of the era, especially during Middle Silurian and Devonian. Well developed Silurian reefs are found in the Wenlock limestone areas of Shropshire and in Wisconsin, Illinois, Iowa, and Indiana, where hundreds of species of corals abounded. These included many colonial-type corals such as *Favosites*, *Catenipora*, *Syringopora*, and *Halysites*. Cup corals include *Strombodes* and the strange *Goniophyllum*, shaped like a four-sided pyramid and covered by a lid made of four stony plates. The latter is characteristic of northern Europe, but many of the others are believed to have enjoyed a world-wide distribution.

Palaeozoic corals.
A. *Alveolites goldfussi,* Middle Devonian, eastern North America.
B. *Favosites favosus,* Silurian, North America and Europe.
C. *Favosites niagarensis,* Middle Silurian, North America.
D. *Catenipora microporus,* Middle Silurian, eastern North America.
E. *Pachphyllum woodmani,* Late Devonian, Iowa.
F. *Lithostrotionella castelnaui,* Mississippian, eastern United States.

In the Devonian Period, cup corals reached their peak of development. Some species reached great size (2 feet long by 4 inches in diameter). *Calceola,* a cone-shaped species with a stony lid, was common in European Devonian strata. Colonial corals include such genera as *Hexagonaria (H. percarinata,* the famous 'Petroskey' stone, *Aulopora, Alveolites* and *Pachyphyllum.* A Devonian reef extends across the Ohio River at Louisville. Kentucky, where a zone of silicified corals form the rapids called the 'Falls of the Ohio'.

The corals decreased in variety during the closing periods of the Palaeozoic but some distinctive species abound. Genera represented in the Carboniferous are *Lithostrotion, Lithostrotionells, Chaetes,* and *Lophophyllidium.* The latter is a common Pennsylvanian cup coral with a spike-like central axis protruding from the centre of the cup.

In contrast to the great coral reefs of the Silurian, Devonian and post-Palaeozoic and Recent periods, the Capitan limestone reef in western Texas and southern New Mexico was built mainly by primitive algae, aided by calcareous sponges, bryozoans, and some brachiopods and corals.

HYDROZOA. The Hydrozoa, first known from the Pre-Cambrian, are probably the ancestral stock of all coelenterates. Because these free-swimming, soft-bodied animals (medusae) had no hard parts they are rarely preserved as fossils. However, several species are known from Ordovician, Devonian, and Carboniferous (Pennsylvanian) formations. The Stromatoporoidea reached their peak in the Devonian, building large colonies, some of which form extensive beds or typical reefs. They range from the Ordovician through the Devonian to problematical fossils of the Cretaceous which may also belong to this group.

GRAPTOLITES. The extinct graptolites have often been called Coelenterates. Their fossils resemble marks made upon the rocks with a pencil, because their chitinous skeletons leave only a black film on the split surfaces of shales and slates. Because of their widespread occurrence most grapolites make excellent index fossils.

The graptolites were colonial animals; the individuals were 'polyp'-like organisms, each secreting a chitinous tube or cup, called a theca. The thecae were arranged in long, straight rows called stipes; the colonies were often attached to seaweed or some floating object or to the sea bottom. Some, such as *Glossograptus ciliatus* (Middle Ordovician, New York) were fastened to a central disc supported by a balloon-like device which floated a whole cluster of colonies. Because of their floating habits, the graptolites are found on all continents. Other species are *Didymograptus bifidus* and *Cyrtograptus kirki* (Silurian), and *Tetragraptus fructicosus* (Lower Ordovician). The Dendroidea were branching, net-like colonies which lived from Late Cambrian to the Mississippian. *Dictyonema flebelliforme* is a characteristic species found in Lower Ordovician rocks in many parts of the world.

ACALEPHAE (lobed jellyfish). Although these are frequently of considerable size they are entirely without hard parts and are thus unfitted for preservation as fossils. Impressions of these delicate organisms are sometimes preserved.

**Vermes** (Worms). These are bilaterally symmetrical animals which may be unsegmented or uniformly segmented. They usually have elongated bodies with a distinct body cavity. The phylum is divided into two classes: Chaetopoda (earthworms and annelids) and Gephyrea (marine annelids).

Because they lack hard parts, the worms have left a scant fossil record. Of more than two hundred genera of fossil worms, less than a third are based on actual body parts. Two-thirds have been applied to fossil tracks, trails, burrows, casts, and tubes. About forty genera are based on worm teeth (scolecodonts), and only about thirty assigned to worm bodies or body parts. Trails are found in some of our oldest fossiliferous rocks (Early Proterozoic) and, except for some algae, the worms are probably the oldest of all known fossils.

Biologically, the worms are so structurally diverse that they have been assigned at different times to at least nine phyla. Even now, classification of many forms is difficult. It is thought by some that the worms are the ancestral stock of the Arthropoda, the largest invertebrate phylum. The segmented worms are the only ones to have left a fossil record of any consequence. That they have existed longer than almost any other form of life is truly amazing because these animals are poorly equipped for preservation in changing environments or for defence against their enemies. Their survival is therefore a remarkable one.

Few pause to consider the importance of worms, living as well as fossil. Yet their numbers are considerable. Some soils hold more than fifty thousand earthworms an acre. These pass eighteen tons of the soil material through their bodies, aerating the soil mantle and changing it physically and chemically. Lobworms living in the sand flats along the coast of Northumberland, England, devour about 3,000 tons per acre of the upper two feet of the shore. Their castings indicate a population density of about eighty-five thousand to the acre.

Similar records of prolific life have been found in the fossil record. For example, in the mountains of British Columbia a small outcrop of the dark grey Middle Cambrian shale near Burgess Pass has given up tens of thousands of fossils. Many are the remains of soft-bodied animals such as algae, sponges, jellyfish, sea cucumbers, and worms. Today these animals are nothing but carbonaceous films between layers of shale. Nevertheless, many show the fine hair-like bristles rarely preserved as fossils (*Canadia spinosa*). Other genera from the Burgess shale are *Worthenella*, *Ottoia*, and *Selkirkia*. A long-ranging genus which built a shell *(Spirorbis)* is known from the Ordovician and is often found attached to brown seaweed washed ashore along our beaches. *Scolithus* tubes, found in Cambrian sandstones over much of the world, were made by worms that lived until Late Devonian time.

**Echinodermata.** These are almost exclusively marine organisms although a few do inhabit brackish water. They have a well developed exterior skeleton composed of calcereous plates and showing radial symmetry of pentameral (fivefold) form. The skeleton may be immovable or movable within narrow limits and is frequently provided with movable appendages (spines). The organisms can be distinguished from the Coelenterates by the presence of a true digestive canal, a distinct body cavity, a more perfectly developed nervous system and, except in certain starfishes, an exclusively sexual mode of reproduction.

They live in every conceivable environment in the seas, from shallow coastal zones to the ocean deeps, in quiet water and in the turbulent waves and currents of the littoral zones. Some live on the muddy bottom, while others prefer the clean sandy or rocky sea floors. The larvae of most echinoderms are free-swimming forms without hard skeletal parts. Many attach themselves to objects on the sea floor and there grow to maturity. Some, however, soon break away to become free-swimming forms. The phylum is subdivided upon the mode of life assumed by the adult form: those which live in a fixed position (Pelmatozoa, or stem-bearing animals) and those which are free to move about (Eleutherozoa). The Eleutherozoa are made up of the sub-phyla Asterozoa and Echinozoa.

Except for the crinoids (sea-lilies), all pelmatozoans are extinct. All classes of the eleutherozoans are living today except a primitive starfish, *Somasteroidea* (Ordovician), the primitive echinoid-like *Bothriocidaroidea* (Ordovician) and box-like eleutherozoan, *Ophiocystia* (Ordovician to Devonian). They are thought to have been derived from a crustacean ancestry, possibly through the Cirripedae.

CYSTOIDEA are represented in the Cambrian by poorly preserved forms of rather doubtful affinites *(Protocystites, Macrocystella, Eocystites, Lichanoides, Trochocystites)*. Their maximum development was in the Ordovician and Silurian, after which they suddenly diminished in number and disappeared in the early Carboniferous. Of the 250 species which have been described, hardly a dozen are found later than the Silurian. They are plentiful in Ordovician rocks in the Leningrad area of Russia, and in Silurian rocks of Oland, Gotland, Sweden, Wales and Bohemia, and in the Chazy and Treton limestones of Canada, New York, Ohio and Indiana. The Silurian limestones of Dudley and Tividale, England show such forms as *Pseudocrinites, Apiocystites, Echinoencrinus* and *Anomylocystites*.

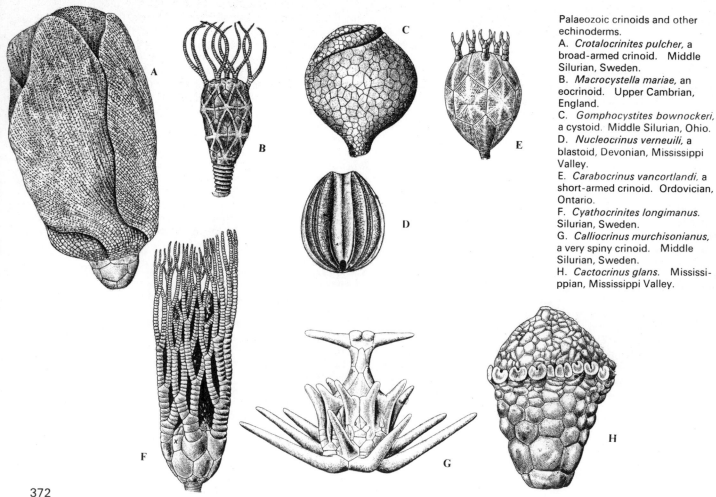

Palaeozoic crinoids and other echinoderms.
A. *Crotalocrinites pulcher*, a broad-armed crinoid. Middle Silurian, Sweden.
B. *Macrocystella mariae*, an eocrinoid. Upper Cambrian, England.
C. *Gomphocystites bownockeri*, a cystoid. Middle Silurian, Ohio.
D. *Nucleocrinus verneuili*, a blastoid, Devonian, Mississippi Valley.
E. *Carabocrinus vancortlandi*, a short-armed crinoid. Ordovician, Ontario.
F. *Cyathocrinites longimanus*. Silurian, Sweden.
G. *Calliocrinus murchisonianus*, a very spiny crinoid. Middle Silurian, Sweden.
H. *Cactocrinus glans*. Mississippian, Mississippi Valley.

Similar forms (*Lepidocrinus, callocystites* and *Caryocrinus*) are found in the Niagaran (Silurian) of North America. Only one form of Cystoidea—*Lepicodiscus*—has been found in the Carboniferous.

BLASTOIDEA are known by early forms in the Ordovician which probably represent transitions from ancestors of cystid type. Several genera occur in the Devonian of Europe and America; for example, in strata of the Kaskaskia group. They are sparse in the Upper Carboniferous and Permian of America and the Island of Timor; above this horizon, no traces of blastoids have been discovered.

CRINOIDEA, or sea-lilies, attained their maximum development in the Palaeozoic. Usually they have a particularly local distribution but occasionally detached fragments occur so profusely that they build up strata many feet thick. The Palaeozoic members of this class lived in shallow water and often occur in the vicinity of coral reefs. The earliest crinoids occur in Cambrian rocks where they are represented by stem joints only. Stem fragments are also known from Ordovician rocks in Britain and are associated with well preserved head portions (or calices) of *Hybocrinus* and *Baerocrinus* from the Ordovician of Russia. In North America, the Trenton and Hudson River limestones are particularly rich in crinoid remains. The Silurian limestones of Dudley, England, and of Gotland in Sweden are famous for the abundance and high state of preservation of their crinoids. The Devonian localities are Eifel, Rhineland, Nassau, Westphalia; the Ardennes and Mayenne, France; Asturias, Spain; Michigan, New York, and the Ohio Falls river, North America. Crinoids are known from the Lower Carboniferous limestones of Tournay and Visé, Belgium, and from Britian, Ireland and Russia (near Moscow). The most famous Lower Carboniferous locality, however, is the Mississippian limestone of Burlington, Iowa and Crawfordsville, in Indiana, which has gained a world-wide celebrity. The Pennsylvanian of Kansas City, Missouri is a notable locality for Upper Carboniferous crinoids, which are also well known from Australia. The most remarkable crinoid fauna of Permian times was discovered in Timor.

The popular name for these organisms, sea-lilies, describes them admirably. It is difficult to imagine anything less like an animal than the crinoid with its tall, slender calcareous stem, expanding at the top into a bulb crowned with delicately divided petals. That they were animals has been proved by the discovery of almost identical organisms living today on the floor of clear sunlit tropical seas.

ASTEROZOA (starfishes and brittle stars) are known from the early Cambrian right through to the present day. They rarely occur as fossils and are found chiefly in slate, calcareous or sandy rocks of shallow water origin. The two classes are well represented in Ordovician and Silurian times by forms which do not differ much from those of today.

ECHINOZOA (sea urchins) are first known in the Ordovician by a single genus, *Bothriocidaris*. In the Silurian of Britain, the class is represented by the *Echinocystoidea*; in America, a form called *Koninckocidaris* is the first known of the *Lepidocentridae*. In the Devonian of Europe, there is one possible cidarid; a number of genera are known from rocks of this age in North America. The cidarids have one species, *Miocidaris*, found in the Lower Carboniferous of Europe.

HOLOTHUROIDEA (sea cucumbers) are best known from the Middle of British Columbia, where some complete specimens of *Eldoria* are found. Elsewhere the fossil record is poor and the class is represented only by fragments in Carboniferous and Permian rocks of Britain and northern Europe.

**Molluscoidea.** Typically, organisms of this phylum secrete a calcareous shell. The respiratory organs lie anterior to the mouth and are in the form of tentacles or fleshy spiral appendages. The mouth conducts into a closed alimentary canal. The nervous system is highly organised. Reproduction is sexual, though in the Bryozoa, it also takes place by budding. The organisms are water inhabitants; the bryozoans are largely marine and the brachiopods wholly marine. The phylum is subdivided into Bryozoa and Brachiopoda.

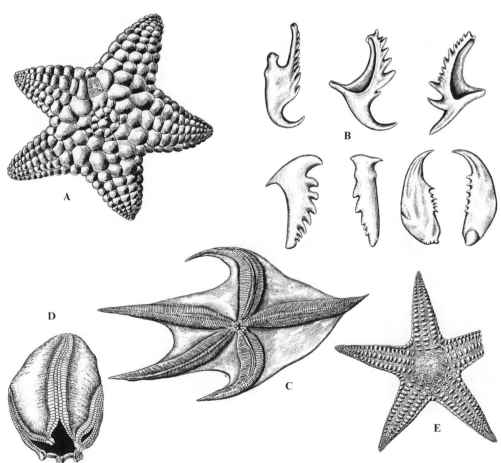

Palaeozoic starfish and annelid worms.
A. *Hudonaster incomptus*, Upper Ordovician, Ohio Valley.
B. *Ildraites*, Middle Silurian. *Eunicites* and *arabellites*, Upper Devonian. Some typical scolecodonts.
C. *Loriolaster mirabilis*, Devonian, Germany.
D. *Cheiropteraster giganteus*, Devonian, Germany.
E. *Devonaster eucharis*, Middle Devonian, New York and Ontario.

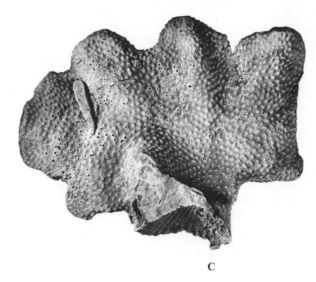

Palaeozoic bryozoans.
A. *Hallopora ramosa,* Upper Ordovician, Ohio Valley.

B. *Archimedes wortheni,* Mississippian, Mississippi Valley.

C. *Monticulipora molesta,* Upper Ordovician, Ohio Valley.

THE BRYOZOA are sometimes called 'moss-animals' because of the encrusting habits of many genera. Of more than 7,000 species, about one-sixth are living at the present time and are not known to occur as fossils; of the fossil varieties, the great majority have been found in rocks of all periods of the Palaeozoic. Probably the most distinctive form is *Archimedes,* with a screw-like axis supporting lace-like fronds; it is found in the Lower Carboniferous (Upper Mississippian) of the Middle States of America, Alaska, north Russia and Spitzbergen.

BRACHIOPODA are among the most useful fossils, ranging from Pre-Cambrian rocks into Recent sediments. Their importance is greatest in the Palaeozoic where they serve as excellent stratigraphic markers and where their abundant fossils offer excellent material for the study of evolutionary development in the phylum. More than 3,500 species have been found in the Palaeozoic formations of North America. Although only a few of 200 species are living today the brachiopods furnish a near-complete record for the study of the environment, habits, and their own development into animals covered with intricately ornamental shells. After reaching the peak of their development in the Middle Palaeozoic, they became again small, relatively simple-shelled forms during more recent times.

The brachiopod shell is composed of two parts, or valves. A line

A small slab of brachiopod shells, *Resserella meeki.* Upper Ordovician of the Ohio Valley.

drawn through the beak divides the shell into two equal parts which are mirror images, each resembling the other. The living and fossil brachiopods are divided into two groups according to the way in which the two valves are held together. In the smaller group, the inarticulates (Atremata and Neotremata), the shells are attached at the beak by muscles. In the larger group, the articulates (Protremata and Teleotremata), the shells are joined along a line that has teeth that fit together (or articulate) to strengthen the closure. The shells are opened and closed by inside muscles. The class got its name (*brachy* — arm, *poda* — foot) because there was a mistaken idea that the animal used its 'arms' as a foot for locomotion, much as the molluscs do. In actual fact, the brachia support the ciliated mantle, the edge of which secretes the shell. By movement of the cilia they circulate food-bearing water through the shell.

Most Cambrian brachiopods are small, simple shells, less than half an inch long. More than 500 species have been found; one genus, *Lingula,* is known to have lived from the Lower Cambrian to the present. It is probably the oldest-known living fossil form, existing through a span of more than 400 million years. *Lingula,* like other genera of the inarticulates, are poor index fossils because of the long range of their species through geologic time, and many are difficult to identify specifically. In Ordovician rocks brachiopods with calcareous shells developed rapidly. All the principal divisions of the group are represented, and the hinged articulates succeeded the inarticulates in abundance as well as variety. The Orthida, of which *Dinorthis pectinella, Platystrophia ponderosa,* and *Herbertella sinuata* are common species, reached their peak in the Ordovician.

Another important group, the strophomenids, that first appeared in the Middle Ordovician, declined after the Devonian, but a few species still survive. *Strophomena planumbona* is found in large numbers in Richmond rocks (Late Ordovician) near Cincinnati, Ohio. Scores of other species are found in the genera *Rafinesquina, Leptaena, Stropheodonta,* and *Schuchertella.* The productids, *Chonetes, Productella,* and *Dictyoclostus,* are found from the Ordovician to the Permian.

The Rhynchonellida, beginning in the Middle Ordovician, and the Terebratulida, first known in the Silurian, lived beyond the Palaeozoic to reach their climax in the Mesozoic (Jurassic and Cretaceous). The former includes the small nut-shaped, short-hinged, strongly plicated *Camarotechia* from the Devonian and Mississippian. The terebratulids are important in Devonian faunas, where *Beachia* and *Rensselaeria* are found.

Possibly the most distinct order of brachiopods is the Spiriferida, characterised by a long hinge line, a wing-shaped outline of the shell, and by the spiral form of the brachia. They were confined mostly to the Palaeozoic, especially the Late Palaeozoic, when many genera of the suborder Spiriferacea flourished. The genus *Spirifer* is essentially confined to the Lower Carboniferous. Other common genera are *Ambocoelia, Delthyris, Syringothyris, Cyrtospirifer, Mucrospirifer,* and *Neospirifer.*

**Mollusca.** The molluscs (the 'soft ones') from a well-characterised and remarkably homogeneous group of invertebrates which has existed since the earliest recognised advent of life upon the globe. They possess bilaterally symmetrical, unsegmented bodies; there is a larval shell gland from which a hard shell (or exoskeleton) is secreted though it is not always permanently retained. They have a well-developed mouth, intestinal canal and excretory system, and a closed but partly lacunary circulation, assisted by a heart with one or more auricles containing usually colourless body fluid (or haemolymph). The nervous system is well developed. Reproduction is sexual, respiration by gills, locomotion by a muscular organ called a 'foot', or by swimming. The organs are typically paired and protected by a mantle. The classification (into five classes) is based on the respiratory organs and the structure of the foot (whether adapted for swimming or not). The reproductive organs and the structure of the heart and nervous system are also important factors in classification, and for separating smaller groups shell characters are largely employed. The classes are: Pelecypoda (clams, or bivalves), Scaphoda (tusk shells), Amphineura (chitons), Gastropoda (snails, or univalves), Cephalopoda.

Unlike many invertebrates, most molluscs move about freely and have become swimming, crawling, and burrowing animals. Some of the burrowing types are carnivorous, even boring through hard shells to get at the animal inside. Others burrow in sand or wood, and some literally excavate holes in the solid rock in which they live. The notorious and destructive shipworm is not a worm at all but a boring clam.

Although the chitons and scaphopods have lived since the Ordovician and Silurian, they are relatively unimportant as fossils because few species have developed. The simple shells of the scaphopods have few distinguishing features to make them suitable for use as index fossils.

PELECYPODA. Marine pelecypods are poorly represented in the Palaeozoic. *Mediolopsis* is known from the British Silurian but as the sea withdrew westwards, its margins became fringed with brackish estuarine waters in which few marine creatures survived and, after carrying on a precarious existence for a while, even these degenerated and disappeared. In Early Carboniferous times, when marine conditions were re-established, forms such as *Posidonia* (or *Posidonomya*) became established: a comparable form is *Caneyella* which is found in the Caney shale of Oklahoma. In America, such forms as *Actinopteria, Nuculites* and *Lunulicardiom* were developed in marine Devonian strata.

Non-member pelecypods, better known as lamellibranchs, are important fossils in Upper Carboniferous times and have been used as 'indicators' in the sub-division of the Coal Measures of Britain. They are sometimes so abundantly present in thin layers in the coal-bearing rocks that they are referred to as 'mussel bands', the name reflecting the close resemblance of their shells to the common fresh-water mussel. The fossils are smaller and more varied than the freshwater specimens and are known by a number of different generic names, such as *Carbonicola, Anthraconaia* and *Naiadites.* There were incursions of the sea from time to time during the Coal Measures period and a typical marine lamellibranch of the Upper Carbonifeorus is *Pterinopecten.*

GASTROPODA. About half the 80,000 living species of molluscs are gastropods, or snails. They can live in almost any environment, from the ocean depths to mountains three miles high. Most of them feed on vegetation, but some are carnivorous, either scavenging for dead animals or killing live ones; a few are sessile, that is, they live in colonies much like corals, are sluggish or inactive, and depend upon their shells for protection. Some species cover the opening of their shells with a small calcareous plate called an operculum.

The fossil record of the gastropods extends from the Cambrian: although not so prominent as trilobites and brachiopods, they form an appreciable element of the fauna of at least the later Cambrian strata. For example, in the Durness limestone of Scotland (Upper Cambrian) gastropods are the most abundant fossils and such forms as *Maclurea, Ophileta, Hormotoma,* and *Raphistoma* are known. An early limpet-like gastropod, *Scenella,* occurs in Cambrian rocks in Norway and Sweden. *Bellerophon,* a free-swimming form, occurs in the Silurian rocks of Wales. In the Lower Carboniferous, gastropods are not abundant, but the flat coiled *Euomphalus* and the low turreted *Pleurotomaria* are among the more important.

Palaeozoic Molluscs.
A. *Straparollus pernodosus,* a snail. Pennsylvanian, Mississippi Valley.
B. *Bellerophon troosti,* a snail, Middle Ordovician, eastern United States.
C. *Holopea symmetrica,* a snail. Middle Ordovician, eastern North America.
D. *Myalina subquodrata,* a mussel. Pennsylvanian, south-western United States.
E. *Platyceras niagarense,* a snail. Middle Silurian, central United States.
F. *Horotoma whitevesi,* a snail. Middle Silurian, New York.
G. *Hexameroceras herzeri,* a cephalopod. Middle Silurian, Ohio.
H. *Maclurites magnus,* a snail. Lower Ordovician, eastern North America.
I & J. Two straight-shelled cephalopods. I. *Mactinoceras glenni,* Middle Ordovician, Tennessee. J. *Spyroceras bilineratum,* Middle Ordovician, eastern half of U.S.A.

CEPHALOPODA. Many of the so-called 'lower' invertebrates had simple bodies and followed sessile habits, living in one place all or most of their life. Their food was borne to them by waves and currents. Not so with the highly developed, free-swimming cephalopods, which include the present-day squids, octopuses, and the pearly nautilus. These complex animals form excellent index fossils in Palaeozoic and Mesozoic rocks.

The cephalopods have well developed eyes and other sense organs, and secrete shells into which they can withdraw for protection. They are predators, seeking out their food and grasping it with tentacles that have many suction discs. When the nautilus or squid wishes to move, it takes water into its mouth and ejects it through a small funnel-shaped tube with sufficient force to move it backwards at considerable speed. The jet age for Man began in the twentieth century; cephalopods have been travelling by this method for about 400 million years. Nautiloid shells are camouflaged on the upper side with a series of stripes which conceal the animal as light plays on the rippling water. When crowded by enemies, further protection is furnished by a brown ink-like substance (sepia) ejected in the form of a dense cloud.

The cephalopods formed a great variety of shells, some straight (Orthoceras or Michelinoceras) or slightly bowed (Valcouroceras); others were openly coiled (Bickmorites) or tightly coiled (Goniatites). Some, like the living squid and cuttlefish, have single internal shells, depending upon their swift movement and sepia for protection. A few—the octopus is one—have no shell at all.

If we examine a nautilus shell, the interior cavity is found to be divided into a series of chambers by slightly curved or disc-shaped partitions (septa). These chambers, very small near the beginning of the coil, gradually increase in size as the animal grows older and larger. As the animal builds new living chambers it leaves behind a tube or siphon which extends back through the septa to the last chamber. The siphon secretes a covering of shelly material, the siphuncle. The shape of the siphuncle and the line of juncture of the septa with the shell wall, called the suture, vary greatly from species to species. They are the features that form the basis for classification of these animals.

The extinct forms have been divided by palaeontologists into two sub-classes: the nautiloids, which have simple septa, and the ammonoids, which have fluted septa. The detailed classification is very complicated; the ammonoids, all extinct have been divided into 14 orders and 75 families, while genera and species are numbered by the thousand.

NAUTILOIDS. The nautiloid record began in the Upper Cambrian where the simple horn-shaped Plectronoceras cambria is found. The sub-class expanded rapidly with extremely diverse forms in the Middle and Upper Ordovician. Straight Lituites litmus from the Baltic, curved Augustoceras shidleri from Kentucky, and coiled Ophioceras nackholmensis from Norway are representative species. Unusual forms common in the Silurian are Ascoceras and Billingsites which, after developing as a slender, slightly curved horn, suddenly expanded into a much larger chamber and cast off the older part of the shell. Other common Silurian nautiloids are Phragmoceras broderipi from Czechoslovakia, Hexameroceras hertzeri from Ohio, and Kionoceras cancellatum from New York. They flourished, but on a diminishing scale, throughout the Upper Palaeozoic, where Centroceras marceliense and Ovoceras oviforme are found in the Devonian of New York. Liceras litratum, Stenopheras dumbeli, Soleno cheilus springeri, and many others are found in the Carboniferous and Permian of the Mid-continent region of the United States.

AMMONOIDS. The ammonoids made their appearance in the Early Devonian. They flourished only moderately until the Mesozoic Era, when they reached their greatest development in the Jurassic seas. Their period of greatness was short-lived, for by the end of Cretaceous time they became extinct. The genus Goniatites, a common Mississippian form, had a typical ammonoid suture composed of lobes and saddles representing the arches and troughs in the septa. Later ammonoids developed more and more complicated folds which, by the end of the Permian, had become quite complex. All Devonian and Mississippian ammonoids have a simple 'goniatitic' suture, but Prodomites gorbyi from the Lower Mississippian (Kinderhook) of the upper Mississippi Valley is one of the first species to have secondary convolutions in its septa, known as the ceratitic suture. Early in the Permian the suture pattern became more complicated as the convolutions of the suture developed second- and third-order lobes: the ammonite suture.

The goniatitic suture provides an excellent opportunity to study the life history of the ammonites, as the later whorls enclose and protect the immature portions of their shells. The first sutures indicate the ancestral character of the animal because it is possible to correlate them with the mature sutures of its ancestor. For example, Perronites (Middle Permian) can be traced back through Pennsylvanian genera to its ancestor (Goniatites) from the Upper Mississippian.

The oldest known ammonoids are from the lower Devonian of Germany. Other forms have also been found in the Middle and Upper Devonian, some of which have world-wide distribution. These include the Middle Devonian Anarcestes and Maenioceras, and Manticoceras, Cheiloceras, and Platyclymenia of the Upper Devonian. In Europe and North America, Protocanites and Beyrichoceras of the Lower Mississippian are succeeded by Goniatites and Eumorphoceras.

The Permian witnessed the culmination of many of the Carboniferous and early stocks; typical Permian genera are Properrinites, Waagenoceras and Cyclolobus.

**Arthropoda.** About 75 per cent of all known animal species belong to this phylum and some three-quarters of a million arthropod species are known. They are segmented organism with appendages which are usually locomotive in function but may be modified to subserve special functions such as the seizure of food, respiration, sensation, copulation, oviposition and fixation. These modifications and the more or less complete fusion of the segments into groups may result in the differentiation of three distinct body regions, the head, the thorax, and the abdomen. There is a well developed brain and central nervous system. The eye may be simple or compound. Respiration in smaller forms is through the general surface of the body whereas larger forms have specialised areas such as gills (branchiae) or lungs (trachae). There is a true mouth, alimentary canal and anus. A dorsal heart, enclosed in a sac, is present. The entire biological organisation of the Arthropoda indicates a close relationship with Vermes.

Many kinds of arthropods exist in incredible numbers—plagues of grasshoppers in flight often blot out of the sun like dense clouds, and 'clouds' of ostracods often crowd the water so thickly that, seemingly, there is not enough water for them to swim in. Undoubtedly the invertebrate peak of evolution has been attained by some orders of the arthropods. They can exist on all sorts of food substances, both plant and animal, they are well equipped for offence and defence against enemies, and they have a high order of social development.

TRILOBITES. The trilobites are the only extinct group of the Arthropoda, yet they are the most important fossils of the Early Palaeozoic. To judge from the degree of biological development in Early Cambrian trilobites, these creatures must have had a long Pre-Cambrian history, though no record has yet been discovered. The sudden appearance of four of the six independent orders of trilobites in the Lower Cambrian requires that we should refer them back to ancestors without hard parts capable of fossilisation. The other two orders must have developed their protective coverings later in the Cambrian and in the Lower Ordovician.

The trilobites, as the name suggests, had a body divided into three parts. Normally this threefold division might be head (cephalon), body (thorax), and tail (pygidium). Here, however, it refers to a longitudinal division of the cephalon, thorax and pygidium into three lobes a central or axial lobe, with a marginal or pleural lobe on each side. Although they vary widely in size (from a fraction of an inch to over two feet long) most species range from one to about three inches long.

The anatomical features of the cephalon, namely the free and fixed cheeks, the 'glabella' or central part, the eyes, and particularly the facial suture, are important elements used to bring together the

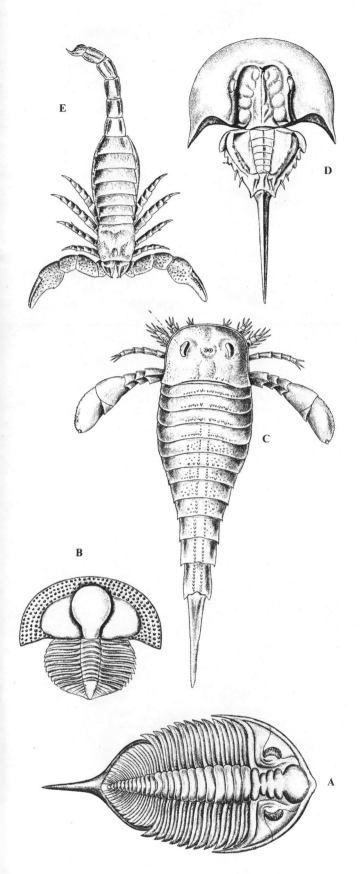

family groups into five of the six orders. The trilobites are subdivided as follows, with the name of the group, geologic range, typical genera, general character, and geographic distribution:

1. Protoparia, Lower Cambrian. *Olenellus.* Large cephalon, eyes crescent-shaped, sutures marginal, many thoracic segments, often with a pair elongated into spines, and a small pygidium, some with a telson or spine-like tail. North America, Africa, Europe, and western Asia.

2. Proparia. Middle Cambrian to Devonian. *Dalmanites, Phacops.* Eyes large, glabella bulb-like, sutures curve inward to glabella, thence along edge of eyes to margin, leaving eyes on the free cheek; fewer thoracic segments than Protoparia. All continents

3. Ophistoparia. Lower Cambrian to Permian. *Olenoides, Isotelus, Paradoxides.* Suture runs from posterior edge of cephalon to anterior edge, eyes often strongly elevated, some carapaces relatively featureless, others spiny. Often grew to large size (30 inches). Largest and longest ranging trilobite order.

4. Hypoparia. Lower Ordovician to Devonian. *Cryptolithus, Paraharpes.* Sutures marginal, cephalon bordered by a pitted fringe, eyes small or missing, thoracic segments few. North America, Europe.

5. Eodiscida. Lower and Middle Cambrian. *Pagetia.* Small (about ½ inch). cephalon and pygidium about equal in size, few (2 or 3) thoracic segments, eyes near margin and small. North America, Asia, Europe, and Australia.

6. Agnostida. Lower Cambrian to Ordovician. *Hypagnostus, Peronopsis.* No suture or eyes. Very small (less than ⅓ inch), two thoracic segments. All continents.

**Encrustacea.** OSTRACODS. The small, bivalved crustaceans called ostracods are geologically important fossils, especially in the Palaeozoic. Although the oldest fossils are found in the Lower Ordovician the sub-order has lived on to become, next to the copepods, the most abundant crustacean in the seas today. Ostracod shells are found in such profusion in some formations that they are useful stratigraphic correlations. Because of their small size (0·5 to 4 mm.) they are frequently recovered from wells when drilling for petroleum.

The ostracod covering is composed of two almost equal shells, joined together at the dorsal margin by a tooth-and-socket hingement. The shells are ovate in outline and may be smooth (*Leperditia fabulites,* Ordovician; *Bairdia curta,* Carboniferous), or diversely ornamented by rounded or elongate nodes (*Hollina insolens,* Devonian; *Verrucosella golcondensis,* Mississippian), or vertically ridged (*Tetradella cicatricosa,* Devonian), horizontally ridged (*Glyptopleura alata,* Mississippian), or finely veined (*Venula striata,* Mississippian). Others are coarsely pitted (*Halliella pulchra,* Devonian) or finely pitted (*Amphissites centronotus,* Permian); some are spiny (*Hollinella radiata,* Upper Mississippian), and still others are characterised by a wide band or frill along much of the free margins (*Dibolbina cristata,* Silurian and *Hollinella*). Many show a combination of several kinds of shell marking. For example, *Hollinella* has a frill, spines, and nodes.

OTHER CRUSTACEANS. Although less important than the trilobites and ostracods as Palaeozoic fossils, several other groups of crustaceans, because of their abundance today, should be mentioned. Three of the four orders of branchiopods are known as fossils, two of which are from Palaeozoic ricks. The orders *Notostraca,* with a single shell, and *Chonostraca,* a bivalve, range from the Cambrian to Recent. The valves of the latter are often mistaken for small clam shells (*Cyzicus ortoni,* Pennsylvanian).

The Cirripedae include the barnacles, once thought to be molluscs. However, after the third moult, a young barnacle secretes a bivalved shell rather like that of an ostracod. Then, attached to a rock or shell by its hard parts, it moults the two valves and secretes a series of strong overlapping plates. The oldest barnacle is probably a species of *Hercolepas,* an acorn-like barnacle from the Silurian.

The best known crustaceans today are members of the Order *Malacostraca,* which includes the land-living sow bugs, the crayfish of the freshwater streams, and the marine crabs, lobsters, shrimps. They are classified according to the nature of the shell which encloses the head and most of the thorax. The phyllocarids (*Colpocaris*

Palaeozoic trilobites and other arthropods.

A. *Dalmanites limulurus.* Middle Silurian, eastern United States.

B. *Cryptolithus bellulus.* Middle Ordovician, central and eastern United States.

C. *Eurypterus remipes,* a eurypterid. Upper Silurian, New York.

D. *Palaeolimulus avitus,* a relative of horse-shoe crabs. Permian, Kansas.

E. *Palaeophonus nuncius,* a scorpion. Upper Silurian, of Sweden.

*elytroides*, Mississippian), except for one genus in the Triassic, are confined to the Palaeozoic. They have a large, shield-shaped shell covering only the anterior part of the animal. However, most of the other groups, are younger than the Permian.

INSECTS. The insects exceed other divisions of the arthropods in number, variety, and adaption to varied environments. They are characterised by three distinct, segmented parts of the body — the head, thorax, and abdomen. The segments of the head are fused to form a single rigid part. The thorax has three segments, each of which carries one pair of legs, and the abdomen has ten to twelve segments. After hatching, the eggs pass through several growth stages, some of which may be spent in water. The adults are exclusively air breathers.

Despite the fact that few insects pre-date the Pennsylvanian, the fossils from these beds, as well as those of the Permian, are noted for their diversity of form and large size. There are huge cockroaches (4 inches) and dragonflies (30-inch wing spread) several times as big as living species. Some living roaches and crickets have come down through nearly a quarter of a billion years essentially unchanged.

The most primitive known insect, a wingless form *(Rhyniella praecursor)*, was found in the Rhynie chert of Middle Devonian age in Scotland. The oldest winged insects, however, came from Pennsylvanian rocks where, in spite of the poor chances for the fossilisation of terrestrial forms, large numbers of well-preserved specimens have been found in several places. Pennsylvanian and Permian insects have been collected from the Mazon Creek beds of Illinois, Lower Permian strata near Abilene, Kansas, the Saartbrucken beds of Germany, and various formations in France, New South Wales and the Soviet Union.

OTHER ARTHROPODS. The arthropods of sub-phylum Arachnida differ from trilobites and other crustaceans because they have only one pair of pre-oral appendages, which carry pincers. They have no antennae, so they are unlike insects, crustaceans, and myriads (centipedes). Three sub-classes are known: the Merostomata which includes the eurypterids and king crabs, known from the Cambrian to Recent; the Embolobranchiata, or scorpions and spiders, from the Silurian to Recent; and the sea spiders, from the Devonian to Recent.

The merostomes are aquatic animals with gills for breathing and well-developed mouth parts. They include the spike-tailed xiphosurians *(Prestwichianella danae* of Illinois and *Belinurus reginae* of Ireland, of Pennsylvanian age) and the eurypterids, the largest of all fossil arthropods *(Eurypterus remipes,* Upper Silurian and *Stylonurus excelsior,* Middle Devonian, both from New York). *S. excelsior* reached a length of $4\frac{1}{2}$ feet.

The Embolobranchiata are a distinct class that includes the spiders and scorpions. They have a separate head and antennae, no gills, lack the wings of insects, and have eight legs. Modern scorpions and the fossil species are similar in appearance and in body parts. Characteristic species include the Upper Silurian *Palaeophonus nuncius* of Sweden and *Eoscorpius dunlopi* from the Upper Carboniferous of Scotland. The spiders have well defined dorsal and ventral abdominal segments and the anterior, median, and posterior segments are separated by flexible membranes.

The sea-spiders are not important as fossils; only two genera are known from the Palaeozoic, both from the Lower Devonian of Germany.

The myriapods, the thousand-legged 'worms', have a meagre fossil record, so little is known about their arthropod ancestors. The two fossil classes include the millipedes and the centipedes. The oldest species, a millipede from the Silurian of Scotland, is thought to have been an air breather, like the living species, but there is no known record of the change in habitat from their sea-dwelling ancestors. Some of these fossils have been found in the Pennsylvanian shales at Mazon Creek, Illionois.

**Vertebrata.** The phylum is divided into five classes: Pisces, Amphibia, Reptilia, Aves, and Mammalia.

The early history of the vertebrates is lost because the oldest known fossils do not tell a connected story covering the period of

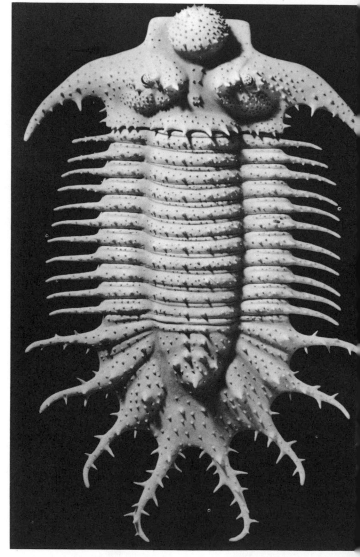

*Top*
*Cephalaspis (front)* a jawless fish covered with bony armour; length about 7 inches.
*Pterichthys (rear)* a true fish with jaws and very thick bony armour on the forepart of its body. These creatures are found in different parts of the Devonian Old Red Sandstone of western Europe. Their close relatives are known from Asia and America.

*Bottom*
Restoration of *Terataspis,* Central Hall, Buffalo Museum of Science. The largest known trilobite, it lived on the Middle Devonian coral reef near New York, 250 million years ago.

transition from invertebrate to vertebrate. The absence of such fossils has been ascribed to the first vertebrates' lack of hard parts to fossilise. Vertebrate history probably started with primitive chordates, similar perhaps to the present-day lancelet *Amphioxus* found in shallow, salt water.

The most primitive vertebrates such as the agnaths, or 'jawless ones', are represented by the hagfish and lampreys. These modern agnaths are distinguished by circular mouths without jaws but set with sharp teeth, and by long eel-like bodies. Many ancient forms had bony armour. The oldest vertebrate fossils are fragments of such protective armour. They were found in the Middle Ordovician Harding sandstone of Colorado and Wyoming, and in Lower Devonian rocks of the Midland Valley of Scotland and in the Welsh border country. There is some disagreement as to whether these deposits are of marine or freshwater origin.

AGNATHA. The first complete agnath specimens were found in Late Silurian and Early Devonian rocks. Known as ostracoderms, these were from a few inches to two feet in length, and almost all were covered by an armour of scale and bony plates. Typical of the group was *Cephalaspis*, which had a fish-like body covered with rows of elongated scales, a tail in which the backbone bent upward (heterocercal), two scale-covered flaps that may have been primitive pectoral fins, and a dorsal fin. The head was covered with a thick bony shield with marginal 'horns' in front of the pectoral fins. The eyes, near the centre of the shield, were directed upward with a plate between them carrying an opening for a pineal 'eye'. The nostril was located in a slit in front of this plate, in a position similar to that in living lampreys. The internal skeleton was cartilaginous. The ostracoderms died out at the end of the Devonian. Many excellent fossils have been found in the British Isles, eastern Greenland, Norway, and Spitzbergen.

PLACODERMS. The placoderms, known as *Gnathostomes* (jaw-mouthed creatures), and also confined to the Palaeozoic (Upper Silurian to Lower Permian), were the first fish to have both jaws and paired fins. They include the so-called 'spiny sharks' or acanthodians, which were primitive placoderms a few inches to a foot long. They possessed up to seven pairs of paired fins, each fin supported by a strong spine. *Climatius*, a typical genus from the Late Silurian and Early Devonian, was a freshwater dweller.

In the Devonian seas also lived the arthrodires, or fishes with jointed necks. Except for heavy plates covering the head and neck, they had lost their armour. We find *Coccosteus*, about two feet long, in the Old Red Sandstone of England and Scotland and equivalent formations in North America and Novaya Zemlya; in the Upper Devonian there were the giants *Dinichthys* and *Titanichthys*, which were as much as 30 feet long. They are particularly well known from the Cleveland black shales of Ohio. Except for broad plates of bone over the head and thorax, these monsters had cartilaginous skeletons. The bevelled plates that covered their jaws made excellent substitutes for teeth when it came to cutting prey to pieces.

Another group of placoderms are the antiarchs, typified by the 'wingfish', *Pterichthyodes*. This creature was so heavily armoured that for a time it was mistaken for a crab. Its pectoral fins developed into elongate, bone-covered structures which gave it the name of 'wingfish', though they were used to help it to crawl. A close relative was *Bothriolepis*, with folds of skin in the pelvic region that may have served as fins. The flippers were articulated in a manner resembling the arthropod limb. This fish also had traces of an internal skeleton in the shoulder region as well as external armour. When stranded on tidal flats, *Bothriolepis* may have breathed air into two pockets that branched from the pharynx, thus surviving until the water returned, perhaps foreshadowing the development of an animal that could live equally well on land and in the water, the amphibian.

CARTILAGINOUS FISHES. Placoderms developed bone internally as well as on the surface of their bodies. Sharks, rays and their kin lost this bone-building capacity and reverted to skeletons constructed of cartilage. Although cartilage is almost never preserved in fossil form, we know enough about these fish from

*Top*
*Holoptychius* (above,) a late Devonian lobe-finned fish which ranged round the world. Its teeth resembled those of amphibians. *Cheirolepis*, a primitive fish with bony rays in its fins, like those of such modern fish as the perch. These drawings are not to scale: *Cheirolepis* was 9 inches long; *Holoptychius* was 30 inches.

*Centre*
*Eusthenopteron*, an air-breathing fish whose fins have bones that could develop into feet and legs.

*Bottom*
*Ichthyostega*, a fin-tailed amphibian from Upper Devonian rocks of Greenland. About 4 feet in length, this animal is clearly transitional between fish and typical amphibians.

*Top*
*Seymouria,* about 20 inches long,
was a link between amphibians
and reptiles. From the Lower
Permian redbeds of Texas.

*Bottom*
*Eryops.* This massive amphibian
appeared in late Pennsylvanian
times but reached its greatest
size in the early Permian. This
species, about 7 feet long, was
found in the lower Permian redbeds of Texas.

petrified teeth and body impressions to be able to date them at least to the Devonian and say that they constituted an important part of the marine fauna of the Devonian, Mississippian, and Pennsylvanian (Carboniferous) periods. They declined during the Early Mesozoic but increased again in numbers during the Cretaceous and are still important part of the marine fauna.

The Cleveland shale in Ohio is the source of some of the earliest sharks, which shared a marine environment with the great *Dinichthys.* There were also ancient freshwater sharks such as the *Pleuracanthus* of the Pennsylvanian and Permian.

BONY FISHES. The bony fishes were a higher form with improved jaw structure and a bony skeleton. They also had an air bladder which functioned as a lung in many primitive and some specialised types, as well as a body covered with scales. Their fins give us a basis for dividing these important fishes into two groups: the crossopterygians, or lobe-fins, and the actinopterygians, or ray-fins.

LOBE-FINS. *Osteolepis* from the Devonian Old Red Sandstone of England was less than a foot long, had a blunt head, and was covered with small rhomboid scales. The late Devonian *Holoptychius,* three times as big as *Osteolepis,* had a stocky body and rounded scales, and ranged over much of the world. *Rhizodus,* of the Mississippian, grew to nearly 15 feet in length. *Megalichthys* is a typical Permian lobe-fin. This group is characteristic of the Mesozoic as well as of the Upper Palaeozoic. A related group, the coelacanths, lived in fresh water during the Carboniferous and Permian but became adapted to a marine habitat during the Cenozoic Era. The coelacanths were thought to be extinct until a representative, *Latimeria,* was caught in the ocean off South Africa in 1938. This surviving lobe-fin, almost six feet long, is much larger than its ancestors. A related group, the lungfish, became common in Late Devonian times, when it inhabited shallow seas. Today lungfish survive in South America, Africa, and Australia, those of the last-named continent being the most primitive.

RAY-FINS. They ray-fins were of primitive types and first appeared in Devonian rocks with shark-like tails and diamond-shaped scales. One of these, *Cheirolepis,* was about 12 inches long and had a well developed heterocercal tail, a slender body, blunt-nosed head, and small scales.

AMPHIBIANS. Amphibians (with a few specialised exceptions) hatch from eggs laid in fresh water and spend their early lives breathing by means of gills. Later (again with exceptions), the creatures become air breathers and spend much of their lives on land. Newts and salamanders are long-tailed members of this group; toads and frogs are short-tailed.

Amphibians did not develop from lungfish, which were highly specialised by Middle Devonian times, but sprang from crossopterygians late in the Devonian, a time when streams and small bodies of water seasonally dried up under arid conditions prevailing over north-western Europe, Greenland, and perhaps other parts of the world. During seasonal droughts, the lobe-finned crossopterygians fared better than others because they were probably able to make their way across land to permanent bodies of water by walking on their stubby fins.

Many fossils have been found that show the stages of amphibian evolution. The crossopterygian pectoral fins already contained the bones which later became the framework of feet and legs. In the latest Devonian or earliest Mississippian, an amphibian called *Ichthyostega* left its bones in Greenland. *Ichthyostega,* four to five feet long, had four legs and an ample, amphibian skull, though scales covered the skin and the tail retained fins.

*Ichthyostega* gave rise to the group of labyrinthodont amphibians, so called because of the intricately folded or crumpled layers of enamel, in their teeth, a characteristic inherited from their crossopterygian ancestors. They were important in the Late Palaeozoic and into Mesozoic times. Many continued to live most of the time in water, but some developed thick limbs and became essentially land dwellers. Typical of these was the carnivorous *Eryops* (Early Permian) with a broad, heavy body supported by short powerful limbs, a broad flat skull, and long, narrow tail. Another labyrinthodont, the heavily-armoured *Cacops,* apparently had enormous eardrums. Many of the older amphibians had disappeared by Middle Permian times and the labyrinthodonts are not known after the Triassic.

REPTILES. The rise and fall of the reptiles is one of the most interesting stories told by the fossil record. Starting in the Pennsylvanian when they sprang from the amphibians, they became the most important group of vertebrates after the end of the Palaeozoic, when they included many spectacular and bizarre animals. Reptiles reached their maximum importance in the Mesozoic, often called the 'Age of Reptiles'. With the end of the Cretaceous many lines, including the dinosaurs, died out.

Reptiles were able to divorce themselves from water because of important developments in their eggs. Fish and amphibian eggs are laid in water, essential in keeping the embryo moist. The reptile egg is encased in a waterbag, or amnion, which keeps the embryo moist. A second membrane, the allantois, lies inside the amnion and serves as a respiratory organ and a receptacle for waste products. Food for the embryo is found in a generous yolk. These features enabled reptiles to lay their eggs on land. Even though the adult life of some species may be spent in the water, reptilian eggs are usually laid on land and the young are sufficiently well developed when hatched to fend for themselves. Some early reptiles, however, and some living today are ovoviviparous, that is, the egg is retained in the body of the female and hatched there, so that the mother gives birth to a brood of living young.

The few reptile fossils that have been found in the Pennsylvanian have come from swamp deposits. There must have been some reptiles living on drier lands at higher altitudes, but there is no record of them until the Permian, when both primitive and advanced types were numerous. Although the amphibians remained plentiful, the reptiles competed strongly with them throughout the Permian, and by early Triassic became dominant. By the end of the Triassic, almost every known reptilian stock had appeared.

Some years ago the bones of a primitive four-footed animal were found in Early Permian rocks near Seymour, Texas. This animal, called *Seymouria babylorensis,* has a skull which displays many features of the amphibian, but many other skeletal features which are distinctly reptilian. The presence of lateral-line canals on its skull, similar to those of fish, seems to indicate that it lived in the water as an adult. It can be classified as a primitive cotylosaur.

Cotylosaurs are often called 'stem reptiles' as it is thought that the reptilian stocks evolved from them. They were common in the Late Pennsylvanian and Early Permian but few were left by the close of the Permian. Besides *Seymouria,* the group also includes *Diadectes,* about five feet long and of stocky build. The largest cotylosaurs are found in the Middle and Upper Permian of Europe and Africa.

Some early reptiles, like *Dimetrodon* and *Edaphosaurus* from late Pennsylvanian and Permian red beds of Texas, had a row of long, slender bony spines on the back, covered with a sail-like web of skin. Although the precise function of this 'sail' is not known, it may have been a combination of radiator and heater. Since reptiles must get their body heat from their environment, they could have been warmed by turning their 'sail' to the sun and pumping blood through the blood vessels under the skin. If too hot, the reptile could have found a shady spot and the circulating blood would then have been cooled in the same way. These animals reached eight to eleven feet in length and weighed 600 pounds or more.

Turtles—the chelonians—are not known before the Triassic. Their ancestor was probably the small Middle Permian reptile *Eunotosaurus* from South Africa. At the back of a small first rib, this fossil has eight very broad ribs that almost touch at their edges. Although the true turtle shell is a separate ossification from the ribs, the broad ribs of *Eunotosaurus* are thought to be a step toward the development of a protective covering.

Other reptiles besides turtles deserted their place of origin, the land, and took to the water, where they became fish-eaters. The slender *Mesosaurus* was a little over a foot long, had a long, narrow snout set with sharp, close-set teeth, and developed from a stem reptile, one of the cotylosaurs. *Mesosaurus* and the related *Stereosternum* have been found only in the Pennsylvanian and Early Permian of South Africa and Brazil, a fact used to support the theory that these continents were once united.

Reptiles represent the highest group of animals living in the Palaeozoic Era. From their beginning in the Pennsylvanian, the group evolved rapidly, almost explosively. In the Permian the reptiles outnumbered the amphibians, and the class was well on its way to the extreme diversification that occurred in the Mesozoic, when one branch, the dinosaurs, came to include the largest land animals that have ever been known to have existed on the Earth.

*Dimetrodon,* a 'fin-backed' reptile of late Pennsylvanian and early Permian times. This species, 10 to 11 feet long, was discovered in Lower Permian red beds of Texas.

# Life in the Mesozoic

As we have seen, the Palaeozoic era passed quietly into the Mesozoic. In Britain there is no sharp line dividing Permian from Triassic rocks and the two are often referred to as though they belonged to one system: the Permo-Trias (New Red Sandstone). Desert conditions prevailed and the adverse climatic environment is indicated by the paucity of the fossil record. In the Mediterranean, however, marine conditions continued through the Permo-Trias and all the main groups of animal life were maintained from Carboniferous times. Evolutionary changes can be demonstrated, as in the corals: the Palaeozoic type (Tetracoralla) in which the chamber is divided into four quadrants seems to have vanished completely and is replaced by the Hexacoralla in which the structure has six divisions. The echinoderms declined almost to the point of extinction in early Permo-Trias times but recovered later. Graptolites disappeared completely. Brachiopods, so conspicuous in the Palaeozoic, declined markedly, though a few genera survived into Mesozoic times. Of the molluscs, the lamellibranchs and gasteropods, which were not prominent in the Palaeozoic, continued to develop, reaching a climax in the Jurassic. Cephalopods of the nautilus type declined until only one genus remained; the goniatite type continued, giving rise to large numbers of ammonites. Some of these became extinct at the end of the Trias and only two genera survived to continue the line into the Jurassic and Cretaceous.

Among the arthropods, trilobites finally disappeared. Aquatic arachnids survived in the king crab and the scorpions. Other crustaceans developed into lobsters and crabs. Terrestrial insects and spiders were but little affected by the transition into the Mesozoic.

Among the vertebrates, the Amphibia declined from their dominant position and became insignificant. Reptiles developed rapidly and became adapted to all modes of life: before the end of the Mesozoic, however, many forms became extinct. Some types of fish completely disappeared—for example, the mudfish, *Ceratodus;* others survived and evolved into various types of enamel-scaled fishes.

## FLORA

There was a considerable difference between the trees of a Carboniferous forest and the trees of a Jurassic forest. In the Jurassic there were no more *Calamites,* sigillarias or lepidodendrons to be found. Ferns however, remained and to them were added more and more modern plants, such as the gymnosperms and later, in the Cretaceous, angiosperms or flowering plants.

**Gymnosperms.** A typical Jurassic forest included cycads—trees with straight or beehive-shaped, woody trunks with large pith cavities. The tufts of large leaves at the end of the trunks suggest those of date palms. Some of their descendants still grow in the tropics. Between their leaves hung huge, flower-like organs as much as six inches long. The forest also included ferns, ginkgos and conifers. The only living species of ginkgo tree is a native of China. Specimens have been widely acclimatised in public gardens and elsewhere in the Occident; many have been transplanted to North America, where the tree is known as the maidenhair tree. Fossil leaves of the ginkgo are preserved by the hundreds in Oregon. They have also been found in the Mesozoic strata of the Arctic, Africa and Australia.

The conifers include araucarians, pines, firs, cedars and sequoias. In North America, in the Petrified Forest National Monument of northern Arizona, thirty-eight species of plants have been identified from the Triassic Chinle formation. One is the spectacular conifer *Araucarioxylon.* Petrified logs found in the Monument indicate that some of these trees were as much as four feet in diameter and 120 feet high. The nearest living relative of *Araucarioxylon* is *Araucaria,* or monkey-puzzle pine.

Similarly, near Whitby in northern England, fossil tree-stumps have been found in Upper Jurassic sand-stones. Stumps and fallen trunks of conifers can be seen also in the Purbeck beds of Portland Bill and the Isle of Wight, and a specimen of the species *Dadoxylon* is at the time of writing on show in the British Museum.

**Angiosperms.** The oldest known remains of flowering plants, angiosperms, from the Jurassic were found in the fossil pollen from a coal bed in Scotland. True flowering plants with seeds enclosed in fruits appeared in the Cretaceous. Willows and poplars began to grow along the river banks, and vast areas were covered by forests of oaks, birch and beech. All these, nowadays typical of temperate regions, were then mixed with tropical or sub-tropical species such as palms, figs, oleanders, magnolias, eucalypti and breadfruit trees. Climate zones had not yet come into being.

## FAUNA

**Protozoa.** The shells of these tiny, one-celled animals, some of which are so minute that they are only just visible to the naked eye, have been studied in detail because of their value in correlating strata in oilfields all over the world. Their small size makes it possible to recover them from drill cuttings and cores without damages. The Foraminifera were abundant in the Mesozoic, particularly in Cretaceous times, where they make up the bulk of the chalky limestones so characteristic of this formation in Europe. Typical genera are *Opthalmidium, Spiroloculina, Nodosaria* and *Lenticulina.*

They were accompanied by Radiolaria (protozoans having a shell made of glassy silica) and by Coccoliths, minute calcareous discs which are usually found embedded in the fine matrix composed of chalk.

**Coelenterata and Enchinodermata.** It is convenient to consider these two phyla together because of the frequent association of corals (Coelenterata) with crinoids and sea-urchins (Echinodermata) in Mesozoic seas. Coral-forming polyps seem to live abundantly only in warm, shallow and very clear water. The abundance of corals in the Oolite Limestone, which stretches from Dorset to Lincolnshire, suggests that during Jurassic times such conditions prevailed over much of south and central England. Typical forms are *Montlivaltia* and the *Thecosimilia* and *Thamnastrea.*

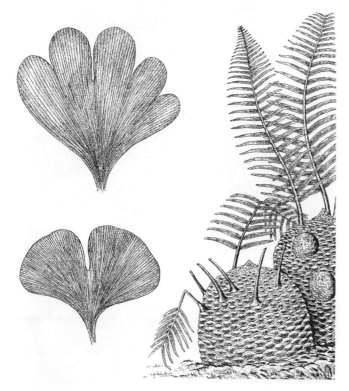

On the left are ginkgo leaves from the Jurassic of Oregon *(above)* and the Cretaceous of Alaska *(below)*. On the right, a restoration of the cycad *Cycadeoidea,* from the Jurassic of South Dakota.

Sea-urchins were associated with the corals. Exceptionally good exposures are to be seen at Calne in Wiltshire where the Corallian Oolite (mid-Upper Jurassic) outcrops. Typical genera are *Cidaris, Hemicidaris, Acrosalenia* and *Stomechinus.* Many kinds of radially symmetrical or regular echinoids are found in Jurassic rocks; associated with them are sea-urchins which have a more or less well-developed bilateral symmetry, known as irregular echinoids. Of these *Holectypus* is a good example. In Upper Cretaceous times sea-urchins lived in large numbers in the clear Chalk sea: typical genera are *Holaster, Micraster, Salenia, Discoidea, Echinocorys* and *Conulis.* Associated with these forms are the sea-lilies *Unitarinus, Marsupites* and *Bourgueticrinus.* Typical of the asteroid, or starfish, division of the echinoderms is the *Pentagonaster* group.

There are few corals in the Mesozoic rocks of North America because the seaways were restricted until the Cretaceous. On the other hand, coral reefs were abundantly developed in the Jurassic seas of central and southern Europe. An archipelago with its barrier-reefs, fringing reefs and atolls stretched from the south of Russia to North Africa and England. In France, the Côte de Meuse was an ancient barrier-reef lying seaward from an island formed by the Vosges Mountains. It was built by branching or massive stony corals, with their companion fauna of sea-lilies, long-spined sea-urchins and tick-shelled molluscs. Along the silted-up axis of the Côte d'Or the reefs were lacking, but they recurred in the Yonne and then continued along the edge of the island of the Massif Central as far as Berry. Other less important reefs have been discovered along the coast of Normandy. Towards the end of the Jurassic period reefs disappeared from the Paris basin and the northern Jura because the waters became muddy and there was, too, a general lowering of temperature. They were then restricted to the central Jura (Valfin reefs).

At the end of the Jurassic there was a further retreat and the reefs became restricted to the southern Jura, where they formed part of the immense barrier-reef which was the boundary of the Alpine seas. They can be seen today in four large areas: Solenhofen in Bavaria; Cérin in southern Jura; Mont Salève, near Geneva; and Bec de l'Echaillon, near Grenoble. In France, reefs of the same age are found in Languedoc and Provence.

Reefs continued to diminish in importance during the Cretaceous, giving way to limestones made up of mollusc shells. Corals

migrated south of a line running from the Pyrenees to Provence. During the Tertiary they withdrew still further south until they became restricted to the tropics, where all except a few species are found today.

**Molluscoidea.** The brachiopods of the Mesozoic can be referred to two groups: *Terebratula* and *Rhynchonella,* which first appeared in Palaeozoic times. The terebratulids are generally circular to oval in shape and have a smooth shell. The rhynchonellids are more pointed and usually the margin of the shell is wavy in outline. Although most of the Jurassic brachiopods belong to one or other of these two main groups, others, such as *Spiriferina,* represent the last remaining examples of the spirifers which reached their climax in the Palaeozoic. The best known brachiopods from Cretaceous times are *Rhynchonella cuvieri* and *Terebratula lata,* used as index fossils for the Middle Chalk.

**Mollusca.** Molluscs were abundant throughout the Mesozoic, reaching their climax in the Jurassic: all the classes are represented: pelecypods, gastropods, belemnites and, above all, the cephalopods.

PELECYPODS. Not very plentiful during the Palaeozoic, the pelecypods increased abundantly in Mesozoic times, often forming limestone beds of considerable thickness. One of the most striking of such limestones is that known as the Rudist Limestone, found in Spain, southern France, the Appenines and Carpathians, as well as north Africa and south Asia. The earliest rudist genus, a sessile (or sedentary) lamellibranch, is *Duceras* which appears in the middle part of the Upper Jurassic. The valves are twisted into two close spirals and one of them, usually the left valve, is fixed to some external support. *Valletia* is a typical example from the Upper Jurassic and Early Cretaceous. Other genera are *Toucasia* and *Requienia,* in which the fixed left valve is coiled while the free right valve has become a more or less flattened lid. Later Cretaceous forms are *Carprina* and *Hippurites.*

Most pelecypods live in the soft mud or sand of the sea-floor. Some, such as the scallop *Pecten,* are capable of swimming; shells of this from occur abundantly in the Liassic ironstone of Cleveland, Yorkshire, where they are known as 'Pecten seams'. Genera such as *Pholadomya* and *Lithophagus* from Jurassic rocks seem to have been burrowing types, while the Jurassic mussels *Modiola* and *Pteria* had a sedentary mode of existence. Other Jurassic genera such as *Ostrea, Chlamys* and *Lima* were associated with the building up of coral reefs.

In the Cretaceous, the pelecypods are represented by many genera but more especially by forms akin to *Pecten.* Others, such as *Plicatulla* and *Spondylus,* were of sedentary habit. In the Chalk beds the *Trigonia* and *Inoceramus* genera are the most typical forms.

GASTROPODS. These were not very important in Mesozoic times. In the Early Jurassic, *Cerithium* and *Natica* became established; later, forms such as *Littorina* and *Turritella* are found in many of the world's coral reefs. *Fusinus* and *Neptunea* appear in the Cretaceous.

CEPHALOPODS. The ammonites, directly related to the goniatites of the Palaeozoic, were abundant in Mesozoic seas. For a long time their fossilised shells were taken for rolled-up petrified snakes. In the fifteenth century an ammonite was embedded in the lintel of one of the doors of Bayeux cathedral with the inscription: 'Believe in divine miracles: this serpent was changed into stone'. Later, ammonites were called 'Ammon's horns' because of their resemblance to the horns with which the artists of antiquity adorned the brow of Jupiter Ammon.

Ammonites play an immense role in palaeontology and in stratigraphy because of their great geographical range. They had floating larvae which were readily disseminated by ocean currents; they evolved rapidly and their species were not long-lived; certain of them are strictly characteristic of particular marine environments (cold water, warm water, deep water, shallow water). An ammonite often shows the precise geologic age of the strata from which it was collected and conditions under which rocks were deposited.

*Top*
A large prawn, *Panaeus speciosus.* Lower Jurassic of Germany.

*Bottom*
*Crioceras,* a coiled ammonite from the Cretaceous of Europe.

*Top*
*Lepidotus minor,* a fish from the Jurassic of Germany.

*Bottom*
*Hippurites,* a clam from the Upper Cretaceous of Western Europe.

There are thousands of species of ammonites. They vary in size from that of a small pea to that of a wagon wheel, and they vary in the manner which the shell is coiled. The successive whorls may just touch, as in *Arietites* (Lower Lias) and *Hildoceras* (Upper Lias), or they may overlap each other so that sometimes only one whorl remains visible from the outside, as in *Oxynoticeras* (related to *Arietites*) and *Hyperlioceras* (related to *Hildoceras*). The last chamber (called the body chamber) may have a circular, oval, quadrangular or arched cross-section. The orifice of this chamber may be a single opening or it may be partly divided to form orifices for the mouth, the eyes and the tentacles. They vary in the shape of the suture, the line along which an internal division joins the chamber wall. After the Palaeozoic goniatites with simple sutures came the ceratites of the Triassic whose sutures are festooned, and then the true ammonites of the Jurassic with crinkled sutures. Finally, in the Cretaceous, the sutures again became festooned. At the same time the shell uncoiled, as in *Macroscaphites* or *Hamites,* and eventually straightened, as in *Baculites.*

The ammonites disappeared completely at the end of the Cretaceous; the reason for their relatively rapid extermination is unknown.

BELEMNITES. These cephalopods are allied to the present-day cuttle-fish; they do not appear as fossils until the Triassic and reached a climax in the Lower Cretaceous. They have a most unusual and highly characteristic cigar-shaped form; before their true nature was recognised, they were believed to be petrified thunderbolts. Typical genera are *Acroteuthis, Hibolites* and *Neohibolites.* They are particularly abundant in the Lower Cretaceous where they are used as index fossils for the identification of subdivisions of these rocks. The belemnites became extinct at the end of the Cretaceous but were replaced by allied types — ancestors of the cuttle-fish and octopus.

### Arthropoda

INSECTS. The evolution of flowering plants is linked with the development of insects feeding on nectar and pollen. It is not surprising, therefore, to find that the Mesozoic landscapes were full of life and noisier than the dismal quiet of the Mississippian and Pennsylvanian forests. Dragonflies, grasshoppers, cockroaches, and termites—many of which remain little changed to this day—were gradually joined by bees, flies, and their allies, whose larvae could develop in a wide variety of environments. Beetles appeared in the Permian. Weevils bored into seeds; dung-flies fed on the excrement of the large reptiles; bees and flies flitted from flower to flower gathering nectar.

### Vertebrata

AMPHIBIANS. Triassic amphibians were largely of the labyrinthodont group known as Stereopondyls. They represented a secondary return to fresh water. They developed a simplified vertebral column and, as often happens with aquatic animals, grew to a large size. Indeed, some were the largest amphibians known. Bizarre species arose; *Buettneria* of the Upper Triassic of the south-western United States, for example, had a large flat head with a heavy skull, a flat body, and weak limbs and feet. Stereospondyls became extinct at the end of the Triassic.

Labyrinthodont amphibians had, from the Pennsylvanian onward, given rise to the evolutionary line of the anurans, or frogs and toads. The Pennsylvanian *Miobatrachus* had a skull and skeleton that foreshadowed the later anurans. The Triassic *Protobatrachus* of Madagascar shows further developments along these lines and displays the skull of a frog on an animal that still had a long tail. By the Jurassic recognisable, tailless ancestors of the present-day frogs and toads had developed.

REPTILES. At the beginning of the Mesozoic a sudden expansion took place in the world of reptiles with the appearance of dinosaurs (Greek *deinos* — terrible, *sauros* — lizard). They are the most widely known of all fossil vertebrates. Dominating terrestrial life for more than one million years, their rule began in the Early Triassic with the appearance of small, lightly built dinosaurs adapted for running on their hind legs, perhaps in pursuit of large insects on which they may have fed.

The dinosaurs are classified in two orders: the Saurischia (reptile pelvis) and the Ornithischia (bird pelvis). The Saurischia, whose pelvis is three-pronged, include all carnivorous dinosaurs as well as many herbivorous ones. The Ornithischia, with a two-pronged pelvis, remained herbivorous throughout their existence. Both orders began as bipedal forms, but many members of each reverted to a four-footed posture. A spectacular example of this is the brontosaur of the Jurassic *(see illustration)*, one of the gigantic plant-eating forms which inhabited the swamps of that time. A relative, *Brachiosaurus* of East Africa and North America, was over 80 feet long and probably weighed 50 tons. The sauropods living in the immense swamps which covered the site of the Rocky Mountains during the Jurassic must have eaten about 700 pounds of vegetation a day. Almost defenceless, they were preyed upon by theropods, a general term applied to fierce carnivorous dinosaurs such as *Allosaurus*. Thirty-four feet long, this beast was dwarfed by the great plant-eating sauropods, but it was equipped with tremendous jaws and teeth, as well as strong hind legs. The front limbs, as in many carnivorous dinosaurs, were reduced in size to the point of uselessness.

Also developed during the Jurassic were the stegosaurs, which attained a length of about 30 feet. Along their backs ran a double crest of broad plates which passed into formidable spines at the caudal end. Their mouths and teeth indicate a herbivorous diet. Although quadrupeds, stegosaurs were ornithischians, as was their bipedal contemporary, *Iguanodon*, best known from Belgium. A group of seventeen skeletons of these dinosaurs was found in a pocket of clay in the coal basin of Bernissart.

In the Cretaceous, as in the Jurassic, there appeared new and spectacular varieties of herbivorous dinosaurs, both heavy quadrupedal forms and the more lightly-built bipedal ones. There were also new bipedal carnivores. The trachodonts, or duck-billed dinosaurs *(see illustration, p. 386)*, had little in the way of defensive armament. Their beak-like jaws bore a mosaic of some 2,000 small teeth for grinding vegetation.

A four-footed herbivore, *Triceratops*, with the ornithischian type of pelvis, was better defended. Its massive body, 18 to 25 feet long including the tail, possessed a large skull which extended back over the neck in a bony shield; two front horns over the eyes pointed forward, while a third horn surmounted the nose.

Mesozoic ammonites and two belemnites.
A. *Belemnitella americana*, Upper Cretaceous, New Jersey to Texas.
B. *Macroscaphites ivani*, Lower Cretaceous (Noocomian) Europe.
C. *Engonoceras pierdenale*. Lower Cretaceous, Texas.
D. *Schloenbachia cristata*. Lower Cretaceous of Europe.
E. *Lytoceras immane*. Upper Jurassic, Europe.
F. *Helicoceras stevensi*. Upper Cretaceous (Pierre), western United States.
G. *Dufrenoya justinae*. Lower Cretaceous, America.

Cretaceous fossils.
A. *Exogyra costa*, a relative of the oyster. Lower Cretaceous, eastern and southern United States.
B. *Rhaphidonema farringdonense*, a calcareous sponge. Cretaceous (Aptian), England.
C. *Scaphites quadrangularis nodosus*. Ammonite showing pearly lustre. Upper Cretaceous. Plains of the western United States.
D. *Cidaris coronata*, an echinoid or sea urchin. Upper Jurassic, Germany.
E. *Pecten quinquinarius*. Upper Cretaceous, U.S.A.

Hip ganglion

Brain

**Top**
*Stegosaurus,* a Jurassic 'bird-hipped' dinosaur that lived near the giant *Apatosaurus.* Stegosaurus, 18 to 25 feet long, is noted for the 'second brain' above the hips.

**Centre left**
The duck-billed dinosaur *Anatosaurus,* also called *Trachodon.* It lived among late Cretaceous swamps of the western United States and may have ranged across the continent. It was probably a good swimmer.

**Centre right**
*Iguanodon,* a plant-eating dinosaur about 15 feet long. Found in Lower Cretaceous coal deposits of Europe.

**Bottom**
*Scolosaurus,* a broad bodied, heavily armoured dinosaur of late Cretaceous times. Length about 18 feet.

*Top*
*Tyrannosaurus rex,* about 50 feet long, the largest carnivorous dinosaur known. In walking, it balanced its body by means of its tail. It was certainly the monarch of the reptile world, and had teeth that were eight inches long. Upper Cretaceous of Montana and nearby regions.

*Bottom*
*Brachiosaurus (left)* and Apatosaurus, the brontosaur *(right)* were two of the largest 'lizard-hipped' dinosaurs. They lived in Colorado and nearby regions in latest Jurassic times.

Certainly the monarch of the reptile world was the carnivorous *Tyrannosaurus* of the Late Cretaceous, which measured some 50 feet long. Its skull, about four feet long, was armed with crenulated teeth eight inches long.

*Protoceratops,* an Asiatic relative of *Triceratops,* was a small dinosaur, five to six feet long, what a skull that bears a curious resemblance to the beak of a parrot. An American expedition to Mongolia discovered a clutch of its eggs, 75 million years old; each was about six inches long and contained an embryo. The skeleton of another reptile nearby, *Oviraptor,* suggests that this animal stole the eggs for food.

The powerful skeletal and muscular development of the dinosaurs contrasts strongly with the puny growth of their brains. The brain of the largest dinosaur was no bigger than that of a sheep. A second 'brain', really a large ganglion, lay in the region of the hips.

For reasons we do not fully understand the dinosaurs became extinct at the end of the Cretaceous. The most important factors in their decline may have been the increasing severity of the climate, the decreasing areas of swamp, and increasing aridity. Being cold-blooded, they could not survive periods of protracted cold. Highly specialised, they needed a highly specialised environment. Their limited brains may also have hindered adaptation to new conditions. Competition with the new, more highly developed and more adaptable mammals may also have been a factor leading to their demise.

SWIMMING REPTILES. Swimming reptiles, like terrestrial reptiles, do not form a natural group. Their superficial resemblances depend solely on their common adaptation to an aquatic way of life. The best known of them all, *Ichthyosaurus,* shows that its body was adapted for rapid swimming. Its long, large muzzle with pointed jaws carried numerous teeth. The large eyes were surrounded by bony rings. The feet were flipper-shaped for swimming. In addition to the flippers there was a dorsal fin and

*Left*
*Pterodactylus (left),* a short-tailed Jurassic pterosaur of Europe. This species was about as large as a pigeon. On the right is *Rhamphorhynchus,* 24 inches long, including his tail.

*Below*
*Archaeopteryx,* the oldest known bird, had teeth, toes on his wings, and a lizard-like tail covered with feathers. Fossils of this bird were discovered in the Jurassic Solenhofen beds of Germany.

Skull

a caudal fin, the lower lobe of which contained the end of the vertebral column (unlike sharks, where the end of the vertebral column lies in the upper lobe). Ichthyosaurs were apparently ovoviviparous, that is, the female retained the eggs inside the body until birth, and skeletons have been found with up to half-a-dozen well-preserved foetal skeletons in the body cavity.

Ichthyosaurs ranged to 10 feet in length, and the long-necked, long-tailed pleiosaurs up to 40 feet. The great aquatic lizards, *Tylosaurus* and *Mosasaurus,* on the other hand, had elongated bodies and attained lengths of 20 to 30 feet; a species found in New Jersey reached 45 feet.

FLYING REPTILES. During the Jurassic and Cretaceous flying reptiles resembling bats developed. Here again, adaptation to the same means of locomotion tended to produce similar characteristics. The primitive flying reptiles had, like most reptiles, identical teeth and long tails. In later forms, teeth were lost and the tail shortened. Like bats, they possessed a membranous fold of skin supported by enlarged hands and attached to the thighs and the foot of the tail. One digit (the fourth), instead of four as in the bats, was longer than the rest. Like birds, they had a mobile neck, a breast bone (though not keeled as in birds), and their skeletons were made up of pneumatic or hollow bones.

*Rhamphorhynchus,* the one closest to its bipedal terrestrial ancestors, had a number of fine, pointed teeth and a long tail that ended in a small diamond-shaped 'sail'. Later *Pterodactylus* arose; it differed from *Rhamphorhynchus* in the smaller tail and the relatively larger head. Neither genus was as large as an eagle. More extraordinary was *Pteranodon* from the North American Cretaceous, with a wing span of 25 feet. There was a considerable number of flying reptiles during Cretaceous times. The largest one on the wing was the *Pteranodon.* Noticeable was the shape of the head, which had a long, backward-projecting crest—presumably a counterpoise to the long, toothless beak. Despite its large wingspread, however, *Pteranodon's* body was no larger than that of a turkey.

## THE FIRST BIRD

The oldest known bird is the crow-sized *Archaeopteryx* (Greek *archaios* — very old, and *pteryx* — wing) from the Jurassic lithographic limestone of Solenhofen (Bavaria). The fine sediment of this stone has preserved both feathers and bones. Unlike those of modern birds, its bones were not hollow. The tail was long and reptile-like but had a row of feathers down each side. The feathered wings were feeble and possessed three-clawed fingers.

# Life in the Cenozoic: Tertiary

Earth movements at the end of the Mesozoic and during the Early Tertiary gave rise to a widespread withdrawal of the sea in America and northern Europe. Tertiary sediments are dominantly estuarine or freshwater but minor incursions of the sea took place from time to time; as a consequence conditions varied rapidly and are reflected in the fossil record.

In the Tertiary deposits, as in all earlier systems, molluscs filled a prominent place in the fauna. Among the cephalopods, the ammonoids after reaching a climax in the Jurassic had disappeared; belemnites had also gone; nautiloids survived only in one genus, *Nautilus*. The gastropods and pelecypods, however, went on evolving and producing new genera and species even into the present epoch: *Conus* in the Eocene, *Trochus* in the Miocene, and *Nassa* in the Pliocene. The Protozoa were represented by a large increase in the number of Foraminifera. Among the vertebrates, the bony fishes became dominant and the enamel-scaled types declined. Reptiles, having reached a climax in the Mesozoic, declined markedly. The Tertiary saw the establishment of the mammals as the dominant vertebrate stock.

Early Tertiary fossils have been discovered in all continents, but genera and species were not as widespread as in the older periods. However, certain nautiloids *(Hercoglossa)*, gastropods *(Turritella)*, pelecypods *(Venericardia)*, and foraminifers *(Globigerina* and *Globotruncana)* had world-wide distribution.

## VEGETATION AND CLIMATE

The Tertiary flora, like present flora, was typified by the predominance of flowering plants over all other types. There were three vegetation groups in Europe: the conifers of the cold and temperate regions, the deciduous trees of the temperate regions, and the broad-leaved evergreens of the tropical areas. The vegetation zones were not as clearly delimited as they are today, but tropical elements were being forced farther and farther south.

In France, for example, the Early Tertiary travertine of Sézanne near Rheims, the green band of the *calcaire grossier* near Paris, and the marls of the ancient lake at Aix-en-Provence show a mixture of tropical plants (palms, banana trees, jujube trees), and plants that still flourish in France today (walnuts, oaks, birches, poplars, and willows).

Towards the middle of the Tertiary, conifers and deciduous trees began to outnumber the tropical elements, as is shown by fossils from the lake beds of Oeningen (Switzerland). The vegetation at this time included pine, fir, birch, oak, poplar, willow, cherry, plum, and almond. Beneath the trees there was a tangle of ferns. Only in places well exposed to the sun did southern plants mix with them. The forests were broken up by prairies, where herds of herbivorous animals grazed by lakes and rivers inhabited by crocodiles. In the water various rushes, reeds, bullrushes, irises, and water lilies flourished. Finally, the fossil flora of the Late Tertiary travertine of Meximieux, Ain, shows the further retreat of plants typical of hot regions. The modern climatic zones were becoming established. Palms no longer reached farther north than the central Mediterranean coasts.

There was throughout the Tertiary a progressive decline in temperature. By comparing plant species from various beds with present-day floras botanists can determine what the temperature may have been at different periods in the Tertiary. They estimate, for example, that in France the mean annual temperature was 77°F. in the Eocene (Paris), 72°F. in the Oligocene (Aix-en-Provence), 64°F. in the Miocene (Oeningen, Switzerland), 59°F. in the Pliocene (Meximieux), as contrasted with 50°F. today. Thus the equatorial climatic zone gradually contracted in favour of the temperate zones, which in turn contracted in favour of the polar zones. Under these conditions, plants of the northern hemisphere migrated southward during the Tertiary — a migration analogous to that of the coral reefs during the Mesozoic.

In North America the sequence from a tropical Eocene climate to a cooler climate in the later Tertiary has been worked out for the John Day basin of Oregon. Also, isotherms for each of the Tertiary epochs have been determined for the area from Lower California to Alaska. In the Eocene the forests in central Oregon were like the rain forests of present-day South America. Then pine, spruce, maple, elm, birch, cottonwood, walnut, sedges, willow, and the dawn redwood *(Metasequoia)* grew on St Lawrence Island, between Alaska and Siberia. *Metasequoia* is unusual in that fossils of it were found and described in 1863, some eighty years before the living tree was found in 1941.

The Plains environment in middle North America was present for the first time in the Early Miocene; seed husks from grasses have been found in deposits of this age. These spear grasses, *Stipidium,* spread widely over the area shortly after their appearance and their fossils occur abundantly in the sandy stream deposits of the Miocene and Pliocene.

Other well-known Tertiary floras have been collected from the Latah formation of Washington and the Wilcox and Claiborne of Alabama (Eocene), from the Bridge Creek of Oregon (Oligocene), the Florissant beds of Colorado (Miocene)—noted also for their fossil insects—and from the Citronelle formation of the Gulf Coast.

As flowering and fruit-bearing plants became important in the Tertiary, flower-loving and honey-gathering insects (beetles, flies, bees, and butterflies) increased in numbers. More than 6,000 species of insects have been recovered from the fossiliferous beds of Sézanne and Aix-en-Provence, from the phosphorites of Quercy, Monte Bolca (Italy) and Oeningen (Switzerland), and from the amber of the Baltic. This last has provided the best specimens. Amber is fossil resin which oozed out of pines and firs, trapping and preserving all sorts of insects, millipedes, spiders, feathers, hairs, leaf fragments, flowers and seeds.

At Oeningen the fossil-bearing beds consist of alternating marine and freshwater deposits of Oligocene and Miocene age. These sediments consist of grey to green, calcareous or feldspathic sandstones and conglomerates which settled in a large lake flanked by grasslands where herds of ruminants and horses grazed. The animals had their own associated and characteristic insects which, on falling into the streams, were carried to the lake and covered by fine-grained sediment. These insects included gad flies and

*Uintatherium*, also called *Dino-ceras*. An Eocene herbivore about 5 feet tall, with six blunt horns. Western United States.

tsetse flies which plagued the animals with their stings, and maggots, flies and dung flies which lived mainly on their droppings.

## FAUNA

**Foraminifera.** These shell-bearing Protozoa, in spite of their generally minute size, have often assumed the role of rock builders. They were not confined to the Tertiary; Palaeozoic and Mesozoic seas contained so many that several deposits of these eras, notably the Cretaceous Chalk, are made up almost entirely of their shells. In the Tertiary, however, the Foraminifera reached their acme, principally in such forms as miliolids and nummulitids. Miliolids are about as big as grains of millet. The fine freestone (a type of sandstone) from the neighbourhood of Paris contained about three million to the cubic yard and owes its granular texture to them.

Rather more complex are the nummulitids, whose shells are made up of a large number of small chambers communicating with one another in a spiral about an axis. Externally they resemble coins, and the terms 'coin stone' and 'nummulite' have been given to the nummulitic limestone of the Paris district. The pyramids of Egypt were constructed of a similar limestone.

Nummulites vary in diameter from a quarter of an inch to three inches. In most species there is a small form with a large central chamber and a large form with a small central chamber. The two species probably represent an alternation of generations, that is, the generation with a large central chamber gave rise to a generation with a small central chamber, which in turn gave rise to a generation with a large central chamber, and so on. Nummulites were characteristic of the warm seas of the Early Tertiary.

The abundance and wide distribution of the foraminifers have made it possible to recognise rocks of the same age in different parts of the world by the species in them. A rock layer or group of layers characterised by one or more typical species is called a 'zone'. In the San Francisco Bay region sixteen zones of foraminiferous rocks can be recognised in an almost uninterrupted sequence.

**Molluscs.** The ammonoids completely disappeared before the Tertiary. The squid-like belemnites had been exceedingly plentiful from the Late Triassic to the Late Cretaceous, but then declined rapidly and disappeared during the Eocene. Pelecypods were of great importance in the Tertiary, but it was the gastropods which became the pre-eminent molluscs of this period.

The principal pelecypods were those types which had perfected hinge lines and possessed a respiratory tube (siphon) that enabled them to bury themselves fairly deep in the sand or mud. These species were comparable with the cockles, venus clams and razor clams of modern seas.

A siphon also characterised many typical Tertiary gastropods such as *Cerithium*. These molluscs had a grooved shell aperture for the passage of the siphon. In contrast, there was a group with no siphon and no groove. Some members of these were widespread, such as *Turritella*, a slim, graceful shell; *Natica*, a fat snail; and *Palundina*, a freshwater form. Finally, a group of gastropods called *pulmonata* developed in which the bronchial chamber had become adapted to air breathing. Examples from this group were *Physa, Lymnaea, Planorbis* (all freshwater) and the land snails.

Large pelecypods were abundant, such as *Venericardia*, a clam, and *Ostrea*, an oyster. Oysters of California were as much as a foot-and-a-half long during the Miocene. Following the extinction of the ammonoids at the end of the Mesozoic the related cephalopods, the nautilods — which still survive — became abundant in

Africa, Trinidad, Peru, and Alabama. Echinoids were important in Cenozoic seas; one of these, the sand dollar, often inhabited the sandy bottoms in large colonies (*Langanum,* France; *Linthia,* New Jersey; *Oligopygus,* Florida).

**Fishes.** Many Tertiary fishes belonged to the same genera as modern fishes, but as the climate in many regions was different they often show different geographical distribution. Tertiary marine sands and limestones contain numerous sharks' teeth, the only part of these cartilaginous fish that can be fossilised. Some were small and of simple triangular shape, while others, huge and with crenellated edges, belonged to a giant species of *Carcharodon,* the man-eating shark, which may have been 60 feet long. Sharks' teeth are abundant in many North American Tertiary formations, such as the Calvert Miocene of the Atlantic coast. Another important horizon is the phosphatic chalk deposit in Morocco, Algeria, and Tunisia where more than sixty different species have been recognised.

The dominant fishes of the Tertiary, however, were the Teleostei, or bony fishes. Archaic forms of the Mesozoic were replaced in the Tertiary by the more evolved members of the group: carp, pike, perch, cod, tuna, mackerel, tench, gudgeon, and flatfish. Their skeletons, well preserved in limestones and shales, can be studied stage by stage to learn the sequence of their evolution. Good specimens come from the Eocene Green River shales of Wyoming, the Eocene beds of Monte Bolca (Italy), the Oligocene of Aix-en-Provence (France) and Glaris (Switzerland), the Miocene of Oeningen (Switzerland), Licata (Sicily), and Oran (Algeria). Aberrant deep-water forms and others with luminous organs are also known.

**Amphibians.** Tertiary amphibians were much like those of today. Frogs and toads impregnated with calcium phosphate have been preserved in the phosphorites of Quercy, France. The lepospondyls had a different type of vertebral column from the other amphibians discussed earlier. They were never large, and they were never numerous. They arose in the Mississippian, culminated in the Permian, and their descendants linger on today as the salamanders and tropical coecilians. A typical Pennsylvanian genus was *Microbrachis* of the order Microsauria. Another order, the Nectridia, may have culminated in *Diplocaulus* from the Texas Permian. It had small jaws, weak legs, and a flat body, but an extraordinarily broad, flat-pointed skull.

**Reptiles.** The reptiles that survived into the Tertiary, apart from the eosuchians, which soon disappeared, represented the four orders known today: rhynchocephalians, turtles, crocodilians and lizards.

The lizard-like rhynchocephalians, of which only the tuatera of New Zealand survives, have been unimportant since the Triassic.

Turtles go back to the Triassic, *Proganochelys* being a characteristic genus. Their adaptations to fresh and salt water, moist woods, hot deserts, and oceanic islands have been excellent. They are among the most ancient of present-day quadrupeds. Oligocene fossil tortoises have been found in the South Dakota badlands and in Egypt. The Pleistocene *Colossochelys* of Asia was one of the largest land genera. The ancient sea-turtle *Archelon* had a twelve-foot flipper spread; *Meiolania* had a two-foot-wide horned skull.

*Protosuchus* of the Triassic was an ancestral crocodilian. The geosaurs were marine crocodilians of the Jurassic and Early Cretaceous. The eosuchian line, to which our present alligators and crocodiles belong, appeared in the Cretaceous. It culminated late in that era when the 40-foot-long Texan *Phobosuchus* with a six-foot skull and some of the most powerful jaws ever known subsisted, in part at least, on dinosaurs.

Snakes and lizards both belong to the order Squamata which is believed to be derived from the Triassic eosuchian *Prolacerta.* Later, in the Mesozoic, such varanid lizards as the mosasaurs grew large and took to the sea. Today most lizards are small; none is marine. The origin of snakes is obscure; although it is assumed that they were derived from lizards, the fossil record is scanty and uncommunicative on this point.

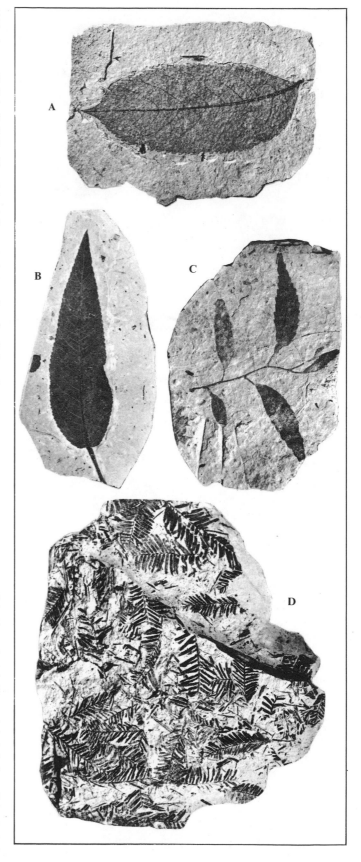

Leaves of tertiary trees.
A. *Juglans magnifica* (walnut). Oligocene of Colorado.
B. *Populus americana* (poplar). Oligocene and Miocene, Western United States.
C. *Zelkova drymeja.* Oligocene and Miocene, western United States.
D. *Metasequoia occidentalis* (dawn redwood). Palaeocene to Miocene, western North America.

391

*Above*
Some Cenozoic molluscs.
A. *Venus tridacnoides,* a 'hard-shell' clam. Interior of left valve. Miocene, eastern United States.
B. *Venus tridacnoides.* Exterior of left valve.
C. *Ostrea titan,* an oyster as much as 50 inches long. Miocene, California.
D. *Lyropecten jeffersonius.* Miocene, eastern United States and California.
E. *Fusus,* a snail partly covered by barnacles. Miocene, eastern United States.

*Below*
Two slender shark teeth from the Pliocene Red Crag formation of England.

**Birds.** In the Cenozoic birds with toothless beaks succeeded the Mesozoic's primitive toothed birds. Many modern groups of birds arose in the Eocene. In North American formations, ancestral or early bitterns, eagles, falcons, grouse, cranes, rails, sandpipers, and owls have been reported. Grebes, geese, vultures, pheasants, and turkeys go back to the Oligocene. Miocene formations contain fossils of shearwaters, boobies, gannets, cormorants, kites, buteos, plovers, murres, and parrots. Pliocene deposits have yielded tree-ducks and pigeons.

In Europe, the phosphorites of Quercy, France, contain mostly birds of prey. At Saint-Gerand-le-Puy, aquatic birds are more common; during the Oligocene, a large lake existed here, comparable with those of central Africa today.

Large running, flightless birds now confined to the southern hemisphere were scattered over other parts of the world in the Tertiary. In the Meudon conglomerate (Eocene), the ostrich-like *Gastornis* was found. *Diatryma,* a heavy, big-billed, flightless, running predator, seven feet tall, lived in Wyoming in the Eocene. In the Miocene, the five-foot-tall *Phororhacos,* with a skull thirteen inches long, was a formidable predator in Patagonia. Ostriches dating from the Pliocene are known from Europe, Asia, and Africa.

## MAMMALS

Mammals are the most adaptable of living creatures. They have a constant body temperature which enables them to withstand both cold and heat. Since they are viviparous they are not subject to the dangers which beset birds and many reptiles during the incubation period. Their relatively large brain gives them a special advantage in the struggle for existence.

**Primitive Mammals.** We know little about the mammals of the Mesozoic. Most were small, and probably both marsupials and placentals were represented. Marsupials were widely distributed in the Cretaceous but aggressive competition from placentals in the Cenozoic permitted them to flourish only in the Australian and South American continents, which by then were isolated from the rest of the world by water. In those two areas marsupials underwent adaptive specialisation and occupied most of the biologic niches that the placentals did elsewhere. Thus in South America there developed the wolf-like *Borhyaena* in the Miocene, and *Thylacosmilus,* a marsupial sabre-tooth, in the Pliocene. In Australia evolution through the Tertiary produced a marsupial 'mole', 'mouse', 'wolf', 'squirrel' (phalanger), 'rabbit' (bandicoot), and 'woodchuck' (wombat), and kangaroos fill in part the biologic niche of ungulates elsewhere.

Marsupials have varied in size from that of a rat to that of a hippopotamus. Extreme examples include the opossum of Montmartre, little bigger than a man's hand, which Cuvier, the early French zoologist, found in a bed of gypsum; and the Australian *Diprotodon,* bigger than a rhinoceros.

In contrast to the primitive mode of gestation of the young in a marsupial pouch, there is in the more advanced placental mammals' gestation in the womb and feeding of the embryo by means of a placenta which joins it to its mother and ensures a well-developed state at birth. Placentals which lived at the beginning of the Tertiary were still fairly primitive, but they constituted the root stock from which present-day forms gradually came.

**Placental Mammals.** Eocene placentals in general had five toes on each foot. Their dentition was complete, the molars had short roots and blunt surfaces, and the brain was small and smooth. Insectivores are regarded as being closest to the ancestral stem. Typical of other early mammals is *Phenacodus,* a primitive ungulate the size of a fox, which lived in Europe and North America in the Early Tertiary. Although early, it is too large and occurs too late to be ancestral to more specialised forms. Other Eocene fossils include ancestral horses, rhinoceroses, carnivores, insectivores, and even lemurs.

While preserving much of their primitive character certain of these early root-stock mammals evolved during the Eocene to produce large specialised forms such as *Coryphodon* of North America and Europe, and *Uintatherium* of the United States.

Amber from Oligocene sands of the Baltic coast. Much of this substance has been collected for use in jewellery and is of interest for the small animals trapped alive and preserved in a perfect state. *Top left:* millipede; *top right:* flies; *bottom left:* spider; *bottom right:* ants. Museum national Histoire.

This second was about the size of a rhinoceros and had three pairs of horns and long dagger-like canine teeth. Its brain was smaller, in proportion, than that of almost any other mammal and was scarcely more evolved than that of Mesozoic reptiles.

## ANCESTRY OF MODERN MAMMALS

Bats go back to the Eocene and are presumed to be derived from insectivores, although fossil evidence is missing. Edentates arose from the Palaeocene palaeanodonts. They culminated in the Pleistocene in South America in such genera as the giant armadillo *Glyptodon* and the huge ground sloth *Megatherium,* contemporaneous with and possibly kept captive by early man.

Rodents go back to the squirrel-like Palaeocene *Paramys.* Beavers go back to the Oligocene; the largest — the size of a small bear — was the giant beaver, *Castoroides,* of the Pleistocene. Rabbits date back to the Palaeocene; the *Eurymylus* of Mongolia was close to the ancestral stock. Whales appear first in the Eocene in *Eocetus* and *Zeuglodon* which were already well adapted for a marine life. They filled the biologic niche of the extinct Mesozoic mosasaurs and ichthyosaurs. Whales must have evolved rapidly as the intermediate record between them and their land ancestors is missing.

**South American Ungulates.** South America was separated from the rest of the world by the sea from the Palaeocene to the Late Pliocene. During this period of time the ungulates in that continent underwent an adaptive radiation similar to that of the more advanced ungulates (hoofed mammals) in the rest of the world. Didolodonts, derived from the ancient condylarths of North

*Top*
*Coryphodon,* an early Eocene mammal about 8 feet long, from the western United States. It may have inhabited swamps, living much like a hippopotamus.

*Centre*
*Hyracotherium,* commonly called *Eohippus.* This four-toed ancestor of horses ranged from western Europe to Colorado and Wyoming. Length about 18 inches.

*Bottom*
*Palaeotherium,* a horse-like mammal about 4 feet high, from Eocene of Europe.

*Right*
*Baluchitherium,* a hornless rhinoceros 18 feet high at the shoulder from the Oligocene of Southern Asia.

America, arrived in South America in the Palaeocene; they gave rise to five orders — the notoungulates, the litopterns, the astrapotheres, the xenungulates, and the pyrotheres.

The notoungulates were the most important and ranged widely in size. The largest, *Toxodon,* resembled a hornless rhinoceros; it lingered into the Pleistocene.

The litopterns included the South American 'horses' and primitive 'camels'. The Pliocene *Thoatherium* had only one toe on each foot, and a hoof like a horse. The Pleistocene *Macrauchenia* was the last of this primitive camel-like line. The astrapotheres and xenungulates developed in a manner parallel to the uintatheres of North America. The pyrotheres suggest proboscideans. When land connection was re-established with North America in the Late Pliocene South America was invaded by more advanced mammals from the north; their competition put an end to the last South American ungulates during the Pleistocene.

**Elephants.** As modern representatives of the order Proboscidea (Greek *proboskis* — trunk) elephants can be recognised by their great size, mobile and prehensile trunk, and massive feet terminating in five digits. Their dentition consists of two tusks or incisors and four enormous molars, the crown of which displays a succession of layers of enamel alternating with ivory and tooth cement. These molars function as rasps with the front-to-back movement of the mandible. They are replaced by new molars five or six times in a lifetime. The skull, of considerable volume, is made lighter by enormous sinuses and contains a well developed brain elephants being the more intelligent of present-day animals.

Against the almost world-wide range of their ancestors the elephants of Asia and Africa are today rare and restricted. Their oldest known ancestor *Moeritherium,* lived in the Eocene near Lake Moeris in Egypt. It was the size of a large wild boar and had two short tusks in each jaw. Its molars, six at the top and six at the bottom, had two transverse crests, each with two large tubercles. It is thought that this animal led an amphibious existence in the lakes and rivers of North Africa and fed on aquatic plants.

*Moeritherium* is replaced in the Oligocene of Egypt by a proboscidean of greater size, *Palaeomastodon.* Different species ranged from three to six feet in height. They had short trunks and short tusks on both jaws. The molars increased in size and had low crowns with blunt-tipped crests. In the Miocene a new lengthening of the trunk and the incisors led to the four-tusked mastodons which lived in Europe, Asia, and North America. As a result of the great development in the lower tusks — almost equal in size to the upper ones — the mandible was elongated and the whole head much longer than that of the modern elephant. The molars generally had four transverse rows of tubercles.

In the Pliocene the four-tusked mastodons were less important than the two-tusked forms. At the same time the true elephants appeared, probably descended from Miocene mastodons. The lower tusks of these Pliocene proboscideans disappeared and the mandible became greatly shortened. The number of the tubercle rows on the surface of the molars increased from four to five; in each row the tubercles fused together into a dental ridge; and at the same time the forehead was raised and the skull enlarged.

This evolution was accompanied by a migration across the world. From Africa, the cradle of the Proboscidea, the mastodons invaded Europe, Asia, North and South America; their last

397

*Left*
Progress among early proboscideans
A. *Moeritherium* stood 2 feet or more high at the shoulder and was the earliest proboscidean that is known. It lived in northern Africa during Eocene and early Oligocene times.
B. *Phiomia*, about 4 feet 5 inches in height, was an Oligocene proboscidean that also lived in northern Africa. It had short tusks on both the upper and the lower jaws, and may have been the ancestor of all later mastodons.
C. *Palaeomastodon*, 3 to 6 feet in height, also lived in northern Africa during early Oligocene times. Its skull and lower jaw were very long.
D. *Gomphotherium* stood 5 to 8 feet high. Its skull was as much as 3 feet long. This was the first proboscidean to wander from Africa to Europe, Asia, and North America. Varied species lived during the Miocene and Pliocene epochs.

*Right*
True mastodons appeared in in Europe during the Miocene epoch. This species, the American mastodon, lived during the Pleistocene Ice Age and died out no more than 10,000 years ago. The animal ranged from 7 feet to 9 feet 6 inches in height.

representatives were contemporaneous with the first men. At the end of the Pliocene and the beginning of the Quaternary mammoths and elephants came into being. The imperial mammoth reached 14 feet in height and had curved tusks. Molars of some species of late proboscideans had as many as fifteen ridges, the intervening spaces of which were filled with cement. As the crown wore away it revealed the central ivory.

In addition to these direct ancestors of the elephant, an aberrant proboscidean, *Dinotherium* (Greek *deinos* — terrible, and *therion* — a wild beast) arose in the Miocene and lived into the Pleistocene. It had two downward-curving tusks on the lower jaw that may have been used for digging. The molars numbered five on each half jaw and had two transverse ridges, rather like the teeth of tapirs. The largest specimens were about the size of a large African elephant.

**Rhinoceroses.** From ungulates with five digits to those with only three, and from animals with trunks to those with nasal horns — in other words, to the rhinoceroses. In the Oligocene there were ancestral rhinoceroses whose smooth, thin nasal bones plainly did not support horns. Incisors and canines were present in both jaws, and the molars exhibited folds of enamel in a fairly simple pattern. The back feet had three digits; the front ones still had four, though one was in progress of regression.

In the Miocene, the true horned rhinoceroses were represented by a considerable number of species in which the nasal bones gradually thickened and on the upper jaw (maxillary) were even joined together by a septum. The incisors and canines disappeared except in certain modern species in Asia. The folds of enamel on the molars became more complicated. The front feet lost their fourth digit and became similar to the back ones. The most spectacular genus was the Asiatic; it was 18 feet tall at the shoulder, and its head (without a horn) was carried on a long neck.

**Tapirs.** The most primitive tapirs, the small lophiodonts, appeared during the Eocene. They had four toes on the front feet and three toes on the hind feet. These animals gave rise in the Oligocene to *Protapirus*, in the Miocene to *Miotapirus*, and in the Pliocene to *Tapirus*, the genus still living today. Evolution of the tapirs involved, principally, an increase in size, increasing complexity in the pre-molars, and a retraction of the nasal bones associated with the development of a small flexible proboscis. Modern tapirs, however, are perhaps the most primitive odd-toed ungulates living.

Tapirs were widely distributed during the Pleistocene but today they are restricted to Central and South America and Malaya. This is a classic example of discontinuous distribution and shows that tapirs once occupied the intervening areas—a fact supported by the fossil record.

**Horses.** Of all mammals, the horse is the best organised for running over plains and eating grass. Its adaptation to running was effected by elongation and straightening of the limbs and by reduction in the number of toes. Adaptation to a herbivorous environment is indicated by the lengthening of the neck for grazing, the disposition of the incisors into sharp blades for cutting grass, the flat form of the molars for grinding, and finally the elongation of the molars so as to last throughout the life of the animal. These features did not appear all at once, but were produced in the course of an evolution whose successive stages have been discovered in the Tertiary formations of the western United States.

Without making a study of all the ancestors of the horse, we can at least state that in height they varied from that of a small dog to that of the modern horse; that their feet, which at first had three or four digits, lost their lateral digits until only the median was preserved; that molars with blunt tubercles gave way to molars with ridges — the molars with 'wearing plates' similar to those in the horses of today.

In the beginning, horses with four and three digits were adapted to walking on the soft ground of a climate that was humid. Teeth with tubercles or ridges correspond to a diet of buds, branches, and leaves — forest-type vegetation. There is a considerable

analogy between these forest-living horses and the wild boars and tapirs of the present day.

From the Miocene onward, dryness succeeded humidity, and the forests gave way to immense grasslands. The molars of horses evolved into rasps suitable for grinding the tough stalks of the grasses. Three-toed feet gradually evolved into single-toed feet better suited for running over hard ground. It is remarkable that after a long and perfect evolution in the North American continent horses disappeared from there for many centuries, to be reintroduced later from Europe by the Spanish conquerors.

While the series of the horse's ancestors is complete in North America and includes *Hyracotherium*, *Orohippus*, *Mesohippus*, *Parahippus*, *Protohippus*, and *Pliohippus*, the Old World possessed only those branches which, detached from the main trunk, had migrated by the Asiatic route across the isthmus that occupied the site of the present Bering Strait from the Pliocene to the Pleistocene.

*Palaeotherium*, discovered in the gypsum of Montmartre, and famous since its bones were used by Cuvier to establish the foundations of vertebrate palaeontology, corresponded approximately to primitive American horses, but was of heavier build. Its restoration shows an animal resembling the tapir in the general shape of its body. It had three digits on each foot, and its molars had a 'wearing plate' of fairly simple design. The palaeotheres died out in the Oligocene.

*Hipparion* spread widely from the end of the Miocene onward in Europe, Asia, Africa and North America; its remains are abundant. It became extinct in the Pleistocene.

**Porcines and Ruminants.** Among the perissodactyls (odd-toed ungulates), the median digit is always the largest. The porcines and ruminants differ in this respect: two digits (digits III and IV, counting the thumb as the first digit) remain more or less equally developed and may or may not be accompanied by digits II and V.

The porcines have four developed digits, a complete set of teeth, molars with blunt tubercles, and a simple stomach not suited to rumination. They include the pigs, wild boars, and hippopotami. Their Tertiary ancestors were for the most part amphibious. Among these were the anthracotheres that are thought to have lcd a semi-aquatic existence in swamps.

Ruminants have only two digits, well developed and resting on the ground, although a survival of the lateral digits can be seen today in the musk-deer and the Cervidae (deer with antlers). Their dentition is incomplete; living representatives lack upper incisors and canines. The molars are high and with use develop a flat crown with a 'wearing plate'. They have a multiple stomach suitable for rumination. With the exception of this last feature, which cannot be observed in the fossilised state, all other characteristics are known to have evolved gradually since the beginning of the Tertiary era.

The primitive North American oreodonts (Eocene-Pliocene) had four digits on each foot (like pigs), teeth that formed an almost continuous series, and molars with crescentic cusps.

The small Eocene *Protylopus* was an early camel, a group that increased in size until the giant Pliocene *Alticamelus* stood ten feet above the ground. The llamas are modern South American camels, distinguished from the extinct primitive camel-like *Macrauchenia* mentioned earlier. The small *Archaeomeryx* of Mongolia was an ancestral Eocene ruminant. One branch developed into the deer, of which group the giant Irish elk had the largest antlers. It became extinct in the Pleistocene. The closely related giraffes branched from the deer line in the Miocene; the okapi represents a primitive type. The sivatheres of the Pliocene and Pleistocene were heavy-bodied giraffes with broad bony horns; they may have been known to the Sumerians.

The bovids — sheep, goats, bison, antelopes, cattle — arose in the Miocene. *Eutragus* was an early representative. Some became huge, such as *Bison crassicornis* of the North American Pleistocene. The Miocene *Merycodus* is ancestral to the North American pronghorn. Africa is the present centre of abundance and variety in members of this group, but with the advance of civilisation their future is in doubt.

*Left*
Fossil skeleton of *Megatherium*.

*Right*
*Smilodon,* a Pleistocen sabre-toothed cat of North America. Its remains are common in tar pits in Los Angeles, California. Height at shoulder about 40 inches.

**Carnivores.** The earliest flesh-eating mammals in the Palaeocene, Eocene, and Oligocene were the small creodonts or primitive carnivores (for example, *Pterodon*), and they had a full set of teeth. The modern carnivores, or Fissipedia, also arose in the Palaeocene.

In the creodonts, the molars were beginning to be specialised for seizing prey and breaking bones; the feet were adapted for gripping, the brain was small. Examples of these are *Hyaenodon* and *Pterodon*, abundant in the phosphorites of Quercy, France, and also in North America. The shape of their teeth suggests that *Hyaenodon* ate fresh meat while *Pterodon* ate carrion, breaking the bones with the flat crushing 'heel' of its molars.

In contrast with the creodonts, modern carnivores have a larger brain and developed a sharp cutting tooth, a modified molar — the carnassial — on each side of each jaw. The creodont-like but larger-brained miacids (dog-like creatures) arose in the Palaeocene and gave rise in turn to the modern carnivores. *Cynodictis* of the Late Eocene was an early dog. *Dinictis* of the Oligocene was an early cat. The Oligocene *Hoplophoneus* was the first of the sabre-tooth cat line which later became extinct with *Smilodon* in the Pleistocene. *Phlaocyon* was the first raccoon, *Plesictis* the first weasel. The Eocene *Stenoplesictis* was an early civet differing little from the modern genet, regarded as close to the ancestral carnivore stem. *Ictitherium* of the Miocene was the first hyena.

Seals and walruses, known collectively as pinnipeds, first appear as fossils in the Miocene, though they may well have arisen earlier. Their adaptation to water is less complete than is that of the whales or sea-cows, for the pinnipeds must come out on land or ice to breed.

**Apes and Lemurs.** The apes, monkeys, and lemurs constitute the order of primates characterised by relatively enormous brain development. Their limbs have hands with opposable thumbs and flat nails. Their dentition is of a primitive type corresponding to an omnivorous diet. The lemurs, the most primitive primates, at present restricted to Madagascar and Indo-Malaya, were formerly widespread in Europe and North America. *Plesiadapis* is a typical Palaeocene genus. The Eocene *Notharctus* is not widely different from the modern lemur. From the Eocene onward we might find among early lemurs the root stock from which the human species has arisen.

Monkeys populated most of the continents before being limited to Asia, Africa, and South America. An early genus was the small Oligocene *Parapithecus*. *Propliopithecus,* a primitive ape, appeared in the Oligocene in Egypt. Evolution in the apes has been characterised by an increase in size of the body and of the brain. The Miocene of Africa has yielded *Limnopithecus*, regarded as ancestral to the gibbons, and *Proconsul*, held to be ancestral to the Eurasian *Dryopithecus* and the other apes. The Miocene of France has given us the tailless apes *Pliopithecus* and *Dryopithecus* with a dentition close to that of the great apes of today.

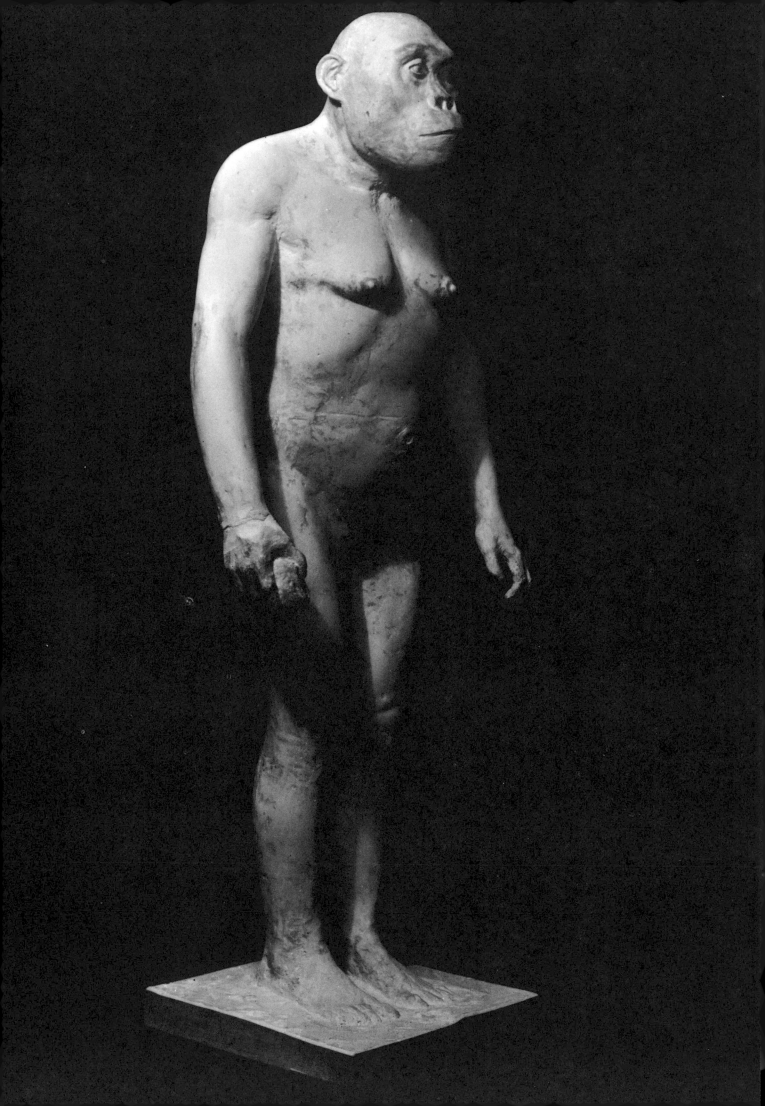

# Life in the Cenozoic: Quaternary

## ANIMAL AND PLANT MIGRATIONS

The essential difference between the faunas and floras of the Pleistocene and those of today is their distribution throughout the world. It is clear that the alternation of glacial and interglacial periods must have influenced both vegetation and animal population. Cold floras and warm floras alternately prevailed. Finally the glaciers drew back to their present-day limits and the present climate was established. Today's fauna and flora have had to adapt themselves to these new conditions in the course of the ten to fifteen thousand years since the last glaciation.

The periodic changes in the distribution of plants and animals were brought about by extensions of range and by migration. When, year by year, the cold became more marked it encouraged the spread of plants and animals typical of the northern regions, while at the same time it induced the retreat of others to southern regions. When the climate began to warm up again the process was reversed. At the present time we are in a period of warming up; already, for example, cod and herring do not travel as far south as they did a hundred years ago, while sardines, mackerel, and tuna are gradually moving north.

If the past floras and faunas of France, where postglacial biotic changes have been closely studied, are compared with those found there today, the species can be divided into five groups: those which have stayed on (as the red deer); those which migrated toward the north (as the reindeer); those which took refuge in the high mountains following the line of the retreating alpine glaciers (as the chamois); those which emigrated to the south (as the hippopotamus); and those which finally disappeared (as the mammoth).

**Warm Faunas.** The warm faunas were those of the interglacial periods. In the first interglacial several species of elephants, rhinoceroses, and hippopotami appeared, to which horses and deer were added later. This fauna was a legacy of that which inhabited France during the Tertiary. It contained the formidable sabre-tooth cat, *Machairodus*, not found in any subsequent interglacial. Also present was the early mammoth, *Mammuthus meridionalis*. This animal was about 12 feet high at the shoulder and had long curved tusks and molars made up of parallel layers of enamel.

Before the mammoth had disappeared the straight-tusked *Loxodonta antiqua* arose. This elephant was more imposing; it measured 14 feet at the shoulder and 15 feet 6 inches to the top of the head. It seems to have been ancestral to the modern African elephant. In Europe other descendants of *Loxodonta antiqua* were the dwarf elephants, varying in height from 3 to 6 feet. Their fossil bones have been found in the Mediterranean islands of Sicily, Malta, Cyprus, and Crete. Their occurrence in island areas implies the former existence of bridges of land joining them to Europe or North Africa.

**Cold Faunas.** With the return of the glaciers the warm fauna migrated to the south and a cold fauna took its place. This was made up of those elements of the warm fauna able to adapt themselves to the cold, and of new elements which moved southward ahead of the glaciers.

Among these were the 10-foot-high woolly mammoth, *Mammuthus primigenius*, an elephant adapted to living in a cold climate. Its tusks and molar teeth are found in great numbers in many Pleistocene terrains. Certain localities are veritable mammoth cemeteries. The frozen lands of Siberia contain so many of their remains that fossil ivory has long been worked there. On the Siberian coast and in the valleys of the Ob, Yenisei and Lena deposits are continually uncovered by river erosion.

When well-preserved mammoths have been discovered intact in frozen soil, dogs and wolves have been fed on the flesh of their corpses. Many such discoveries have been made and a number of natural history museums exhibit pieces of skin and the soft parts of such mammoths. Serological studies of the refrigerated blood have shown that the woolly mammoth was more closely related to the present-day Asian elephant than to the African elephant. This is also confirmed by the mammoth's concave forehead and small ears, and by the arrangement of the layers of enamel on its molars. Two fatty lumps on its head may have been used as sources of reserve energy in times of food shortage. Traces of a thick fleece of grey wool mixed with reddish hairs can be distinguished on fragments of skin. Examination of stomach contents shows that the mammoth fed on pine needles and cones, leaves and seeds of herbaceous willows, dwarf birches, and many other cold, humid-type plants.

The mammoth lingered into the post-glacial stage in Eurasia and east-central North America. It became extinct in Europe during the time of the Magdalenian culture (topmost Quaternary). It was a contemporary of early man and was well known to the prehistoric inhabitants of the valley of the Vezère. Drawings of it are not uncommon on cave walls; the cave of Combarelles in the Dordogne area of France contains a superb representation in which all the mammoth's characteristics can be seen—its long recurved tusks, its occipital hump, and its woolly hair. The reasons for the creature's extinction are not known; man may have had something to do with it. Whatever the reasons, it disappeared not long after retreat of the glaciers.

The woolly rhinoceros was another member of the last glacial fauna. It became extinct in Aurignacian times. It is known from many fossils, from at least one carcass found frozen in the ice of Siberia, and from one preserved in a sort of fossil paraffin in Poland. It was about the same height as the modern rhinoceros of South Africa but, like the mammoth, was covered with thick wool mixed with hairs. Unlike the woolly mammoth, the rhinoceros did not reach North America.

With the increasing cold of oncoming glacial stages many creatures, including man, took refuge in the caves of limestone

*Australopithecus.* Tool-making hominid, one of the oldest known human beings, lived in Africa over 500,000 years ago. Height about 4 ft. Restoration by Maurice Wilson based on female skeleton from Sterkfontein, Transvaal.

districts. As a result deposits accumulated; the caves of France, for example, have delivered up the bones of numerous animals: lions, hyenas, and cave bears. The lion and hyena were related to species now living in Africa. The cave bear was the biggest and most widespread of the cold fauna carnivores and its bones have accumulated by the thousands in caves. Other animals that lived in France during the last glacial period now live in northern Europe and America: they include the reindeer, moose, musk ox, wolf, arctic fox, and wolverine.

South of the ice-caps lay a bleak, frozen tundra where only lichens and short plants grew; it was an area inhabited by wolves and reindeer. Farther from the ice came the taiga, the forests of conifers and birch, inhabited by the mammoth and woolly rhinoceros. The fact that their carcasses are today found entombed in the frozen swamps of the tundra suggests either that individual beasts wandered there and perished, or that the present tundra occupies the site of a former taiga. In North America mastodons were swamp dwellers.

The existence of glaciers demanded a cold, humid climate with plenty of snow. As the climate became drier or warmer the glaciers began to retreat, and as the retreat continued the cold faunas of tundra and taiga were succeeded by a steppe fauna. Winds from the ice-cap and its outwash plain would blow clouds of glacial dust or loess over the steppes each summer. On this fine fertile silt new growths of grass would flourish each spring, attracting herds of horses, mammoths, and bison to graze on the abundant herbage.

Other inhabitants of the steppes were the taiga antelope, the hamster, and the lemming. The steppe fauna was followed by a forest fauna which included the red deer, fallow deer, wild boar, fox, badger, squirrel, rabbit, beaver, European bison, and wild ox.

Beyond the forests a humid climate favoured the development of peat-bogs. From Pleistocene peat we know of *Megaceros*, an Irish elk, a giant open-country deer with antlers up to 13 feet wide, the largest of any known species. For the most part its remains have been found in the peat-bogs of Ireland.

## THE PLEISTOCENE IN SOUTH AMERICA

Before it became re-united to North America in the Pliocene by the isthmus of Panama, South America was inhabited by strange animals, different from those living elsewhere at that time. Notable among these were the edentates. They had simple teeth without enamel, or no teeth at all; claws at the end of their toes were used for digging.

The greatest of all, *Megatherium*, was truly the 'great beast' that its name implies. It was a giant sloth, about 20 feet long and weighing 5 tons, with massive hind-quarters, a heavy tail, and thick-set bones; it probably often stood upright.

The related *Mylodon* was no taller than an ox. It differed from its larger cousin in having a relatively longer tail and a skin that contained little pieces of bone called ossicles. Fragments of its skin have been found in Patagonia; they are nearly an inch thick, covered

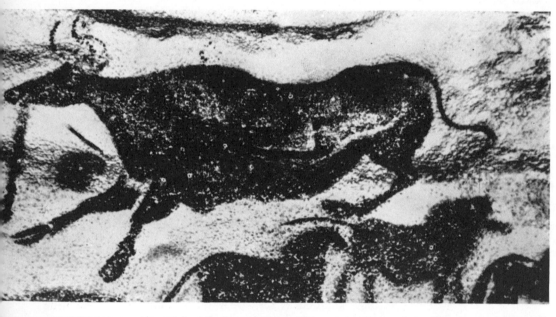

*Top*
A rock painting from the caves at Lascaux.

*Bottom*
The woolly rhinoceros of Ice-Age Europe and Asia. It was hunted by early man, who also painted its picture in caves.

**Top**
*Megatherium,* largest of the giant ground sloths, and two kings of glyptodont. These animals lived in South America during the Pleistocene. *Megatherium* and *Glyptodon* wandered into North America.

**Centre**
*Loxodonta meridionalis,* an early Pleistocene elephant of Europe, was closely related to the present-day African species.

**Bottom**
The woolly mammoth ranged Europe, northern Asia and North America during cold portions of the Pleistocene Ice Age. It was sometimes trapped and preserved in ground ice. In North America this and other species of mammoth lived until 10,000 or 11,000 years ago.

with long hairs, and contain ossicles arranged in a mosaic. A
closely related genus was kept in confinement by early man, and
fragments of skin were found in a cave whose entry was restricted
by a wall. The soil forming the floor of the cave contained the
remains of cooking, the excrement of a herbivorous animal, and
straw for its fodder. In a niche in the wall was a human skeleton.
The folklore of the Tehuelche Indians, the present-day inhabitants
of Patagonia, describes an animal which may well have been
this giant sloth.

Another fascinating edentate from South America, *Glyptodon,*
a giant cousin of the armadillo, was also a contemporary of early
man. Its body was hidden by a rigid carapace about five feet long,
slightly resembling that of a tortoise. The head and tail were also
protected by plates and rings of bony armour. One glyptodont, the
group to which *Glyptodon* belonged, had a spiny ball at the end of the
tail. Charcoal, cinders, worked flints, and calcined bones of
*Glyptodon* itself, broken open to yield the marrow, have been found
beneath *Glyptodon* carapaces. This genus also roamed parts of
North America.

Until the seventeenth century the elephant-bird, *Aepyornis,*
lived in the swampy regions of the central Madagascan Plateau.
A short-winged bird about as tall as an ostrich but weighing three
times as much, it laid eggs of two-gallon capacity—the equivalent of
six ostrich eggs or 130 hen's eggs. The inhabitants knew it by the
name Vouroupokra and may have helped in its extermination.

The largest bird that ever existed was *Dinornis maximus,* the
giant moa of New Zealand. It attained a height of more than 10 feet
and had completely lost its wings. Its feathers were thin and stringy.
For defence it lashed out with its huge feet with their three large
toes, each bearing a thick claw. According to Maori legends describing the hunting of the moas, these birds lived until about the
fourteenth century.

## FOSSIL MAN

The physical development of man from an ape-like ancestor involved
the development of a large brain. This was associated with the
development of a forehead and chin, and a 'straight-line' face, rather
than one with jutting jaws. Increased brain weight could best be
supported on an upright body; this involved a shifting of internal
weight, stresses and supports, and the use of only two limbs for
progression. This in turn freed the arms and hands for other uses,
such as the wielding of weapons and tools. Of exceptional importance was the development or retention of an opposable thumb,
which enables man to grasp and hold objects. Upright posture
enabled him to see farther, and reliance on the sense of sight increased
while reliance on the sense of smell declined. Important, too, was
the development of excellent co-ordination between eye and hand.

The special attribute of man was and is his remarkable brain,
which has enabled him to seize dominion over the rest of the living
world. In a number of respects—ten fingers, ten toes, and his
dentition—man remains relatively unspecialised. This proved an
advantage in his struggle for dominance, for even a huge brain in
a specialised body with flippers like a whale, hoofs like a horse, or
atrophied eyes like a mole, would never have allowed him to build a
house, draw a bow, or learn to write.

These physical developments are often traced to the arboreal
existence of man's anthropoid ancestors. A life of swinging through

trees was aided by retention of four fingers and an opposable thumb; by long, strong arms; and by acute vision that co-ordinated with the use of the arms and hands.

That man's ancestors forsook such an existence is attributed, in part at least, to climatic changes that replaced much of the forest with grasslands. In such an environment only a ground-walking or ground-running animal could survive. Structurally man is relatively defenceless and cannot outrun predators, so his survival hinged on sighting enemies from afar (erect posture and keen vision), on his craft in eluding them and his skill in securing prey by the use of hand-thrown or hand-held weapons or traps.

The more immediate ancestors of modern man are sought among the advanced Pliocene apes such as *Dryopithecus*. One Pliocene anthropoid that has attracted attention is the structurally primitive *Oreopithecus*, fossils of which have been found in coal measures at Bassinello in central Italy. *Oreopithecus* had a vertical face, and no muzzle; its teeth were more man-like than ape-like, and the males were larger than females. In size it was a little larger and heavier than a gibbon. One school at least is convinced that it belongs in the family Hominidae, and regards it as an early, aberrant member of the family of Man.

With the advent of the Pleistocene, man-like apes gradually gave way to ape-like men. The earliest fossil that can be classified as an ape-man is that of *Australopithecus* of South Africa; the remains date back about one million years. His brain-case was about 600 cc. in size, just a little larger than that of a gorilla. He walked upright, produced and used 'tools of bone and horn, and lived in caves. *Paranthropus* lived in the same region at about the same time and may, or may not properly be classified in a different genus.

*Australopithecus* was first named in 1925 from the skull of child of six found at the limeworks at Taung, near Kimberley, South Africa. His position as a hominid, a member of the family Hominidae that includes Man, was disputed by authorities for a long time but was finally accepted. In 1936 and 1938 a brain-case and parts of another skull were found at Sterkfontein and at Kromdraai, both in South Africa. These fossils were secured by Dr Robert Broom, who found that they differed from the Taung skull and named them *Paranthropus*. After World War II, other *Paranthropus* skulls were found in a nearby deposit at Swartkrans which yielded parts of thirty-five different australopithecines. In 1947 in a limestone cave in the valley of Makapansgat, near Potgietersrust, about 150 miles from Sterkfontein, part of another australopithecine skull was found, the first of many to come from there.

Structurally, the australopithecines could well be directly ancestral to man. *Paranthropus* looks like a somewhat specialised offshoot. *Australopithecus* was a meat-eater or was omnivorous, while *Paranthropus* may have been a vegetarian. In 1959 Dr and Mrs L.S.B. Leakey found in the Oldoway Gorge, Tanzania, in a formation associated with crude stone tools, a larger australopithecine which they named *Zinjanthropus*.

A significant advance in human evolution is displayed by *Pithecanthropus*, or Java Man, whose fossils were found in 1892 near Trinil, along the Solo River in Java, in Middle Pleistocene deposits estimated to be about 350,000 years old. The brain-case of this Java Man was 850 cc. in size, a substantial increase over that of *Australopithecus*. He was about the size of modern man, walked upright, had a skull, no forehead, beetling brows, projecting face, big jaws, and large teeth. Pithecanthropine remains have been found in China, Algeria, Tanzania, and in Java, and it may well be that *Pithecanthropus* occupied the entire Old World.

Pekin Man was originally named *Sinanthropus pekinensis* from fossil teeth taken from a cave at Choukoutien, about twenty-seven miles from Peking, China. The first full skull was found in 1929. Before World War II put an end to the excavations, fossilised parts of more than forty individuals had been taken from the site.

Many scientists now place Pekin and Java Man in the same genus. The name of the former, claim many anthropologists, should be *Pithecanthropus pekinensis*. Pekin Man lived during the Middle Pleistocene or second interglacial. He walked upright, had a thick skull, beetling brows, a retreating but more pronounced forehead than Java Man, and a brain-case of 1,075 cc. He also had a shorter jaw than Java Man, and his teeth were smaller and more human. Pekin Man was omnivorous and sometimes cannibalistic; he used

fire, stone and bone tools. He belonged to the Chellean culture.

Heidelberg Man, originally described as *Homo heidelbergensis*, lived some 250,000 years ago. He was first known from a lower jaw found in 1907 in the Mauer district, six miles from Heidelberg, Germany. The jaw was large, ape-like, and chinless, but contained small, human-type teeth.

Solo Man, perhaps not too distantly removed from Heidelberg Man, comes from a later deposit near Trinil, quite close to the place where *Pithecanthropus erectus* was first found. In 1931 and 1932 these sediments yielded various other early human skulls with brain-cases larger than Java Man. The skulls had apparently been broken so that the brain could be removed and eaten. It is assumed therefore that Solo Man was a cannibal.

Also similar to Heidelberg Man is *Telanthropus*, found at Swartkrans, South Africa, in the same sediments with *Paranthropus*. The lower jaw resembles that of Heidelberg Man; the teeth suggest Java and Pekin Man. Other Middle Pleistocene jaws, found at Ternifine, in Algeria, by Professor Camille Arambourg in 1954 and 1955, resembled those of Pekin and Java Man but were associated with western hand-axe tools, not the chopper tools of the East.

Rhodesian Man discovered in 1921 at Broken Hill, Zambia, had a cranial capacity of 1,300 cc. His skull, several other bones, and another upper jaw were found. The skull lacked the thickness of that of Java, Pekin, or Solo Man but it had deep brow ridges. A similar skull was found several years later at Saldanha Bay, 80 miles north of Cape Town. It is dated as Gamblian, an African stage contemporaneous with the fourth glacial of the northern hemisphere. Associated tools included late hand-axes.

Neanderthal Man, the next main human type, appeared about 150,000 years ago, at the beginning of the Upper Pleistocene. He is distinguished by short stature, thick body, big projecting face, a low, large, backward projecting skull, prominent brow ridges, a receding forehead and chin, and brain size similar to that of modern man. He knew the use of fire, wore clothes, lived in caves, used a wide variety of tools, and buried his dead. He appeared during the last interglacial and lived on through the last glacial stage. His kind was widely scattered over the Old World, from Germany to Cape Town and from Spain to Java. Important sites are found in Israel, Iran, and Uzbekistan. He disappeared before or at the end of the Ice Age.

Neanderthal Man takes his name from the valley of the Neander, near Düsseldorf, Germany, where the first bones were found. Since their discovery in 1856, many others have been uncovered in Europe, Asia and Africa, mostly in cave deposits. The cave at Moustier in France has given its name to the Mousterian stage, as the Neanderthal culture is called.

Cro-Magnon Man, the prototype of our present *Homo sapiens*, appeared about 50,000 years ago. The average size of his brain-case was 1,450 cc. He was generally tall, walked upright, had a well-developed forehead and chin, and lacked prominent brow ridges. His earliest culture is classified as Aurignacian. Discoveries made in France, Italy, Egypt, and Morocco have established that modern man was contemporary with Neanderthal Man 50,000 years ago. They have established, too, that he may have been responsible for the elimination of Neanderthal Man by interbreeding, destruction, or competition that was too aggressive. Skulls found at Swanscombe, England, and at Steinheim may have come into existence about the time when both the Neanderthal and Cro-Magnon men appeared, or shortly afterwards. Both skulls are suggestive of modern man; both are from deposits of the second interglacial stage.

In the course of prehistoric time *Homo sapiens* was represented by four different types: Grimaldi, Cro-Magnon, Chancelade and brachycephalic.

The Grimaldi type had a long skull (dolichocephalic), and was so named because it was discovered during excavations carried out by the Prince of Monaco in the caves of Grimaldi near Menton. Two skeletons were found there, one of a woman, the other of a teenage boy. The somewhat bulbous forehead and the projections of the tooth-bearing, bony gum suggest that they were negroid.

The Cro-Magnon type, also long-skulled, was thus named because the first five skeletons were unearthed during excavations

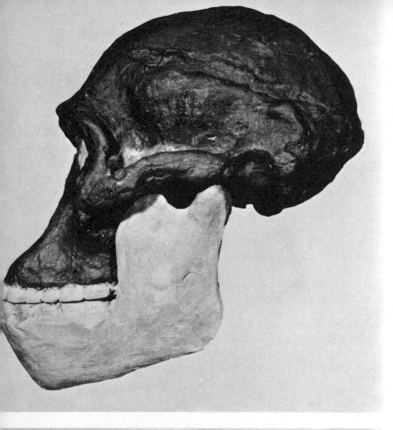

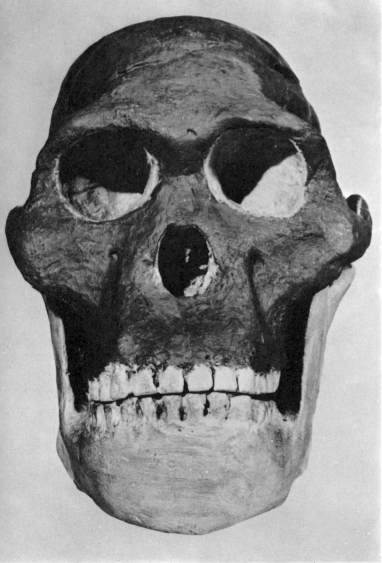

Skulls of *Australopithecus transvaalensis*, found at Sterkfontein.

of the rock shelter at Cro-Magnon, France. Similar types have since been found all over the Dordogne, in Charente, Saône-et-Loire, Belgium, England, and in Germany. Although they appeared earlier than the Grimaldi type, the Cro-Magnon types persisted in Europe for a much longer period. Cro-Magnon Man was distinguished by high stature (6 feet), athletic body, a thin nose, robust chin, long limbs, and prominent heels.

The Chancelade type is also long-skulled, named after Chancelade, near Perigueux (France), where the skull was discovered. Its short stature (5 feet 3 inches), high, wide face, and prominent cheek bones were once thought to show fairly close affinities with the modern Eskimos but this theory is now questioned.

A type with a round skull (brachycephalic) came into Europe from Asia about 6,000 years ago. This type, less artistic, had a higher level of industry than the dolichocephalic types. It brought in new ways of life, notably the use of tools and weapons of polished stone. Later still, about 3,500 B.C., other brachycephalic peoples brought with them a knowledge of the working of metals.

Since that time one group of dolichocephalic people (Cevenols) has been preserved almost unchanged in certain mountainous regions; another spread into Germany and Scandinavia (Nordic types); a third became established in the south of France, Italy, Spain, and Portugal (Mediterranean types). Others have either disappeared or contracted alliances with the newcomers. The mixture of brachycephalics and dolichocephalics produced an intermediate group of mesocephalics, who also contribute to the present complexity of the human species.

Some scientists think that early man can be classified more simply. Ernst Mayr, an eminent palaeontologist and systematist of Harvard University, said in 1956 that if the same standards of form and structure were used in classifying man as are used in classifying the rest of the animal kingdom, all men could be put into three species: *Homo transvaalensis*, which would include *Australopithecus, Paranthropus* and similar forms; *Homo erectus*, which would include *Pithecanthropus, Sinanthropus*, and similar forms; and *Homo sapiens*, which would include Neanderthal and Cro-Magnon Man, together with all the other races.

## MAN'S PLACE OF ORIGIN

If, as most palaeontologists think, man derived from a small group of Miocene apes, *Dryopithecus* or others, the problem then is to pin-point the place of his origin. Following the discoveries of *Pithecanthropus* and *Sinanthropus,* many thought that man originated in the Far East. The subsequent discoveries of *Australopithecus* and *Paranthropus* focused attention on Africa, where the oldest palaeolithic industries are found. Professor Camille Arambourg, the French anthropologist, wrote:

'The Black Continent instead of Asia could well have played a foremost rôle in the evolution of humanity. It must not be forgotten that this very old continent has always been one of the most important centres of evolution. To it we owe the appearance and development of numerous zoological groups. In particular we have seen that the anthropoids arose there, and that all the first representatives of the branch with human affinities lived there at the beginning of the Miocene.

'It would appear logical, if not certain, to suppose that *Homo sapiens,* lacking a protective cover of hair, could only be born in a relatively warm climate or, at least, one sufficiently temperate. It was the use of clothes which later enabled him, as he migrated farther from his place of origin, to face the more rigorous climate of the glacial period... We are not far wrong in thinking that it is perhaps on the southern shores of the Mediterranean, encouraged by the warm climate and the abundance of nature, that modern man was able to break away from his ancestral ties. Thus the basin of the 'divine sea with its sonorous shores' from which, one day, must have come out of a mixture of races the Oriental mystics and philosophies and our Hellenic culture, could also have been the crucible where, at the dawn of time, the elaboration of thought was achieved.'

**Man in America.** Although man's history in America goes far beyond the ten thousand years once considered the continent's maximum, it still does not equal the antiquity of the fossils and implements found in the early Pleistocene deposits of Europe, Africa and Asia. There are very few skeletal remains of America's earliest inhabitants, and most knowledge of his presence has been gained from weapons and tools. Particularly significant are flint implements used by Folsom (New Mexico) Man, distinguished by the longitudinal groove or fluting on each side of the points. They are contemporaneous with extinct species of bison *(Bison figginsi)* and mammoth. Carbon-14 tests show that these objects are eight to ten thousand years old.

At Lindenmeier, north of Fort Collins, Colorado, two thousand Folsom-type flints were found associated with the burned bones of many bison. Several other sites in North America have yielded important discoveries of early man. Vero, on the east coast of Florida, has produced some human foot bones and parts of a human pelvis in association with bones of an extinct elephant, mastodon, horse, and ground sloth, some of which show marks made by flint knives, presumably made when the flesh was being cut off. Nearby Melbourne, in Florida, has also yielded human bones and artefacts, including pottery, as well as a fauna similar to that found at Vero. In a gravel pit near Frederick, Oklahoma, flint implements have been found with the bones of extinct tapirs, elephants, camels, glyptodonts, and ground sloths.

The prehistoric culture of Incas, Mayas, and others in Central and South America was preceded by a different, older culture. This may have been developed by the first human inhabitants of the western hemisphere, who arrived there from Asia by way of a land bridge over the Bering Strait about 25,000 years ago. At that time the sea level had been lowered perhaps 300 feet when the water froze in the glacial ice during the Pleistocene. Actually a drop of only 100 feet would have created a land bridge across Bering Strait between Alaska and Siberia.

## THE HOAX OF THE PILTDOWN MAN

In 1908 unusual flints were discovered in gravel being used for road building near Piltdown, Sussex. The finder, Charles Dawson, cautioned the road workers to be on the lookout for other material. The men soon came up with a bone fragment, part of a skull. A few years later other fragments, apparently parts of the same skull, were found. Then a piece of jawbone turned up; all were described as *Eoanthropus dawsoni,* or the 'dawn man'. To the credit of many palaeontologists and anthropologists of the time, the authenticity of the finds was argued at length because the skull appeared to belong to a much more advanced type than the contemporary pithecanthropoids, and the jaw much less so.

The answer to the riddle was found in the 1950s in the results of radioactive and fluoride tests. It was discovered that the bones and artefacts had been treated with chemicals and stains to stimulate an antique appearance. The older, genuine bones and teeth of extinct animals have a high fluorine content, whereas younger bones have no fluorine. The tests proved the jawbone of *Eoanthropus dawsoni* to be that of an orang-outan, a living species. Thus ended one of the classic hoaxes in palaeontology.

## PREHISTORIC INDUSTRY

An impressive demonstration of the inventive mind of man was the fashioning of tools and weapons. At the outset our most distant ancestors had used clubs and sticks broken from trees; later they used stones which they shaped and then polished. Other materials, such as bone, horn, and ivory, were also used. Finally, at the dawn of historical times, metals were introduced for the making of primitive implements.

The earliest example of hominid industry is found in the bone and horn culture of the australopithecines. Exhaustive study and research by Dr Raymond A. Dart and his associates in South Africa over many years have shown that these ape-men used bone and horn long before man started making stone tools and weapons.

Anthropologists divide the subsequent stone industry times into the Palaeolithic or Old Stone Age, and the Neolithic or New Stone Age. The first is characterised by chipped stone implements of a primitive culture, the second by polished stone implements.

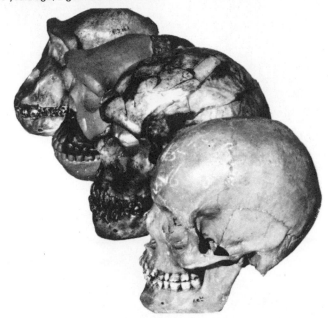

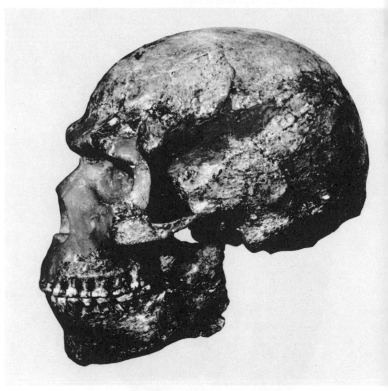

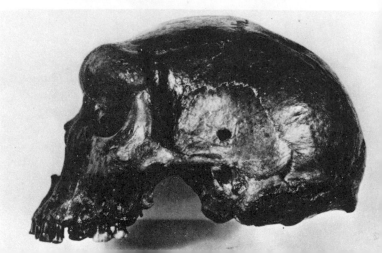

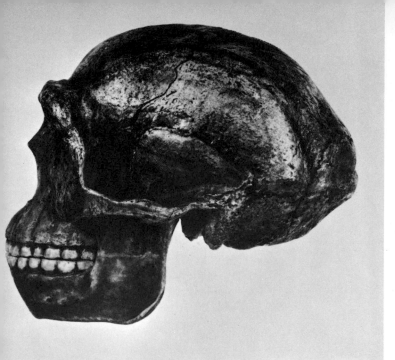

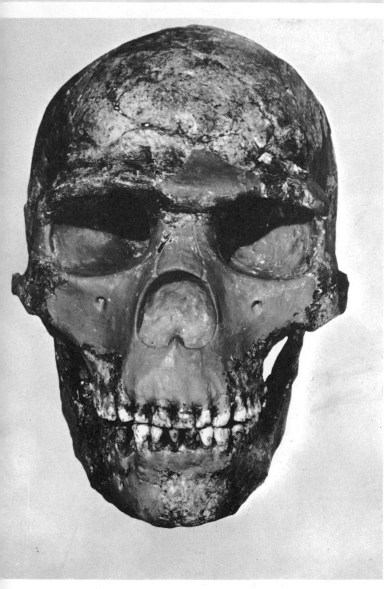

In the Palaeolithic men worked with stone and bone. Their first materials were flint, sandstone, quartzite, obsidian, chalcedony, jasper, and other hard rocks. These they split by heating in a fire, or chipped by hitting them with other stones. Palaeolithic industry lasted a long time, perhaps 700,000 years. It has been possible to subdivide the Pleistocene on an archaeological basis, but it is not always possible to correlate these archaeological sub-divisions with the geological epochs or ages. Different populations living at the same time were not necessarily at the same cultural stage. However, we can say that all men during the last Ice Age, and many of them before that time, were in the Palaeolithic cultural stage. Some men continued in the Palaeolithic after the retreat of the ice; the Neolithic did not occur anywhere until sometime after the ice had retreated from Europe. The Palaeolithic can be subdivided as follows: Chellean, Abbevillian, Acheulian, Mousterian, Aurignacian, Solutrean, Magdalenian.

The earliest and most crudely-fashioned stone tools of the Chellean culture are associated with Pekin Man. This culture is sometimes called Abbevillian, after the town of Abbeville where the high terrace of the Somme yielded early prehistoric tools in the form of almond-shaped blocks of stone trimmed by removing large flakes from both sides. Some of these hand-axes measure 11 inches long by 6 inches wide and weigh more than 4½ pounds; they have an irregular, sinuous outline. As they cannot be fitted with a handle they must have been held in the middle of the hand. Similar tools have been found in the high terraces of the Seine, Marne (Chelles), Charente, and Garonne in France, and in the 150-foot terrace of the Thames in England. They are always found in river alluvium, often mixed with the bones of elephants, rhinoceroses, and hippopotami. It is assumed that they were fashioned by men of the *Pithecanthropus* type during the first interglacial period and that these men, essentially nomadic, followed the valleys during their migrations.

The next stage, the Acheulian, is so called after Saint Acheul in France, a type station for stone implements that have also been found along the Seine, Charente, and Garonne, and in the south of England and Italy. Acheulian fist-axes were found with portions of a human skull in the Middle Gravels of the 100-foot Thames terrace at Swanscombe, Kent. Several features show that the Acheulian is younger than the Abbevillian: the cutting edges of the stone tools are smaller and better dressed than those of the preceding culture; the flakes are smaller and the outline is more regular; smaller implements in the shape of scrapers and points are also mixed with them; finally, these implements are in the middle, not the high terraces, and in the older loess which covers them. The climate was at first warm enough for elephants, rhinoceroses and hippopotami, but it was followed by a cold fauna in which mammoths and woolly rhinoceroses appeared for the first time. Thus the Acheulian corresponds to the second interglacial and the third glacial period. The human beings were in all probability of the *Pithecanthropus* type.

With the coming of the third stage, the Mousterian, stone implements changed entirely. Earlier types persisted but were associated with an increasing number of magnificent points and scrapers made from flakes which had been retouched on one face only. The block chosen for treatment was first given a preliminary trimming which converted it into a 'core-tool'. A blow on the upper surface detached flakes which were then retouched or dressed by striking against a stone, a piece of wood, or a bone. Magnificent core-tools from Grand Pressigny, in France, were prepared from flint so useful that it was carried by nomads and traders all over Europe.

The oldest Mousterian implements are found in low river terraces and in the recent loess, while the youngest is found in rock shelters or caves. There are well-known cave deposits in the Dordogne (notably Le Moustier) and in the Charente, Yonne, Corrèze, and the Pyrenees. There is thus a valley Mousterian and a cave Mousterian. The former is contemporary with a warm fauna (third interglacial), the latter with a cold fauna (fourth glacial). The human element was Neanderthal Man.

While the combined Abbevillian and Acheulian endured for

*Above*
Skulls of a 50,000-year-old Neanderthaloid man. The skull was found early this century on Mount Carmel.

*Right*
Site of the famous Laura cave paintings, 100 miles from Cooktown, in Queensland, Australia.

over 700,000 years, the Mousterian lasted scarcely 80,000 years, and the three subsequent stages combined — Aurignacian, Solutrean and Magdalenian — did not last more than 60,000 years. Human cultural evolution was speeding up. Neanderthal man was replaced by *Homo sapiens* with his high forehead, inventive mind, and artistic taste. Earlier stone tools were replaced with a profusion of scrapers, borers, 'saws' and engraving needles. Retouching was now effected on both sides of the tool. Each cultural stage had its own industries. For example, there are the magnificent laurel-leaf blades and willow-leaf points belonging to the Solutrean. A number of Solutrean workshops, tools, cores and rejected implements have been found. One of the most famous workshops is at Grand Pressigny. The most interesting, perhaps, is at Roche-moure (Ardèches) where debris scattered over a circular area reveals the working position of different workmen, each of whom had his own speciality. Some made scrapers, others needles.

The bones of animals associated with Aurignacian, Solutrean, and Magdalenian implements indicate that they date from the last glacial period and the beginning of the postglacial. The mammoths and woolly rhinoceroses gradually gave way to reindeer, which were replaced in their turn by animals of the steppe and forest. Because the climate was generally cold, man remained a cave dweller, as in the Mousterian, and the material for his implements was found inside or around caves and rock shelters. Type localities are Aurignac (Haute-Garonne), Solutre (Saône-et-Loire), La Madeleine (Dordogne). Kent's Cavern near Torquay, Devon, has yielded examples of Aurignacian and Magdalenian implements; other British localities include the Mendips, Cat's Hole, Gower, Ffynnon Beuno in the Vale of Clwyd, and the Creswell Crags in Derbyshire.

**Polished stone industry.** The Neolithic or New Stone Age is distinguished by the use of polished stone tools, weaving, and fired pottery, and by the domestication of plants and animals. This age in Europe dates back 7,000 years. The long-headed (dolichocephalic) races of *Homo sapiens* mixed with the round-headed (brachycephalic) races from Asia; polished stone replaced chipped stone; the axe became the principal implement. Dressing was by percussion or by abrasion with sand. Such methods were used by the Mexican Indians at the time of the Spanish conquest. The faces of the piece were then smoothed by pressure flaking and, finally, were polished on fixed grindstones or on polishers. Axes were drilled with a hole for shaftings. Pieces at different stages of the drilling process as well as the flint drills used in making the hole are found in the workshops. Certain polished jade or nephrite axes add lustre to the collections in the larger museums.

**Bone industry.** Bone tools had been used as far back as australopithecine times, long before stone. It was still used, along with stone, throughout the Palaeolithic. Bone relics become increasingly important from the Mousterian onward. The tips of stag antlers, for instance, were shaped into points for daggers, and bone splinters were used for a variety of purposes.

In the Aurignacian, Solutrean and Magdalenian cultures, the bone industry sprang into renewed life. Bone from horses, horn from reindeer and deer, and fossil ivory from the mammoth were used for spear points, daggers, spatulas, barbed harpoons, fish-hooks, and eyed needles — all made with consummate skill. Bracelets, pendants, and necklaces of ivory or of balls of ivory alternating with teeth and shells are often found on human skeletons. One curious object may have been a spear straightener. It is made from a piece of reindeer horn taken from the junction of the beam and the points and is variously ornamented and perforated.

**Metal industry.** Although metals have been used in Egypt for 9,000 years, their use in Europe goes back only 5,000 years and was spread by the brachycephalic peoples. The first metals utilised were gold and copper, which needed only hammering to produce useful shapes. The process of smelting and alloying was not devised until much later. The alloying of copper and tin gives bronze, which first appeared in western Europe about 2,000 B.C. Later still, man discovered metallurgy, learning that an iron ore smelted with charcoal gives liquid iron which can then

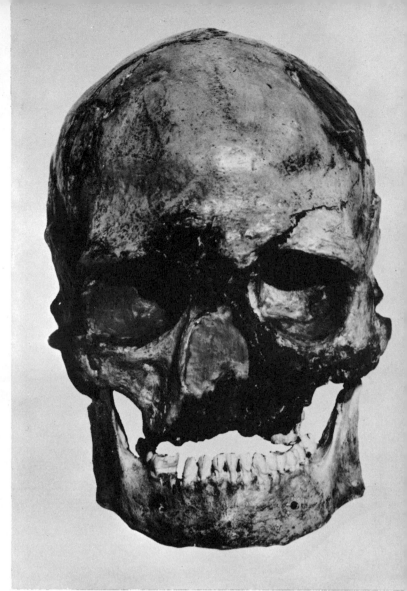

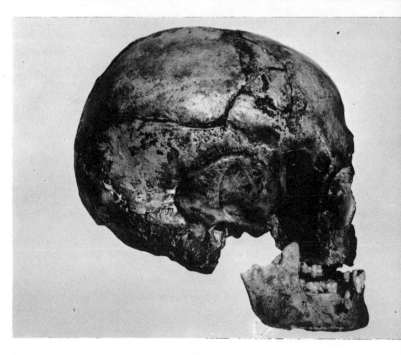

*Left*
The Sistine Chapel at Lascaux. The picture shows a detail from the left wall of the spacious Bull's Chamber; between the first and second bull can be seen a herd of deer and also the silhouette of a horse.

*Above*
Cheddar Man, a British representative of Cro-Magnon Man, from the Upper Palaeolithic of 15,000 years ago.

413

Early Java Man *(Pithecan-thropus modjokertensis)*.

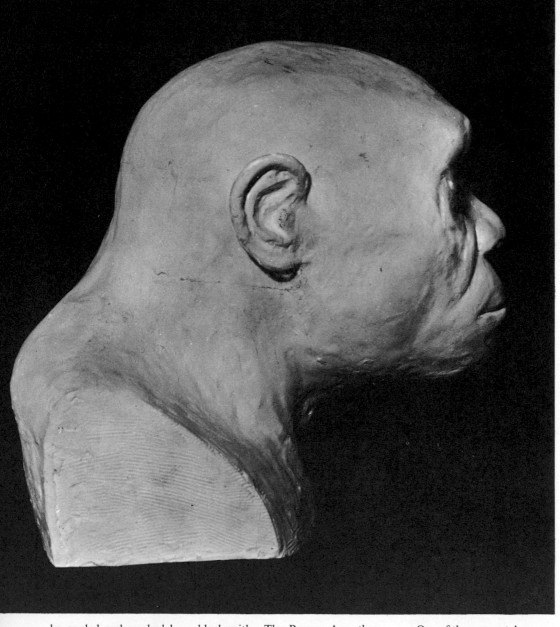

be cooled and worked by a blacksmith. The Bronze Age, then, was succeeded by the Iron Age. The use of this new metal came into Europe about 1,000 B.C.

## PREHISTORIC DWELLINGS

The earliest men, and probably most Neanderthal men, were at first nomadic and possessed no permanent dwellings. The fact that their flint implements are found largely in river shoreline deposits and in the loess which covers them seems to indicate that men tended to live along rivers and follow the river valleys which offered easier means of travel than deep forests.

Pekin Man was a cave dweller; during the last glacial stage (Würm) Neanderthal Man, seeking refuge against the cold, became a cave dweller. He still lived an open-air life, but sheltered beneath a rocky overhang. Although he was free to change his rock shelter after a few days, his earlier nomadism gradually became tempered into a non-migrant existence.

The permanent occupation of European caves dates from the second half of the Würm glaciation and was effected by men of the Grimaldi, Cro-Magnon, and Chancelade races. Fires were lit at cave entrances where today heaps of cinders, charcoal, and the bones of animals killed for food reveal their early function. Unwanted objects would be thrown out of the cave mouth to form a talus slope; the excavation of such slopes is of great interest. In the cave itself and the times far from the entrance, percolating waters carried along fallen debris which finally became encrusted by lime deposits. Here, too, there is much that is of interest. The cave walls and ceiling were often decorated with engravings drawings and paintings.

One of the caves at Arcy-sur-Cure, in France, provided speleologists and prehistorians with a floor of kitchen debris and hammered stones exactly as it was left when a rock fall made the cave inaccessible some 50,000 years ago.

The construction of dwellings is of more recent origin, and took place after the Ice Age. The first dwellings were huts made of wattles covered with clay, and had roofs of straw of reeds. In Alsace and Germany the foundations of raised huts have been found, with a kitchen hearth, a pit for refuse, and a cowshed, all built over a cellar. Huts were usually grouped together in villages or fortified camps in the valleys, which were still the great communication routes; some were built on piles in lakes.

Especially interesting are the lake villages of Lakes Geneva, Annecy, Chalian, and Clairvaux, and those of the Swiss and Italian Alps. Such lake villages as those of Robenhausen (Switzerland) and Bikuspin (Poland) may have had anything up to 100,000 stakes or piles to support the platforms on which the huts were built. The architecture gradually improved, and bolts, doors, and windows were made, and beams were joined by mortices and tenons. Bridges and catwalks connected the village with the land. Breakwaters protected it against the waves; palisades and watch-towers protected it from invaders. Today, at the bottom of the lakes, the bases of the piles, implements, pottery, and fragments of material can still be found. Many bone harpoons and hooks, wood floats, and hemp fishing-lines have also been found. Modern lake dwellings are found today in various parts of both Africa and south-east Asia.

**Agriculture and stock-raising.** Long before stock-raising and agriculture were practised, man got his food by hunting and fishing. Armed with bows and arrows, spears, slings, and daggers, he hunted

414

Rhodesian Man (*Homo rhodesiensis*). Restorations by Maurice Wilson.

the mammoth, cave bear, bison, wild ox, reindeer, deer, roebuck, horse, and wild boar, depending on the period. One of the most famous hunting places was that of Solutre, near Mâcon. At the foot of a cliff below a rocky spur the remains of more than 100,000 horses have been found. The horses must have been chased up the 'spur' and forced to jump from the top of the cliff. Elsewhere, pits that were used as traps have been discovered.

From the Neolithic onwards, domestication and stock-raising included cattle, goats, sheep, and pigs. Most of these animals were of Asiatic origin and came to Europe with the short-skulled men. In the piles of Neolithic kitchen waste complete skeletons of domestic animals have been found, but of wild animals only those bones that would be found in the pieces brought back by the hunters.

The Neolithic kitchen middens of the Danish coast are heaps of shells, oysters and mussels, sometimes 10 feet high, 50 yards wide, and 300 yards long; they were left by the coastal peoples. Similar shell deposits have been found at other points along the coasts of western Europe.

Neolithic Man was also a farmer, skilled in handling the hoe, the primitive plough, and the flint sickle. Grains of wheat, rye, barley, and oats, and the remains of flaxen textiles have been found in lake village deposits.

## PREHISTORIC ART

The Upper Palaeolithic Aurignacian and Magdalenian culture levels are characterised by extraordinary sculptures, engravings and paintings found in many of the once inhabited caves. Between 15,000 and 25,000 years ago there was a sudden blossoming of art in south-west France, and north and east Spain (also called the Spanish Levant). Of the scores of cave paintings examined so far, possibly those of Lascaux in France, Altamira in north Spain, and Parpallo in the Spanish Levant are the most outstanding. Paintings in styles similar to those of Parpallo are known in Africa, from the Sahara to the Republic of South Africa and have also been known in India and Australasia.

**Sculpture.** It is probable that sculpture was, chronologically, the first of the arts. Because it is three-dimensional, it is usually less abstract than engraving or painting. It might be supposed that man first noticed the chance resemblance between certain stones and animal or human forms, then passed from these stone figures which he could retouch and perfect to works entirely his own.

Female figures, such as the 'Venus' of Laussel and Lespugne, are the most common, and show a development that suggests the steatopygia of the modern Hottentots. Certain parts of the body may have been exaggerated to make the statuettes symbols of fertility. Later, Chancelade and Cro-Magnon men fashioned many portable objects, ornaments, arms, implements for display, batons of authority, amulets, and statuettes representing different animals. Well-known examples are a horse's head carved in reindeer horn from Mas d'Azil, and a statuette of a mammoth carved in mammoth ivory discovered at Lespugne.

Sculpture passed from portable art to mural art. A projecting piece of limestone wall in a cave may first have suggested the form of an animal which the artist gradually brought out into relief. One of the finest series of sculptures known is that of the rock shelter at Angles-sur l'Anglin. There the Magdalenian artist produced a three— tiered frieze, where bison, chamois, ibex, horses, and three 'Venuses' with slender graceful forms are drawn up in line. Clay in

415

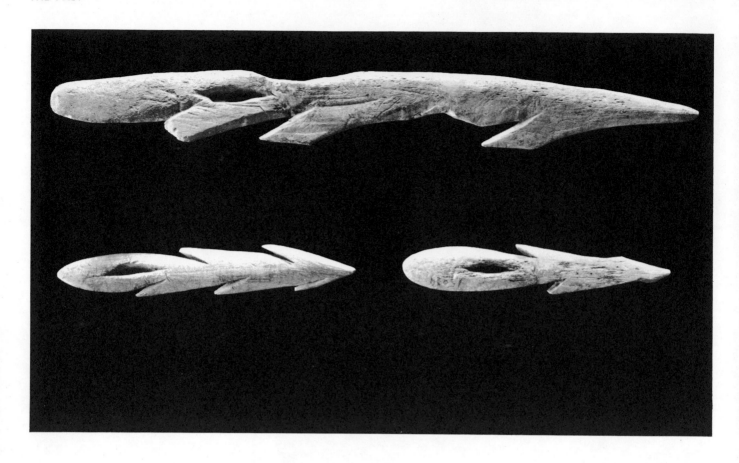

the Pyrenean caves of Tuc d'Audoubert and Montespan was used for making models, notably a group of bison resting on the ground.

**Engraving and painting.** Prehistoric pictorial art was first discovered in 1880 in the cave at Altamira, Spain. The discovery in 1940 of the cave at Lascaux raised admiration for prehistoric animal art to a high pitch. Usually mural engravings and paintings are found in remote and dark places in the caves, suggesting that the artist must have provided illumination for himself from torches or stone lamps burning animal fats. His palette consisted of three fundamental colours: the black of manganese, yellow ochre, and red ochre or red hematite. Stone saucers and shell palettes have been found in which these colours were ground with fat.

The animals portrayed are mostly mammoths, woolly rhinoceroses, bison, reindeer, and horses. Some frescoes depict with impressive realism several animals represented in different attitudes. Hunting and dancing scenes with human figures are rarer; the painting at Lascaux, however, shows a bison wounded by a long arrow with the hunter falling in front of the animal. It is interesting to note that human figures shown in attitudes of movement in the cave paintings at Cogul, in east Spain, are precisely the same as those depicted at Saltadora in the Levant and Domboshawa in Rhodesia. Similarly, animal drawings in the famous bison fresco of Altamira are comparable with those of eland at Khosta, Lesotho.

In addition to its realism, this mural art is distinguished by its simplicity; there are no useless, trivial details. It is also remarkable for the intensity of movement and life expressed. A man who painted or engraved a running deer with such accuracy must previously have observed the animal with an acuteness of which few men today are capable or accustomed.

In its decline Palaeolithic art produced such things as the curious coloured stones of Mas d'Azil. They are pebbles on which the artist has painted mysterious signs in red ochre, sometimes a cross and stars possibly representing the sun, or undulating lines perhaps symbolising waves or serpents.

**Utilitarian art.** Today it is thought unlikely that art existed for art's sake among the Palaeolithic people. All sculpture, engraving, and painting must have had a utilitarian rôle. The works have been compared to modern totemic symbols, some of which were preserved originally in tabooed localities in the belief that they would bring good luck to the hunt. We can imagine the scenes of magic and sorcery in which this ancient art had a central place if we consider the superstitions of primitive peoples still living who follow what may be similar rites.

Eminent prehistorians agree that art of the period was used for magic and religious purposes. It is therefore no longer surprising to discover drawings, paintings, and even sculptures in parts of the caves that are difficult of access. Profiles in several caves of masked and sometimes completely disguised men are taken to be those of sorcerers presiding over scenes of magic and incantation. The works at Labastide, Hautes-Pyrénées, for example, suggest that it was a sanctuary where prehistoric sorcerers devoted themselves to mysterious ceremonies, sometimes very complex and undoubtedly inherited from an ancient cave-dwelling tradition.

## BURIAL PLACES AND MEGALITHS

Funeral rites appear to have begun in the Magdalenian. From then on the dead were buried, aligned along certain cardinal directions, often with their knees bent up to the chin. Various ornaments, such as bracelets, necklaces, diadems and crowns, have been found with them.

In the Neolithic, man abandoned his caves and began living in huts. The memory of cave burials may have led him to construct monuments, like the dolmens common over much of France, particularly in Brittany. A dolmen (*dol* — table and *men* — stone) is a table formed from a huge flat stone resting horizontally on two vertical stones. One of the biggest is the Merchants' Table at Locmariaquer, Morbihan. In Great Britain dolmens are found at places as far apart as Cornwall and Stennis, Orkney. Where

*Left*
Magdalenian harpoons made of bone. Various sorts of hunting and fishing weapons came into use during Magdalenian times.

*Right*
A Middle Magdalenian pebble engraving showing bison. From Bruniquel, in France.

several dolmens have been built in line to form a covered way, serveral aligned skeletons may have rested. Originally, dolmens and covered ways were buried beneath artificial mounds or tumuli of considerable proportions. Those at Carnac are 390 feet long, 195 feet wide, and 39 feet high. Such burial mounds are analogous to the pyramids of Egypt, where the dead were buried with their animals and household objects.

Neolithic Man also covered parts of Britain and France with standing stones, or menhirs (*men* — stone, and *hir* — long), often also called megaliths. In France there are more than 6,000 menhirs, of which 4,750 are in Brittany. They are 5,000 to 6,000 years old, though a few may be younger. The biggest, now broken, is that of Locmariaquer, which is 68 feet long. Near Carnac, the menhirs form avenues, leading to a cromlech (*crom* — circle and *lech* — place), which was probably a kind of altar where priests officiated. The stone circles at Avebury and Stonehenge are the largest cromlechs in Britain and the biggest megalithic monuments in the world.

The avenues at Menec, Kermario and Kerlescan run in an east-west line for $2\frac{1}{2}$ miles. At Menec there are 1,099 stones in 11 parallel rows which can be followed for half a mile to a semi-circular cromlech of 70 menhirs. At Kermario there are 982 stones in 10 lines, also half a mile long, while farther east at Kerlescan there are only 540 menhirs in 13 rows, and 39 stones in the terminal cromlech. The avenue at Stonehenge is $1\frac{1}{2}$ miles long:

In each row the tallest stones are at the western end and the smallest at the eastern; the cromlech, when it exists, is always next to the biggest menhirs. Stones placed transverse to the rows indicate certain astronomical directions: at Kermario, that of the direction in which the sun rises at the solstices; at Menec and Kerlescan that of sunrise at the equinoxes. Similarly, the 'Altar Stone' at Stonehenge and a stone at Woodhenge are aligned on the rising sun at the summer solstice.

All these facts suggest that these are religious monuments. The avenues were sacred paths; the cromlechs were sanctuaries where priests officiated. Each temple had a particular purpose: Kerlescan was consecrated to the feast of seedtime; Kermario to the harvest. Neolithic peoples were essentially farmers and shepherds. The sun, plough, and ears of wheat are represented on many of their monuments. The bones of oxen and horses lie in their tombs, side by side with human ashes. Carnac was probably a sacred city during this time just as Jerusalem and Mecca are sacred cities today.

Like the cromlechs in Brittany, peat-bogs in Scandinavia seem to have been considered sacred. The bogs were places where people were engulfed and disappeared. Investigations of these bogs in the Jutland peninsular have brought to light some surprising facts about the Late Quaternary Danish religions. Bodies in a perfect state of preservation have been found there. Some were executed criminals or enemies who had been thrown into the peat with their hands tied and their eyes blindfolded; others were priests who were sacrificed so that the goddess of the soil would look favourably on the community. With them have been found bronze helmets with long horns, copper vases, trumpets, bracelets, ornaments and, in one case at least, a silver basin and a ceremonial chariot.

Neolithic life in Europe was similar in many ways to Neolithic life in other parts of the world. Few cultures, however, are ever exactly alike, and man achieved his various cultural stages unevenly in different parts of the world. Around the Mediterranean man was enjoying a Neolithic culture, while on the fringes of the retreating ice sheet and in other areas remote from the main group of mankind Palaeolithic cultures still prevailed. The pattern remains the same in some measure; there are still Stone Age dwellers in New Guinea, Australia, and the Kalahari Desert of South Africa, while scientists in the most advanced industrial nations are sending satellites into outer space.

With Neolithic Man our study of the origin and history of the Earth, life and prehistoric man ends. His descendants soon developed the art of writing. At that point prehistory drew to a close and the era of recorded human history began.

# Acknowledgments

Chuck Abbott 366; Aerial Photography 109; Air France 143, 157B; Altigraph 60T; American Museum of Natural History 99T, 381, 390, 405T, 156BL; Atlas Photos 79; Australian News and Information Bureau 222T, 301B; Austrian State Tourist Department 330; G. Douglas Boulton 409; Bourgin 59C, 74, 138TL, 138R, 144R, 145T; Bourgin and Perrit 70; British Museum (Natural History) 402, 408, 409, 410, 413,414, 415; British Official Photograph (Crown copyright) 233; British Petroleum Company 185TR, 185B, 269; Buffalo Museum of Science, New York 378B; Bundesbildstelle, Bonn 248; Camera Press 14, 153, 239, 316T; Canadian Government Travel Bureau 208; J. Allan Cash 76T, 96L, 108, 110L, 119, 120T, 123, 130, 132BR, 139T, 148, 150, 172TL, 194, 205, 230, 327, 335B; Central Office of Information (Crown copyright) 167C; Charbonnages de France 249R, 257, 259B; James A. Cunningham 75; De Beers Consolidated Mines 217T, 219, 221; Decca Radar 36B; Department of Transport, Canada 90; English Electric 298C; E.S.S.A. 47T; Esso Petroleum Company 174, 182, 274B, 275, 279, 280, 281, 285, 287; Federal Information Department Rhodesia 97T; C. L. Fenton 152, 161, 235B, 236T, 325T, 335T, 343, 358, 359, 360, 364T, 364B, 368, 369, 371, 372, 373, 374, 375, 376, 377, 378T, 379, 380, 383, 384TL, 386, 387, 388, 391, 392, 395, 396, 397, 398, 399, 401, 404B, 405C, 405B, 406TL; M. A. Fenton 40B, 43R, 147; Field Museum of Natural History 406BL, 147; French Government Tourist Office 339, 347T; Richard Gee 362; Geological Society of North America 107; Donald C. Good 237; Douglas Glass 220; G.P.O. 36CR; Ice-

# Index

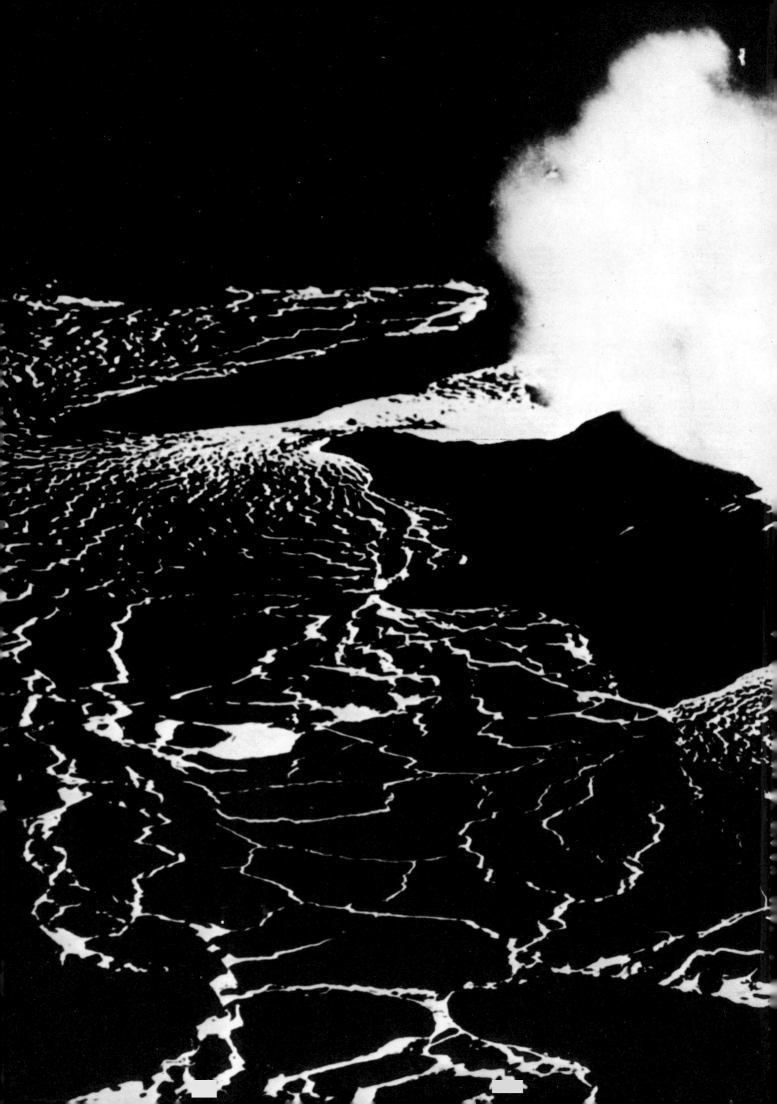